Treatise on Invertebrate Paleontology

Part E

Porifera

Revised

Volume 2:
Introduction to the Porifera

J. Keith Rigby, Coordinating Author

by R. M. Finks, R. E. H. Reid, and J. K. Rigby

Prepared under Sponsorship of
The Geological Society of America, Inc.

The Paleontological Society
The Palaeontographical Society
SEPM (Society for Sedimentary Geology)
The Palaeontological Association

Raymond C. Moore
Founder

Roger L. Kaesler
Editor

Jill Hardesty, Jane Kerns
Michael Cormack
Assistant Editors and Editorial Staff

The Geological Society of America, Inc.
and
The University of Kansas
Boulder, Colorado, and Lawrence, Kansas
2003

Library of Congress Catalogue Card Number 53–12913
ISBN 0–8137–3130–5

Distributed by the Geological Society of America, Inc., P.O. Box 9140, Boulder, Colorado 80301, www.geosociety.org, from which current price lists of parts in print may be obtained and to which all orders and related correspondence should be directed. Editorial office of the *Treatise:* Paleontological Institute, The University of Kansas, 1475 Jayhawk Blvd., Room 121, Lawrence, Kansas 66045-7613, www.ku.edu/~paleo.

Citation information: Kaesler, R. L., ed. 2003. Treatise on Invertebrate Paleontology, Part E (Revised), Porifera, vol. 2. The Geological Society of America & The University of Kansas. Boulder & Lawrence. xxvii + 349 p., 135 fig., 10 tables.

The *Treatise on Invertebrate Paleontology* has been made possible by (1) funding principally from the National Science Foundation of the United States in its early stages, from The Geological Society of America through the bequest of Richard Alexander Fullerton Penrose, Jr., and from The Kansas University Endowment Association through the bequest of Raymond C. and Lillian B. Moore; (2) contribution of the knowledge and labor of specialists throughout the world, working in cooperation under sponsorship of The Geological Society of America, the Paleontological Society, the SEPM (Society for Sedimentary Geology), the Palaeontographical Society, and the Palaeontological Association; (3) acceptance by The University of Kansas of publication without any financial gain to the University; and (4) generous contributions by our individual and corporate sponsors.

PART E, Revised

PORIFERA

VOLUME 2: INTRODUCTION TO THE PORIFERA

R. M. Finks, R. E. H. Reid, and J. K. Rigby

CONTENTS

INFORMATION ON *TREATISE* VOLUMES

Parts of the *Treatise* are distinguished by assigned letters with a view to indicating their systematic sequence while allowing publication of units in whatever order each is made ready for the press. Copies can be obtained from the Publication Sales Department, The Geological Society of America, 3300 Penrose Place, P.O. Box 9140, Boulder, Colorado 80301, www.geosociety.org.

PUBLISHED VOLUMES

Part A. INTRODUCTION: Fossilization (Taphonomy), Biogeography, and Biostratigraphy, xxiii + 569 p., 169 fig., 1979.

Part C. PROTISTA 2 (Sarcodina, Chiefly "Thecamoebians" and Foraminiferida), Volumes 1 and 2, xxxi + 900 p., 653 fig., 1964.

Part D. PROTISTA 3 (Protozoa: Chiefly Radiolaria, Tintinnina), xii + 195 p., 92 fig., 1954.

Part E. ARCHAEOCYATHA and PORIFERA, xviii + 122 p., 89 fig., 1955.

Part E, Revised. ARCHAEOCYATHA, Volume 1, xxx + 158 p., 107 fig., 1972.

Part F. COELENTERATA, xx + 498 p., 358 fig., 1956.

Part F. COELENTERATA, Supplement 1 (Rugosa and Tabulata), Volumes 1 and 2, xl + 762 p., 462 fig., 1981.

Part G. BRYOZOA, xiii + 253 p., 175 fig., 1953.

Part G, Revised. BRYOZOA, Volume 1 (Introduction, Order Cystoporata, Order Cryptostomata), xxvi + 625 p., 295 fig., 1983.

Part H. BRACHIOPODA, Volumes 1 and 2, xxxii + 927 p., 746 fig., 1965.

Part H, Revised. BRACHIOPODA, Volume 1 (Introduction), xx + 539 p., 417 fig., 40 tables, 1997.

Part H, Revised. BRACHIOPODA, Volumes 2 and 3 (Linguliformea, Craniiformea, Rhynchonelliformea [part]), xxx + 919 p., 616 fig., 17 tables, 2000.

Part H, Revised. BRACHIOPODA, Volume 4 (Rhynchonelliformea [part]), xxxix + 768 p., 484 fig., 3 tables, 2002.

Part I. MOLLUSCA 1 (Mollusca General Features, Scaphopoda, Amphineura, Monoplacophora, Gastropoda General Features, Archaeogastropoda, Mainly Paleozoic Caenogastropoda and Opisthobranchia), xxiii + 351 p., 216 fig., 1960.

Part K. MOLLUSCA 3 (Cephalopoda General Features, Endoceratoidea, Actinoceratoidea, Nautiloidea, Bactritoidea), xxviii + 519 p., 361 fig., 1964.

Part L. MOLLUSCA 4 (Cephalopoda: Ammonoidea), xxii + 490 p., 558 fig., 1957.

Part L, Revised. MOLLUSCA 4, Volume 4 (Cretaceous Ammonoidea), xx + 362 p., 216 fig., 1996.

Part N. MOLLUSCA 6 (Bivalvia), Volumes 1 and 2 (of 3), xxxvii + 952 p., 613 fig., 1969.

Part N. MOLLUSCA 6 (Bivalvia), Volume 3, iv + 272 p., 153 fig., 1971.

Part O. ARTHROPODA 1 (Arthropoda General Features, Protarthropoda, Euarthropoda General Features, Trilobitomorpha), xix + 560 p., 415 fig., 1959.

Part O, Revised. ARTHROPODA 1 (Trilobita: Introduction, Order Agnostida, Order Redlichiida), xxiv + 530 p., 309 fig., 1997.

Part P. ARTHROPODA 2 (Chelicerata, Pycnogonida, Palaeoisopus), xvii + 181 p., 123 fig., 1955 [1956].

Part Q. ARTHROPODA 3 (Crustacea, Ostracoda), xxiii + 442 p., 334 fig., 1961.

Part R. ARTHROPODA 4, Volumes 1 and 2 (Crustacea Exclusive of Ostracoda, Myriapoda, Hexapoda), xxxvi + 651 p., 397 fig., 1969.

Part R. ARTHROPODA 4, Volumes 3 and 4 (Hexapoda), xxii + 655 p., 265 fig., 1992.

Part S. ECHINODERMATA 1 (Echinodermata General Features, Homalozoa, Crinozoa, exclusive of Crinoidea), Volumes 1 and 2, xxx + 650 p., 400 fig., 1968 [1967].

Part T. ECHINODERMATA 2 (Crinoidea), Volumes 1–3, xxxviii + 1,027 p., 619 fig., 1978.

Part U. ECHINODERMATA 3 (Asterozoans, Echinozoans), xxx + 695 p., 534 fig., 1966.

Part V. GRAPTOLITHINA, xvii + 101 p., 72 fig., 1955.

Part V, Revised. GRAPTOLITHINA, xxxii + 163 p., 109 fig., 1970.

Part W. MISCELLANEA (Conodonts, Conoidal Shells of Uncertain Affinities, Worms, Trace Fossils, Problematica), xxv + 259 p., 153 fig., 1962.

Part W, Revised. MISCELLANEA, Supplement 1 (Trace Fossils and Problematica), xxi + 269 p., 110 fig., 1975.

Part W, Revised. MISCELLANEA, Supplement 2 (Conodonta), xxviii + 202 p., frontis., 122 fig., 1981.

THIS VOLUME

Part E, Revised. Porifera, Volume 2 (Introduction to the Porifera), xxvii + 349 p., 135 fig., 10 tables, 2003.

VOLUMES IN PREPARATION

Part B. PROTISTA 1 (Chrysomonadida, Coccolithophorida, Charophyta, Diatomacea, etc.).

Part E, Revised. PORIFERA (additional volumes).

Part F, Revised. CNIDARIA (Scleractinia).

Part G, Revised. BRYOZOA (additional volumes).

Part H, Revised. BRACHIOPODA (additional volumes).

Part K, Revised. MOLLUSCA 3 (Nautiloidea).

Part L, Revised. MOLLUSCA 4 (Ammonoidea) (additional volumes).

Part M. MOLLUSCA 5 (Coleoidea).

Part O, Revised. ARTHROPODA 1 (Trilobita) (additional volumes).

Part Q, Revised. ARTHROPODA 3 (Ostracoda).

Part R, Revised. ARTHROPODA 4 (Crustacea Exclusive of Ostracoda).

Part T, Revised. ECHINODERMATA 2 (Crinoidea).

Part V, Revised. GRAPTOLITHINA.

Part W, Revised. TRACE FOSSILS.

EDITORIAL PREFACE

Roger L. Kaesler
[The University of Kansas]

From the outset the aim of the *Treatise on Invertebrate Paleontology* has been to present a comprehensive and authoritative yet compact statement of knowledge concerning groups of invertebrate fossils. Typically, preparation of early *Treatise* volumes was undertaken by a small group with a synoptic view of the taxa being monographed. Two or perhaps three specialists worked together, sometimes co-opting others for coverage of highly specialized taxa. Recently, however, both new *Treatise* volumes and revisions of existing ones have been undertaken increasingly by teams of specialists led by a coordinating author. This volume, Part E Revised, Porifera, Volume 2, has been prepared by such a team. In the early stages of the work, R. M. Finks and R. E. H. Reid worked together on the volume. Final preparation of this volume was coordinated by J. K. Rigby, working with manuscript that was submitted previously by both Finks and Reid. Editorial matters specific to this volume are discussed near the end of this editorial preface.

ZOOLOGICAL NAMES

Questions about the proper use of zoological names arise continually, especially questions regarding both the acceptability of names and alterations of names that are allowed or even required. Regulations prepared by the International Commission on Zoological Nomenclature (ICZN) and published in 1999 in the *International Code of Zoological Nomenclature,* hereinafter referred to as the *Code,* provide procedures for answering such questions. The prime objective of the *Code* is to promote stability and universality in the use of the scientific names of animals, ensuring also that each generic name is distinct and unique, while avoiding unwarranted restrictions on freedom of thought and action of systematists. Priority of names is a basic principle of the *Code;* but, under specified conditions and by following prescribed procedures, priority may be set aside by the Commission. These procedures apply especially where slavish adherence to the principle of priority would hamper or even disrupt zoological nomenclature and the information it conveys.

The Commission, ever aware of the changing needs of systematists, revised the *Code* in 1999 to enhance further nomenclatorial stability, specifying that the revised *Code* should take effect at the start of 2000. In spite of the revisions, the nomenclatorial tasks that confront zoological taxonomists are formidable and have often justified the complaint that the study of zoology and paleontology is too often merely the study of names rather than the study of animals. It is incumbent upon all systematists, therefore, at the outset of their work to pay careful attention to the *Code* to enhance stability by minimizing the number of subsequent changes of names, too many of which are necessitated by insufficient attention to detail. To that end, several pages here are devoted to aspects of zoological nomenclature that are judged to have chief importance in relation to procedures adopted in the *Treatise,* especially in this volume. Terminology is explained, and examples are given of the style employed in the nomenclatorial parts of the systematic descriptions.

GROUPS OF TAXONOMIC CATEGORIES

Each taxon belongs to a category in the Linnaean hierarchical classification. The *Code* recognizes three groups of categories, a species-group, a genus-group, and a family-group. Taxa of lower rank than subspecies are excluded from the rules of zoological nomenclature, and those of higher rank than superfamily are not regulated by the *Code.* It is both natural and convenient to discuss nomenclatorial matters in general terms first and then to consider each of these three recognized groups separately. Especially important is the provision that within each group the categories are coordinate, that is, equal in rank, whereas categories of different groups are not coordinate.

FORMS OF NAMES

All zoological names can be considered on the basis of their spelling. The first form of a name to be published is defined as the original spelling (*Code,* Article 32), and any form of the same name that is published later and is different from the original spelling is designated a subsequent spelling (*Code,* Article 33). Not every original or subsequent spelling is correct.

ORIGINAL SPELLINGS

If the first form of a name to be published is consistent and unambiguous, the original is defined as correct unless it contravenes some stipulation of the *Code* (Articles 11, 27 to 31, and 34) or unless the original publication contains clear evidence of an inadvertent error in the sense of the *Code,* or, among names belonging to the family-group, unless correction of the termination or the stem of the type genus is required. An original spelling that fails to meet these requirements is defined as incorrect.

If a name is spelled in more than one way in the original publication, the form adopted by the first reviser is accepted as the correct original spelling, provided that it complies with mandatory stipulations of the *Code* (Articles 11 and 24 to 34).

Incorrect original spellings are any that fail to satisfy requirements of the *Code,* represent an inadvertent error, or are one of multiple original spellings not adopted by a first reviser. These have no separate status in zoological nomenclature and, therefore, cannot enter into homonymy or be used as replacement names. They call for correction. For example, a name originally published with a diacritical mark, apostrophe, dieresis, or hyphen requires correction by deleting such features and uniting parts of the name originally separated by them, except that deletion of an umlaut from a vowel in a name derived from a German word or personal name unfortunately requires the insertion of *e* after the vowel. Where original spelling is judged to be incorrect solely because of inadequacies of the Greek or Latin scholarship of the author, nomenclatorial changes conflict with the primary purpose of zoological nomenclature as an information retrieval system. One looks forward with hope to further revisions of the *Code* wherein rules are emplaced that enhance stability rather than classical scholarship, thereby facilitating access to information.

SUBSEQUENT SPELLINGS

If a subsequent spelling differs from an original spelling in any way, even by the omission, addition, or alteration of a single letter, the subsequent spelling must be defined as a different name. Exceptions include such changes as an altered termination of adjectival specific names to agree in gender with associated generic names (an unfortunate impediment to stability and retrieval of information); changes of family-group names to denote assigned taxonomic rank; and corrections that eliminate originally used diacritical marks, hyphens, and the like. Such changes are not regarded as spelling changes conceived to produce a different name. In some instances, however, species-group names having variable spellings are regarded as homonyms as specified in the *Code* (Article 58).

Altered subsequent spellings other than the exceptions noted may be either intentional or unintentional. If "demonstrably intentional" (*Code,* Article 33), the change is designated as an emendation. Emendations may be either justifiable or unjustifiable. Justifiable emendations are corrections of incorrect original spellings, and these take the authorship and date of the original spellings. Unjustifiable emendations are names having their own status in nomenclature, with author and date of their publication. They are junior, objective synonyms of the name in its original form.

Subsequent spellings, if unintentional, are defined as incorrect subsequent spellings. They have no status in nomenclature, do not enter into homonymy, and cannot be used as replacement names.

AVAILABLE AND UNAVAILABLE NAMES

Editorial prefaces of some previous volumes of the *Treatise* have discussed in appreciable detail the availability of the many kinds of zoological names that have been proposed under a variety of circumstances. Much of that information, while important, does not pertain to the present volume, in which authors have used fewer terms for such names. The reader is referred to the *Code* (Articles 10 to 20) for further details on availability of names. Here, suffice it to say that an available zoological name is any that conforms to all mandatory provisions of the *Code.* All zoological names that fail to comply with mandatory provisions of the *Code* are unavailable and have no status in zoological nomenclature. Both available and unavailable names are classifiable into groups that have been recognized in previous volumes of the *Treatise,* although not explicitly differentiated in the *Code.* Among names that are available, these groups include inviolate names, perfect names, imperfect names, vain names, transferred names, improved or corrected names, substitute names, and conserved names. Kinds of unavailable names include naked names (see *nomina nuda* below), denied names, impermissible names, null names, and forgotten names.

Nomina nuda include all names that fail to satisfy provisions stipulated in Article 11 of the *Code,* which states general requirements of availability. In addition, they include names published before 1931 that were unaccompanied by a description, definition, or indication (*Code,* Article 12) and names published after 1930 that (1) lacked an accompanying statement of characters that differentiate the taxon, (2) were without a definite bibliographic reference to such a statement, (3) were not proposed expressly as a replacement (*nomen novum*) of a preexisting available name (*Code,* Article 13.1), or (4) for genus-group names, were unaccompanied by definite fixation of a type species by original designation or indication (*Code,* Article 13.2). *Nomina nuda* have no status in nomenclature, and they are not correctable to establish original authorship and date.

VALID AND INVALID NAMES

Important considerations distinguish valid from available names on the one hand and invalid from unavailable names on the other. Whereas determination of availability is based entirely on objective considerations guided by articles of the *Code,* conclusions as to validity of zoological names may be partly subjective. A valid name is the correct one for a given taxon, which may have two or more available names but only a single correct, hence valid, name, which is also generally the oldest name that it has been given. Obviously, no valid name can also be an unavailable name, but invalid names may be either available or unavailable. It follows that any name for a given taxon other than the valid name, whether available or unavailable, is an invalid name.

One encounters a sort of nomenclatorial no-man's land in considering the status of such zoological names as *nomina dubia* (doubtful names), which may include both available and unavailable names. The unavailable ones can well be ignored, but names considered to be available contribute to uncertainty and instability in the systematic literature. These can ordinarily be removed only by appeal to the ICZN for special action. Because few systematists care to seek such remedy, such invalid but available names persist in the literature.

NAME CHANGES IN RELATION TO GROUPS OF TAXONOMIC CATEGORIES

SPECIES-GROUP NAMES

Detailed consideration of valid emendation of specific and subspecific names is unnecessary here, both because the topic is well understood and relatively inconsequential and because the *Treatise* deals with genus-group names and higher categories. When the form of adjectival specific names is changed to agree with the gender of a generic name in transferring a species from one genus to another, one need never label the changed name as *nomen correctum.* Similarly, transliteration of a letter accompanied by a diacritical mark in the manner now called for by the *Code,* as in changing originally *bröggeri* to *broeggeri,* or eliminating a hyphen, as in changing originally published *cornu-oryx* to *cornuoryx,* does not require the designation *nomen correctum.* Of course, in this age of computers and electronic databases, such changes of name, which are perfectly valid for the purposes of scholarship, run counter to the requirements of nomenclatorial stability upon which the preparation of massive, electronic databases is predicated.

GENUS-GROUP NAMES

Conditions warranting change of the originally published, valid form of generic and subgeneric names are sufficiently rare that lengthy discussion is unnecessary. Only elimination of diacritical marks and hyphens in some names in this category and replacement of homonyms seem to furnish basis for valid emendation. Many names that formerly were regarded as homonyms are no longer so regarded, because two names that differ only by a single letter or in original publication by the presence of a diacritical mark in one are now construed to be entirely distinct (but see *Code,* Article 58).

As has been pointed out above, difficulty typically arises when one tries to decide whether a change of spelling of a name by a subsequent author was intentional or unintentional, and the decision has to be made often arbitrarily.

FAMILY-GROUP NAMES

Family-Group Names: Authorship and Date

All family-group taxa having names based on the same type genus are attributed to the author who first published the name of any of these groups, whether tribe, subfamily, or family (superfamily being almost inevitably a later-conceived taxon). Accordingly, if a family is divided into subfamilies or a subfamily into tribes, the name of no such subfamily or tribe can antedate the family name. Moreover, every family containing differentiated subfamilies must have a nominate subfamily (*sensu stricto*), which is based on the same type genus as the family. Finally, the author and date set down for the nominate subfamily invariably are identical with those of the family, irrespective of whether the author of the family or some subsequent author introduced subdivisions.

Corrections in the form of family-group names do not affect authorship and date of the taxon concerned, but in the *Treatise* recording the authorship and date of the correction is desirable because it provides a pathway to follow the thinking of the systematists involved.

Family-Group Names: Use of *nomen translatum*

The *Code* (Article 29.2) specifies the suffixes for tribe (-ini), subfamily (-inae), family (-idae) and superfamily (-oidea), the formerly widely used ending (-acea) for superfamily having been disallowed. All these family-group categories are defined as coordinate (*Code,* Article 36.1): "A name established for a taxon at any rank in the family group is deemed to have been simultaneously established for nominal taxa at other ranks in the family group; all these taxa have the same type genus, and their names are formed from the stem of the name of the type genus [Art. 29.3] with appropriate change of suffix [Art. 34.1]. The name has the same authorship and date at every rank." Such changes of rank and concomitant changes of endings as elevation of a subfamily to family rank or of a family to superfamily rank, if introduced subsequent to designation of the original taxon or based on the same nominotypical genus, are *nomina translata*. In the *Treatise* it is desirable to distinguish the valid alteration in the changed ending of each transferred family-group name by the term *nomen translatum,* abbreviated to *nom. transl.* Similarly for clarity, authors should record the author, date, and page of the alteration, as in the following example.

Family HEXAGENITIDAE
Lameere, 1917

[*nom. transl.* DEMOULIN, 1954, p. 566, *ex* Hexagenitinae LAMEERE, 1917, p. 74]

This is especially important for superfamilies, for the information of interest is the author who initially introduced a taxon rather than the author of the superfamily as defined by the *Code.* For example:

Superfamily AGNOSTOIDEA M'Coy, 1849

[*nom. transl.* SHERGOLD, LAURIE, & SUN, 1990, p. 32, *ex* Agnostinae M'COY, 1849, p. 402]

The latter is merely the individual who first defined some lower-ranked, family-group taxon that contains the nominotypical genus of the superfamily. On the other hand, the publication that introduces the superfamily by *nomen translatum* is likely to furnish the information on taxonomic considerations that support definition of the taxon.

Family-Group Names: Use of *nomen correctum*

Valid name changes classed as *nomina correcta* do not depend on transfer from one category of the family group to another but most commonly involve correction of the stem of the nominotypical genus. In addition, they include somewhat arbitrarily chosen modifications of endings for names of tribes or superfamilies. Examples of the use of *nomen correctum* are the following.

Family STREPTELASMATIDAE
Nicholson, 1889

[*nom. correct.* WEDEKIND, 1927, p. 7, *pro* Streptelasmidae NICHOLSON in NICHOLSON & LYDEKKER, 1889, p. 297]

Family PALAEOSCORPIDAE
Lehmann, 1944

[*nom. correct.* PETRUNKEVITCH, 1955, p. 73, *pro* Palaeoscorpionidae LEHMANN, 1944, p. 177]

Family-Group Names: Replacements

Family-group names are formed by adding combinations of letters, which are prescribed for all family-group categories, to the stem of the name belonging to the nominotypical genus first chosen as type of the assemblage. The type genus need not be the first genus in the family to have been named and defined, but among all those included it must be the first published as name giver to a family-group taxon. Once fixed, the family-group name remains tied to the nominotypical genus even if the generic name is changed by reason of status as a junior homonym or junior synonym, either objective or subjective. Seemingly, the *Code* requires replacement of a family-group name only if the nominotypical genus is found to have been a junior homonym when it was proposed (*Code,* Article 39), in which case ". . . it must be replaced either by the next oldest available name from among its synonyms [Art. 23.3.5], including the names of its subordinate family-group taxa, or, if there is no such synonym, by a new name based on the valid name . . . of the former type genus." Authorship and date attributed to the replacement family-group name are determined by first publication of the changed family-group name. Recommendation 40A of the *Code,* however, specifies that for subsequent application of the rule of priority, the family-group name ". . . should be cited with its original author and date (see Recommendation 22A.2.2), followed by the date of its priority as determined by this Article; the date of priority should be enclosed in parentheses." Many family-group names that have been in use for a long time are *nomina nuda,* since they fail to satisfy criteria of availability (*Code,* Article 11.7). These demand replacement by valid names.

The aim of family-group nomenclature is to yield the greatest possible stability and uniformity, just as in other zoological names. Both taxonomic experience and the *Code* (Article 40) indicate the wisdom of sustaining family-group names based on junior subjective synonyms if they have priority of publication, for opinions of the same worker may change from time to time. The retention of first-published, family-group names that are found to be based on junior objective synonyms, however, is less clearly desirable, especially if a replacement name derived from the senior objective synonym has been recognized very long and widely. Moreover, to displace a widely used, family-group name based on the senior objective synonym by disinterring a forgotten and virtually unused family-group name based on a junior objective synonym because the latter happens to have priority of publication is unsettling.

A family-group name may need to be replaced if the nominotypical genus is transferred to another family group. If so, the first-published of the generic names remaining in the family-group taxon is to be recognized in forming a replacement name.

SUPRAFAMILIAL TAXA: TAXA ABOVE FAMILY-GROUP

International rules of zoological nomenclature as given in the *Code* affect only lower-rank categories: subspecies to superfamily. Suprafamilial categories (suborder to kingdom) are either not mentioned or explicitly placed outside of the application of zoological rules. The *Copenhagen Decisions on Zoological Nomenclature* (1953, Articles 59 to 69) proposed adopting rules for naming suborders and higher taxa up to and including phylum, with provision for designating a type genus for each, in such manner as not to interfere with the taxonomic freedom of workers. Procedures were outlined for applying the rule of priority and rule of homonymy to suprafamilial taxa and for dealing with the names of such taxa and their authorship, with assigned dates, if they should be transferred on taxonomic grounds from one rank to another. The adoption of terminations of names, different for each category but uniform within each, was recommended.

The Colloquium on Zoological Nomenclature, which met in London during the week just before the 15th International Congress of Zoology convened in 1958, discussed thoroughly the proposals for regulating suprafamilial nomenclature, as well as many others advocated for inclusion in the new *Code* or recommended for exclusion from it. A decision that was supported by a wide majority of the participants in the colloquium was against the establishment of rules for naming taxa above family-group rank, mainly because it was judged that such regulation would unwisely tie the hands of taxonomists. For example, a class or order defined by an author at a given date, using chosen morphologic characters (e.g., gills of bivalves), should not be allowed to freeze nomenclature, taking precedence over another class or order that is proposed later and distinguished by different characters (e.g., hinge teeth of bivalves). Even the fixing of type genera for suprafamilial taxa would have little, if any, value, hindering taxonomic work rather than aiding it. Beyond mere tidying up, no basis for establishing such types and for naming these taxa has yet been provided.

The considerations just stated do not prevent the editors of the *Treatise* from making rules for dealing with suprafamilial groups of animals described and illustrated in this publication. Some uniformity is needed, especially for the guidance of *Treatise* authors. This policy should accord with recognized general practice among zoologists; but where general practice is indeterminate or nonexistent, our own procedure in suprafamilial nomenclature needs to be specified as clearly as possible. This pertains especially to decisions about names themselves, about citation of authors and dates, and about treatment of suprafamilial taxa that, on taxonomic grounds, are changed from their originally assigned rank. Accordingly, a few rules expressing *Treatise* policy are given here, some with examples of their application.

1. The name of any suprafamilial taxon must be a Latin or Latinized, uninominal noun of plural form or treated as such, with a capital initial letter and without diacritical mark, apostrophe, diaeresis, or hyphen. If a component consists of a numeral, numerical adjective, or adverb, this must be written in full.

2. Names of suprafamilial taxa may be constructed in almost any manner. A name may indicate morphological attributes (e.g., Lamellibranchiata, Cyclostomata, Toxoglossa) or be based on the stem of an included genus (e.g., Bellerophontina, Nautilida, Fungiina) or on arbitrary combinations of letters (e.g., Yuania); none of these, however, can end in -idae or -inae, which terminations are reserved for family-group taxa. No suprafamilial name identical in form to that of a genus or to another published suprafamilial name should be employed (e.g., order Decapoda LATREILLE, 1803, crustaceans, and order Decapoda LEACH, 1818, cephalopods; suborder Chonetoidea MUIR-WOOD, 1955, and genus *Chonetoidea* JONES, 1928). Worthy of notice is the classificatory and nomenclatorial distinction between suprafamilial and family-group taxa that are named from the same type genus, since one is not considered to be transferable to the other (e.g., suborder Bellerophontina ULRICH & SCOFIELD, 1897 is not coordinate with superfamily Bellerophontoidea MCCOY, 1851 or family Bellerophontidae MCCOY, 1851).

3. The rules of priority and homonymy lack any force of international agreement as applied to suprafamilial names, yet in the interest of nomenclatorial stability and to avoid confusion these rules are widely applied by zoologists to taxa above the family-group level wherever they do not infringe on taxonomic freedom and long-established usage.

4. Authors who accept priority as a determinant in nomenclature of a suprafamilial taxon may change its assigned rank at will, with or without modifying the terminal letters of the name, but such changes cannot rationally be judged to alter the authorship and date of the taxon as published originally. A name revised from its previously published rank is a transferred name (*nomen translatum*), as illustrated in the following.

Order CORYNEXOCHIDA Kobayashi, 1935

[*nom. transl.* MOORE, 1959, p. 217, *ex* suborder Corynexochida KOBAYASHI, 1935, p. 81]

A name revised from its previously published form merely by adoption of a different termination without changing taxonomic rank is a *nomen correctum.*

Order DISPARIDA Moore & Laudon, 1943

[*nom. correct.* MOORE in MOORE, LALICKER, & FISCHER, 1952, p. 613, *pro* order Disparata MOORE & LAUDON, 1943, p. 24]

A suprafamilial name revised from its previously published rank with accompanying change of termination, which signals the change of rank, is recorded as a *nomen translatum et correctum.*

Order HYBOCRINIDA Jaekel, 1918

[*nom. transl. et correct.* MOORE in MOORE, LALICKER, & FISCHER, 1952, p. 613, *ex* suborder Hybocrinites JAEKEL, 1918, p. 90]

5. The authorship and date of nominate subordinate and supraordinate taxa among

suprafamilial taxa are considered in the *Treatise* to be identical since each actually or potentially has the same type. Examples are given below.

Subclass ENDOCERATOIDEA Teichert, 1933

[*nom. transl.* TEICHERT in TEICHERT & others, 1964, p. 128, *ex* order Endoceroidea TEICHERT, 1933, p. 214]

Order ENDOCERIDA Teichert, 1933

[*nom. correct.* TEICHERT in TEICHERT & others, 1964, p. 165, *pro* order Endoceroidea TEICHERT, 1933, p. 214]

TAXONOMIC EMENDATION

Emendation has two distinct meanings as regards zoological nomenclature. These are alteration of a name itself in various ways for various reasons, as has been reviewed, and alteration of the taxonomic scope or concept for which a name is used. The *Code* (Article 33.1 and Glossary) concerns itself only with the first type of emendation, applying the term to intentional, either justified or unjustified changes of the original spelling of a name. The second type of emendation primarily concerns classification and inherently is not associated with change of name. Little attention generally has been paid to this distinction in spite of its significance.

Most zoologists, including paleontologists, who have emended zoological names refer to what they consider a material change in application of the name such as may be expressed by an importantly altered diagnosis of the assemblage covered by the name. The abbreviation *emend.* then must accompany the name with statement of the author and date of the emendation. On the other hand, many systematists think that publication of *emend.* with a zoological name is valueless because alteration of a taxonomic concept is introduced whenever a subspecies, species, genus, or other taxon is incorporated into or removed from a higher zoological taxon. Inevitably associated with such classificatory expansions and restrictions is some degree of emendation affecting diagnosis. Granting this, still it is true that now and then somewhat more extensive revisions are put forward, generally with a published statement of the reasons for changing the application of a name. To erect a signpost at such points of most significant change is worthwhile, both as an aid to subsequent workers in taking account of the altered nomenclatorial usage and to indicate where in the literature cogent discussion may be found. Authors of contributions to the *Treatise* are encouraged to include records of all especially noteworthy emendations of this nature, using the abbreviation *emend.* with the name to which it refers and citing the author, date, and page of the emendation. Examples from *Treatise* volumes follow.

Order ORTHIDA Schuchert & Cooper, 1932

[*nom. transl. et correct.* MOORE in MOORE, LALICKER, & FISCHER, 1952, p. 220, *ex* suborder Orthoidea SCHUCHERT & COOPER, 1932, p. 43; *emend.*, WILLIAMS & WRIGHT, 1965, p. 299]

Subfamily ROVEACRININAE Peck, 1943

[Roveacrininae PECK, 1943, p. 465; *emend.*, PECK in MOORE & TEICHERT, 1978, p. 921]

STYLE IN GENERIC DESCRIPTIONS

CITATION OF TYPE SPECIES

In the *Treatise* the name of the type species of each genus and subgenus is given immediately following the generic name with its accompanying author, date, and page reference or after entries needed for definition of the name if it is involved in homonymy. The originally published combination of generic and trivial names of this species is cited, accompanied by an asterisk (*), with notation of the author, date, and page of original publication, except if the species was first published in the same paper and by the same author as that containing definition of the genus of which it is the type. In this instance, the initial letter of the generic name followed by the trivial name is given without repeating the name of the author and date. Examples of these two sorts of citations follow.

Orionastraea SMITH, 1917, p. 294 [**Sarcinula phillipsi* MCCOY, 1849, p. 125; OD].

Schoenophyllum SIMPSON, 1900, p. 214 [**S. aggregatum;* OD].

If the cited type species is a junior synonym of some other species, the name of this latter is given also, as follows.

Actinocyathus D'ORBIGNY, 1849, p. 12 [**Cyathophyllum crenulate* PHILLIPS, 1836, p. 202; M; =*Lonsdaleia floriformis* (MARTIN), 1809, pl. 43; validated by ICZN Opinion 419].

In some instances the type species is a junior homonym. If so, it is cited as shown in the following example.

Prionocyclus MEEK, 1871b, p. 298 [**Ammonites serratocarinatus* MEEK, 1871a, p. 429, *non* STOLICZKA, 1864, p. 57; =*Prionocyclus wyomingensis* MEEK, 1876, p. 452].

In the *Treatise* the name of the type species is always given in the exact form it had in the original publication. Where mandatory changes are required, such as removal of diacritical marks or hyphens, these are introduced later in the text, typically in the description of a figure.

Fixation of Type Species Originally

It is desirable to record the manner of establishing the type species, whether by original designation (OD) or by subsequent designation (SD). The type species of a genus or subgenus, according to provisions of the *Code,* may be fixed in various ways in the original publication; or it may be fixed subsequently in ways specified by the *Code* (Article 68) and described in the next section. Type species fixed in the original publication include (1) *original designation* (in the *Treatise* indicated by OD) when the type species is explicitly stated or (before 1931) indicated by n. gen., n. sp. (or its equivalent) applied to a single species included in a new genus, (2) defined by use of *typus* or *typicus* for one of the species included in a new genus (adequately indicated in the *Treatise* by the specific name), (3) established by *monotypy* if a new genus or subgenus has only one originally included species (in the *Treatise* indicated as M), and (4) fixed by *tautonymy* if the genus-group name is identical to an included species name not indicated as the type.

Fixation of Type Species Subsequently

The type species of many genera are not determinable from the publication in which the generic name was introduced. Therefore, such genera can acquire a type species only by some manner of subsequent designation. Most commonly this is established by publishing a statement naming as type species one of the species originally included in the genus. In the *Treatise* such fixation of the type species by subsequent designation in this manner is indicated by the letters SD accompanied by the name of the subsequent author (who may be the same person as the original author) and the publication date and page number of the subsequent designation. Some genera, as first described and named, included no mentioned species (for such genera established after 1930, see below); these necessarily lack a type species until a date subsequent to that of the original publication when one or more species is assigned to such a genus. If only a single species is thus assigned, it becomes automatically the type species. Of course, the first publication containing assignment of species to the genus that originally lacked any included species is the one concerned in fixation of the type species, and if this publication names two or more species as belonging to the genus but did not designate a type species, then a later SD designation is necessary. Examples of the use of SD as employed in the *Treatise* follow.

Hexagonaria GURICH, 1896, p. 171 [**Cyathophyllum hexagonum* GOLDFUSS, 1826, p. 61; SD LANG, SMITH, & THOMAS, 1940, p. 69].

Mesephemera HANDLIRSCH, 1906, p. 600 [**Tineites lithophilus* GERMAR, 1842, p. 88; SD CARPENTER, herein].

Another mode of fixing the type species of a genus is through action of the International

Commission of Zoological Nomenclature using its plenary powers. Definition in this way may set aside application of the *Code* so as to arrive at a decision considered to be in the best interest of continuity and stability of zoological nomenclature. When made, it is binding and commonly is cited in the *Treatise* by the letters ICZN, accompanied by the date of announced decision and reference to the appropriate numbered opinion.

Subsequent designation of a type species is admissible only for genera established prior to 1931. A new genus-group name established after 1930 and not accompanied by fixation of a type species through original designation or original indication is invalid (*Code,* Article 13.3). Effort of a subsequent author to validate such a name by subsequent designation of a type species constitutes an original publication making the name available under authorship and date of the subsequent author.

HOMONYMS

Most generic names are distinct from all others and are indicated without ambiguity by citing their originally published spelling accompanied by name of the author and date of first publication. If the same generic name has been applied to two or more distinct taxonomic units, however, it is necessary to differentiate such homonyms. This calls for distinction between junior homonyms and senior homonyms. Because a junior homonym is invalid, it must be replaced by some other name. For example, *Callophora* HALL, 1852, introduced for Paleozoic trepostomate bryozoans, is invalid because Gray in 1848 published the same name for Cretaceous–Holocene cheilostomate bryozoans. Bassler in 1911 introduced the new name *Hallophora* to replace Hall's homonym. The *Treatise* style of entry is given below.

Hallophora BASSLER, 1911, p. 325, *nom. nov. pro Callophora* HALL, 1852, p. 144, *non* GRAY, 1848.

In like manner, a replacement generic name that is needed may be introduced in the *Treatise* (even though first publication of generic names otherwise in this work is generally avoided). An exact bibliographic reference must be given for the replaced name as in the following example.

Mysterium DE LAUBENFELS, herein, *nom. nov. pro Mystrium* SCHRAMMEN, 1936, p. 183, *non* ROGER, 1862 [**Mystrium porosum* SCHRAMMEN, 1936, p. 183; OD].

Otherwise, no mention is made generally of the existence of a junior homonym.

Synonymous Homonyms

An author sometimes publishes a generic name in two or more papers of different date, each of which indicates that the name is new. This is a bothersome source of errors for later workers who are unaware that a supposed first publication that they have in hand is not actually the original one. Although the names were published separately, they are identical and therefore definable as homonyms; at the same time they are absolute synonyms. For the guidance of all concerned, it seems desirable to record such names as synonymous homonyms. In the *Treatise* the junior of one of these is indicated by the abbreviation *jr. syn. hom.*

Not infrequently, identical family-group names are published as new names by different authors, the author of the name that was introduced last being ignorant of previous publication(s) by one or more other workers. In spite of differences in taxonomic concepts as indicated by diagnoses and grouping of genera and possibly in assigned rank, these family-group taxa, being based on the same type genus, are nomenclatorial homonyms. They are also synonyms. Wherever encountered, such synonymous homonyms are distinguished in the *Treatise* as in dealing with generic names.

A rare but special case of homonymy exists when identical family names are formed from generic names having the same stem but differing in their endings. An example is the family name Scutellidae RICHTER & RICHTER, 1925, based on *Scutellum* PUSCH, 1833, a trilobite. This name is a junior homonym of Scutellidae GRAY, 1825, based on the echinoid genus *Scutella* LAMARCK, 1816.

The name of the trilobite family was later changed to Scutelluidae (ICZN, Opinion 1004, 1974).

SYNONYMS

In the *Treatise,* citation of synonyms is given immediately after the record of the type species. If two or more synonyms of differing date are recognized, these are arranged in chronological order. Objective synonyms are indicated by accompanying designation *obj.,* others being understood to constitute subjective synonyms, of which the types are also indicated. Examples showing *Treatise* style in listing synonyms follow.

Mackenziephyllum PEDDER, 1971, p. 48 [**M. insolitum;* OD] [=*Zonastraea* TSYGANKO in SPASSKIY, KRAVTSOV, & TSYGANKO, 1971, p. 85, *nom. nud.; Zonastraea* TSYGANKO, 1972, p. 21 (type, *Z. graciosa,* OD)].

Kodonophyllum WEDEKIND, 1927, p. 34 [**Streptelasma Milne-Edwardsi* DYBOWSKI, 1873, p. 409; OD; =*Madrepora truncata* LINNE, 1758, p. 795, see SMITH & TREMBERTH, 1929, p. 368] [=*Patrophontes* LANG & SMITH, 1927, p. 456 (type, *Madrepora truncata* LINNE, 1758, p. 795, OD); *Codonophyllum* LANG, SMITH, & THOMAS, 1940, p. 39, obj.].

Some junior synonyms of either the objective or the subjective sort may be preferred over senior synonyms whenever uniformity and continuity of nomenclature are served by retaining a widely used but technically rejectable name for a genus. This requires action of the ICZN, which may use its plenary powers to set aside the unwanted name, validate the wanted one, and place the concerned names on appropriate official lists.

OTHER EDITORIAL MATTERS

BIOGEOGRAPHY

Purists, *Treatise* editors among them, would like nothing better than a stable world with a stable geography that makes possible a stable biogeographical classification. Global events of the past few years have shown how rapidly geography can change, and in all likelihood we have not seen the last of such change as new, so-called republics continue to spring up all over the globe. One expects confusion among readers in the future as they try to decipher such geographical terms as U.S.S.R., Yugoslavia, or Ceylon. Such confusion is unavoidable, as books must be completed and published at some real time. Libraries would be limited indeed if publication were always to be delayed until the political world had settled down. In addition, such terms as central Europe and western Europe are likely to mean different things to different people. Some imprecision is introduced by the use of all such terms, of course, but it is probably no greater than the imprecision that stems from the fact that the work of paleontology is not yet finished, and the geographical ranges of many genera are imperfectly known.

Other geographic terms can also have varying degrees of formality. In general, *Treatise* policy is to use adjectives rather than nouns to refer to directions. Thus we have used *southern* and *western* in place of *South* and *West* unless a term has been formally defined as a geographic entity (e.g., South America or West Virginia). Note that we have referred to western Texas rather than West Texas, which is said to be not a state but a state of mind.

NAMES OF AUTHORS: TRANSLATION AND TRANSLITERATION

Chinese scientists have become increasingly active in systematic paleontology in the past two decades. Chinese names cause anguish among English-language bibliographers for two reasons. First, no scheme exists for one-to-one transliteration of Chinese characters into roman letters. Thus, a Chinese author may change the roman-letter spelling of his name from one publication to another. For example, the name Chang, the most common family name in the world reportedly held by some one billion people, has been spelled more recently Zhang. The principal purpose of a bibliography is to provide the reader with entry into the literature. Quite arbitrarily, therefore, in the interest of information retrieval, the *Treatise* editorial staff has decided to retain the roman spelling that a Chinese author has used in each of his publications rather than attempting to adopt a common spelling of an author's name to be used in all citations of his work. It is entirely possible, therefore, that the publications of a Chinese author may be listed in more than one place under more than one name in the bibliography.

Second, most but by no means all Chinese list their family name first followed by given names. People with Chinese names who study in the West, however, often reverse the order, putting the family name last as is the Western custom. Thus, for example, Dr. Yi-Maw Chang, formerly of the staff of the Paleontological Institute, was Chang Yi-Maw when he lived in Taiwan. When he came to America, he became Yi-Maw Chang. In the *Treatise,* authors' names are used in the text and listed in the references as they appear in the source being cited.

Several systems exist for transliterating the Cyrillic alphabet into the roman alphabet. On the recommendation of skilled bibliographic librarians, we have adopted the American Library Association/Library of Congress romanization table for Russian and other languages using the Cyrillic alphabet.

MATTERS SPECIFIC TO THIS VOLUME

Some languages, in this volume most notably the Polish and Czech languages, are enriched with the use of diacritical marks that provide enhanced alphabetical diversity. While celebrating diversity, we have nevertheless elected to omit such marks from Polish and Czech geographical terms used in the *Treatise*. We continue to insert diacritical marks in authors' names. Two factors have led us to this editorial decision. First, we in the *Treatise* editorial office typeset electronically all the pages, and such diacritical marks must be inserted by hand into the final computer-prepared pages. This is a costly and time-consuming operation that is fraught with the possibility of introducing errors. Second, in the burgeoning information age of the new millennium, databases and schemes for information retrieval will be of critical importance in managing paleontological information. Stability and uniformity of terminology are requisites of database-management systems, and the use of diacritical marks and computer technology are likely to remain incompatible for some time to come. We hope that linguistic purists will be tolerant of this transgression, which we have undertaken solely in the interest of expediency, consistency, and information retrieval.

In this volume we have taken special pains to acknowledge authorship of chapters and subsections. Readers citing the volume are encouraged to pay close attention to the actual authorship of a chapter or subsection.

Stratigraphic ranges of taxa have been compiled from the ranges of lower taxa. In all instances, we have used the *range-through* method of describing ranges. In instances, therefore, where the work of paleontology is not yet finished, some ranges of higher taxa will not show gaps between the ranges of their subtaxa and may seem to be more complete than the data warrant.

ACKNOWLEDGMENTS

The Paleontological Institute's Assistant Editor for Text, Jill Hardesty, and the Assistant Editor for Illustrations, Jane Kerns, have faced admirably the formidable task of moving this volume through the various stages of editing and into production. In this they have been ably assisted by other members of the editorial team including Jack Keim with photography and computer graphics, Mike Cormack with his outstanding computer skills, Mary Huyck with her work on illustrations, and Denise Mayse and Jean Burgess with general support. Jill Krebs, the remaining member of the Paleontological Institute editorial staff, is involved with preparation of PaleoBank, the paleontological database for future *Treatise* volumes, and has not been closely involved with this volume.

Many of the illustrations in this volume were prepared a number of years ago by the late Roger B. Williams, who served the Paleontological Institute as assistant editor for illustrations until shortly before his death. We remain indebted to him for his dedication to the *Treatise* project, his very high standards, and his synoptic view of the *Treatise*.

This editorial preface and other, recent ones are extensive revisions of the prefaces prepared for previous *Treatise* volumes by former editors, including the late Raymond C. Moore, the late Curt Teichert, and Richard A. Robison. I am indebted to them for preparing earlier prefaces and for the leadership they have provided in bringing the *Treatise* project to its present status.

Finally, on behalf of the members of the staff of the Paleontological Institute and the other authors of this volume, I am pleased to have the opportunity to extend our sincere thanks to Professor J. Keith Rigby for helping us bring this volume to publication. Throughout his career, his work has been marked by scrupulous scholarship, complete dedication to the task, and unwavering attention to detail. He has brought the same approach to this project, and we are grateful to him for all he has done.

REFERENCES

International Commission on Zoological Nomenclature. 1999. International Code of Zoological Nomenclature, 4th edition. International Trust for Zoological Nomenclature. London. 306 p.

Moore, R. C., and C. Teichert. 1978. Treatise on Invertebrate Paleontology. Part T, Echinodermata 2(1). The Geological Society of America and The University of Kansas. Boulder & Lawrence. 401 p.

Robison, R. A., and C. Teichert. 1979. Treatise on Invertebrate Paleontology. Part A, Introduction. The Geological Society of America and The University of Kansas. Boulder & Lawrence. 569 p.

Roger L. Kaesler
Lawrence, Kansas
October 27, 2003

STRATIGRAPHIC DIVISIONS

The major divisions of the geological time scale are reasonably well established throughout the world, but minor divisions (e.g., subseries, stages, and substages) are more likely to be provincial in application. The stratigraphic units listed here represent an authoritative version of the stratigraphic column for all taxonomic work relating to the revision of Part E. They are adapted from the International Union of Geological Sciences 2000 International Stratigraphic Chart, compiled by Jürgen Remane, Chairman of the International Commission on Stratigraphy (ICS), with the collaboration of all ICS Subcommissions. A copy of the chart can be obtained at the following website: http://www.iugs.org/iugs/pubs/intstratchart.htm.

Cenozoic Erathem
- **Quaternary System**
  - Holocene Series
  - Pleistocene Series
- **Neogene System**
  - Pliocene Series
  - Miocene Series
- **Paleogene System**
  - Oligocene Series
  - Eocene Series
  - Paleocene Series

Mesozoic Erathem
- **Cretaceous System**
  - Upper Cretaceous Series
  - Lower Cretaceous Series
- **Jurassic System**
  - Upper Jurassic Series
  - Middle Jurassic Series
  - Lower Jurassic Series
- **Triassic System**
  - Upper Triassic Series
  - Middle Triassic Series
  - Lower Triassic Series

Paleozoic Erathem
- **Permian System**
  - Lopingian Series
  - Guadalupian Series
  - Cisuralian Series
- **Carboniferous System**
  - Pennsylvanian Subsystem
  - Mississippian Subsystem
- **Devonian System**
  - Upper Devonian Series
  - Middle Devonian Series
  - Lower Devonian Series
- **Silurian System**
  - Pridoli Series
  - Ludlow Series
  - Wenlock Series
  - Llandovery Series
- **Ordovician System**
  - Upper Ordovician Series
  - Middle Ordovician Series
  - Lower Ordovician Series
- **Cambrian System**
  - Upper Cambrian Series
  - Middle Cambrian Series
  - Lower Cambrian Series

COORDINATING AUTHOR'S PREFACE

J. Keith Rigby
[Brigham Young University]

This volume is the first to revise and update the Porifera section of the 1955 pioneer volume of *The Treatise of Invertebrate Paleontology, Part E,* by M. W. de Laubenfels. That significant work helped initiate renewed interest in the sponges among the next generation of researchers, an interest also prompted, in part, by the focused studies of N. D. Newell and G. A. Cooper and their students and coworkers on the faunas and facies of the Permian reefs of Texas and New Mexico.

The current work includes this introductory volume and, to follow soon after, a systematic part concerned largely with those forms that have been traditionally included in the Porifera. An additional volume is in preparation documenting those groups relatively newly included in the phylum, such as the stromatoporoids, chaetetids, and sclerosponges. Some minor overlap of these two efforts is inevitable, particularly where taxonomic limits are blurred in the gray area established between categories.

Sponges are relatively simple animals with a long and incomplete geologic record, for many taxa likely left no readable record of their organic skeleton or skeletons of easily disassembled elements, although they contributed to cherty units and spiculites throughout the geologic record. Sponges are common animals in modern seas, yet relatively few paleontologists have focused research efforts on fossil sponges and their geologic record, possibly because of their perceived limited value as time-stratigraphic index fossils and also because their selective preservation of biologic information has made use of classifications of living forms difficult for fossils in some instances because of limitation of what can be seen in the fossils. As a result, records of sponges with solid skeletons, such as the lithistid demosponges or groups of calcareous sponges, have been overplayed because of their ease of preservation compared to the probably large and diverse populations of sponges with easily disarticulated or organic skeletons that are undocumented. Nonetheless more than 1,200 genera of fossil sponges have been described and illustrated here, representing numerous families and orders in each of the traditional classes of the Porifera.

Investigations of fossil Porifera have had cycles of interest when numbers of investigators and, as a consequence, numbers of publications increased and decreased, as documented in the extensive bibliography compiled for the volumes. Peaks of activity were recorded in the 1870–1890s and 1960–1990s, but comparatively few papers were published in the mid-1800s, the 1900–1920s, and during the immediate past few years.

This revision of the Porifera volume is largely the result of three specialists: Robin E. H. Reid, Robert M. Finks, and J. Keith Rigby. Reid and Finks began their compilations and manuscript preparation in the 1970s at the invitation of R. C. Moore, with Reid focusing on Mesozoic and Cenozoic hexactinellid and demosponge taxa and Finks on Paleozoic taxa and the heteractinid and calcareous sponges. Both of them completed manuscripts on several chapters of the introduction and made preliminary compilations on some systematic sections. Reid completed manuscripts on Mesozoic and Cenozoic demospongiid and hexactinellid taxa by the late 1970s but shifted his research interests some time later to work on vertebrate fossils. I became officially involved in 1987 after it became apparent that preparation of manuscripts and illustrations for the fossil sponges needed renewed efforts to complete and update earlier compilations and descriptions, and to include descriptions and illustrations of the various taxa added to the geologic record since the period of active involvement of the first two authors. Numerous colleagues have contributed much and willingly helped with literature and locality and distribution data. They have assisted in helping to settle many problems satisfactorily.

Appreciation is expressed to colleagues for their assistance in preparation of the manuscript and illustrations for this volume of the *Treatise of Invertebrate Paleontology.* In particular I would like to thank Baba Senowbari-Daryan, Institute für Paläontologie, Universität Erlangen-Nürnberg; Andrzej Pisera, Polska Akademia Nauk, Warsaw; Diego Garcia-Bellido Capdevila, Universidad Complutense, Madrid; Loren Babcock, Ohio State University; Dorte Mehl-Janussen, Forschungsinstitut und Naturmuseum Senckenberg, Frankfurt am Main; Fan Jiasong, Institute of Geology, Beijing; Wu Xichun, Chengdu Institute of Geology, China; Rachel Wood, University of Cambridge; Andrey Yu. Zhuravlev, Palaeontological Institute, Russian Academy of Sciences, Moscow; V. J. Goryansky, Leningrad; Barry D. Webby, University of Sidney; and John Pickett, Geological Survey of New South Wales. Appreciation is also extended to Robert B. Blodgett, Oregon State University; Robert E. Sloan and Penny Krosch, University of Minnesota; Wilbert R. Danner, University of British Columbia; Karl W. Flessa, University of Arizona; Fred D. Bosworth, Johns Hopkins University; Carl W. Stock, University of Alabama; Colin W. Stearn, McGill University; Robert J. Elias, University of Manitoba; Françoise Debrenne, Laboratoire de Paléontologie, Paris; Björn E. E. Neuman, University of Bergen; Tomasz Wrzolek, Silesian University, Poland; Daniel C. Fisher, Museum of Paleontology, University of Michigan; Thomas E. Bolton, Geological Survey of Canada, Ottawa; and Desmond H. Collins, Royal Ontario Museum.

I thank the faculty and staff of the Department of Geology, Brigham Young University, for their support and continuing interest, and thank personnel of the Interlibrary Loan Office of the Harold B. Lee Library, also at Brigham Young University, for their invaluable assistance. I also express gratitude for secretarial help in preparation of final verisons of manuscripts at Brigham Young University, some of which was partially funded by a grant from The Paleontological Institute, University of Kansas, which also funded translation by Dmitri Slinkov of critical Russian literature for the compilation. The continued interest and support of editors and the staff of The Paleontological Institute, University of Kansas, through the long process of preparation and production of the manuscript and illustrations, editing the compilations, and, finally, publication of the volume is also much appreciated.

REPOSITORIES AND THEIR ABBREVIATIONS

Abbreviations and locations of museums and institutions holding type material, which are used throughout the volume, are listed below.

AGE: Archiv für Geschiebekunde, Geologisch-Paläontologisches Institut, Hamburg, Germany
AI: Institute of Geological Sciences, Polish Academy of Sciences, Kraków, Poland
AMNH: American Museum of Natural History, New York City, New York, USA
AM or AMu: Australian Museum, Sydney, Australia
BM: Berlin Museum, Berlin, Germany
BMNH: British Museum (Natural History), London, United Kingdom
BMS: Buffalo Museum of Science, Buffalo, New York, USA
BSPGM: Bayerische Staatssammlung für Paläontologie und historische Geologie, München, Germany
BYU: Geology Department, Brigham Young University, Provo, Utah, USA
CCG: Chengdu College of Geology (now Chengdu University of Technology), Chengdu, Sichuan, China
CEGH-UNC: Cátedra de Estratigrafía y Geología Histórica, Universidad Nacional de Córdoba, Córdoba, Argentina
CSGM: Central Siberian Geological Museum, United Institute of Geology, Geophysics, & Mineralogy, Siberian Branch of the Russian Academy of Sciences, Novosibirsk, Russia
CU: University of Cincinnati, Cincinnati, Ohio, USA
CUG: Colgate University, Geology Department Collections, Hamilton, New York, USA
CPC: Bureau of Mineral Resources, Canberra, Australia
CRICYT: Centro Regional de Investigaciones Científicas y Tecnológicas, Mendoza, Argentina
FEGI: Far East Geological Institute, Russian Academy of Sciences, Vladivostok, Russia
FM: Field Museum (Natural History), Chicago, Illinois, USA
GII: Institut für Geologie und Paläontologie der Universität Innsbruck, Innsbruck, Austria
GIK: Geologisch-Paläontologisches Institut, Universität zu Köln, Köln, Germany
GPIMH: Geologisch-Paläontologisches Institut und Museum der Universität Hamburg, Hamburg, Germany
GSC: Geological Survey of Canada, Ottawa, Canada
GSM: British Geological Survey (formerly Geological Survey Museum; Institute of Geological Sciences, London), Keyworth, Nottinghamshire, United Kingdom
GSS: Geological Survey of Scotland, Edinburgh, United Kingdom
GSWA: Geological Survey of Western Australia, East Perth, Australia
HM: Hunterian Museum, University of Glasgow, Glasgow, United Kingdom
IGASB: Institute of Geology, Academia Sinica, Beijing, China
IGPTU: Institut und Museum für Geologie und Paläontologie, Tübingen Universität, Tübingen, Germany
IPFUB: Institut für Paläontologie, Freie Universität, Berlin, Germany
IPPAS: Institute of Palaeobiology, Polish Academy of Sciences, Warsaw, Poland
IPM: Institut de Paléontologie du Muséum national d'Histoire naturelle de Paris, Paris, France
IPUB: Institüt für Paläontologie, Universität Bonn, Bonn, Germany
IPUM: Instituto di Paleontologia, Università di Modena, Modena, Italy
IRSNB: Institut Royal des Sciences naturelles de Belgique, Brussels, Belgium
ISM: Illinois State Geological Survey, Urbana, Illinois, USA, formerly at Illinois State Museum, Springfield, Illinois, USA
IU: Indiana University, Bloomington, Indiana, USA
JPI: Jianghan Petroleum Institute, Jingsha, Hubei, China
KUMIP: University of Kansas, Lawrence, Kansas, USA
LGI: Leningrad Mining Institute, Leningrad, Russia
MCCA: Museo Comunale in Cortina d'Ampezzo, Cortina d'Ampezzo, Italy
MCZ: Museum of Comparative Zoology, Harvard University, Cambridge, Massachusetts, USA
MFGI: Museum Far Eastern Geological Institute, Vladivostok, Russia
MHGI: Museum of the Hungarian Geologic Institute, Budapest, Hungary
MIGT: Museum, Institute of Geology, Dushambe, Tajikistan
MMMN: Manitoba Museum of Man and Nature, Winnipeg, Canada
MMF: Geological and Mining Museum, Sydney, Australia
MNCN: Museo Nacional de Ciencias Naturales, Madrid, Spain
MNHN: Muséum National d'Histoire Naturelle de Paris, Paris, France
MUZ IG: Museum of the State Geological Institute, Warsaw, Poland
NIUPGAS: Nanjing Institute of Geology and Paleontology, Academia Sinica, Nanjing, China
NMV: National Museum of Victoria, Melbourne, Australia
NRM: Naturhistoriska Riksmuseet (Swedish Museum of Natural History), Stockholm, Sweden
NYSM: New York State Museum, Albany, New York, USA

ODM: Old Dominion College, Norfolk, Virginia, USA
OSU: Ohio State University, Department of Geology, Columbus, Ohio, USA
OUZC: Ohio University Zoological Collections, Athens, Ohio, USA
PDMNH-P: Paleontological Department of the National Museum, Museum of Natural History, Prague, Czech Republic
PIUB: Paleontological Institute of the University of Bonn, Bonn, Germany
PIUFB: Paläontologisches Institut, Freie Universität Berlin, Berlin, Germany
PIUW: Paläontologichen Instituts, Universität Wien, Vienna, Austria
PIUZ: Paleontological Institute, University of Zürich, Zürich, Switzerland
PIW: Institut für Paläontologie der Universität Würzburg, Würzburg, Germany
P-MD: Provincial Museum of Danzig, Danzig, Germany
PRM: Peter Redpath Museum, Montreal, Canada
PU: Princeton University, Princeton, New Jersey, USA
ROM: Royal Ontario Museum, Toronto, Canada
SAM: South Australian Museum, Adelaide, Australia
SGIP: Sammlung des Geologisch-Paläontologichen Institutes der Universität Palermo, Palermo, Italy
SMF: Natur-Museum und Forschungs-Institut, Senckenberg, Germany
SPIE: Sammlung des Institut für Paläontologie, Universität Erlangen-Nürnberg, Erlangen, Germany
SPIML: Sammlung des Paläontologischen Institutes der Universität Marburg, Lahn, Germany
SPIT: Sammlung des Paläontologischen Institutes der Universität Tübingen, Tübingen, Germany
SSPHG: Staatliches Sammlung für Paläontologie und historische Geologie, München, Germany
SSSBGF: Stratigraphische Sammlung der Sektion Geowissenschaften der Bergakademia Freiberg, Freiberg, Germany
SUP: Sydney University, Department of Geology, Sydney, Australia
TMM: Texas Memorial Museum, University of Texas, Austin, Texas, USA
TsNIGER: Ts NIGER Museum, Russia
UA: University of Alberta, Edmonton, Alberta, Canada
UAF: University of Alaska, Fairbanks, Alaska, USA
UC: University of Cincinnati, Cincinnati, Ohio, USA
UCC: Chicago Natural History Museum, formerly in Walker Museum, Chicago, Illinois (see also FM), USA
UCM: Universidad Complutense de Madrid, Madrid, Spain
UG: University of Göttingen, Göttingen, Germany
UL: Lodz University, Institute of Geography, Lodz, Poland
UM: University of Minnesota, Minneapolis, Minnesota, USA
UMG: University of Montana, Department of Geology, Missoula, Montana, USA
UNE: University of New England, Armidale, New South Wales, Australia
UPLGS: Université de Paris, Laboratoire de Géologie de la Sorbonne, Paris, France
U-SK: Universitäts-Sammlung zu Kiel, Germany
UTBEG: University of Texas, Bureau of Economic Geology, Austin, Texas, USA
VK: Theo Van Kemper Collection, Amsterdam, The Netherlands
WAGS: Western Australia Geological Survey, Perth, Australia
WAM: Western Australia Museum, Perth, Australia
WIF: Wadi Institute of Himalayan Geology, Dehra Dun, India
WMC: Woodwardian Museum, University of Cambridge, Cambridge, United Kingdom
WMNM: Wesfälisches Museum für Naturkunde, Münster, Germany
YaFAN: Institute of Geology, Yakut Branch, Siberian Division AN SSR, Yakutsk, Russia
YPM: Yale Peabody Museum, New Haven, Connecticut, USA
ZPAL: Institute of Paleobiology, Warsaw, Poland

GENERAL FEATURES OF THE PORIFERA

R. E. H. Reid

[formerly of Department of Geology, The Queens University of Belfast]

INTRODUCTION

Sponges are simple, multicellular, sedentary, marine, and nonmarine aquatic animals that feed and respire by drawing water through their bodies. Circulating water is propelled by the uncoordinated flagella of cells termed choanocytes, unknown in true metazoans and resembling individual choanoflagellates. An internal skeleton is usual and typically consists of calcareous or siliceous **spicules;** but siliceous spicules may be accompanied or replaced by an organic skeletal material (**spongin**). A few modern sponges have no skeleton.

Modern and most fossil sponges generally have been referred to four or five classes: Calcarea Bowerbank, Demospongea Sollas, Hexactinellida Schmidt, Heteractinida Hinde, and Sclerospongia Hartman & Goreau. The Calcarea are sponges with a skeleton of calcareous spicules. The Demospongea have soft parts like those of the Calcarea, but their skeletons consist of siliceous spicules, of siliceous spicules and spongin together, or of spongin fibers only; or sometimes a skeleton may be absent. The Hexactinellida also have siliceous spicules but are distinguished from the Demospongea by the form of the spicules, by absence of spongin, and by histological differences. The recently rediscovered Sclerospongia have been considered by some to represent an additional class that includes modern forms and fossils, such as the stromatoporoids and chaetetids. Other workers include these sponges in the Demospongea. Modern sclerosponges have a skeleton of siliceous spicules, spongin, and basal massive calcareous structures. A small extinct group with distinctively shaped spicules of uncertain but probable calcareous original composition is placed herein into a separate class, the Heteractinida Hinde. Its stratigraphic range is Lower Cambrian to Lower Permian.

ANATOMY AND HISTOLOGY

The general organization of sponges is centered on their water circulation. There are many variations in detail, but all types can be explained as having some modification of a simple basic pattern. This basic sponge has a simple cup-shaped or tubular body attached to a substratum by a closed base but is open at the top (Fig. 1). The part of the body, termed the **lateral wall,** that bounds the central cavity above the fixed base is pierced by many small pores. The central cavity, called the **paragaster** or **spongocoel,** is lined with **choanocytes;** their beating flagella cause water to be drawn in through the small lateral pores and expelled through the large open top, the osculum. Food particles are ingested by choanocytes as the water passes through; digestion is not extracellular, and respiration occurs also.

This very simple pattern is not known as an adult condition in modern sponges, although the simplest Calcarea form branching tubes with simple porous walls and develop from an unbranched tubular larva, the **olynthus.** In most Calcarea and all Demospongea and Hexactinellida, choanocytes do not line the paragastral surface but are restricted to lateral **flagellated chambers.** These are sometimes simply free diverticula of the lateral wall but are usually located within it and not visible externally. Water circulation is more complex, correspondingly, with two main patterns (see below). There are also two main types of histological organization: one in Calcarea and Demospongea, the other in Hexactinellida.

In Calcarea and Demospongea, the lateral wall (Fig. 2) has a gelatinous internal groundmass or **mesenchyme,** which is

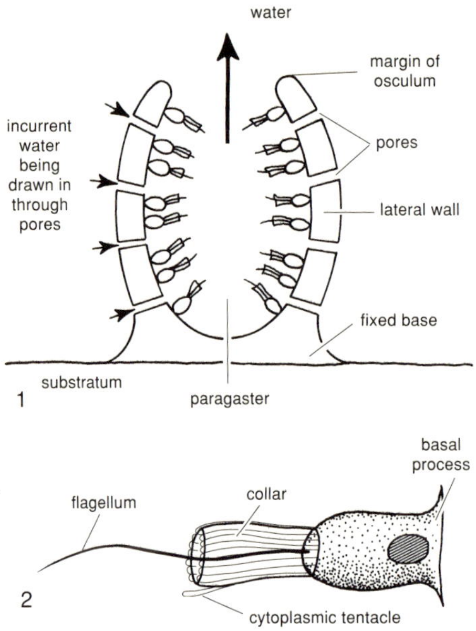

FIG. 1. Basic morphology and water circulation of sponges; *1,* longitudinal section through a simple asconoid sponge to show basic morphological features and water circulation; the section does not correspond with the adult state of any modern sponge, although represented briefly by the olynthus larvae of some Calcarea, and initially asconoid rhagons of a few Demospongea; *2,* single choanocyte, with collar made of cytoplasmic tentacles, and long flagellum (new).

typically coated by flattened dermal cells or **pinacocytes** where choanocytes are absent. The mesenchyme contains cells of various types, some of which are able to move within it, such as the generalized **archaeocytes.** It also includes the cells that secrete the spicules (**scleroblasts** or **sclerocytes**) or spongin (**spongioblasts** or **spongocytes**).

The Calcarea have three grades of circulatory organization (Fig. 3). In ascons, the adult body forms branching, thin-walled tubes, with choanocytes on the paragastral surface. The walls are perforated by intracellular pores that pierce cells termed **porocytes.** Circulating water enters the paragastral cavity directly by way of the pores in this simple condition, which is termed **asconoid** (Fig. 3.1). In **sycons,** choanocytes are restricted to flagellated chambers arranged radially around an axial paragaster (Fig. 3.2). In the simplest or **sycettoid** type, the chambers are separate, lateral diverticula of a central, paragastral tube. In **grantioid** sycons, they are internal spaces in a compact lateral wall, which appears to represent fusion of the separate chambers. Radial passages for ingress of water are left between the chambers and are called **inhalant canals** or **prosochetes;** together, they comprise a **canal system.** The condition of these sponges, in which chambers discharge to the paragaster directly, is called **syconoid.** In **leucons,** internal flagellated chambers are grouped around exhalant canals, the **apochetes,** which carry the excurrent water to the paragastral cavity, and this type of circulation is **leuconoid** (Fig. 3.3). All the known Demospongea are leucons as adults, although some develop from a larval rhagon with internal chambers arranged in the syconoid manner and no canals of either sort. In SOLLAS's (1887) pre-*Challenger* account of the Porifera, the leuconoid type of canal system was described as the rhagon type, because some of the leuconoid Demospongea develop from rhagons. But a rhagon as such (SOLLAS, 1887, 1888) is the larval form only, with chambers,

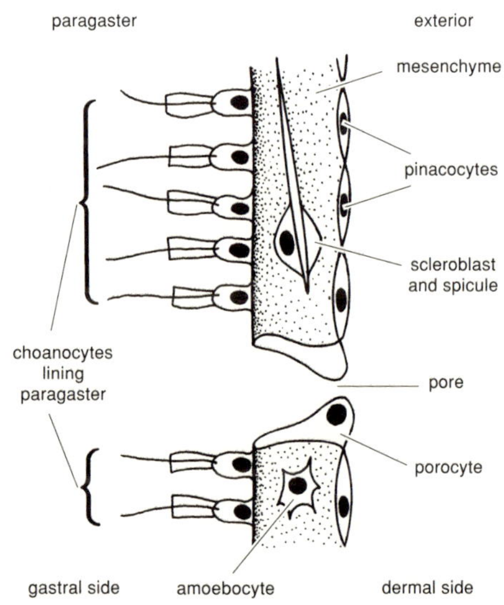

FIG. 2. Longitudinal section of lateral wall in an asconoid member of class Calcarea showing cytological and other features; choanocyte layer at left corresponds with thick lines of Figures 3–4 (new).

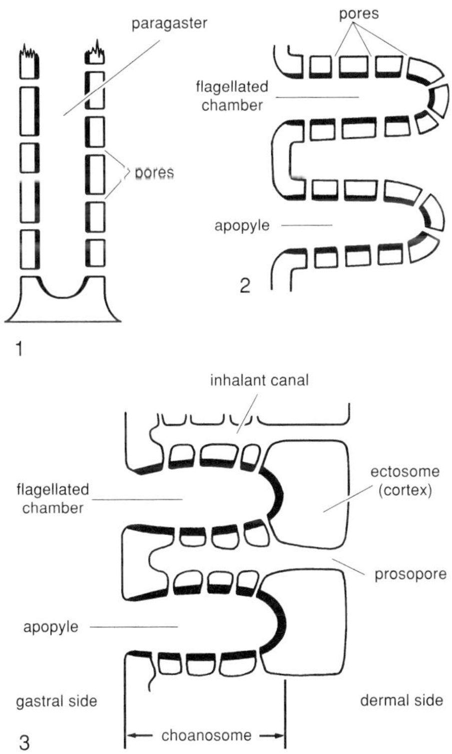

FIG. 3. Asconoid and syconoid Calcarea, in diagrammatic longitudinal sections; *thick lines* refer to choanocyte layer (Fig. 2, left side); *1,* tubular asconoid sponge; *2,* radial diverticula of a sycettoid sycon; *3,* lateral wall of a grantioid sycon, with an ectosome developed as a thick cortex (new).

but without a canal system; and other leucons do not develop from rhagons.

In most Calcarea and Demospongea with enclosed inhalant canals, the external or dermal surface of the body is formed by a compact chamberless layer, the ectosome (Fig. 4–7). The internal part containing the chambers is then called the **choanosome.** An ectosome may be either a thin, skinlike dermis or a thicker, rindlike cortex. Inhalant canals may extend through an ectosome but more commonly begin from larger subdermal spaces underlying it, which the water enters through pores or canals in the ectosome. A similar structure on the exhalant or **gastral** side of the wall is the **endosome.**

This term was proposed as the name of an ectosome-like stratum on the exhalant side of the choanosome, with the inner trabecular network of Hexactinellida (SOLLAS, 1887) and a demosponge structure (SOLLAS, 1888) as examples. Accordingly it must not be used as equivalent to choanosome (e.g., see BURTON, 1963 for correct usage).

Larvae of Hexactinellida are much like those of some Demospongea, but the adults have no mesenchyme and no covering pinacocytes. Flattened choanocytes are connected together syncytially by lateral processes to form a **reticulate** choanocytal membrane, diverticular outgrowths of which form the flagellated chambers (Fig. 5–6). The chambers are supported from both sides by a three-dimensional network of syncytial filaments, the **trabeculae,** comprising a trabecular network. The outer (inhalant) and inner (exhalant) surfaces are formed by **dermal** and **gastral membranes** composed of trabeculae netted and flattened in the plane of the surface. Interspaces of the trabecular network are filled with water from outside, which presumably circulates through them. The spicules are formed in the trabeculae by multinucleate **scleroblast-syncytia.** The arrangement of chambers may be syconoid (Fig. 5) or leuconoid (Fig. 6), with the latter more usual. Simple examples have neither inhalant nor exhalant canals, so that water must reach and leave the chambers entirely

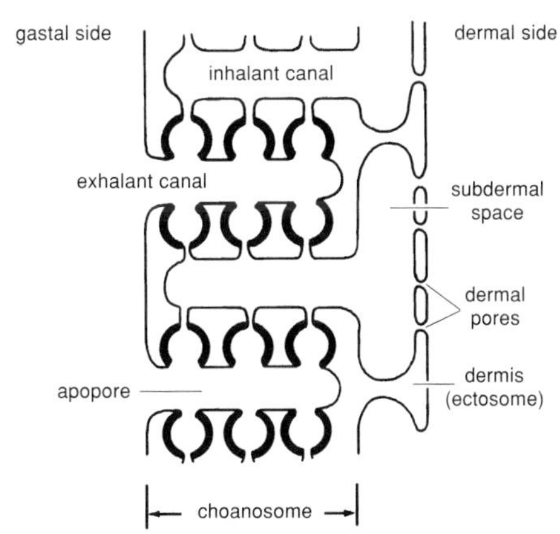

FIG. 4. Diagrammatic longitudinal section of wall of a leuconoid sponge, with ectosome developed as thin, porous dermis covering subdermal spaces (new).

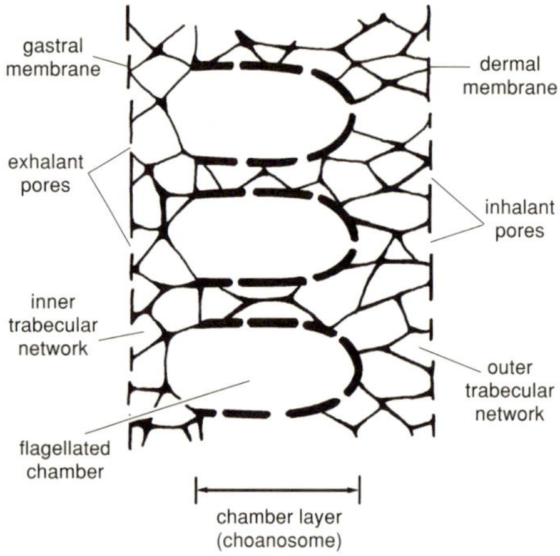

FIG. 5. Diagrammatic longitudinal section of lateral wall in Hexactinellida; choanocytal membrane (*thick lines*) forms radial flagellated chambers; in adult modern Hexactinellida this simple pattern is modified by diverticulation of chambers or folding of chamber layer (see also Fig. 6) (new).

through spaces between trabeculae; in the others, chambers are grouped around exhalant canals, and inhalant canals may also be present between them. Both types of canals are produced by special enlargement of the ordinary intratrabecular spaces and are, thus, not directly related to the similar prosochetes and apochetes of Calcarea and Demospongea. Inhalant canals begin under the dermal membrane or start from larger subdermal spaces underlying it; the **exhalant canals** end under the gastral membrane, end in larger subgastral spaces underlying it, or open through the membrane directly due to secondary fenestration. The external membranes are thus comparable morphologically to the ectosome and endosome of Calcarea and Demospongea, although different histologically.

The internal anatomy of sponges is not significantly modified by differences in **habitus** (i.e., form of the body), but leuconoid types may have modifications of the basic pattern pictured above. In the basic type of sponge body, the inhalant and exhalant (or dermal and gastral) surfaces are external and internal respectively, with the latter forming the lining of a paragastral cavity (Fig. 7.1). In several conditions there is no paragaster. First, expansion of a basically cuplike body can produce a plate- or mushroomlike growth, with the upper surface being homologically gastral and the underside dermal. Water discharged from exhalant canals thus leaves the body directly instead of through an axial paragaster. If the openings or **apopores** of exhalant canals are prominent features of the surface they are often then called **oscula** and function as such, although not homologous with a primary osculum. In **flabellate** sponges, the body is fanlike, tonguelike, or bladelike, with opposite surfaces being dermal and gastral respectively. In other forms the paragastral cavity is suppressed so that no true gastral surface exists. Exhalant canals then open at the top of the body or at various points in the sides. In encrusting forms, water is usually discharged through numerous oscula in an otherwise inhalant upper surface (Fig. 7.2).

It is worth noting here that the term **pore** can have several meanings when applied to Porifera. Pores of the soft parts may be (1)

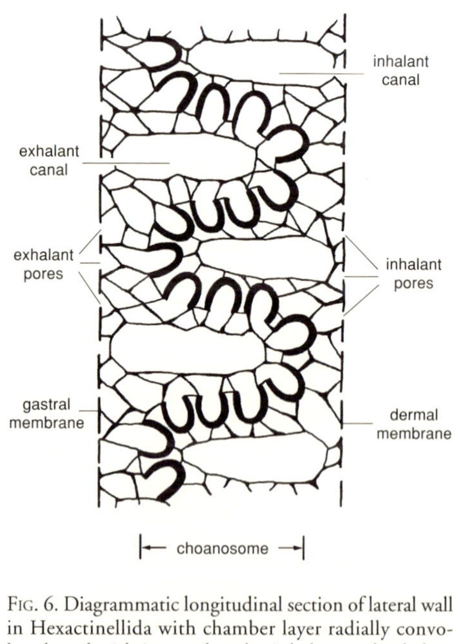

FIG. 6. Diagrammatic longitudinal section of lateral wall in Hexactinellida with chamber layer radially convoluted, and with intertrabecular inhalant and exhalant canals (new).

intracellular perforations of the walls of ascons; (2) perforations in the walls of flagellated chambers, through which water is received from inhalant canals or intertrabecular spaces; (3) external openings (**prosopores**) of inhalant canals or apertures (apopores) of exhalant canals; (4) perforations in the surface of an ectosome or endosome, not related directly to canals of the choanosome; or (5) perforations in the dermal and gastral membranes of Hexactinellida. The same name or variants may also be used for features of skeletal surfaces, in the form of special apertures, where flagellated chambers or canals have entered a skeletal framework. In other words, this term is vernacular and is used to mean any small aperture or perforation without homological implication. It is important not to confuse the different features for which it may be used.

In the first edition of *Treatise Part E* (DE LAUBENFELS, 1955), skeletal pores and canals were called by the same names as the true canals (prosochetes, apochetes) and their apertures (prosopores, apopores). There are several objections to this terminology. First, the skeletal features are not the canals of the soft parts and represent them only imperfectly. In some Hexactinellida, skeletal apopores or apochetes are related to flagellated chambers (e.g., in *Cyclostigma* SCHRAMMEN, *Aphrocallistes* GRAY) instead of to exhalant canals (i.e., apochetes). In genera with no modern species, they could represent either. A skeletal framework may have no canalization in a sponge with both prosochetes and apochetes in the soft parts. Last, many fossils have features of uncertain function. The present authors, therefore, use skeletal pore and skeletal canal as general designations and when function is unknown. Similarly the terms **ostia** and **epirhyses** are used as they were by RAUFF (1893, 1894, 1895), SCHRAMMEN (e.g., 1910, 1912), and IJIMA (1927) when such specific terms can be used.

SPICULES

Typical sponge spicules are produced by secretion of calcite or opaline silica around

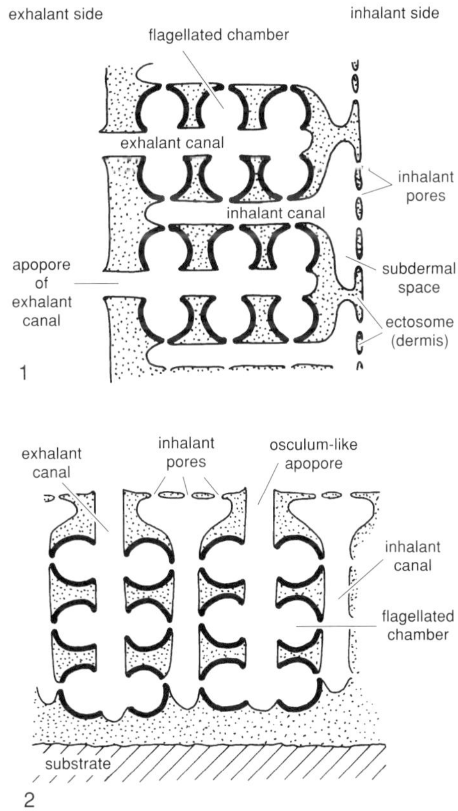

FIG. 7. Diagrams of eurypylous types of canal system; *1,* sponge with separate (and opposite) inhalant and exhalant surfaces; *2,* encrusting form with inhalant pores and apopores on one surface; both examples shown with ectosome developed as a porous dermis covering subdermal spaces from which inhalant canals run inwardly; *thick lines* represent choanocytal linings of flagellated chambers; other parts *finely outlined* and *stippled;* canals, shown here diagrammatically as straight and simple, are normally branched more or less extensively in practice (new).

organic **axial filaments** and are either needlelike or pinlike with a single coring axial filament, or they are stellate bodies consisting of three or more radiating rays cored by filaments that radiate from a spicular center (Fig. 8.4, 8.5, 8.9). **Rays** formed in this manner are **actines,** and their mode of formation is actinal. In other types, initial actinal bodies are modified in various ways by **nonactinal secretion,** producing, for example, regular or irregular outgrowths without organic axes. Some spicules are formed wholly by nonactinal secretion. A raylike

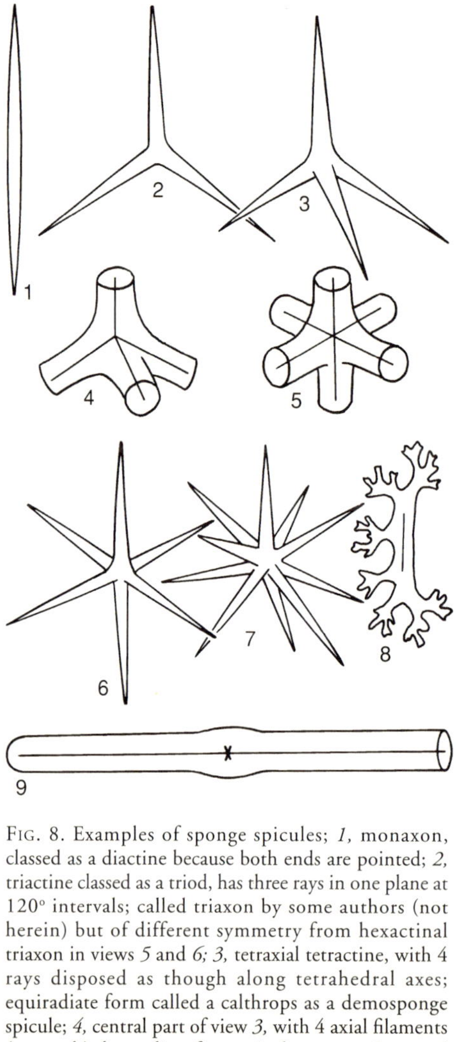

FIG. 8. Examples of sponge spicules; *1,* monaxon, classed as a diactine because both ends are pointed; *2,* triactine classed as a triod, has three rays in one plane at 120° intervals; called triaxon by some authors (not herein) but of different symmetry from hexactinal triaxon in views *5* and *6; 3,* tetraxial tetractine, with 4 rays disposed as though along tetrahedral axes; equiradiate form called a calthrops as a demosponge spicule; *4,* central part of view *3,* with 4 axial filaments (or canals) that radiate from spicular center; *5,* central part of triaxial hexactine, with central axial cross formed by 6 axial filaments, the central part of view *6; 6,* triaxial hexactine with 6 rays that follow 3 axes that intersect at right angles; triod in view *2* is not a variant of this type of spicule; *7,* polyactine, also a polyaxon; *8,* irregularly shaped spicule (desma of a lithistid demosponge), with monaxial character shown by single axial canal; *9,* central part of triaxial diactine from hexactinellid sponge (also called a rhabdodiactine), having triaxial center; 4 aborted rays represented by axial rudiments only (new).

structure without an axial filament is a **pseudoactine,** and individual outgrowths or complete spicules without organic axes are **anaxial.** Actinal calcite or silica secretion often begins while the axial filament is still growing so that this filament projects terminally from a mineral sheath. The end (or each end) is finally covered after ceasing to grow. After death, decay of an axial filament produces a hollow **axial canal,** which is often enlarged by internal solution of the surrounding mineral matter. Such canals may be open at the ends, due to secondary perforation or to growth of the spicule having been incomplete. In fossils, the canals may be lost in diagenesis. Internal solution may also produce misleading **pseudoaxial canals** in originally solid anaxial structures.

The length or diameter of spicules that occur within the body is usually between several micrometers and several centimeters, although larger spicules may occur. Much larger sizes, however, are reached by spicules that some sponges protrude from their bases to form anchoring structures, in which one ray imbedded in the body continues to grow. In the largest known example, the modern hexactinellid *Monoraphis chuni* SCHULZE is supported by a single basal needle reaching more than two meters long and a centimeter thick.

Modern nomenclature of spicules is based generally on that of the authors of the *Challenger* reports (POLÉJAEFF, 1883; RIDLEY & DENDY, 1887; SCHULZE, 1887a; SOLLAS, 1888), with additions by various later authors (e.g., RAUFF, 1893, 1894, 1895). The names used are of various types as described below.

a. Names ending **-actine** (or -act or -actin in some literature) refer usually to the number of rays. Stellate spicules are **triactines, tetractines, pentactines, hexactines,** and others, according to number of rays present, or they are simply called polyactines if the number is more than six (Fig. 8). Pinlike or needlelike spicules are called **monactines** or **diactines** according to whether they are sharp at one end only or both, assuming that this implies growth in one direction or two. Adjectives referring to the number of rays are formed similarly with the termination -actinal, for example, monactinal or hexactinal.

In some types of spicules, the number of axial filaments and the number of rays are identical; but in others, some axial filaments cease to grow shortly after inception, so that no corresponding rays are formed. The rudimentary axial filaments persist at the spicular center as **axial rudiments.** The spicule takes its name from the number of rays developed and not from the number of axial filaments (Fig. 8.9), for example as in the **octactines** that characterize some **heteractinid** sponges where axial filaments are not known.

b. Names ending **-axon** refer to the orientation of rays and axial filaments, which are pictured as following a varying number of growth axes. There are four main types.

1. **Monaxons** are spicules with a single axial filament, in which actinal growth occurs at one end only or both and thus along one axis in one or both directions (Fig. 8.1).

2. **Tetraxons** have four axial filaments arranged as though following tetrahedral axes or in some distorted form of this pattern and have four or fewer rays (Fig. 8.3).

3. **Triaxons** have axial filaments following three axes that intersect at right angles and have six or fewer rays (Fig. 8.5–8.6).

4. **Polyaxons** comprise all spicules in which rays grow in five or more unrelated directions (Fig. 8.7). **Pentiradiate** and **sexiradiate** spicules occur in the heteractinid sponges and are spicules in which five and six rays, respectively, radiate from a center and occur in a single plane.

The term **diaxon** is also used for less important spicules in which rays or axial filaments follow two axes only, which intersect or meet at an angle. This aspect of the character of spicules is expressed adjectivally with terms ending -axial, for example, **monaxial** or **tetraxial.**

The use of terms based on hypothetical growth axes is partly conventional, not literal. In the strictest sense, tetraxons and triaxons are four- and six-rayed spicules, in which the rays follow the ideal tetrahedral and rectangular patterns, but the same terms are applied to distorted variants of these types. Thus, an anchorlike variant of the regular tetraxon, with three rays bent back toward a fourth, is still called a tetraxon, and an irregular variant of a regular triaxial hexactine is similarly still a triaxon. But triaxons and tetraxons do not comprise simply any types of spicules in which growth of rays can be said to follow three or four axes. In particular, spicules with three rays in one plane at equal angles (120°) are literally triaxial, but they are not variants of the six-rayed rectangular triaxon and are called triods, not triaxons.

The status of spicules as monaxons, tetraxons, and so on is not altered by suppression of rays if the axial filaments persist or by formation of secondary anaxial outgrowths that may alter their overall form. For instance, many Hexactinellida contain needlelike spicules with the outward form of monaxons, in which a **spicular center** at some point between the two ends has a six-rayed and triaxial axial cross, with four rays represented by axial rudiments (Fig. 8.9). Such a spicule is not a monaxon but a diactinal triaxon or **rhabdodiactine.** In a group of Demospongea called **lithistids,** articulated spicules that form a skeletal framework may begin their development as obvious monaxons or anaxial corpuscles but come to resemble tetraxons or polyaxons through formation of solid, raylike outgrowths (Fig. 8.8); these spicules are **monaxial** and **anaxial,** not **tetraxial** or **polyaxial.**

Arrangement of spicules according to the number of axes has often been given first place in their classification. Herein the class of Porifera in which particular spicules occur is taken as the first consideration, because similar morphological types can have different relationships in different classes.

c. Many names of individual types of spicules are based on various aspects of their form, including resemblances to other objects. A regular tetraxon with four equal rays is called a **calthrops,** from its resemblance to a caltrop; but a tridentlike tetraxon, with one ray long and three short, is a triaene. The

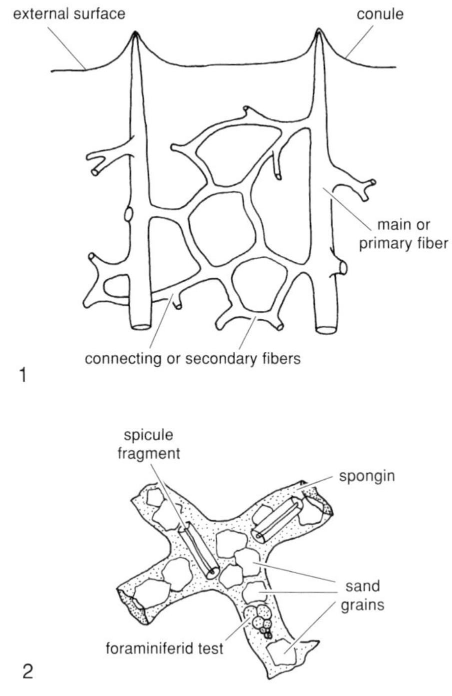

FIG. 9. Keratose skeletons of dictyoceratid type; *1,* section showing primary (main) and secondary (connecting) fibers, and conules where surface is raised by ends of primary fibers; *2,* fibers cored by foreign bodies, here shown as mainly sand grains but also includes spicule fragments and a foraminifer test; spongin *stippled* (new).

monaxon-like triaxial diactines of Hexactinellida are rhabdodiactines, in contrast to **orthodiactines** in which two rays meet at right angles. A cross-shaped tetractine is a **stauractine.** Various starlike spicules are called **asters** and given individual names ending -aster. In monaxons, nomenclature is based on the form of the ends; for instance, a monaxon is an **oxea** if both ends are sharp, a **style** if one is sharp but one blunt, and a **strongyle** if both are blunt. The form of the rays in other spicules may be shown by adding prefixes (e.g., oxy-, sharp, in oxyhexactine; **acanth-**, in **acanthostyle,** implies a spiny ornament).

d. In some instances, different names have come to be used for similar spicules of different classes. In particular, triactines and tetractines of Calcarea are often called **triradiates** and **quadriradiates** but are similar morphologically (though not ontogentically) to spicules called **triods** and **triaenes** in Demospongea. Some terminology used herein is different from that of the first edition of *Treatise Part E* (DE LAUBENFELS, 1955), for example, **triaxial** and **triaxon.** The term triaxial was introduced by SCHMIDT (1870) to distinguish from demosponge tetraxons the spicules of the Hexactinellida in which up to six rays follow three axes at right angles. This is the meaning of triaxial and triaxon in the *Challenger* volumes (SCHULZE, 1887a; SOLLAS, 1888) and in most later literature (e.g., RAUFF, 1893, 1894, 1895); but earlier VON LENDENFELD (1887) and SOLLAS (1887) used triaxon to include triods. This was an error pointed out by SOLLAS (1888) himself but copied by DE LAUBENFELS (1955). Herein we give the term triaxon its usual meaning. The triod and triaxon are both literally triaxial but need to be distinguished as having entirely different symmetries; the triod is not found in Hexactinellida, and those of Calcarea and Demospongea show relationship to tetraxons.

Tetraxons, as defined by the *Challenger* authors (SCHULZE, 1887a; SOLLAS, 1888), who followed SCHMIDT (1870) and VOSMAER (1882, 1883, 1884, 1885, 1887), and by nearly all later authorities (e.g., RAUFF, 1893, 1894, 1895), comprise spicules of the types of the demosponge calthrops and triaene, with four rays arranged as though following tetrahedral axes. They do not include the cross-shaped stauractines of Hexactinellida, which are properly triaxons or diaxons according to whether there are axial rudiments of the two additional rays of a hexactine (SCHULZE, 1887a; RAUFF, 1893, 1894, 1895; IJIMA, 1927).

SPONGIN

Spongin is a protein material secreted by some Demospongea but not found in other Porifera. Forms whose principal spicules are monaxons may have them cemented together by spongin or imbedded in reticulate spongin fibers. In **keratose** sponges only spongin is secreted, and the skeleton consists of reticulate spongin fibers or of branching fibers that are separate except at a common base (Fig. 9). The two types of spongin skeleton are distinguished as **dictyoceratid** and **dendroceratid.** The secreted skeleton is sometimes reinforced with foreign material, comprising a **xenoskeleton.**

Spongin may also occur in nonskeletal roles as fine filaments or membranes.

OTHER SKELETAL CHARACTERISTICS

Spicular skeletons of sponges vary widely in character and in various ways. Different types of spicules occur in different classes, and not all the types that occur are found in any one genus. Spicules are sometimes all similar but commonly vary in form, size, or both. When two or more types are present in one skeleton, they are often distributed differently. The spicules of some sponges are simply embedded in the soft parts without other connection, but others have spicules cemented, articulated, or fused together to form a rigid skeletal framework. Flagellated chambers or a canal system may then cause interruption of skeletal meshwork and sometimes other modifications. Some sponges have special types of spicules protruded from their surfaces, and these may take part in formation of anchoring structures.

When spicules of a sponge are of two or more sorts that differ in form, size, function, or some combination of these factors, each sort present is a category of spicules. As simple examples, the spicules may consist of monaxons and tetraxons of similar or different sizes or of monaxons of two or three sizes that do not intergrade. Herein names of the categories are those of the spicules themselves. Further terms are used where spicules differ in distribution and function. For instance, a spicule may occur in all parts of the body or be restricted to the choanosome or to an ectosome; it is then correspondingly somal, choanosomal, or ectosomal. Protruded spicules forming anchoring structures are classed as **radical** or called **basalia.** Details of the types of spicules present with their distribution and function are termed the **spiculation.**

In most Demospongea and all Hexactinellida, the spicules are differentiated into two main functional categories comprising main supporting elements, or **megascleres,** and accessory elements, or **microscleres.** As implied by the names, these are typically of large and small sizes, respectively. It is difficult to generalize, but the lengths or diameters of many examples fall in ranges above and below about 0.1 mm. Sometimes, however, a spicule that is normally a microsclere grows to megascleric size; it is usually then still called a microsclere homologically, although some authors call it a megasclere simply because of its size. The types of spicules found as megascleres and microscleres vary widely in different sponges. The two may be similar in form but are commonly different, and some types of spicules occur only as megascleres or microscleres. For instance, the demosponge triaene is always a megasclere; on the other hand, many-rayed spicules occur as microscleres in some Demospongea but never as demosponge megascleres. Megascleres are often differentiated further into different types supporting the interior (choanosome) and surface parts (e.g., dermis, dermal membrane), and different types of microscleres may also be differently distributed.

Spicules forming rigid skeletal frameworks are united by calcareous cement (**sclerosome**) in the Calcarea (Fig. 10.1), by articulation (**zygosis**) in the Demospongea (Fig. 10.2), and by fusion in the Hexactinellida (Fig. 10.3). They are always megascleres in the two latter classes, in which distinct microscleres are present, and are typically those of the choanosome, although megascleres supporting the surfaces may also be united. A skeletal framework remains intact after death unless physically broken, although other loose spicules that occur are usually lost with the soft parts unless trapped in the meshes. In forms with loose spicules only, the skeleton collapses as the soft parts

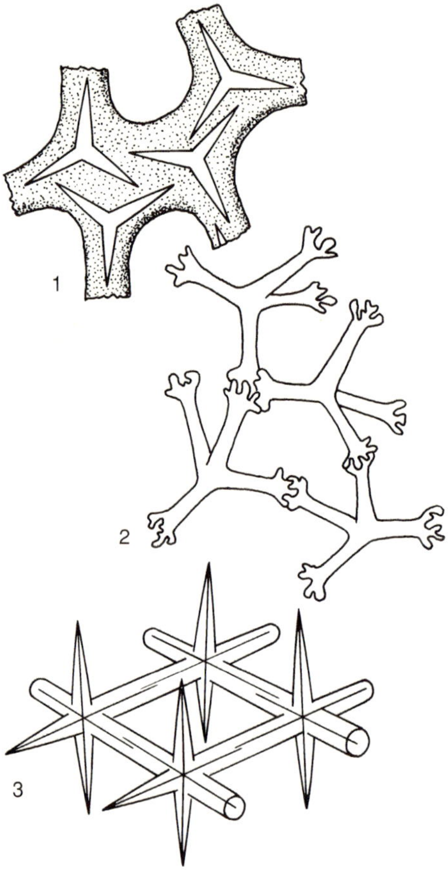

FIG. 10. Rigid spicular skeletons; *1,* triactines of a member of class Calcarea embedded in calcareous cement (sclerosome); *2,* tetraxial desmas of a member of class Demospongea, order Lithistida, united by articulation (zygosis); *3,* parenchymal hexactines of a member of class Hexactinellida, united by fusion during development (ankylosis) to form part of a rigid dictyonal framework; positions of individual dictyonal hexactines indicated by their axial filaments (new).

decay, and the spicules are dispersed by current action. In consequence, most fossil Porifera are forms with **skeletal frameworks.**

In some forms with skeletal frameworks, flagellated chambers and inhalant and exhalant canals are small enough to lie within the meshes of continuous skeletal meshwork. But in other forms canals and sometimes chambers are larger than the normal skeletal meshes and cause interruption of meshwork that develops around them. This modification of the skeleton is skeletal canalization. It is also seen in keratose sponges with reticulate skeletons producing, for instance, the larger holes and channels in a bath sponge. The features produced may be restricted to meshwork at the surface or penetrate the skeleton deeply, with inhalant and exhalant features overlapping within it. These features persist when the soft parts decay and give an indication of their character, although not a complete one. An internal cavity, for instance, may represent only the principal trunk of a canal whose smaller branches pass out through the lining skeletal meshes. Apertures in skeletal surfaces due to canalization are called **skeletal pores** herein, and internal canal-like passages are **skeletal canals.** Skeletal pores and canals are ostia and epirhyses (RAUFF, 1893, 1894, 1895; SCHRAMMEN, 1910, 1912; IJIMA, 1927), if identified certainly as inhalant, and **postica** and **aporhyses,** if certainly exhalant. Some special types of internal features are also distinguished (IJIMA, 1927). But many fossil sponges have skeletal pores or canals that cannot be identified functionally so that use of these terms only is more convenient. Absence of canalization in a fossil skeletal framework does not mean that the sponge lacked flagellated chambers or even a canal system; on the contrary, chambers must always be assumed to have been present, with, for example, inhalant and exhalant canals in any instance in which the modern examples are leucons.

Skeletal pores and canals that arise by internal interruption of growing skeletal meshwork are classed as **intraskeletal.** Accounts that represent canalization in some fossil Hexactinellida (e.g., *Ventriculites* MANTELL) as produced by plication of the wall are incorrect and are badly inconsistent with the nature of the soft parts of modern forms. The skeletal canals of these sponges are true intraskeletal features, produced as in other Porifera. Diverticulation of the choanocytal membrane to form flagellated chambers and of the basic chamber layer producing leuconoid conditions are purely internal occurrences, involving no folding of the wall as a whole. The inhalant and exhalant canals of Hexactinellida are enlarged intertrabecular spaces, produced without folding of any sort.

FORM AND ATTACHMENT

The external form or habitus of sponges varies widely. Simple examples form cups or hollow cylinders or expand into funnel-like or mushroomlike shapes. In others, hollow tubes branch dendritically, divide and anastomose repeatedly, or radiate from the outside of an axial tube or funnel. Longitudinal folding of the wall or a basically funnel-like growth can produce features ranging from closely spaced plications to radiating, finlike flanges. Some forms with no paragaster or spongocoel are massive, taking globular, pyriform, cushionlike or irregular shapes; others are tuberlike or bushlike or are flabellate sponges, of fanlike or similar shapes, with opposite sides inhalant and exhalant, or with exhalant apertures on both sides. Another sponge forms thin crusts, with oscula of a number of separate paragaster-like cavities. Some sponges grow in forms intermediate between various types now described. There is also a variety of minor and special developments. For instance, the main mass of the body may be raised on a stalk, or a lateral wall may be pierced by local apertures called **parietal gaps.** In some sponges outward form is modified by secondary structures, which arise at the end of normal growth; in Hexactinellida, a terminal osculum may be covered by a porous transverse **diaphragm** or **sieve plate,** or a body consisting of dividing and anastomosing tubes may be enclosed in a capsule.

Many sponge species conform to one habitus, although individuals may vary in their detailed development. Others are more variable and may have gradations through two or more of the types that are distinct in other sponges. Variation in form is presumably genetic in basis, but environment may also have an influence, and one species may then take different forms under different conditions (see Variability and Variation, p. 223).

Sponges are attached to the substratum by an encrusting base, by imbedded rootlike outgrowths of the basal part of the body, or by protruded spicules imbedded in an underlying sediment and typically forming a beardlike or ropelike **root tuft.** They are then classed accordingly as **basiphytes, rhizophytes,** or **lophophytes.** A few have no type of attachment and are simply anchored by their weight or capable of drifting (see also Functional Morphology and Adaptation, p. 219–220; and Ecology and Paleoecology, p. 243).

INDIVIDUALITY

Sponges have less distinct individuality than true Metazoa, and what comprises an individual may be problematical. Forms with regular shapes have a generally individual character; but sometimes a group of what are normally distinct individuals are produced from a single common base. Closely spaced sponges of one species may also grow together basally or so as to form a single composite mass when branching, massive, or encrusting. Sponges with more than one osculum are often interpreted as colonies, from an argument (MINCHIN, 1900) that each osculum marks an individual; but multiple oscula can arise in a variety of ways, for example, by transverse constriction and division of a single primary osculum, by lateral budding, or by apopores assuming the status of oscula.

APPENDIX: TERMINOLOGY

Some of the terminology in sections of this volume may be different from that of the first edition of the *Treatise Part E* (DE LAUBENFELS, 1955).

triaxial, triaxon. The term triaxial was introduced by SCHMIDT (1870) to distinguish spicules of the Hexactinellida, in which up to six rays follow three axes at right angles, from demosponge tetraxons. This is the meaning of triaxial and triaxon in the *Challenger* volumes (SCHULZE, 1887a; SOLLAS, 1888) and in most later literature (e.g., RAUFF, 1893, 1894, 1895); but earlier VON LENDENFELD (1887) and SOLLAS (1887) used triaxon to include triods. This was an error, pointed out by SOLLAS himself (1888) but copied by DE LAUBENFELS (1955). The present authors give the term triaxon its usual meaning. The triod and triaxon are both literally triaxial but need to be distinguished as having entirely different symmetries; the triod is not found in Hexactinellida, and those of Calcarea and Demospongea are related to tetraxons.

tetraxon. Tetraxons, as defined by the *Challenger* authors (SCHULZE, 1887a; SOLLAS, 1888), who followed SCHMIDT (1870) and VOSMAER (1882, 1883, 1884, 1885, 1887), and by nearly all later authorities (e.g., RAUFF, 1893, 1894, 1895), comprise spicules of the types of the demosponge calthrops and triaene, with four rays arranged as though following tetrahedral axes. They do not include the cross-shaped stauractines of Hexactinellida, which are properly triaxons or diaxons according to whether there are axial rudiments of the two additional rays of a hexactine (SCHULZE, 1887a; RAUFF, 1893, 1894, 1895; IJIMA, 1927).

rhagon. In SOLLAS's (1887) pre-*Challenger* account of the Porifera, the leuconoid type of canal system was described as the rhagon type because some of the leuconoid Demospongea develop from rhagons. But a rhagon as such (SOLLAS, 1887, 1888) is the larval form only, with chambers but without a canal system; and other leucons do not develop from rhagons.

endosome. This term was proposed as the name of an ectosome-like stratum on the exhalant side of the choanosome, with the inner trabecular network of Hexactinellida (SOLLAS, 1887) and a demosponge structure (SOLLAS, 1888) as examples. Accordingly, it must not be used as equivalent to choanosome and is used correctly, for example, by BURTON (1963).

skeletal pores and canals. In the first edition of *Treatise Part E* (DE LAUBENFELS, 1955), these features of the rigid skeleton were called by the same names as the true canals (prosochetes, apochetes) and their apertures (prosopores, apopores), which are soft part structures in living sponges. There are several objections to this method. First, the skeletal features are not the canals of the soft parts and represent them only imperfectly. In some Hexactinellida, skeletal apopores or apochetes are related to flagellated chambers (e.g., in *Cyclostigma* SCHRAMMEN, *Aphrocallistes* GRAY), instead of exhalant canals (i.e., apochetes). In genera with no modern species, they could represent either. A skeletal framework may have no canalization in a sponge with both prosochetes and apochetes in the soft parts. Last, many fossils have features of uncertain function. Herein, therefore, the terms skeletal pore and skeletal canal are used both as general designations and when function is unknown. In addition, terms (ostia, epirhyses, etc.) used by RAUFF (1893, 1894, 1895), SCHRAMMEN (1910, 1912), and IJIMA (1927) are used herein when specific terms can be used.

CLASSIFICATION

J. K. Rigby

[Department of Geology, Brigham Young University]

The outline classification presented below is principally of fossil sponges treated in the systematic volume of *Treatise Part E (Revised),* vol. 3 (in press). It summarizes taxonomic relationships and presents stratigraphic ranges of the various taxa. Numbers in parentheses indicate the number of recognized fossil genera in each suprageneric group, from family to class, with the number of subgenera included being the second number, listed after the semicolon.

The order of taxa in the outline represents taxonomic relationships and may be different from the order presented in the systematic volume because taxa in those sections are generally separated into Paleozoic and Mesozoic-Cenozoic occurrences. As a result, some major categories may be represented in two or three systematic sections. For example, the section treating hypercalcified sponges documents sponges strictly included in the class Calcarea and also some demosponges (noted by the symbol ** after the stratigraphic ranges) that have calcareous skeletons. In other sections, a family with a long stratigraphic range may be represented in both Paleozoic and Mesozoic sections or chapters where those demosponges or hexactinellid sponges are treated.

SUMMARY OF CLASSIFICATION

Phylum Porifera Grant, 1836 (1,183;30)
- Class Demospongea Sollas, 1875 (615;11)
  - Subclass Clavaxinellida Lévi, 1956 (54)
    - Order Protomonaxonida Finks & Rigby, herein (39)
    - Order Clavulina Vosmaer, 1887 (15)
  - Subclass Choristida Sollas, 1880 (52;2)
    - Order Plakinida Reid, 1968 (5)
    - Order Pachastrellida Reid, herein (13)
    - Order Ancorinida Reid, 1968 (19;2)
    - Order Craniellida Reid, 1968 (2)
    - Order Uncertain (12)
    - Order and Family Uncertain (1)
  - Subclass Tetractinomorpha Lévi, 1953 (30)
    - Order Streptosclerophorida Dendy, 1924 (13)
      - Suborder Eutaxicladina Rauff, 1894 (13)
    - Order Hadromerida Topsent, 1898** (17)
  - Subclass Ceractinomorpha Lévi, 1953 (479;9)
    - Order Dictyoceratida Minchin, 1900 (5)
    - Order Dendroceratida Minchin, 1900 (0)
    - Order Verongida Bergquist, 1978 (3)
    - Order Halichondrida Topsent, 1898 (2)
    - Order Poecilosclerida Topsent, 1928 (16)
    - Order Haplosclerida Topsent, 1898 (18)
    - Order Sigmatosclerophorida Burton, 1956 (4)
    - Order Agelasida Verrill, 1907** (145)
    - Order Vaceletida Finks & Rigby, herein** (51)
    - Order Lithistida Schmidt, 1870 (114)
      - Suborder Orchocladina Rauff, 1895 (114)

Order Tetralithistida Lagneau-Hérenger, 1962 (93;9)
- Suborder Tetracladina Zittel, 1878 (70;9)
- Suborder Dicranocladina Schrammen, 1924 (16)
- Suborder Pseudorhizomorina Schrammen, 1901 (2)
- Suborder Didymmorina Rauff, 1893 (5)

Order Megalithistida Reid, herein (28)
- Suborder Helomorina Schrammen, 1924 (6)
- Suborder Megamorina Zittel, 1878 (22)

Order Axinellida Bergquist, 1967 (1)
Order Uncertain (1)
Order and Family Uncertain (5)
Order Monolithistida Lagneau-Hérenger, 1955 (18)
- Suborder Megarhizomorina Schrammen, 1924 (2)
- Suborder Sphaerocladina Schrammen, 1910 (16)

Order and Suborder Uncertain (1)
Order Spirosclerophorida Reid, 1963 (114;9)
- Suborder Rhizomorina Zittel, 1878 (88;9)
- Suborder Uncertain (26)

Class Hexactinellida Schmidt, 1870 (432;19)
- Subclass Amphidiscophora Schulze, 1887 (160)
  - Order Amphidiscosa Schrammen, 1924 (41)
  - Order Reticulosa Reid, 1958 (118)
  - Order Hemidiscosa Schrammen, 1924 (1)
- Subclass Hexasterophora Schulze, 1887 (272;19)
  - Order Lyssacinosa Zittel, 1877 (36)
  - Order Hexactinosa Schrammen, 1903 (134;8)
  - Order Lychniscosa Schrammen, 1903 (81;11)
  - Order and Family Uncertain (6)
  - Order Uncertain (15)

Class and Order Uncertain (2)
Class Heteractinida de Laubenfels, 1955 (32)
- Order Octactinella Hinde, 1887 (26)
- ?Order Hetairacythida Bedford & Bedford, 1937 (4)
- Order and Family Uncertain (2)

Class, Order, and Family Uncertain (3)
Class Calcarea Bowerbank, 1864 (57)
- Subclass Calcinea Bidder, 1898 (3)
  - Order Clathrinida Hartman, 1958 (???)
  - Order Murrayonida Vacelet, 1981 (3)
- Subclass Calcaronea Bidder, 1898 (54)
  - Order Leucosolenida Hartman, 1958 (0)
  - Order Sycettida Bidder, 1898 (4)
  - Order Stellispongiida Finks & Rigby, herein (32)
  - Order Sphaerocoeliida Vacelet, 1979 (5)
  - Order Lithonida Doederlein, 1892 (12)
  - Order and Family Uncertain (1)

Class and Order Uncertain (29)
Class Uncertain (13)
Unrecognizable Genera (235)

CLASS DEMOSPONGEA: GENERAL MORPHOLOGY AND CLASSIFICATION

R. E. H. Reid

[formerly of Department of Geology, The Queens University of Belfast]

The Demospongea are sponges with mesenchyme and **pinacoderm**, as in Calcarea, but with a skeleton of siliceous spicules, of organic matter (spongin), of both, or with no skeleton. The spicules are monaxial to polyaxial or anaxial, and four-rayed spicules are characteristically tetraxial. Spicules with more than four rays occur normally only as microscleres.

No single skeletal character distinguishes all Demospongea from other sponges. The spicules vary widely, and different kinds occur singly or in numerous different combinations. Most living genera with spicules have either both megascleres and microscleres or megascleres only; but a few have microscleres only or small spicules not regarded as properly either megascleres or microscleres. Megascleres and microscleres may be similar but are often of different types so that megascleres are diactines or tetractines, for example, but microscleres are polyactines. Three main types of skeletons occur in forms containing megascleres. In **choristid** Demospongea, the **choristids**, some or all the megascleres are tetractinal or triactinal, although monaxons (usually diactines) may also be present, and all megascleres are loose in the soft parts without articulation or connection by spongin. In **monaxonids** the megascleres are monaxons only, developed as diactines or monactines, and may be loose in the soft parts, cemented together by spongin, or embedded in spongin fibers.

Lithistids have some or all the megascleres in the form of articulated **desmas**, which form a stony skeletal framework. The desmas are usually developed from initial monaxons or tetraxons but have articulatory parts (**zygomes**) and sometimes raylike arms (**clones**) that are characteristically anaxial. Some desmas are entirely anaxial. The anaxial structures of lithistid desmas do not occur in megascleres of choristid or monaxonid sponges. In keratose sponges with a skeleton composed of spongin only, the skeleton consists of dendritic or reticulate fibers or may also include loose spongin spicules when the fibers are dendritic.

Fossil Demospongea are recognized by occurrence of one of the three main types of megascleric skeletons, by traces of a spongin skeleton, or by dissociated megascleres or microscleres of types only found in Demospongea. Nearly all forms containing megascleres are distinguished from Hexactinellida by the absence of triaxial examples (hexactines and variants), and some are distinguished further by occurrence of tetraxons (calthrops, triaenes, tetraxial desmas), which do not occur in Hexactinellida. Union of megascleres solely by articulation (zygosis) distinguishes lithistid skeletal frameworks from structures formed in Calcarea or Hexactinellida by cementation or fusion of spicules. Very rarely, some megascleres of Hexactinellida may be articulated, but lithistid desmas are never triaxial.

Some fossil sponges are known that may be either Hexactinellida or monaxonid Demospongea. This taxonomic problem arises because the principal megascleres of some hexactinellids are monaxon-like diactines (or rhabdodiactines), which cannot be recognized as hexactinellidan spicules from their outward form only. In living examples other spicules are obvious triaxons (hexactines and others: see class Hexactinellida, *Treatise Part E (Revised),* vol. 3, in press), and an **axial cross** between the ends of most diactinal megascleres has rudimentary axial filaments of two or four undeveloped rays; but most fossils have the main megascleres only, and the character of their axial systems is often not demonstrable. If a sponge with diactinal megascleres of

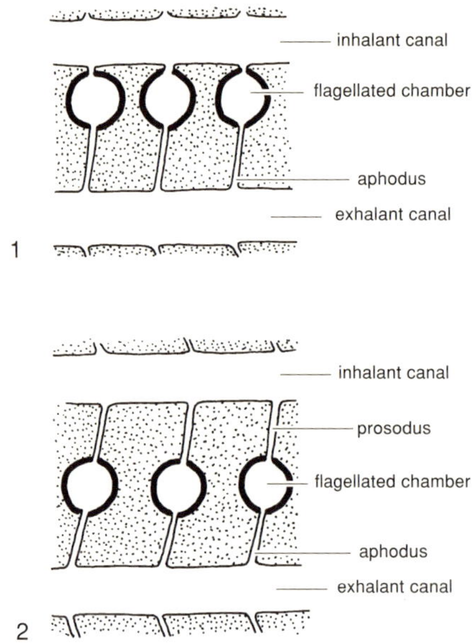

FIG. 11. Aphodal and diplodal canal systems; *1,* aphodal type, in which flagellated chambers discharge to exhalant canals proper by fine ducts termed aphodi; *2,* diplodal type, with aphodi as in *1,* and with additional fine ducts termed prosodi through which water is led to flagellated chambers from inhalant canals proper; as in eurypylous sponges (see Fig. 7), the canals are branching passages in practice, and incurrent and excurrent water may pass through different surfaces of the body or different parts of the same surface; *thick lines* represent choanocytal linings of flagellated chambers; other parts *finely outlined and stippled* (new).

undetermined character has the spicules united by fusion or by lateral **synapticula,** often running like the rungs of a ladder between parallel diactines, it belongs to the Hexactinellida. If, however, the spicules are unconnected, there is strictly no way of deciding which class is represented; but comparisons with forms of known position sometimes suggest that such sponges may be Hexactinellida. No sponge with loose diactines only, however, can be shown conclusively to be a hexactinellid unless the megascleres are demonstrably diaxial or triaxial.

The apparent stratigraphic range of the class Demospongea is Lower Cambrian to recent, although the oldest forms included are supposed monaxonids without certainly diagnostic spicules.

SOFT PARTS

The soft parts of adult Demospongea are broadly similar to those of leuconoid Calcarea, with various special features. As in the Calcarea, a mesenchymal groundmass is present. Choanocytes are restricted to flagellated chambers, which are typically globular, and always receive water from inhalant canals and discharge it through exhalant canals. External and canalar surfaces are coated by a layer of discrete pinacocytes, by a syncytial **epithelioid membrane,** or, in some keratose sponges, apparently by membranous spongin. There is nearly always an ectosome, developed as a thin, skinlike dermis or as a thick and often tough and fibrous cortex. In some forms, a dermislike endosome forms an exhalant surface. In **euryphylous** sponges, the chambers receive water directly from the main trunks or branches of inhalant canals, and their **apopyles** open directly into exhalant canals. **Aphodal** sponges (Fig. 11.1) have fine ducts termed **aphodi** between the chambers and larger exhalant canals proper. **Diplodal** sponges (Fig. 11.2) have similar fine **prosodi** between inhalant canals and chambers. The euryphylous and aphodal conditions are the common ones. An ectosome developed as a **dermis** is typically porous and underlain by subdermal spaces, from which the inhalant canals run into the interior. Inhalant canals may pass inward through a **cortex** from the surface but more often start from subdermal spaces under or within it. If the spaces lie under the cortex, they are **subcortical,** and water reaches them through short **intracortical** canals sometimes called **chones.** Less commonly, subdermal spaces interrupting a cortex are roofed by a thin porous surface layer. In either instance, pores seen at the surface may form concentrations in special poriferous areas. Within a cortex, chones or true canals may have musclelike sphincters for closure.

In some Demospongea, the young sponge developed first from the newly settled larva is a transitory **rhagon** that has flagellated chambers but no canal system and no ectosome. A rhagon takes the form of a thinly walled, hollow cone with an oscular opening at the top (Fig. 12.1–12.2). The lateral wall of the upper part (**spongophare**) contains globular flagellated chambers, whose apopyles open directly into the paragastral cavity; but the basal part (**hypophare**) applied to the substrate has no chambers. Small pores in the external surface lead into the chambers of the upper part. The adult leuconoid condition, which is often correspondingly eurypylous, is produced by complication of the wall, which in some instances is comparable to the diverticulation of sycettoid Calcarea. In one instance (LÉVI, 1957b), an even simpler rhagonlike larva has an asconoid condition, with choanocytes lining the whole paragaster, and no flagellated chambers. In other Demospongea, however, the canals are developed first within a solid larva, and aphodi may be present before an osculum appears. For many, no embryos or larvae are known.

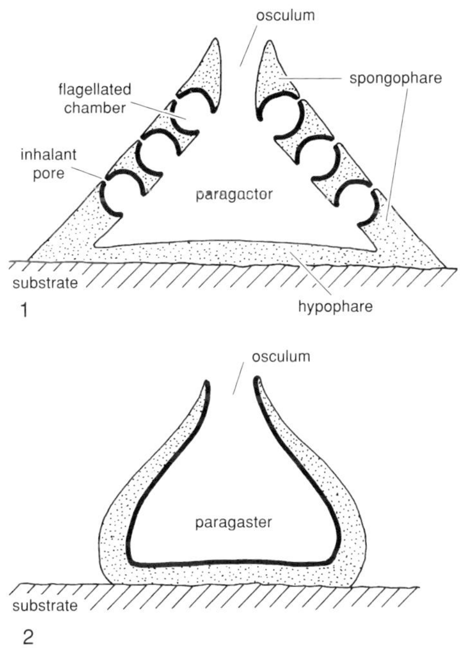

FIG. 12. Rhagons shown in vertical section; *1,* rhagon in original sense (SOLLAS, 1888), with upper part (spongophare) containing flagellated chambers but no canal system, and a chamberless basal part (hypophare); *2,* asconoid larva of *Halisarca dujardini* JOHNSTON, Holocene, also currently called a rhagon; *thick lines* represent choanocytal linings of flagellated chambers or paragastral cavity; other parts *finely outlined and stippled;* substrate *crosshatched* (new).

INFLUENCE OF SOFT PARTS ON THE SKELETON

Two main results of the influence of the soft parts on the skeleton may be noted within fossil material.

First, division of the soft parts into external ectosome and internal choanosome can result in restriction of some spicules to one of these parts of the body or in special arrangement of megascleres in relation to an ectosome, although they do not lie entirely within it. There are two important instances. In choristid sponges, tetractinal megascleres are often developed as triaenes (see p. 20–21), arranged so that three similar rays lie in or under an ectosome, and a fourth layer ray extends into the choanosome. In massive (e.g., globular) choristids with triaenes, this leads to arrangement of megascleres in a radiating (radiate) manner. In lithistids, the desmas of the skeletal framework are characteristically choanosomal megascleres. At the surfaces, however, there may be separate ectosomal megascleres in the form of loose triaenes, loose monaxons, or special desmas different from those of the interior.

Second, growth of skeletal meshwork around canals that are larger than ordinary skeletal meshes results in production of skeletal pores or canals (p. 4). These features occur in many lithistid and keratose sponges, but only the former are common as fossils. The skeletal pores or canals may be outlined by specially enlarged meshes only but involve more typically interruptions equal in width to several or many meshes. In lithistids, the desmas forming meshwork surrounding them are distorted often by curvature or in other ways.

Judging from modern forms, the general construction of connected skeletal frameworks is not normally controlled by the arrangement of canals or flagellated chambers

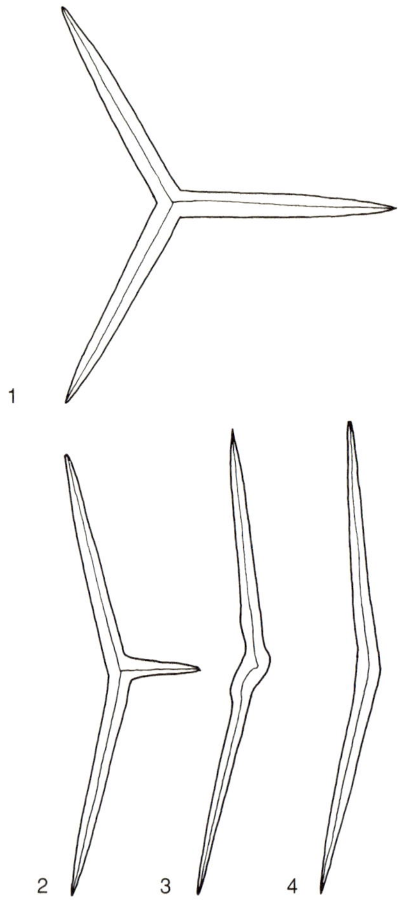

FIG. 13. Spicules of types intergrading from triactine to diactine; *1,* triod, or regular triactine, with three equal rays at 120% intervals; *2,* triactine with two equal rays, which meet at a wider angle than corresponding rays in view *1,* and 1 short ray; *3,* angled diactine, irregularly bent near center and with axial filament kinked centrally at point corresponding with position of short ray in view *2; 4,* more regular diactine with rays still angled; irregular or angled character of diactines (views *3–4*) is thought to imply derivation by reduction of 1 ray of a triactine (views *1–2*); these transitions are seen chiefly in spicules of choristid Plakinidae but also occur in monaxonid *Alectona* CARTER, whose megascleres are normally of types shown in views *3–4* (new).

in the soft parts, as might be imagined, but seems to be essentially independent unless modified by canalization. Correspondingly, various fossil lithistids have similar types of canalization in skeletons that are differently constructed, and formed from different kinds of desmas. Nonetheless, in some fossils apparent skeletal canals were produced by special arrangements of the spicules, although how such features were related to the soft parts is unknown.

SPICULES

Demosponge spicules consist of hydrated (opaline) silica with some content of organic material that is secreted concentrically around organic axial filaments or partly or entirely without control by distinct axial structures. Megascleres may be formed by groups of scleroblasts, but many microscleres arise inside single ones. Axial filaments are usually present in megascleres, although lithistid desmas and some lithistid ectosomal spicules are partly or entirely anaxial. In microscleres, axial filaments occur rarely, except in especially large examples; their evident occurrence in some suggests their presence in the others, but this cannot be asserted.

Variations in form and other characters of demosponge spicules result from a number of developments, found singly or in various combinations. Some developments occur only in megascleres or microscleres, but others occur apparently in both, although, strictly, the rarity of detectable axial filaments in microscleres can make apparently similar features not certainly identical in origin. The small spicules forming the skeletons of some Demospongea (Plakinidae, Thrombidae, Samidae) and noted as microscleres have variations similar to those of some choristid megascleres, although sometimes with special developments.

For further description, the following arrangement is convenient.

1. Number of rays. The number of rays (or, in microscleres, apparent rays) may be varied meristically in a sequence 1, 2, 3, 4, 5, etc. (Fig. 13–18). In the simplest instance, all the rays are similar in length, without occurrence of spicules with some rays partially developed. The angles between the rays vary according to their number and are usually subequal or equal, although exceptions may occur. The symmetries of diactinal,

triactinal, tetractinal, and hexactinal spicules are typically monaxial (with two rays in line), triodal (three rays at 120° intervals in one plane), and tetraxial (four rays following three axes at right angles), respectively. This type of variation occurs mainly in microscleres called **euasters** (p. 30), in which the number of rays may vary from one to more than 100 and is typically more than four; but it also occurs among megascleres and small analogous spicules with two to six rays and then usually among individual variants of a normal tetractine.

There may also be intergradation between monactinal and tetractinal spicules through intermediates in which some rays are relatively short, rudimentary, or represented only by internal axial rudiments. Transitions between diactines, triactines, and tetractines are often accompanied by reorientation of rays, so that, for example, two rays in line in a diactine correspond with two at 120° in a triod (triactine). This style of variation occurs mainly in small, megasclere-like spicules of Plakinidae (Fig. 13) but sometimes in megascleres of typical choristids or the initial bodies (**crepides**) of some desmas. It is usually interpreted as being due to a process of reduction (SCHULZE, 1880), because diactines intergrading with regular triactines can be markedly irregular in their middle parts.

Monaxons are usually considered as having either one or two rays according to whether the two ends are different (e.g., sharp and blunt respectively) or similar; but a so-called monactine can originate by secondary rounding of one end of an initial diactine.

2. Branching rays. Rays may divide into true actinal branches due to branching of the axial filament during growth (Fig. 14). Branching is usually dichotomous, with division occurring once only in each ray affected, but may sometimes result in production of three or more branches at one division or may be repeatedly dichotomous.

Clear actinal branching occurs mainly in loose megascleres and small analogous spicules, which are usually tetractinal and never normally monactinal or diactinal. The number of rays affected may be any number of those present, for example, one to four in a tetractine; or branching may be restricted to rays grouped as a cladome (see point 5 below, triaenose symmetry). Occasionally, branching may occur in the initial body (crepis) of a tetraxial or related type of desma (p. 52) or in axial canals representing it.

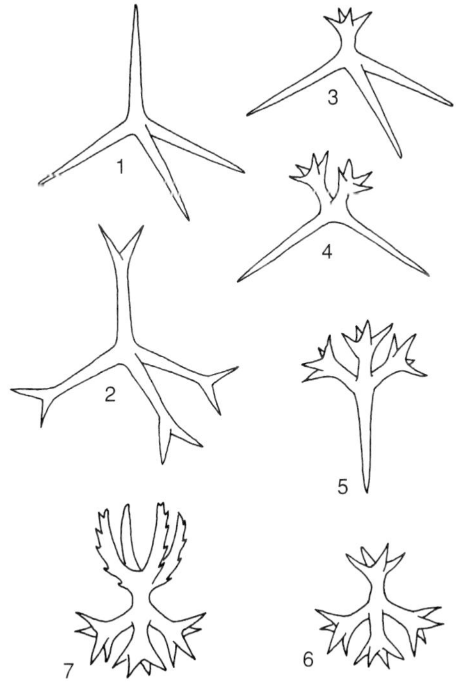

FIG. 14. The calthrops (or regular tetractine) and branched variants; *1,* simple calthrops with unbranched rays; *2,* dichocalthrops with each ray branched dichotomously; *3–7,* lophose tetractines with rays divided into groups of 4 branches; *3,* monolophose tetractine; *4,* dilophose tetractine; *5,* trilophose tetractine; *6,* tetralophose tetractine with all rays developed similarly; *7,* candelabrum, or tetralophose tetractine with branches of 1 ray especially developed (new).

In microscleres, repetitive branching appears to occur in a series of forms (**streptoscleres**, p. 31) that typically have sympodial spiral axes, although axial filaments can be detected rarely.

3. Terminal features. The ends of monaxons or individual rays are generally finely pointed (**oxeote**) but sometimes

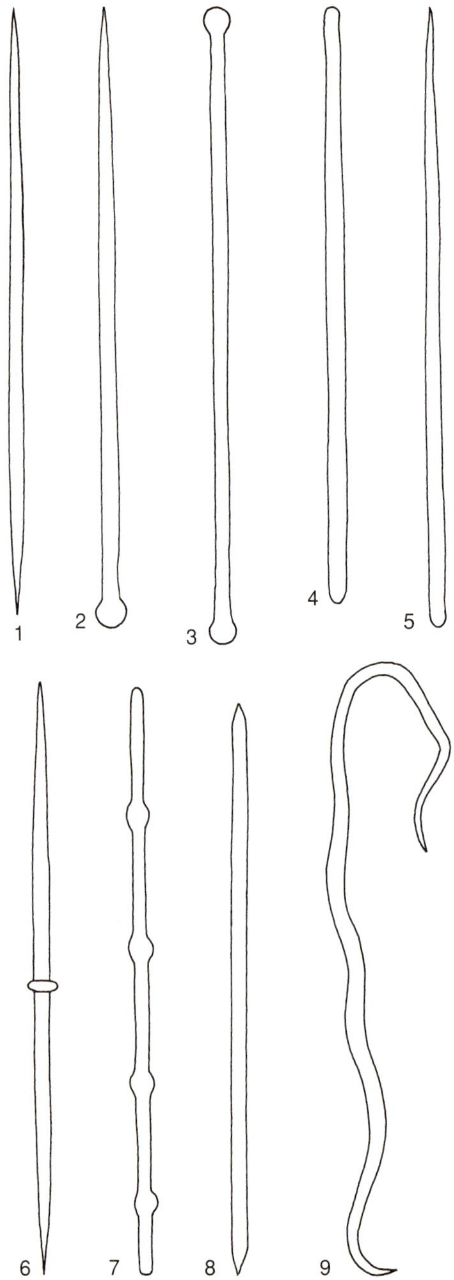

FIG. 15. Nomenclature of demosponge monaxons; *1,* oxea (or, amphioxea), sharply pointed at both ends; *2,* tylostyle, with 1 end sharp, other swollen; *3,* tylote (or amphitylote), with swollen ends; *4,* strongyle (or amphistrongyle), with blunt ends; *5,* style, with 1 end blunt, 1 sharply pointed; *6,* centrotylote oxea, with annular central tylus; *7,* polytylote strongyle; *8,* tornote (or, amphitornote), abruptly pointed at both ends; *9,* ophirhabd with pointed (oxeate) ends (new).

abruptly pointed (**tornote**), rounded (**strongylote**), or spherically knobbed (**tylote**) (Fig. 15). In monaxons, the two ends may be similar or differently developed. In monaxial megascleres with one end only strongylote or tylote, the corresponding end of the axial filament may also have a rounded enlargement; this is usually interpreted as marking the spicular center, that is, the point from which growth began, so that spicules with this feature are regarded as genuine monactines.

4. Curvature and torsion. Curvature of individual rays is a feature of various loose megascleres and small analogous spicules. A monaxial megasclere may be curved along its length, bent centrally, or sometimes irregularly sinuous. Some apparently diactinal microscleres are bent centrally, bow shaped, or rarely bent like a forceps with the two ends converging. Others have continuous torsion in a range from C-shaped or S-shaped to polyspiral and spring or screwlike. These shapes may also occur in the axial parts of microscleres whose final shape is modified further by various types of outgrowths (see point 6 below).

5. Triaenose symmetry. Many loose tetractinal megascleres and some small analogous spicules are developed as triaenes, in which three rays are similar but are distinguished in some way from the fourth (Fig. 16). The three similar rays are called **cladi** and together form a **cladome,** while the fourth ray is the **rhabdome.** This nomenclature was based on the idea (SOLLAS, 1888) that the cladi are branches of a primary monaxon or rhabdus; in practice such spicules may develop from regular tetractines, but the terms have been generally adopted. The cladi may differ from the rhabdome (a) by being shorter or, less commonly, longer; (b) by curvature or displacement away from the regular tetraxial positions, either away from or toward the rhabdome; (c) by branching, which is normally single and usually dichotomous, though trichotomous branching may occur; (d) in lithistids, by production of lateral expansions, which may form leaflike features or coalesce as a sili-

ceous disc; or (e) by some combination of two or more of these developments. The rhabdome is normally a simple straight ray and is commonly longer, sometimes much longer, than the cladi, although shorter or even rudimentary in some sponges. These spicules are characteristically arranged with the cladi toward or in an ectosome or occasionally an endosome and with the rhabdome directed inwardly.

In some sponges, triaenes and occasional variants have meristic variation in the number of cladi, of which one to four may be present (Fig. 17–18). The variants with one, two, or four cladi are called **monaenes, diaenes,** and **tetraenes.** The angles between the cladi are varied in accordance with their number; hence a tetraene occurring as a variant of a triaene with cladi at right angles to the rhabdome (orthotriaene, see p. 25) can have the triaxon symmetry of a hexactinellidan pentactine, although this is very uncommon.

Two other types of loose megascleres or analogous spicules, both uncommon, have a triaene-like symmetry (Fig. 19). **Mesotriaenes** are forms in which the rhabdome is diactinal, with equal or unequal rays in line, and a cladome arising from a point between the ends. Forms with the rhabdal rays equal occur as choanosomal megascleres and appear to be special, five-rayed variants of equiradiate tetractines (calthrops), which they occasionally accompany. The cladi are emitted at right angles to the rhabdome; they may be simple, branched once dichotomously, or repeatedly dichotomous. This type can be distinguished conveniently by SOLLAS's (1888) original name centrotriaene. Mesotriaenes with unequal rhabdal rays occur as occasional variants of some normal triaenes with the rhabdal ray much longer than the cladi; the extra ray may be similar to the cladi in length, shorter, or represented only by an axial rudiment. **Amphitriaenes** are spicules resembling a triaene with a cladome at both ends of the rhabdome. The cladi of the two ends may be simple or dichotomously or trichotomously branched, in any combination. It is not clear how these

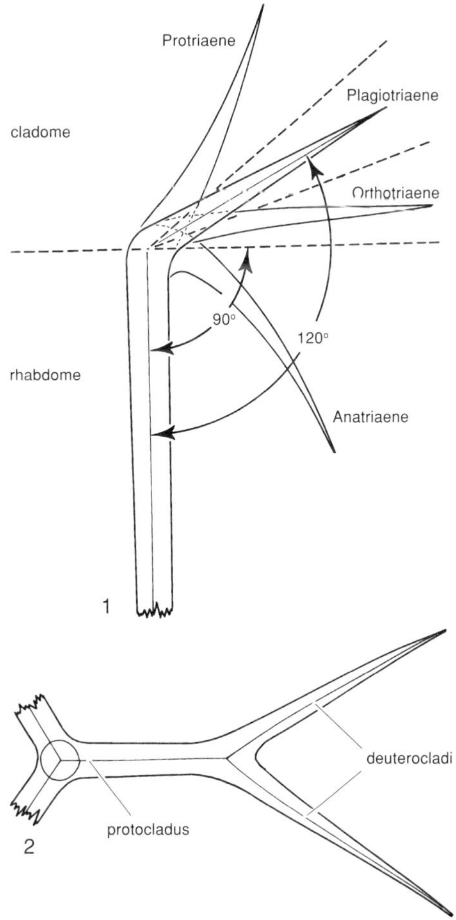

FIG. 16. Nomenclature of triaenes; *1,* nomenclature based on attitude of cladi; figure shows part of rhabdome (long ray) of a long-shafted triaene and 1 cladus of cladome oriented as in plagiotriaenes (strong outline) and protriaenes, orthotriaenes, and anatriaenes (finger outlines); *2,* part of cladome of dichotriaene seen in plan view showing protocladus and deuterocladi of 1 cladus; fine internal lines in both figures are axial filaments (new).

spicules are related to the others. There is sometimes an enlargement at the middle of the axial filament of the rhabdome; if this is regarded as a spicular center, the cladomes should then comprise true branches arising from the ends of a diactine.

In lithistid desmas, the initial body (crepis) is occasionally a triaene, a centrotriaene, or an amphitriaene, although this may not be apparent from the shape of the

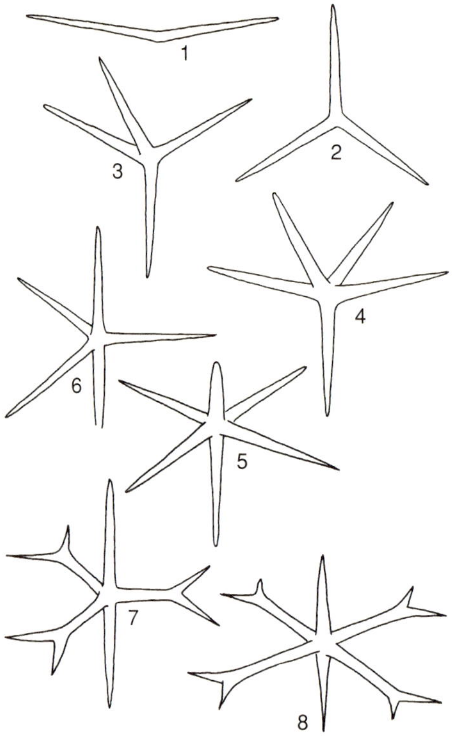

FIG. 17. Calthrops and meristic variants; *1–5,* spicules of *Calthropella* SOLLAS; *1,* diactine with angled rays; *2,* triactine (triod); *3,* calthrops (tetractine); *4,* pentactine, of type in which 2 pairs of rays are set at different angles to a fifth, here pointed downward; this type grades into forms in which the 4 upper rays make equal angles with the fifth and which tend to take orthotetraene shape (see Fig. 18,*4,8*); *5,* hexactine with triaxon symmetry; this example also having 5 equal rays and one that is shorter and terminally blunted; *6–8,* spicules of *Yodomia* LEBWOHL; *6,* pentactine developed as a centrotriaene; compare with view *4,* from which this type may originate by straightening of 2 rays into line; *7,* centrotriaene with simple rhabdal rays (here vertical) but dichotomously branched cladi; *8,* hexactine variant of view *7,* with 4 branched cladi and with triaxon symmetry; this type may be called a centrotetraene or dichocentrotetraene (new).

fully formed desma (see point 9 below). On the other hand, some desmas have triaene-like shapes (e.g., with three raylike arms similar and one differently developed) unrelated to the form of the crepis, which may then be an equiradiate tetractine, a monaxon, or an anaxial corpuscle.

6. Ornament and secondary rays. Ornamentation of rays or of the shaft of a monaxon is a feature of some spicules. In megascleres, the only common ornament consists of small spines or spinules arranged without order, less commonly in regular whorls, and rarely in spiral series. A few megascleres have annular or spiral ridging. Ornamented microscleres are mainly monaxons (or, apparent monaxons) with straight to spiral axes (see point 4 above), having simple to spiral spinulation or various special shapes due to outgrowths that are spinelike, bladelike, or leaflike or form simple or serrated transverse flanges.

In some monaxonids, **spinules** ornamenting megascleres contain axial filaments, which begin some distance from the primary axial filament. These spinules are thus comparable with true cored rays, although clearly secondary additions. Occasionally, megascleres, which are normally spiny monactines (**acanthotylostyles**), are replaced by apparently diactinal to pentactinal spicules, with one to four extra rays arising as outgrowths from one end of an initial monaxon. The extra rays contain axial filaments that begin at some distance from the end of the primary filament instead of meeting it at a center as in true radiate megascleres; these rays are therefore regarded as secondary, corresponding with the cored type of spinule, and presumably evolved by enlargement of spinules at the base of a monactine.

In various microscleres, gradation from spiny monaxon to apparent polyactine suggests analogous developments, but no axial filaments are detectable.

7. Epicentric secretion. In a few instances, spicules are formed or initiated by secretion around a center. Microscleric spheres of some genera consist of silica secreted around a central granule, which seems to be organic. Sometimes a sphere passes morphologically into a short, round-ended monaxon (strongyle), implying its origin from the latter by further shortening unless the opposite is imagined. In some anaxial desmas (**sphaeroclones,** see p. 57) the initial body (crepis) is a siliceous corpuscle, apparently with no organic center, although a granular nucleus may appear during growth.

8. Central and axial thickening. Secondary secretion of silica may occur around the center of a radiate spicule. This is seen mainly in some euaster microscleres, in which secretion around the center of an initial polyactine can produce a spherical mass, with the ends of rays projecting (in **sphaeraster,** see p. 30–31), or solid spherical to discoidal spicules (**sterrasters, aspidasters,** see p. 30–31) with no trace of separate rays but a fine surface granulation.

Analogous secondary thickening may occur along the axial parts of monaxial microscleres, producing changes in form. In extreme instances, forms with a spiral or C-shaped axis are converted into spicules (**sterrospiras, chelasters,** see p. 32–34) resembling solidified euasters.

9. Special features. The character and form of the desmas of lithistids and a few other sponges with partly lithistid characters (**sublithistids,** see p. 49) result from secretion of silica either partly or entirely without control by growing axial filaments. The articulatory parts (zygomes) of the desmas are characteristically anaxial, and some have additional anaxial raylike arms (clones) or are wholly anaxial. Desmas are typically of various irregular shapes with the zygomes being often branching, rootlike outgrowths but sometimes handlike, cuplike, or tonguelike and with clones arranged in various ways when present. The overall form of a desma may be more or less clearly determined by that of an initial body of crepis, which may be tetractinal, triactinal, or monaxial or have little or no obvious relationship to it. In some types, accretion of silica at the ends of a monaxial crepis or around an anaxial initial corpuscle produces spherical masses (centra) from which clones radiate. In addition to features peculiar to a number of main types of desmas (p. 49–51), individual desmas may have further variations related to orientation or location in the skeleton or distortions due to canalization.

Some lithistids also have ectosomal megascleres that are formed either partly by nonactinal secretion or are wholly anaxial. In the former, siliceous outgrowths from an

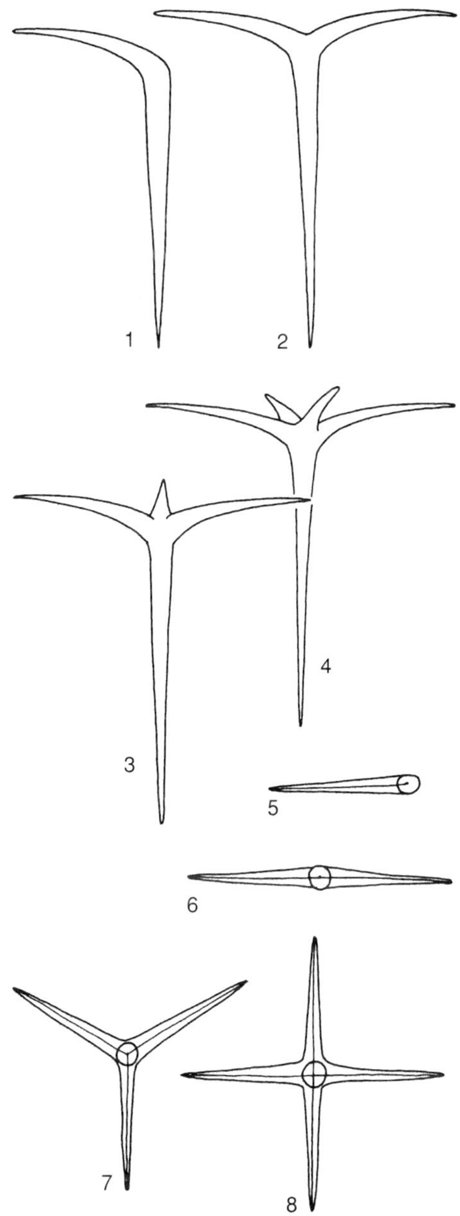

FIG. 18. Nomenclature and form of meristic triaene variants, which differ in number of cladi; illustrated by variants of simple orthotriaene (views *3, 7*); *1–4,* side views, rhabdome downward; *1,* monaene; *2,* diaene; *3,* triaene; *4,* tetraene. *5–8,* views with cladome in plan, and the rhabdome toward front; *5,* monaene; *6,* diaene; *7,* triaene; *8,* tetraene; also illustrates resemblance of demosponge orthodiaenes (views *2, 6*) and orthotetraenes (views *4, 8*) to hexactinellidan tauactines and pentactines (new).

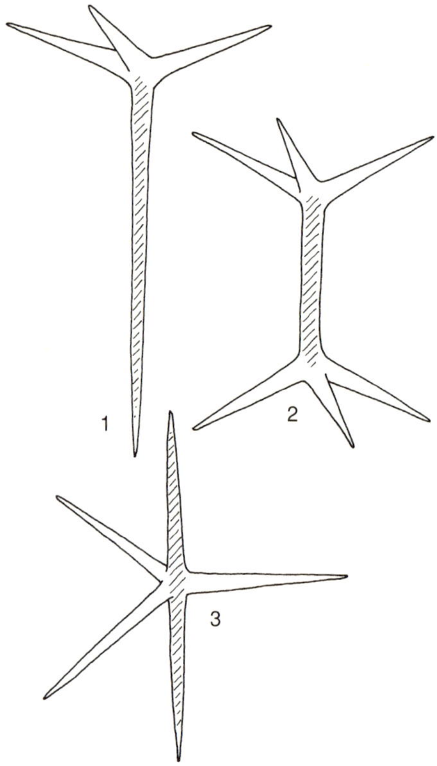

FIG. 19. Megascleres with triaenose symmetry, classed as triaenes *s.l.,* in which rays are contrasted as forming a rhabdome (*crosshatched shading*) and a cladome or 2 cladomes (unshaded rays); *1,* triaene *s.s.,* with a single rhabdal ray and 3 cladi; the rhabdome (rhabdal ray) of this type is typically longer than cladi, as here, and cladi are also often displaced from regular tetraxial positions (here, slightly toward rhabdome); *2,* amphitriaene with central rhabdal shaft bearing cladomes at both ends; this type can be compared with some amphiaster microscleres, to which it was related by TOPSENT (1928b); *3,* mesotriaene of centrotriaene type with 2 opposite rhabdal rays (or, a diactinal rhabdome), and 3 central cladi that meet rhabdal axis at right angles (new).

initial tetractine or monaxon form leaflike expansions or siliceous discs. Anaxial, ectosomal megascleres range from finely branching bodies to scalelike plates.

10. Teratological variants. Individual spicules sometimes have irregular modification by contortion, branching of rays, or addition of extra rays. The axial filaments of extra rays may be continuous with the normal axial system or separate from it. Very rarely, composite spicules may form where two or more have grown together.

All the main types of variations in the character of the spicules are represented in fossils. Only those present in megascleres are seen for the most part, but a few unusual deposits have yielded examples of nearly all the known kinds of microscleres.

The nomenclature of spicules is based simply on various aspects of their form or partly on inceptional characters when changes occur during growth. The last applies specially to lithistid megascleres. Microscleres, desmas, and some other forms are dealt with in sections below, but the following may be noted at this point.

a. Spicules are termed generally as monactines, diactines, triactines, tetractines, etc., according to the number of rays, or simply as polyactines if more than six are present. A regularly tetraxial and equiradiate tetractine is distinguished from triaenes as a calthrops (Fig. 14, 20). A triactine may be a triod, if it has three rays in one plane at 120° intervals, or a tripod, if it has the three arranged pyramidally. Any radiate microsclere or form with raylike outgrowths arising from an axial portion is an aster.

b. The prefixes **dicho**- and tricho- imply dichotomous or trichotomous branching, affecting one to all rays in, for example, a dichocalthrops or only the cladi of a dichotriaene (Fig. 20). Branching producing a regular cluster of three or more branches is lophose, and a lophose tetractine may be monolophose to tetralophose according to how many rays are affected. Repetitive branching is polycladose.

c. Monaxons with the two ends finely pointed (oxeote), abruptly pointed (tornote), rounded (strongylote), or knobbed (tylote) are called oxeas, tornotes, strongyles, and tylotes respectively (Fig. 15–16), or alternatively given the extra prefix amphi-, as in, for example, amphioxea or amphitylote. Those with only one end oxeote are styles if the other is strongylote, **tylostyles** if it is tylote. A strongyloxea has one end oxeote, the other at first tapering as in an oxea but rounded at the tip. A hairlike monaxon is a **raphide.** Some further types take names with the termination -rhabd, e.g., ophirhabd for megascleres that are sinuous oxeas or strongyles.

d. The nomenclature of triaenes is based mainly on the form and attitude of the cladi (Fig. 16). The central type of triaene morphologically is the **plagiotriaene,** with straight or nearly straight unbranched cladi meeting the rhabdome at angles between about 110° and 135°, and regularly tetraxial when the angle is 120°. If the cladi are instead markedly curved toward the rhabdome or grow out at a lesser angle to it, the spicules are **orthotriaenes** until the angle between cladi and rhabdome is 90°, and anatriaenes when it is less than 90°. Triaenes with simple cladi curved markedly away from the rhabdome or meeting it at angles more than 135° are **protriaenes** (Fig. 21.1). Meristic triaene variants with a different number of cladi are named similarly, for example as prodiaenes or anadiaenes. If the cladi are dichotomous or trichotomous, a triaene is a **dichotriaene** or a **trichotriaene;** the primary cladal rays are then the protocladi, and their branches are deuterocladi. If the shape is otherwise similar to that of orthotriaenes or protriaenes, the names orthodichotriaene and prodichotriaene have sometimes been used. Lithistid triaenes with the cladi leaflike or forming a disc are **phyllotriaenes** and **discotriaenes.** The nomenclature of mesotriaenes and amphitriaenes follows similar principles when any special names are given; for instance, a mesotriaene found as a variant of a normal anatriaene can be called an anamesotriaene. Amphitriaenes have also been called homopolar or heteropolar, according to whether the two ends are similar or differently developed.

There is some variation in usage of the names given to simple triaenes. These were based originally by SOLLAS (1888) on the angle between the cladi and the axis of the rhabdome; but later authors (e.g., VON LENDENFELD, 1903; DE LAUBENFELS, 1955) have varied some of his concepts, placing emphasis on curvature of the cladi, or citing different angular relationships. In practice, it is difficult to maintain a consistent

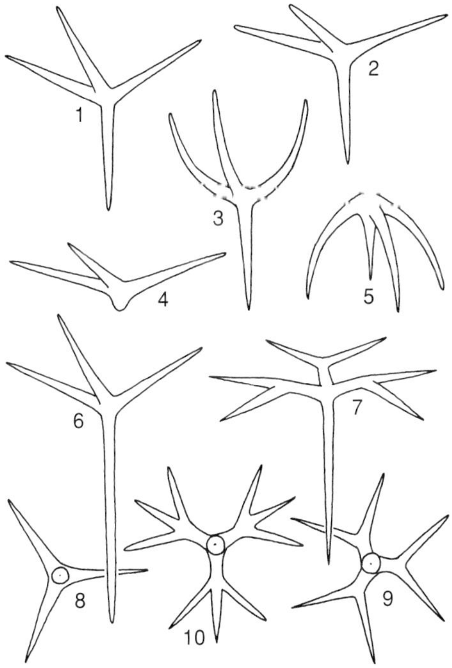

FIG. 20. Short-shafted triaenes, including subtriaenes (views *2–5*); *1,* calthrops oriented as a triaene; included for comparison with others, and because it may intergrade with spicules shown in views *2–6; 2,* equal-rayed (or equiradiate) orthotriaene, with triaenose symmetry occurring only in displacement of 3 rays (at top) toward fourth as compared with true calthrops (view *1*); *3,* equiradiate protriaene, with 3 rays curved away from fourth; *4,* subtriactine, with 3 equal rays arranged as in view *2,* and the fourth (or, rhabdal ray) represented by a rudiment; may occur as a variant of views *1, 2,* or *6; 5,* anatriaene with 3 equal rays (cladi), and a shorter fourth (rhabdal) ray; this type may grade to forms with rhabdal ray reduced to a rudiment, which may be termed anatripods; *6,* typical plagiotriaene, with rays arranged as in a calthrops (view *1*), but with a rhabdal ray longer than 3 others that are taken as cladi; as drawn here, this form is approaching a long-shafted triaene into which it passes gradationally; *7,* dichotriaene, with an unbranched rhabdal ray (pointed downward) little longer than branched cladi; *8–10,* cladome of triaenes (or subtriaenes), as seen with the rhabdal ray toward front; *8,* simple triaene; *9,* dichotriaene; *10,* trichotriaene (new).

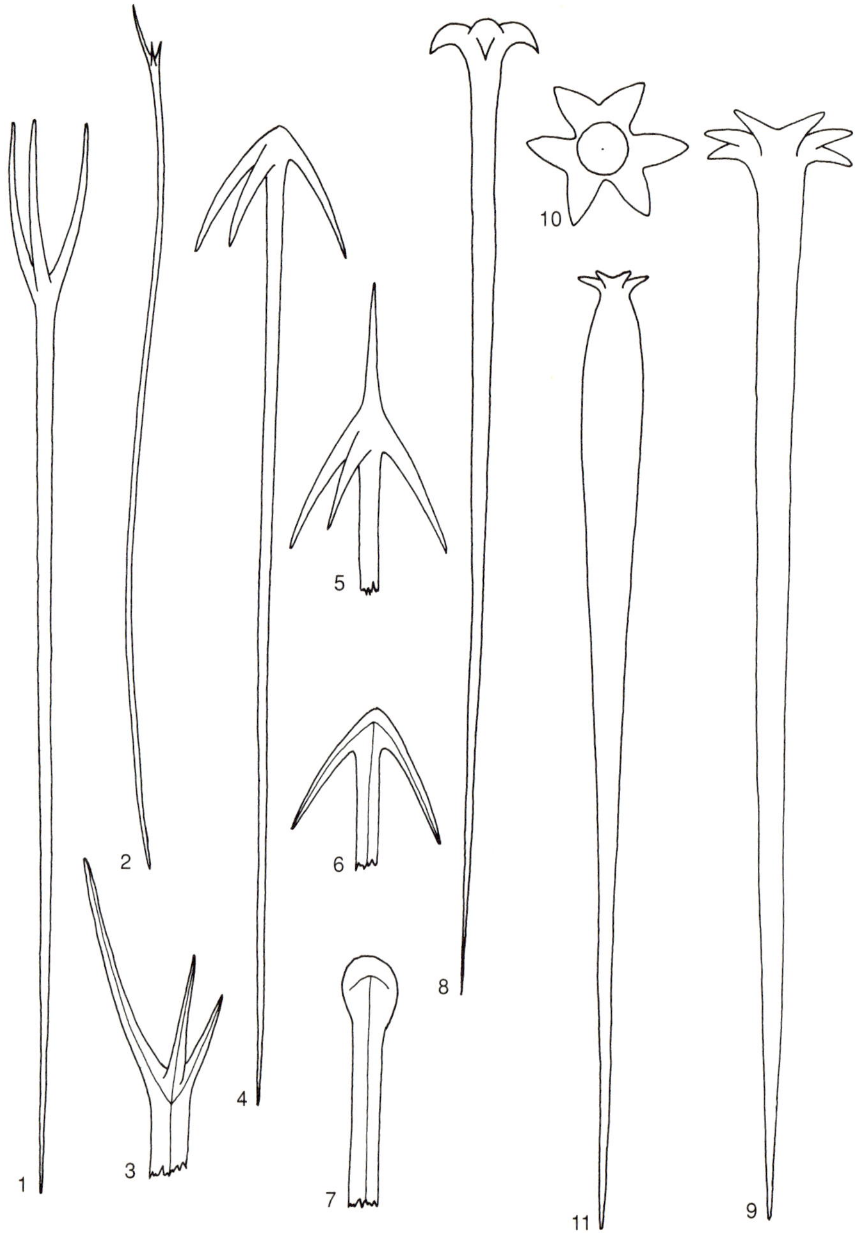

FIG. 21. Long-shafted triaenes and some variants; *1,* long-shafted protriaene, with rhabdome (i.e., long rhabdal ray) about 6 times as long as cladi; *2,* hairlike (trichodal) protriaene of *Tetilla* SCHMIDT with unequal (1 long, 2 short) cladi; *3,* cladal end of same spicule, showing the axial filaments; *4,* typical anatriaene; *5,* mesotriaene variant of view *4* with a second rhabdal ray (pointing upwardly) about same length as cladi; *6,* anadiaene variant of view *4;* cladal end, showing axial filaments; *7,* pseudotylostyle variant of view *6;* cladal (or, tylote) end, showing axial filaments; *8,* long-shafted orthotriaene with recurved cladi, not strictly an anatriaene but approaching that type (see view *4*); *9,* long-shafted dichotriaene, with cladi little longer than maximum diameter of rhabdome; *10,* cladome of view *9* with rhabdal ray (or, rhabdome) toward front; *11,* trachelotriaene, with swollen, clublike rhabdome; cladome as in view *10* but its total diameter little greater than maximum diameter of rhabdome (may be less in some examples) (new).

nomenclature. The problem is partly that the four types distinguished intergrade so that any distinction is arbitrary and partly that the two criteria cited (angle between cladi and the rhabdome; curvature of cladi) vary independently. Some triaenes do not fit readily any named category; for instance, the cladi may begin at an angle of more than 135° to the rhabdome, as in some protriaenes, but curve back toward it or grow out initially as in orthotriaenes but curve strongly away from it. The principal differences are in usage of the terms plagiotriaene and orthotriaene; the writer uses these essentially in the sense they were given by VON LENDENFELD (1903, 1907).

Triaenes are also described as long- or short-shafted, according to the length of the rhabdome; but, again, authors differ in what they mean by long and short. Herein triaenes are called long-shafted if the rhabdome is twice or more than twice as long as the cladi. It is also convenient to make a further distinction between triaenes *s.s.*, with the rhabdome longer than the cladi, and subtriaenes in which it is not longer or is shorter.

e. Spicules with a conspicuous ornament but otherwise simple shapes are named by addition of a prefix to a general name. The main instance is that of spiny spicules, called **acanthoxeas** or **acanthostyles.** The prefix **cric-**, as in **criccalthrops** and **cricostyle,** implies strong annulation, which occurs mainly in some fossil spicules. A strongly annulated monaxon with both ends tylote is a **cricorhabd.**

Another development in monaxons, conveniently treated as ornament, is development of annular to spherical enlargements at the center or at several points between the two ends, without continuous ornamentation. These spicules are **centrotylote** if a single central feature is present and polytylote if there is more than one. The latter type is uncommon and is usually a variant of a tylostyle.

f. The prefix **micro-** designates small spicules, as in microcalthrops or microxea. This usage is restricted here to microscleres but has sometimes been applied to the small spicules of Plakinidae or to small crepides of lithistid desmas.

TETRAXONS AND TRIAXONS

The regular tetraxon, the calthrops, is the central type of demosponge spicule from which all other types have arisen according to a concept due principally to SCHULZE (1880, 1887a) and DENDY (1905, 1916, 1921). A sponge without tetraxons as megascleres or in any form is supposed to have lost them in phylogeny. These ideas are based mainly on the characters of some simple Demospongea with small spicules, comprising the family Plakinidae SCHULZE. The spicules of these sponges are mainly tetractinal but occasionally diactinal, triactinal, or pentactinal, with diactines or triactines predominant in some genera. Diactinal spicules intergrade with triactines or sometimes tetractines through forms in which one or two rays are rudimentary. They may be either angled centrally or markedly irregular when strictly diactinal; this is regarded as showing the origin of monaxons as secondary diactinal derivatives of triactines or tetractines. Tetractinal spicules may be modified by branching but are never true triaenes. In *Plakina* SCHULZE the soft parts are unusually simple: a eurypylous adult develops by diverticulation of a rhagon (SCHULZE, 1880) and may sometimes lack an ectosome (*P. monolopha* SCHULZE). If these sponges are interpreted as primitive, their simple diactinal to pentactinal spicules can be seen as representing the prototypes of all spicules of other Demospongea whether megascleres or microscleres. The primary basis of different types of spiculation then seems to be meristic variation, with all other developments secondary. In megascleres, triaenes are explicable as functional calthrops derivatives that support the ectosome, to which the cladome is typically directed.

In addition, some choristids with tetractinal megascleres in the form of triaenes only have the latter much subordinate to monaxons (usually oxeas) and resemble closely some monaxonids with similar microscleres (euasters or sigmaspires, see p.

19, 30). In a few forms, the triaenes are so inconspicuous that until their discovery a sponge (e.g., *Stellettinopsis* CARTER) has been supposed to be monaxonid. Hence it has been argued that some (SOLLAS, 1888) or all (DENDY, 1916, 1922) monaxonids are forms derived from choristids with triaenes by loss of these spicules. The designation epipolasid has this implication as used by SOLLAS and DENDY though not by DE LAUBENFELS (1936, 1955).

These ideas can provide a complete explanation of demosponge spiculation (see DENDY, 1921) but are not certainly correct. Their basis is comparative study of modern sponges, which are all end forms phylogenetically, without stratigraphic evidence. In fact, the oldest monaxonids are Cambrian in age, but the oldest certain choristids are Late Ordovician. This may be due only to incompleteness of the stratigraphic record, which is very sporadic, or to forms with monaxons being simply the first to develop spicules large enough to be preserved. SOLLAS (1888), however, regarded at least some triaenes as derived from monaxons because long-shafted triaenes can develop ontogenetically from an initial monaxon. This can be thought to be more consistent with the paleontological evidence, although this involves assuming that ontogeny follows phylogeny, which is clearly not always correct; SOLLAS also admitted that some triaenes have no sign of this origin. DE LAUBENFELS (1936) noted loss of triaenes as a possible mode of origin of monaxonids in his order Epipolasida but thought that the calthrops as a megasclere is usually derived from a triaene by shortening of the rhabdome. Even DENDY (1922) held that some subtriaenes seem to be derivatives of normal, long-shafted triaenes (*Paratetilla* DENDY). Further, various authors since VOSMAER (1882, 1883, 1884, 1885, 1887) have seen no close connection, or none in DENDY's sense (e.g., TOPSENT, 1928b), between one group of monaxonids (those with sigmatosclere microscleres, p. 35) and sponges with tetraxons, and LÉVI (1957b) has suggested that their spicules are of independent origin. Last, *Plakina* has embryos of a type (**amphiblastulas**) unknown in any true choristid or any other form possessing megascleres.

As a brief assessment, before full discussion, herein the calthrops is regarded as being acceptable, morphologically, as a central type of spicules in Plakinidae, in most choristids, in monaxonids with euaster microscleres, and in some lithistids; but it does not certainly follow that its symmetry is not secondary or even that all tetraxons are homologous. There are various instances in which the ideas of SOLLAS or DE LAUBENFELS fit the facts at least as well as those of DENDY. Some monaxonids seem to be genuinely allied to choristids, but most have no demonstrable relationship to sponges with tetraxons; though, equally, it does not seem demonstrable that any are of independent origin. More generally, most of the evidence available is from modern forms, which can give no objective indication of the direction of phylogeny, whatever may seem likely. Herein, therefore, the SCHULZE-DENDY picture of phylogeny is not taken as a basic assumption, although parts of it seem justified.

It is also often said that hexactinal triaxons occur only in Hexactinellida and never in Demospongea. This idea is fundamental to SCHULZE's (1887b) picture of phylogeny in siliceous sponges; it is argued of tetraxons and triaxons that neither can give rise to the other, so that each must have arisen independently in separate stocks. The orders Tetraxonia (-ida) and Triaxonia (-ida) used by SCHULZE (1887b), SCHRAMMEN (1912), and others (LAGNEAU-HÉRENGER, 1962; REZVOI, ZHURAVLEVA, & KOLTUN, 1962) express this concept taxonomically.

In fact, triaxial spicules occur in many living Demospongea, although not normally as megascleres. They occur mainly in forms possessing euaster microscleres (p. 30), in which six-rayed euasters are usually triaxial in form. These spicules cannot be dismissed as really anaxial, because large examples may have an observable axial system (e.g., VON LENDENFELD, 1907, pl. 28,*13*). A four-rayed euaster, however, is usually a calthrops and thus a tetraxon. The relationships of tetraxial and triaxial euasters is simply that of four- and six-rayed members of a meristically varying series, with the rays arranged at equal angles that depend on their number. Occasional triaxial megascleres have also been recorded as abnormal hexactinal variants of a normal calthrops (*Calthropella simplex* SOLLAS, 1888) or centrotriaene (*Yodomia perfecta* DENDY, 1916); in addition, a pentactinal triaxon is known as a variant of a normal orthotriaene (*Sphinctrella cribrifera* SOLLAS, 1888). The fossil *Spiractinella* HINDE with mainly triaxial megascleres also seems to be a demosponge, since less common, four-rayed spicules are calthrops and triaenes (REID, 1963c).

It cannot be claimed, therefore, that tetraxons and triaxons must have independent origins or that triaxon spicules are only found in Hexactinellida. The true distinctive feature of hexactinellidan spiculation is that all the spicules present are either hexactinal triaxons or variants of this type with fewer rays, excepting only some monaxons whose relationship to triaxons is not demonstrable. I am unaware of any grounds for asserting the homology of hexactinellidan triaxons with those of Demospongea; the latter seem more likely to have arisen independently in various demosponges, and this must be so if tetraxons are the central type of demosponge spicule.

MICROSCLERES

Because of their bearing on the relationships of choristids, monaxonids, and lithistids, a knowledge of the microscleres is desirable before megaspiculation is considered.

Demosponge microscleres vary widely in form and have many shapes that do not occur in megascleres. For several reasons, they are difficult to treat systematically. A purely morphological treatment is unsatisfactory because of numerous instances in which forms that are similar morphologically appear to be convergent homeomorphs; but a true homological treatment is even more difficult because of the number of examples involved, and because evidence of homologies is often lacking in those examples that are disputable. There are also various artificial problems, due to differences in opinion or usage, resulting, for example, in the same spicule taking different names or a given name having different meanings in the works of different authors.

Some examples of these problems are as follows.

1. In the *Challenger* nomenclature, SOLLAS (1888) grouped all polyactinal and pseudopolyactinal choristid microscleres as asters and divided these further into (a) euasters, with rays (or, apparent rays, emitted from a center), and (b) **streptasters,** with rays arising from an axial part, which is usually spiral (hence the prefix strept-, twisted). A variety of further named types were distinguished within these groups. Later work (DENDY, 1924b) demonstrated that some types included (**plesiasters, metasters,** and **spirasters** of the choristid Pachastrellidae and Theneidae) as streptasters have sympodial axes produced by repetitive branching. A nonspiral type called a **sanidaster** is restricted to the first group, but DE LAUBENFELS (1955, p. 30) defined it as comprising straight spiny monaxons and also included forms with fine spinulation called spiny microrhabds (not streptasters) by SOLLAS (*Halina (Dercitus) bucklandi* BOWERBANK per DE LAUBENFELS, 1955, p. 43, fig. 23,*1b*). SOLLAS's (1888) streptasters thus included two types of spicules, each of which has since been called streptasters in different restricted senses. In addition, the spiraster type of choristid streptaster is homeomorphic with some monaxonid microscleres that appear to be spiral monaxons, called spinispiras by DENDY and BURTON (and herein, p. 32–33) but spirasters by DE LAUBENFELS.

2. In DENDY's (1916, 1917, 1921, 1922) nomenclature, a microsclere regarded as a spiny monaxon or as derived from this type even if euastriform morphologically, is called a pseudaster. This usage has been followed by some authors (e.g., BURTON) but not others

(e.g., DE LAUBENFELS, 1955). The term has also been used for euastriform spicules of various monaxonid sponges in which evidence of homology is lacking, on a basis of subjective assessment of other characters irrespective of the form of the microscleres. Furthermore, even if a seeming euaster intergrades with a monaxon, there is no certainty that either is the prototype phylogenetically, although DENDY's view seems generally more likely.

3. Some choristid sponges (Craniellidae) have twisted monaxial microscleres called **sigmaspires** by SOLLAS (1888), with a spire of about one revolution and appear C-shaped or S-shaped, according to how they are viewed. These were regarded (SOLLAS, 1888) as distinct from similar sigmas of various monaxonid sponges (e.g., Desmacidontidae). Various later authors have equated these spicules as sigmas (e.g., VON LENDENFELD, 1904b; HENTSCHEL, 1909), and DENDY's classification (1916, 1917, 1921, 1922) depends on assuming their homology. But TOPSENT (1928b), since followed by BURTON, rejected the identity of these microscleres and the picture of phylogeny DENDY based on it. DENDY's ideas were rejected by DE LAUBENFELS (1936) but still called both types sigmas. As a further complication, some sigmaspires pass into spicules that would be called chelas if found in Desmacidontidae but were called sigmaspires by TOPSENT (1928b) because of their evident homology with that type.

It is therefore not possible to give an account of the microscleres that conforms with all previous usages. The attempt is made herein to combine the best features of the previous literature with some original views where this seemed desirable. In part, a homological treatment is attempted, in that some types of microscleres are noted as characteristic of various groups of sponges in which they can be thought to be homologous. If, however, a morphological type recurs sporadically in sponges with no evident special relationship, it is usually considered as repetitive.

Fossil Demospongea having the megascleric skeleton contain rarely microscleres, but most of the main types are known from some sedimentary rocks.

a. EUASTERS

Defined morphologically, a euaster is any microsclere in which rays or apparent rays radiate from a center. Euasters occur mainly as the characteristic microscleres of the choristid Calthropellidae, Ancorinidae, and Geodiidae and the monaxonid Coppattiidae and Tethyidae in which they have similar characters. The euasters of these sponges are meristically varying spicules with up to a hundred rays or more and are typically polyactinal, although tetractines or triactines may be present. Diactines or monactines may also occur as further meristic variants, although not strictly euasters morphologically. Tetractinal and hexactinal examples are usually tetraxons (microcalthrops) and triaxons respectively, although they may also have other shapes. The simplest euasters are **oxyasters**, with finely pointed rays and no central swelling. In variants in which the rays have developed differently, the latter are blunt or flat ended in strongylasters, knobbed terminally in tylasters, or spiny in anthasters. The tylaster type is also sometimes called a chiaster, because four-rayed examples with curved rays can be *chi*-shaped. An anthaster with spines developed mainly at the tips of the rays is called an acanthotylaster by some authors. Any of these types may be modified by occurrence of a central enlargement, the centrum, whose diameter may be less or greater than the lengths of the parts of the rays that project from it. Some authors use the term sphaeraster for any such spicules, but others restrict the term to forms in which the centrum is conspicuous (according to SOLLAS, 1888, when the diameter equals or is greater than one third the length of the rays). In specialized sterrasters, restricted to the choristid Geodiidae, a highly polyactinal initial oxyaster solidifies from the center outward, until the rays are marked only by fine surface granulation. A smooth depression, the hilum, on one side of the spicule often marks the position of the secreting scleroblast, within which the sterraster was formed. Some sterrasters are markedly kidney shaped rather than spherical, a type sometimes called a rhax by paleontologists (following RAUFF, 1893). A flattened discoidal variant of the sterraster is an aspidaster. A spicule intermediate between a sphaeraster and a true sterraster is a sterrospheraster.

If DENDY (1921) is followed, these euasters may be seen as comprising meristic variants of a central tetraxon (microcalthrops), with variation mainly in a positive direction though negative variants also occur (triactines, diactines, monactines). In fact, there is no sure indication that any type is central, although four is the maximum number of rays in normal megascleres.

Various further monaxonids have microscleres in the form of polyactinal euasters, without pauciradiate (few-rayed) variants (pentactines, tetractines, triactines, etc.), which have no demonstrable relationship to the euasters described above. Other similar spicules are gradational with intermediates into a monaxial microsclere (e.g., a spinispira, p. 33) or megasclere (e.g., an acanthostyle) and may then be called **pseudoeuasters** if supposed to have been derived from the monaxons. There may be sometimes comparative evidence that a seeming euaster is really a pseudoeuaster, for instance, if an apparently related species contains forms that pass into monaxons.

b. STREPTOSCLERES

This name was adopted (REID, 1963b) for the streptaster microscleres of the choristid Pachastrellidae and Theneidae, also called dichotriacts by DENDY (1924b), because of confusion in usage of the term streptaster and because most of these spicules are not dichotriactines morphologically.

Streptoscleres are streptasters *sensu* SOLLAS (1888) that appear to have meristic variations on a pattern of dichotomous or more complex branching. The simplest typical streptosclere is a spicule that appears to have two branches at the ends of a short central shaft, either in the same plane or in different planes. This type is often accompanied by simple triactines, which in turn grade into oxeas through various irregular intermediates; by comparison, it seems to be a dichotriactine, with one ray branched dichotomously. Other typical streptoscleres have more complex branching (e.g., trichotomous) or repeated heterotomous branching with production of composite (sympodial, DENDY, 1924b) spiral axes. Individual types distinguished are (i) plesiasters, with a short straight axis; (ii) metasters, with a spiral axis making less than one revolution; (iii) spirasters, with an axis making one to several turns; and (iv) amphiasters, with branches in whorls at the ends of a straight central shaft. In practice the picture is more complex, since these types intergrade in all directions. Other complications are (a) that morphological spirasters and amphiasters need not be streptoscleres (see sanidasters, p. 29, and spinispiras, p. 32–33); and (b) that VON LENDENFELD (1907) used metaster as a general name for streptoscleres.

DENDY (1924b) regarded streptoscleres as unrelated to euasters, but VON LENDENFELD (1907) regarded them as aster-derivatives. VON LENDENFELD's view seems more probably correct; microscleric triactines, regarded as the central type (DENDY, 1924b), are widely distributed as pauciradiate euasters; and some forms with streptoscleres (e.g., *Thenea wrightii* SOLLAS) have other oxyasters with up to six rays. There are also several instances in which Ancorinidae have euasters passing into spicules resembling plesiasters (e.g., *Tethyopsis dubia* WILSON), although these forms can also be compared with sanidasters.

By thickening of the axial part, a streptosclere of spiraster or amphiaster type may pass into a spicule resembling a monaxial spinispira (p. 32–33) or a type of amphiaster related to sanidasters. These forms can be identified usually as streptoscleres by association with typical examples but could not be recognized as such if found dissociated.

c. MICRORHABDS, SANIDASTERS, AND VARIANTS

A typical **microrhabd** is a straight monaxial microsclere without raylike spines, although fine spinulation may be present, and is usually an oxea (or microxea) or a strongyle (microstrongyle). These same names may also be used for curved specimens unless there are grounds for regarding these as forms of a sigmaspire (p. 32) or a spinispira (p. 33–34). A short microstrongyle may be almost as thick as it is long or pass into a sphere (p. 36); if spinulate, this type has sometimes been called a phalangaster. A spinulate microrhabd with relatively few, large, raylike spines is a sanidaster or an **amphiaster** if the spines occur in whorls at the ends. The relationships of these spicules may be shown by mutual replacement in species of one genus (e.g., *Sanidasterella* TOPSENT) or intergradation in one species.

In forms with euasters or streptoscleres, a microrhabd may be identifiable as a diactinal euaster or streptosclere. Microrhabds and sanidasters accompanying euasters but not directly comparable to each other are often assumed to be their homologues, and a sanidaster may pass into a euaster through intermediates. But microrhabds also occur in many other sponges without euasters, and comparative evidence may then suggest that, for example, a microxea in one species is equivalent to a megascleric oxea in another. Thus microrhabds appear to have several different origins and have probably arisen independently in many different sponges.

Spiny microxeas and sanidasters are streptasters *sensu* DE LAUBENFELS (e.g., 1955), at least by that author's definition; but he also sometimes used the name for streptoscleres with composite axes. They are not the streptasters of BURTON (1959), which are streptoscleres. A spiny microxea is a streptaster *sensu* DE LAUBENFELS but not in SOLLAS's original (1888) sense.

d. SIGMASPIRES AND VARIANTS

Sigmaspires are arcuate to spiral monaxial microscleres that occur especially in the choristid Craniellidae (Tetillidae auctt.), although indistinguishable spicules also occur in some monaxonids and lithistids. A typical sigmaspire is a blunt-ended, spiral monaxon of about one revolution, appearing C-shaped or S-shaped according to how it is viewed. Many examples are spinulate, although the spinules may be seen only at high magnification. Simple variants range toward a true, flat C-shape or pass into **toxaspires** making rather more than one revolution and appearing bow-shaped in some aspects. A strongly spined spiral sigmaspire has sometimes been called a sigmaspiraster. In *Chrotella* SOLLAS, C-shaped variants have spines in two opposite lateral rows along their length or restricted to the ends; the latter type resembles the chelas of some Desmacidontidae (*Treatise Part E (Revised),* vol. 3, in press) but have still been called sigmaspires (TOPSENT, 1928b) because of their evident homology.

Various authors have termed sigmaspires as **sigmas** (e.g., VON LENDENFELD, 1904c; DENDY, 1924b; DE LAUBENFELS, 1936), although the latter name was based (RIDLEY & DENDY, 1887) on partly comparable microscleres of monaxonid sponges (e.g., Desmacidontidae) with no evident relationship to craniellids. Herein SOLLAS (1888) and TOPSENT (1928b) are followed, as by BURTON (e.g., 1959), who thought these types distinct. Their homology was also rejected by DE LAUBENFELS (1936), although he still called both sigmas (1936, 1955). In VON LENDENFELD's (1904c) nomenclature, the term sigmaspire was expanded to take in the spiraster type of streptosclere, which, however, he later called a metaster (VON LENDENFELD, 1907).

e. SPINISPIRAS AND VARIANTS

Spinispiras and their variants are the characteristic microscleres of the monaxonid Spirastrellidae and Clionidae, which appear to be closely related. Typical spinispiras are blunt-ended (strongylote) monaxons of one to several revolutions bearing spines of varied size that may also have spiral arrangement. They vary from finely microspinulate strongylospires, with many small spinules that are not arranged spirally, to stoutly spined forms homeomorphic with a spiraster streptosclere. Intermediates between these extreme forms have spines of moderate size along the outside of the spiral. Coiling is usually tight in polyspiral examples, although some springlike forms occur; but short forms may resemble a sigmaspire, although usually stouter. In other variations, a spinispira may pass into (i) a smooth strongylospire; (ii) a C-shaped or straight microstrongyle, with small spinules that are not arranged spirally; (iii) a C-shaped form, strongly spined externally, called a sigmaster herein; (iv) a straight monaxon with spines arranged spirally; (v) a straight amphiaster, with spines in whorls at the ends; or (vi) a similar discaster with further whorls between the ends. In *Placospongia* GRAY, a short, long-spined, initial spinispira solidifies, producing a spicule resembling a sterraster, although distinguished by its different ontogeny. Initially mistaken for a genuine sterraster (SOLLAS, 1888), this form is now called a sterrospira (DENDY, 1921, and herein) or a selenaster (DE LAUBENFELS, 1955).

Because coarsely spined spinispiras resemble the spiraster form of streptosclere, all spinispiras are sometimes called spirasters (e.g., DE LAUBENFELS, 1936, 1955); but a finely microspinulate strongylospire is not a spiraster morphologically in the sense of possessing raylike spines. The coarsely spined forms are streptasters *sensu* SOLLAS (1888), but the group does not seem to be related to the streptosclere series (above, p. 31). The spiraster forms are homeomorphic, but the range of variation in the two groups follows different patterns. In particular, the characteristic morphological passage from spiraster to microtriod through intermediate metasters and plesiasters, as in streptoscleres, never occurs in spirastrellids or clionids. There are no other grounds for believing that these families are closely related.

The homology of the spinispira group is debatable. A spinispira is essentially similar to the spinulate sigmaspire, although typically polyspiral; but no sure connection can be demonstrated between the choristid Craniellidae and the monaxonid Spirastrellidae and Clionidae. In *Timea* GRAY, usually classified as a spirastrellid, the microscleres are typically euastriform (usually sphaeraster) but sometimes pass into sigmasters; this was read by DENDY (1921) as meaning that the seeming euasters are pseudasters but could also be taken as suggesting derivation of spinispiras from euasters. Some spinispiras have observable axial filaments, however, and thus seem to be genuine monaxons. A finely spinulate form may also share identical spinulation, a central annulation, or both with a megascleric oxea, which is the most likely prototype.

f. SIGMATOSCLERES

This name was adopted by REID (1963b) for sigmas and other forms that are the characteristic microscleres of the monaxonid Desmacidontidae, also occurring in some other monaxonids and sublithistids.

A typical sigma is a smooth, sharp-ended (oxeote), C-shaped monaxon or a similar form with the ends out of line, and then C-shaped or S-shaped in different views. If the ends are markedly out of line, it is said to be contort. In variant conditions, a sigma may (i) have one end reversed, producing a genuine S-shape; (ii) approach a true spiral shape; or (iii) bear a short external spine centrally or two opposite central spines directed inward and outward. Sigmas of some genera have hooked ends of markedly different sizes or in one instance have sawtooth external serrations on the hooked ends. A rare chiastosigma, apparently related to centrally spined sigmas, resembles two sigmas crossed centrally and is *chi*-shaped in some aspects.

The other types included as **sigmatoscleres** are mainly forms developed from a flat, C-shaped sigma in ontogeny, with the initial shape modified by secondary outgrowths. A **diancistra** has bladelike expansions, the fimbriae, on the inside of the curve and looks like a penknife with a partly opened blade at each end. A **clavidisc** is a similar spicule with the ends grown together, producing an ovate disc with a median slit running lengthwise. A canonchela is like a clavidisc with additional lateral fimbriae growing out from the central parts and is larger on one side than on the other. The rest form a group known as cheloids, comprising chelas and apparently related types, with various types of lateral outgrowths at the ends or sometimes with spines on the convex side. **Chelas** occur in two main forms, called dentate and palmate. A **dentate chela** has grapnel-like ends, with inwardly curving thorn- or toothlike lateral outgrowths arranged in opposite pairs. The ends of the central shaft may also bear small lateral expansions or alae. Terminology of these spicules is sometimes further refined according to the number of terminal flukes, as in tridentate or quinquedentate (the number is odd because the end of the initial sigma forms a single central fluke). A few forms pass from this type of chela into so-called **amphidiscs**, with a symmetrical ring of recurved flukes at each end of a straight shaft. A palmate chela has inwardly facing ends and further lateral teeth when these are present, expanded to form leaflike palms; the alae are strongly developed and sometimes form hoodlike expansions. Either type of chela may have similar ends or one end larger than the other and is then an **isochela** or an **anisochela**, respectively. The two types are also intergrading, through dentate forms with terminal flukes markedly flattened, which are sometimes called arcuate. An unusual spheranchora resembles two clavidiscs intersecting at right angles but develops from a tridentate chela in ontogeny. A placochela has alae meeting in the middle to give the shaft a figure-of-eight shape and palms forming circular expansions on the inward facing ends. A **bipocillus** has lobate leaflike expansions at the ends or spoonlike expansions with the concave sides faced together. Forms often called spiny chelas, herein called **chelasters**, may lack the typical features of chelas but have stout spines developed on the convex side. The spines may occur without order or in partial transverse whorls. Most chelasters are obviously C-shaped, but short stout examples can resemble a sphaeraster in some aspects.

In other developments, a sigma may pass into a simple microxea, with one or more bends along its length, or a toxa with a central bend and the ends curved in the opposite direction. Spicules of these types are, however, found in many other sponges, in which they cannot be considered as sigmatoscleres. A tonglike forceps, found only in the Demospongea, is essentially a toxa with the two ends bent together. Some examples are smooth, but other bear spinules and small terminal expansions.

There are several different views of how these spicules are related. According to DENDY (1921) the central type is the sigma, derived from an oxea by way of a toxa, and the prototype from which the more specialized forms have arisen in phylogeny, as well as in ontogeny. For TOPSENT (1928b), a sigma was supposed to be derived from a simple euaster with the lost rays represented by the central spines of some examples; but a chela was a spicule derived from an amphiaster through an amphidisc. DE LAUBENFELS (1936) suggested that some sigmas are chela derivatives. There is no way of testing these contentions, but DENDY's view seems generally most likely. There is no objective evidence that sigmas and chelas had separate origins. The two types are often found together, apart from the chela arising from a sigma; and a chela may have the two ends out of line, as in the contort type of sigma. Moreover, the general spiculation of the sigma-bearing sponges is normally entirely monaxonid, and the supposed amphiasters of *Samus anonyma* GRAY that TOPSENT (1928b, p. 44) cited in this context are, in fact, amphitriaenes with true actinal cladomes. The same sponge has otherwise no special resemblance to the chela-bearing desmacidontids. Reversion to a sigma from, for example, a chela would certainly be possible, however.

DENDY (1916, 1917, 1921, 1922) also held that the sigma is homologous with the craniellid sigmaspire; but TOPSENT (1928b) and DE LAUBENFELS (1936) both rejected this idea and were followed by BURTON (e.g., 1959). As pointed out by TOPSENT, the two are not identical morphologically: a sigmaspire is typically blunt ended, finely spinulate, and more spiral than C-shaped, but a sigma is typically sharp ended, unornamented and more C-shaped than spiral. There is also no evident relationship between the choristid craniellids and the sigma-bearing monaxonids, whose only close relatives appear to be the keratose sponges. A relationship imagined by DENDY (1922) is, again as said by TOPSENT (1928b), essentially based on preconviction. Sigmas and sigmaspires are, therefore, held herein to be convergent.

In alternative nomenclature, the dentate type of chela is called anchorate (e.g., DE LAUBENFELS, 1955) or is called an **anchora** (ancre in TOPSENT, 1928b) with the term chela then restricted to the palmate type.

g. OTHER PSEUDASTERS

Some monaxonid sponges have further types of pseudastrose microscleres, which seem to have varying homologies.

In **discasters,** sometimes called **discorhabds** (DENDY, 1921), a monaxial shaft bears whorls of separate spines or discoidal flanges between the ends. The best known discaster is the so-called **chessman spicule** of *Latruncalia* DU BOCAGE, in which the shaft typically bears a varying number of marginally serrated flanges, although separate spines may occur. Whorls of spines may also occur at the ends, or one end may be stylote or tylote. In *Sigmosceptrella* DENDY, a discaster with whorls of spines at the ends and between them develops from a sigma-shaped prototype, whose inward facing ends are engulfed by a thickened shaft during development. In *Barbozia* DENDY, a sharp-ended monaxon has two whorls or separate spines between the ends, and *Didiscus* DENDY has two simple discs.

DENDY (1921) interpreted these spicules as homologous and as pointing to the origin of the spinispira series from chelas; but this is not believed by other authors (e.g., TOPSENT, 1928b; DE LAUBENFELS, 1936) nor followed herein. The genera cited have nothing else in common except that all are monaxonids. A spiraster-type spinispira passes into a spined discaster in *Spirastrella corticata* (CARTER); but none of the other discasters pass into spinispiras and no spinispira or any other discaster develops from a sigma. Occasional imperfect variants of the *Latruncalia* discaster, or chessman spicule, are spiny styles or tylostyles and this spicule typically occurs at the surface of the body, where apparently related forms have small ectosomal styles or tylostyles. While not conclusive, this suggests its homology with the latter. The *Sigmosceptrella* discaster may be a chelaster homologically, but that of *Barbozia* is simply a modified oxea, and occurs in a sponge in which abnormal variants of a megascleric oxea may have comparable modification (DENDY, 1921). In *Didiscus,* the young forms of discasters and megascleric oxeas are initially indistinguishable. It seems likely, therefore, that these microscleres include forms with several different origins.

A euastriform microsclere with the form of a polyactinal sphaeraster may intergrade rarely with a spiny monaxial megasclere, which appears to be its prototype unless the opposite is imagined. The megasclere may be a diactinal oxea or strongyle, or a monactinal style or tylostyle, and the linking intermediates are shorter monaxons. This transition seems to correspond with occurrence of short variants of various smooth megascleres, for example, very short strongyles whose length is little more than their thickness, but with shortening carried to the extreme, and persistent spination producing a euastriform spicule. This type of passage from megasclere to microsclere in apparently homologous spicules does not occur in typical microscleres (although spinispiras may share ornament with a megasclere: see p. 32–33), and treatment of the present type as microscleres is essentially arbitrary. In another instance, short, finely spined tylostrongyles, corresponding to forms found as short variants of a typical tylostyle have sometimes been called korynasters (e.g., SCHRAMMEN, 1924a), although they are properly megascleres.

Various freshwater Spongillidae have amphidisc spicules, with a simple or serrated transverse disc at each end of a straight shaft, or amphiaster variants with spines in two irregular clusters. These types may occur with a small acanthoxea and be linked with it by intermediate spicules. In *Dosilia* GRAY, a spiny-shafted amphidisc is accompanied by spicules with passage from acanthoxea to pseudoeuaster, by shortening of the monaxon shaft and growth of many secondary rays from its center, and also by double variants with secondary rays growing from two centers. The amphidisc itself may have variants with secondary rays on the shaft. Spongillid amphidiscs are similar morphologically to those found as uncommon variants of the dentate type of chela (*Iotrochota* RIDLEY; see also p. 34), with which they have sometimes been associated (e.g., DE LAUBENFELS, 1936); but no sure connection can be made between these spicules or the sponges in which they occur.

h. DRAGMAS

Some forms have very fine monaxial microscleres secreted in bundles, each bundle being formed within one scleroblast. These are called **dragmas** and may be straight orthodragmas or take the form of toxas or sigmas. Their hairlike fineness is sometimes described as trichodal. They are not known as fossils.

i. SPHERES

A number of choristid sponges contain microscleric spheres consisting of silica secreted around an initial granule or of two or more such bodies fused together. These forms appear to originate by abortion of a microrhabd, since occasional variants of composite examples may have one component replaced by a normally cored spicular ray. A short microstrongyle may also pass into a sphere by gradation.

DISTRIBUTION OF MICROSCLERES

The distribution of different types of microscleres discussed above, is difficult to summarize completely, but a number of general points are worth restating.

1. The most widespread microscleres are microrhabds, which occur in many choristids, monaxonids, and lithistids. They may occur either alone or with some other type (e.g., euasters or streptoscleres).

2. Euasters, streptoscleres, sigmaspires, spinispiras, and sigmatoscleres are each the characteristic microscleres of particular groups of demosponges, and the type characteristic of each of these groups is either rare or absent in the others. Forms with euasters or sigmatoscleres as the characteristic microscleres have none of the other types, unless sigmas are equated with sigmaspires. Streptoscleres, typical spinispiras, and sigmatoscleres are restricted to the groups they characterize, unless chelalike sigmaspire variants are equated with true chelas.

3. Any of these major types of microscleres may be present in one form only or in two or more differing in size, shape or both. Different forms of one main type may also be differently distributed anatomically; for instance, a sponge with euasters may have small oxyasters in the choanosome and large spherasters or sterrasters packing a cortex.

4. Other pseudasters and further minor types (dragmas, spheres) are distributed sporadically in various demosponge genera. Dragmas that are sigmas are restricted to sigma-bearing sponges.

5. Various Demospongea have no microscleres, although spicules identifiable as megascleres are present. Some of these sponges are otherwise identical or similar to various other forms with microscleres, but some are not.

The occurrence of the five main types of microscleres restricted to some groups of demosponges may be outlined as follows.

6. Typical euasters, in which pauciradiate examples have recognizable meristic variation, occur mainly in the choristid Calthropellidae, Ancorinidae, and Geodiidae, and the monaxonid Tethyidae and Coppatiidae. Simple euasters may occur in some forms with streptoscleres (Theneidae) and some

other sponges. Spherasters and sterrasters have recently been recorded in some fossil lithistids (Megamorina; Hurcewicz, 1966) but may be intrusive since a similar modern form has only spirasters (possible streptoscleres).

7. Streptoscleres occur mainly in the choristid Pachastrellidae and Theneidae but also in some members of two groups of lithistids (Tetracladina, Dicranocladina).

8. Sigmaspires are characteristic of the choristid Craniellidae but also appear to be present in a few monaxonids and lithistids. An apparent sigmaspire, which may not be homologous with these spicules, may also occur as a variant of a spinispira.

9. Spinispiras are restricted to the monaxonid Spirastrellidae and Clionidae, unless spirasters found in some lithistids are thought to be spinispiras.

10. Sigmatoscleres are characteristic of the monaxonid Desmacidontidae and some similar sponges and are otherwise found only in several desmacidontid-like sublithistids.

Occurrence of these five types as the characteristic microscleres of different groups of demosponges has led many since the *Challenger* authors (Ridley & Dendy, 1887; Sollas, 1888) to use them in classification, as either subordinate to the megascleres (see also Rauff, 1893; Schrammen, 1910, 1912; Topsent, 1928b) or providing the principal criteria of classification (e.g., Hentschel, 1909; Dendy, 1916, 1917, 1921, 1922; Burton, 1959; Reid, 1963b). But their use involves interpretative problems, in dealing with homologies (see previous section) and with forms that lack microscleres. The occurrence of five types as characteristic of groups of nonlithistids, which can also be thought to be homogeneous, and distinct in terms of their other characters, is here considered to be acceptable grounds for using these microscleres in the characterization of taxa. It is acceptable provided that such characterization is understood to rest firmly on this basis and not simply on a basis of occurrence of particular types of microscleres. Too little is known about most lithistids for them to be treated in this manner, however. The method is also unsuitable for general paleontology, because microscleres are nearly always lacking in fossil material, and because many fossil lithistids do not resemble modern forms with microscleres or distinctive microscleres.

SPICULAR ONTOGENY AND PHYLOGENY

In description of the microscleres, several instances have been noted if a specialized type of adult spicule is developed from a different prototype in ontogeny by secondary modification of its initial form. Examples include development of a sterraster from a polyactinal oxyaster (p. 30), of a sterrospira from a spinispira (p. 32–33), or of a diancistra or a chela from a sigma (p. 33). It is reasonably likely, although not surely demonstrable, that such changes in ontogeny correspond with an earlier change in phylogeny.

There are, nonetheless, occurrences that have an ontogenetic prototype need not represent a phylogenetic prototype. Because some lithistid desmas develop from initial tetraxons but others from initial monaxons or anaxial corpuscles, it was argued by Schrammen (1910) and Dendy (1921) that these types must have arisen independently. But in *Macandrewia* Gray there is intergradation between normally monaxial desmas and occasional tetraxial variants (Schmidt, 1880; Sollas, 1888; Topsent, 1904). This points to origin of one of these types from the other; moreover, whichever direction is ascribed to phylogeny, the ontogenetic prototype of the phylogenetic end form does not correspond with its phylogenetic prototype. In another example, intergrading phyllotriaenes and discotriaenes of some lithistids may develop from either a dichotriaene or a simple triaene. In turn, a discotriaene with simple and short to rudimentary axial filaments in the cladal disc may pass into a monaxial spicule, in which these filaments are lacking.

The ontogenetic prototypes of some demosponge megascleres, thus, seem to have

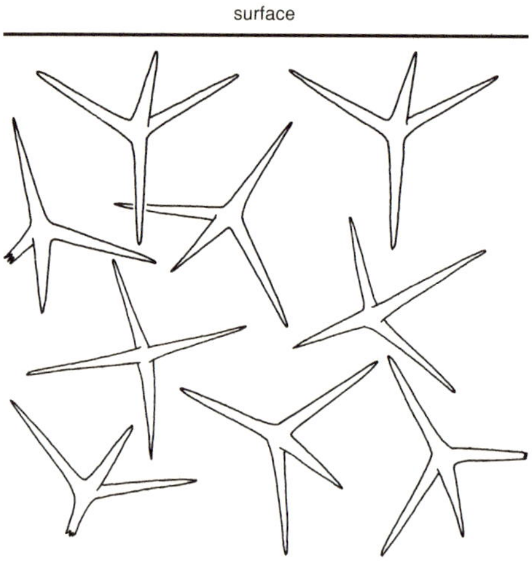

FIG. 22. Choristid skeletons (type 1): all megascleres calthrops, oriented irregularly except that those near surface (top here) may have 3 rays pointing outwardly, 1 inwardly (new).

been subject to caenogenetic alterations in phylogeny, assuming that transitions in these spicules represent evolutionary sequences. This was not understood by SCHRAMMEN and DENDY, although shown earlier by SOLLAS (1888). Presumably the same principle may apply to microscleres. In particular, typical streptoscleres do not appear to develop from triactines, although their adult variations point to this prototype in phylogeny unless an opposite derivation (i.e., spiraster to triactine) is postulated.

MEGASPICULATION: 1. CHORISTIDS

The choristid Demospongea, the choristids, are nonlithistid sponges that either have megascleres with three or more rays, although monaxons may also occur, or have small analogous spicules not of megascleric size. The predominant form in the characteristic spicules is four-rayed (tetractinal) and tetraxial, although this type is sometimes replaced by triactines or rarely outnumbered by spicules with more than four rays. Tetraxial examples may be calthrops, subtriaenes, or true triaenes; and these types may occur alone or in any combination. When triaenes are present as megascleres, one to several different sorts may occur. Monaxial megascleres are usually oxeas, but sometimes strongyles or styles. Additional microscleres may be present or absent; those of modern forms include euasters, streptoscleres, sigmaspires, microrhabds, sanidasters, and spheres but not spinispiras or sigmatoscleres.

The typical choristids are those with megascleres and greatly outnumber the others. It is difficult to generalize, but the megascleres are usually spicules with rays from 0.2 mm long to many times longer. The largest tetraxons are triaenes, in which the rhabdome may reach lengths of 10 to 20 mm or more in the body and may be considerably longer (e.g., 40 to 50 mm) in examples protruded for anchorage. The length of monaxons is comparable with that reached by the rhabdomes of triaenes, though either type may be larger than the other in a given sponge. Calthrops or monaxons may occur without regular arrangement, producing an irregular feltwork of interwoven rays; but triaenes are typically arranged with the cladi toward a dermal or gastral surface and the rhabdome running radially inward. Many massive forms whose megascleres are triaenes and monaxons have generally radiate architecture with all the megascleres following lines that radiate from a point within the body.

To distinguish three intergrading subtypes is useful for further description in which tetraxial megascleres are (i) all calthrops or subtriaenes or both; (ii) the latter intergrading with true triaenes; or (iii) all triaenes. As usual with nonlithistids, these can be illustrated best by reference to living examples, but some fossils are also relevant.

The first group is typified by *Halina* BOWERBANK, *Pachastrella* SCHMIDT, and similar genera. The megascleres may all be forms with three or more rays (*Halina*) or include monaxons (*Pachastrella*). The skeleton is never truly radiate and is sometimes entirely irregular; but calthrops or subtriaenes next to the surface may have three rays directed toward it and the fourth inward, as is usual in triaenes (Fig. 22). Megascleres other than monaxons are most commonly tetractinal calthrops or subtriaenes, with other types (triactines, pentactines) either absent or present as only minor variants; but some forms have only triactinal megascleres or spicules with more than four rays as a characteristic feature. The fossil *Helobrachium* SCHRAMMEN has triactinal but tetraxial megascleres (**helotriaenes**, SCHRAMMEN, 1910), with the fourth ray represented by a buttonlike rudiment and the other rays hooked at the ends; because of these hooked ends the skeleton is loosely coherent, although lithistid features are absent. In the living *Yodomia* LEBWOHL, calthrops or subtriaenes are accompanied by smaller pentactinal centrotriaenes with occasional hexactinal variants.

Sponges with megascleres ranging from calthrops to true triaenes are exemplified by *Poecillastra* SOLLAS, in which they vary continuously from regular calthrops to plagiotriaenes or orthotriaenes, with the rhabdome up to several times longer than the cladi. The skeleton may be more or less irregular or have triaenes arranged radially near the surface but an irregular feltwork of calthrops internally (Fig. 23). If oxeas (monaxons) are present they may form radial bundles, and calthrops of the interior may then occur with one ray aligned with the oxeas and the other three directed outward. This condition approaches that of true radiate choristids, in which all the tetraxons are triaenes. In another development, the fossil *Propachastrella* SCHRAMMEN has irregularly felted calthrops and variants accompanied at the surface by presumably ectosomal dichotriaenes.

The fossil *Spiractinella* HINDE (Fig. 24) appears to have aberrant development of this type of skeleton, with the megascleres mainly of types having five or six rays. The most common megascleres are hexactines, **dichohexactines**, and two kinds of dichotetraenes; but others include calthrops intergrading with simple triaenes and rare **dichotriaenes.** The sponge is known mainly

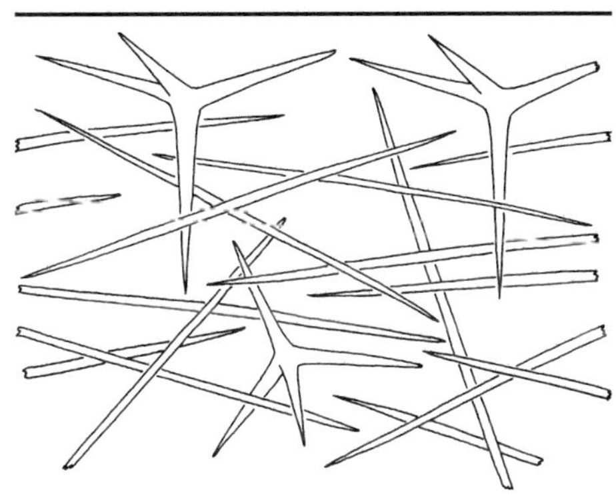

FIG. 23. Choristid skeletons (type 2): example with ectosomal triaenes and choanosome oxeas and calthrops; triaenes are short-shafted orthotriaenes arranged with 3 cladal rays toward surface and fourth (rhabdal) ray pointing inwardly; under their cladomes, presumed here to lie in ectosome, choanosomal skeleton consists of irregular feltwork of oxeas and subordinate calthrops (new).

still from dissociated spicules, but these occur together and share a distinctive spiral ornament. The dichotriaenes and one type of dichotetraene also share a peculiar style of branching, in which one large cladal branch stands roughly at right angles to the rhabdome, and a small branch is directed away from it but parallel with its axis (REID, 1963b). By analogy with later forms, the symmetrical hexactines, dichohexactines, and calthrops were probably spicules of the interior, with the triaenes and tetraenes underlying an external surface. This sponge was long thought to belong to the Hexactinellida, but the calthrops and triaenes require its reference to the class Demospongea. The dichohexactines, although outwardly similar to hexactinellidan **hexasters,** are megascleres (not microscleres) and have genuine actinal branching, as in demosponge dichotriaenes. The genus is also of interest because a similar sponge without tetraxons could not be identified as a demosponge from its spicules. A possible example is the contemporaneous (Carboniferous) *Stromatidium* GIRTY, whose known spicules are **polycladose** dichotetraenes with a strong spiny ornament; in genera from the

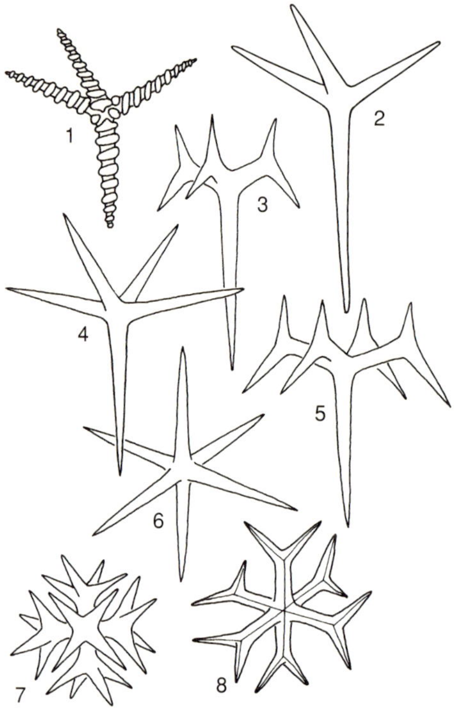

FIG. 24. Unusual choristids: *Spiractinella* HINDE (Lower Carboniferous); *1,* calthrops shows characteristic spiral ornament, omitted in other figures; *2,* short-shafted plagiotriaene, which intergrades with 1; *3,* dichotriaene, with cladal branches (or deuterocladi) arranged in vertical planes; *4,* tetraene, with 2 pairs of opposite cladi (= upper 4 rays as shown) making different angles with the fifth ray; *5,* tetraene with cladi developed as in 3; this type also grades into forms with the branches (deuterocladi) horizontal or recurved toward rhabdome (here pointed downward); *6,* regular hexactine; *7,* dichohexactine, with branches of opposite rays in different planes; *8,* lophose hexactine; this type grades into euaster-like forms by further shortening of primary rays; types 1 and 2 are less common than 5–8, and 3 and 4 are uncommon; because of prevalence of triaxons (view *6–8*), this sponge was mistaken for a member of class Hexactinellida before discovery of other types of spicules (new).

Carboniferous in Ireland, such spicules occur with similarly ornamented calthrops, although no clear relationship is demonstrable. Branching rays are also abnormal in hexactinellidan megascleres. But *Stromatidium* spicules may have union by fusion, which is known in many Hexactinellida but only found teratologically in proven Demospongea.

Typical triaenose choristids have all tetraxons developed as triaenes, which are often long shafted (Fig. 25). The triaenes are typically arranged with the cladome in or under an ectosome and the rhabdome running into the choanosome at right angles to the surface; their arrangement is, hence, radiate in forms of massive habitus. They are sometimes all of one kind but more often of two or three, in the range protriaene to anatriaene or also including dichotriaenes. Some forms have mesotriaenes with a short extra distal ray as variants of a normal triaene (usually an **anatriaene**). Triaenes of different types may also differ markedly in size in one sponge; hence forms of different types and sizes found loose in a sediment need not represent separate species. Accompanying monaxons (usually oxeas) may form an irregular choanosomal feltwork or have radiate arrangement; in the latter instance they may be grouped into radiating bundles that may radiate spirally in some sponges. Fully radiate sponges may have some triaenes in the interior as well as in their typical position. At the surface, either triaenes or oxeas may protrude through an ectosome, which is then **hispidated** (i.e., bristly) externally, and some forms protrude very long-shafted triaenes (usually anatriaenes) to form beardlike rooting structures. The proportion of triaenes and monaxons is also varied. In some forms the triaenes are so few that the sponge at first appears to be monaxonid, and some modern species (e.g., *Craniella cinachyra* (DE LAUBENFELS), *Stellettinopsis corticata* CARTER) were initially described as monaxonids. The triaenes of *S. corticata* are also still more readily overlooked because the length of the cladi is less than the thickness of the rhabdome.

This family has been called Ophiraphiditidae (SCHRAMMEN, 1910); but *Ophiraphidites* CARTER was based on a macerated fragment with **ophirhabds** only, which could represent several monaxonids, see for example *Bubaris* GRAY. The fossil Cephaloraphiditidae (Ophiraphiditidae auctt.) are peculiar choristids, whose choanosomal megascleres are sinuous oxeas or ophirhabds intertwined to form a loosely coherent skeletal framework without true lithistid articulations. Tetraxons found at the surface and presumably ectosomal are mainly subtriaenes, which grade into calthrops or true triaenes with the rhabdome barely longer than the cladi. These forms are unusual in the absence of long-shafted triaenes, which are otherwise normally present when the internal megascleres are monaxial.

Sponges with small spicules analogous with true choristid megascleres are conveniently termed microspiculate. The length of spicular rays is generally less than 0.2 mm, down to less than a tenth of this. The spicules are usually mainly tetractinal but sometimes mainly triactinal (*Plakortis* SCHULZE) or diactinal (*Roosa* DE LAUBENFELS) without including tetractines. The latter, when present, may have variants with more than four rays. In Plakinidae, tetractines may be normal or lophose calthrops but never triaenes. Simple lophose calthrops are typical of *Plakina* SCHULZE and may be monolophose to fully **tetralophose. Candelabrum** spicules of *Corticium* SCHMIDT are special tetralophose calthrops in which the branches of one ray are larger than those of the others and also bear spines on their external faces. In alternative usage, all forms of lophose calthrops are called candelabra. The arrangement of the spicules is generally like that of choristids with calthrops as megascleres, except that a candelabrum may occur with the major branches directed toward a surface or canal. Two other genera of doubtful relationship to plakinids have triaenose spicules. *Thrombus* SOLLAS (Fig. 26.1–26.2) has small spiny triaenes, sometimes varied as diaenes or tetraenes, with cladi unbranched or branched dichotomously or trichotomously. *Samus* GRAY has comparable amphitriaenes with the cladi branched in one cladome only or both. The rhabdome of these spicules may have a central swelling of the axial filament, suggesting that they are really diactines with branching rays.

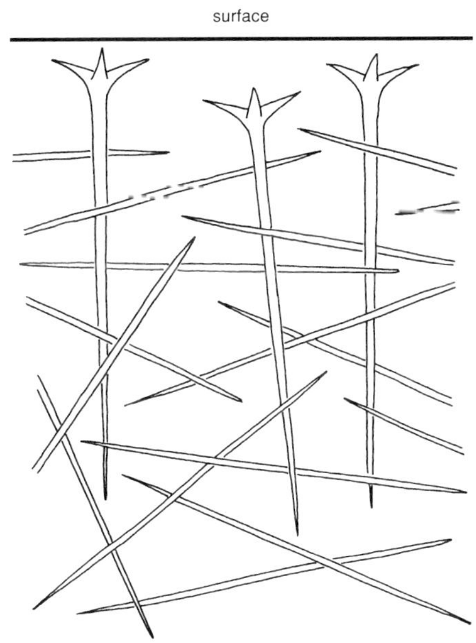

FIG. 25. Choristid skeletons (type 3): with triaenes and oxeas only and radial arrangement of megascleres restricted to triaenes; triaenes are long-shafted orthotriaenes, arranged more or less radially with their cladi in or under an ectosome; choanosomal megascleres are irregularly felted oxeas (new).

Most living choristids, and all with megascleres, have an ectosome developed as a dermis or a cortex. The canal system is eurypylous to diplodal in different forms, with this range in both forms with megascleres and in microspiculate sponges. There may also be further complications, for example presence of chones (p. 16) in corticate sponges. A cortex may be soft or toughly fibrous; when euaster microscleres are present, the cortex may be packed with spherasters, sterrasters, or aspidasters to form a stony rind (Fig. 27).

Living choristids with megascleres include three main groups whose characteristic microscleres are euasters (Calthropellidae, Ancorinidae, Geodiidae), streptoscleres (Pachastrellidae, Theneidae), or sigmaspires (Craniellidae). There are none with spinispiras or sigmatoscleres. The commonest accessory microscleres are microrhabds, although sanidasters, related amphiasters, or

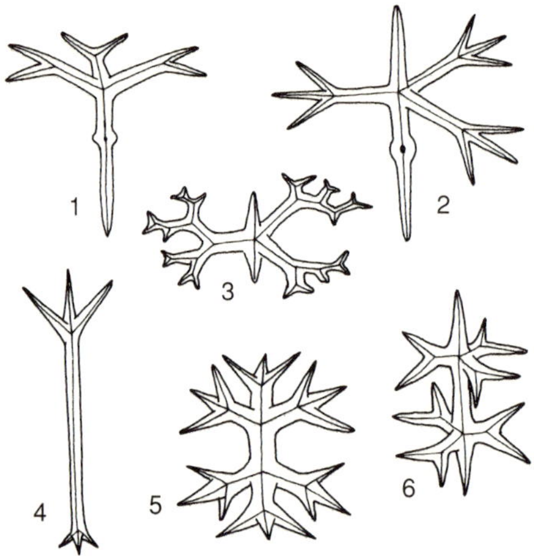

FIG. 26. Unusual choristids: *1–2,* spicules of *Thrombus* SOLLAS (Eocene–Holocene), with rhabdal rays showing a swelling of ray and axial filament that suggests a spicular center; 1, dichotriaene; 2, mesotriaene with trichotomous cladi (or, mesotrichotriaene); spiny ornament omitted in both; 3, centrotriaene of *Triptolemma* DE LAUBENFELS (Eocene–Holocene), with polycladose cladi; 4, heteropolar amphitriaene of *Amphitethya* VON LENDENFELD (Holocene); 5, equipolar amphitriaene of *Samus* GRAY (Holocene) with trichotomous cladi; 6, amphimesotriaene of *Ditriaenella* HINDE (Eocene–Holocene), with dichotomous cladi, and extra rhabdal rays at both ends of central shaft (new).

simple spheres are sometimes present. Some genera have no microscleres or none but microrhabds but are otherwise like forms in which distinctive types are present. The microspiculate Plakinidae have no microscleres, unless all of their spicules are referred to this category; this was done by SOLLAS (1888), but the lophose types do not occur as microscleres in any forms with megascleres. *Thrombus* has microscleres in the form of minute toothed amphidiscs, and *Samus* has sigmaspires.

Some choristids are preserved intact as fossils, but they are more often represented by dissociated megascleres. These should not be made the basis for erection of species or genera, unless they are especially distinctive. As seen from discussions above, several types of tetractines of different sizes may occur in one species; on the other hand, indistinguishable megascleres can be present in members of different families (or orders in some classifications). In addition, many members of three groups of lithistids (Tetracladina, Dicranocladina, Megamorina) have ectosomal dichotriaenes, which may be lost after death, and cannot be distinguished from choristid dichotriaenes; hence isolated dichotriaenes need not represent choristids and are especially open to suspicion in rocks that also yield triaene-bearing lithistids. On the other hand, very few lithistids have triaenes with unbranched cladi, and none have anatriaenes.

The oldest known choristids are Early Carboniferous, but their spicules range from calthrops or dichocalthrops to typical long-shafted triaenes, suggesting a longer history. Sediments containing these spicules may also yield apparent sterrasters, now found only as the characteristic microscleres of Geodiidae. Most fossil choristids resemble forms now living but are difficult to allocate with certainty because similar megascleres can occur in modern genera with different types of microscleres (e.g., euasters or sigmaspires). Exceptionally, specimens with microscleres preserved can be referred to modern genera (e.g., CHAPMAN & CRESPIN, 1934). A few fossil genera (e.g., *Helobrachium* SCHRAMMEN, *Cephaloraphidites* SCHRAMMEN) have megascleres with no modern counterparts.

MEGASPICULATION: 2. MONAXONIDS

The monaxonid Demospongea, the monaxonids, are nonlithistid sponges with megascleres that are all monaxons, except in some instances in which spicules with secondary rays replace a normal monaxon. Accompanying microscleres may be monaxial (microrhabds; sigmatoscleres), triactinal to polyactinal (euasters), or pseudopolyactinal (spinispiras; other pseudasters); or they may be absent. Spongin may be absent or may supplement the megascleric skeleton or largely replace it.

Existing monaxonids have very diverse characters and range generally from sponges that are otherwise almost identical with various choristids with mainly monaxial megascleres to forms in which the skeleton is produced mainly by reticulate spongin fibers and different individuals have a few spicules present or none. Between these extremes fall many other types, having various lesser resemblances to choristid or keratose sponges or with nothing markedly in common with either. The canal system varies from eurypylous to diplodal, as in choristid sponges. There is always an ectosome, which is sometimes a cortex but more often a simple dermis. The monaxial megascleres may be diactines (oxeas, strongyles), monactines (styles, tylostyles), or both; and any type of megasclere may occur in one or more forms. Many genera have smooth megascleres only, but in others some or all are conspicuously spiny. The spiny megascleres again may be diactines, monactines, or both. A few forms have pseudastrose megascleres. The arrangement of megascleres may be radiate, irregularly felted, or follow various patterns related to the presence of spongin. Megascleres of the choanosome and ectosome may be similar or different in form, arrangement, or both.

For further description, modern forms are conveniently divisible into three major groups, as follows:

i. more or less choristid-like sponges with euaster or sigmaspire microscleres, mainly diactinal megascleres, and normally no spongin;

ii. sponges with spinispira microscleres when any distinctive forms are present, with typically monactinal megascleres, and with spongin normally absent although occasionally small amounts occur; and

iii. sponges with sigmatoscleres or no distinctive microscleres, with diactinal or monactinal megascleres and with spongin commonly present and often conspicuous.

The last group also takes in nearly all forms containing spiny megascleres. These groups have various overlaps in the characters of the megascleric skeleton but do not represent intergrading types of organization,

FIG. 27. Choristid skeletons (type 4): example with radiate structure, 2 kinds of triaenes, and cortical armor; spherical bodies beneath surface are sterraster microscleres, produced in choanosome but exported to cortex, where they accumulate to form stony armor; triaenes found under this layer are long-shafted orthotriaenes (4 shown) and smaller anatriaenes (1, at center); choanosomal megascleres are radially oriented oxeas (new).

as with groups distinguished above in description of the choristid sponges. In particular, forms with the contrasted types of microscleres are essentially nonintergrading, and how they are related is unknown.

In the first group, the Coppatiidae and Tethyidae are euaster-bearing monaxonids, of which some are closely similar to various ancorinid choristids apart from the absence of triaenes. The arrangement of megascleres varies from confused, irregularly felted to fully radiate with the spicules grouped in bundles. The megascleres are usually oxeas, although exceptions occur. *Jaspis (?) serpentina* WILSON (1925) has oxeas of normal *Jaspis* species replaced by ophirhabds, like those of the choristid Cephaloraphiditidae. This sponge resembles the fossil *Heteroraphidites* SCHRAMMEN in which, however, absence of tetraxons could be due to their loss before burial. In *Tethya* LAMARCK, with a radiate skeleton, the megascleres are typically strongyloxeas (p. 25), sometimes passing into styles or even tylostyles, which are arranged with the pointed ends outward. Euasters of these sponges are similar to those of ancorinids with a range from simple oxyaster to sterrospheraster, and one sort or more may be present. As in ancorinids, a microrhabd or sanidaster may also occur. The ectosome is commonly a cortex and may be packed with spherasters or sterrospherasters, as in the choristid *Aurora* SOLLAS.

Trachygellius TOPSENT and *Raphidotethya* BURTON are comparable sponges, with oxea megascleres in radiating bundles but with sigmaspire microscleres like those of the choristid Craniellidae. The latter are triaenose sponges with radiate skeletons; among them, *Trachygellius* and *Raphidotethya,* which are both stalked sponges, come closest to *Amphitethya* VON LENDENFELD, which may have triaenes restricted to a stalk. The monaxonid *Stylocordyla* THOMSON, in which microscleres are absent, has a spirally radiate skeleton, reproducing a pattern that occurs in species of the choristid *Craniella* SCHMIDT.

Because such monaxonids can resemble closely various choristids in all characters but the absence of triaenes, some authors have regarded them as forms derived from triaenose choristids by loss of the triaenes. Genera for which this is claimed are sometimes called epipolasid, from SOLLAS's (1888) use of a family Epipolasidae. This name was not based on that of any genus included and is therefore invalid. The genus *Epipolasis* DE LAUBENFELS was established 48 years later, for such forms grouped with choristids in his taxon Choristida. The principal grounds for this idea are the rarity of triaenes in similar triaenose choristids, suggesting partial loss of the ability to produce this type of spicule, which could lead to its being lost completely. There is also good evidence of comparable losses of various other types of spicules, both megascleres and microscleres (e.g., BURTON, 1932). This idea was especially emphasized by DENDY (1905, 1916, 1917, 1921, 1922), who believed that all monaxonid sponges are derivatives of choristid ancestors. On the other hand, DE LAUBENFELS (1936) regarded derivations claimed by DENDY as either simply possible among other alternatives or unacceptable; the latter applies specially to alleged origin of monaxonids with sigmatosclere microscleres from craniellids with sigmaspires (DENDY, 1922), also rejected by TOPSENT (1928b). Close relationship of choristid and monaxonid sponges is probably genuine in the forms described above, although no evidence points objectively and certainly to the direction even of phylogeny. Some choristids with mainly monaxial megascleres and only rare triaenes can be so like the monaxonids cited that identification can depend on an author's observation. A relevant example is *Craniella cinachyra* (DE LAUBENFELS); initially thought to be monaxonid and described as a species of *Trachygellius* (DE LAUBENFELS, 1936), this sponge has since proved to have rare triaenes, even in DE LAUBENFELS's holotype (LITTLE, 1963). Here it seems likely that an otherwise similar sponge with monaxon megascleres only is a genuine relative. But other monaxonids have generally no comparable resemblance to any known choristid, and many have spicules or other features (e.g., spongin fibers) unknown in any choristid. While it is possible that such sponges are choristid derivatives, no general assertion is justified from the evidence available. There is, especially, no clear evidence that forms with sigmatosclere microscleres are related to any choristids except as demosponges.

A second group of modern monaxonids comprises forms with spinispira microscleres (Spirastrellidae, Clionidae) and similar sponges with microrhabds only or no microscleres (Suberitidae, Polymastiidae). In this group, the megascleres are mainly monactinal and most typically tylostyles, although simple styles or oxeas may also occur. A few forms have the latter types only. The arrangement of the megascleres ranges from confused to fully radiate; in the latter instance, monactines are arranged with the pointed ends outward. The canal system varies from eurypylous to diplodal. The ectosome is often a cortex; it may then have a special ectosomal skeleton of small monactines arranged radially to form a cortical palisade. In some forms, the surface is made hispid by protrusion of ectosomal or choanosomal megascleres. In a special development of this condition, the hemispherical to discoidal *Radiale* SCHMIDT has a prominent equatorial fringe of large protruded spicules. A few forms (e.g., some *Polymastia* species) protrude long monactines in which the distal end is tylote, in contrast to normal internal tylostyles in which the outward end is azote. Very rarely, small amounts of spongin may occur cementing megascleres together (e.g., in *Suberites* species), but spongin fibers do not occur.

Typical examples of this group can again have some resemblances to triaenose choristids in radiate arrangement of the megascleres or in the presence of a cortex, but none resembles any known choristid in a manner suggesting origin by simple loss of triaenes. The spinispira microscleres are comparable with the sigmaspires of the choristid Craniellidae, but it cannot be asserted that these types are homologous (see p. 35).

Most other living monaxonids form a third group consisting of sponges with sigmatosclere microscleres and similar forms with only toxas, microrhabds, or no microscleres. These forms are several times more numerous than all other monaxonids together. Sigmatoscleres, when present, may be sigmas, more complex types (e.g., diancistras, chelas), or both. The megascleres may be diactines, monactines, or both, which may be smooth or conspicuously spiny. The canal system varies from eurypylous to diplodal, and the ectosome is usually a thin dermis, although rarely a cortex.

Spongin is commonly present, as either a cementing material or forming skeletal fibers that supplement the megascleric skeleton or largely replace it (Fig. 28). A few forms with fibrous spongin have sparse megascleres in some individuals but none in others, thus passing into keratose sponges. The arrangement of megascleres ranges generally from wholly confused to various special patterns related to the presence of spongin. When a skeleton has distinct meshes enclosed by cemented megascleres or by skeletal fibers, it is reticulate; it is isodictyal if the sides of the meshes are regularly subequal to equal (Fig. 28.1). A radiate arrangement of megascleres occurs occasionally.

When spongin is present, the skeleton may have a variety of conditions. If spongin fibers are absent, the arrangement of megascleres varies from wholly confused to a regular isodictyal pattern, in which individual megascleres are cemented together by their tips to enclose three-sided or four-sided meshes. Distinct spongin fibers are divisible into primary fibers, which radiate to the surface from a central or axial part of the body and secondary fibers that join primary fibers transversely. In forms with skeletal fibers, some of all megascleres are typically either embedded in the fibers or attached to them externally, as either coring spicules that are wholly enclosed within the fibers and aligned longitudinally (Fig. 28) or as **echinating** spicules (Fig. 29) attached externally by one embedded tip and directed obliquely forward. Most commonly, coring spicules are diactinal (oxeas or strongyles), and echinating spicules are monactinal (styles, tylostyles, acanthostyles); but exceptions can occur in both directions. Monactinal echinating spicules occur with the blunt (stylote or tylote) end imbedded in the fiber and the oxeote end outward.

Three types of spicule and fiber combinations are distinguished: (i) chalinid, in which only coring megascleres are present; (ii) axinellid, with echinating megascleres only

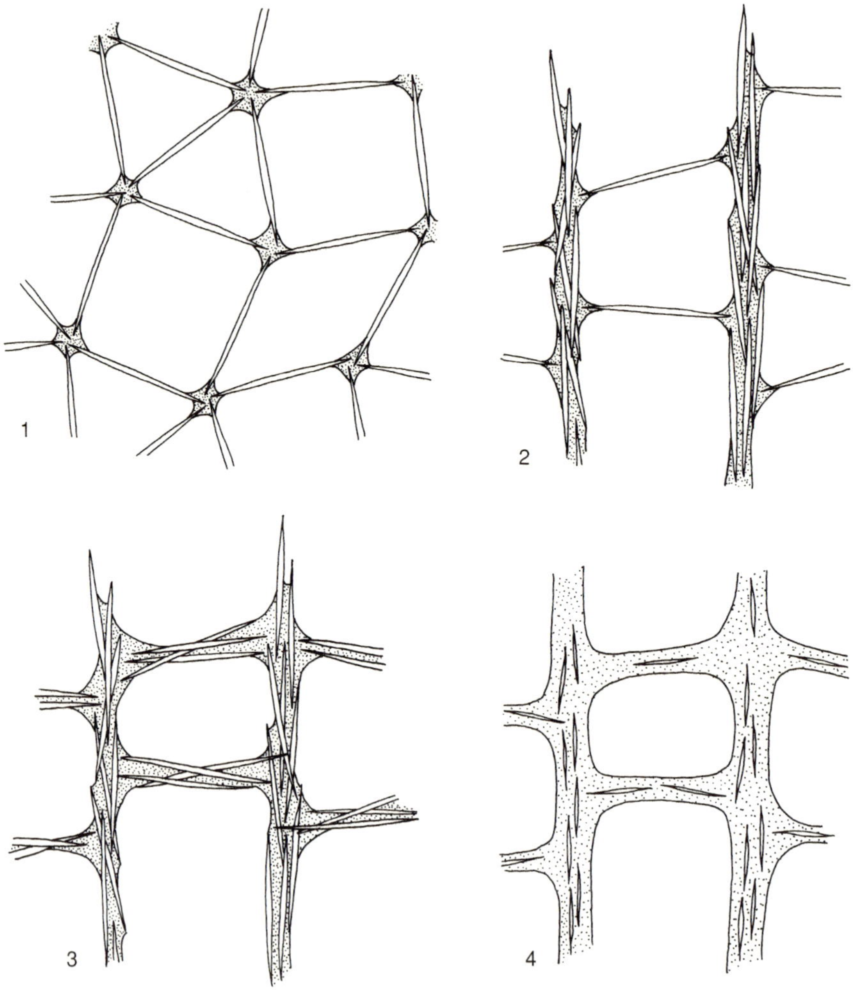

FIG. 28. Reticulate monaxonid structures, in which spicules are united by cementing or fibrous spongin; spicules oxeas in all figures; spongin *stippled; 1,* unispicular latticework of spicules united at their tips, with mainly three- and four-sided meshes; amount of spongin small; *2,* skeleton with multispicular fibers, formed by bundles of cemented spicules, and connected transversely by single spicules whose tips are imbedded in fibers; spongin more important than in view *1* but still subordinate to spicules; *3,* structure similar to view *2* but transverse connections also formed by multispicular fibers; *4,* skeleton formed chiefly by reticulate spongin fibers, which are cored by subordinate spicules; this type grades into forms in which spicules are locally or entirely absent; all structures represented are three-dimensional but are drawn as two-dimensional for simplicity; fibers of chalinid type (new).

(Fig. 30); and (iii) ectyonid, with both coring and echinating megascleres (Fig. 29). The number of spicules can vary considerably. Coring spicules may be present in bundles in both primary and secondary fibers or in the primaries only with the secondaries then containing only single spicules or none; or only a single line of spicules may be present along the axes of both types of fibers or of the primaries only. Echinating spicules vary in numbers from densely crowded examples to occasional isolated spicules, occurring several to many meshes apart in a mainly keratose skeleton. The amount of spongin may also vary; in axinellid fibers it ranges from barely enough to connect the bases of the spicules to so much that most of the spicules are buried in spongin with only the tips of some projecting. In other variations primary fibers may be connected transversely by spicules imbedded at both ends, without true secondary fibers. In axinellid skeletons, free primary fibers radiate often to the surface from a reticulate central or axial skeleton. Such fibers, with small amounts of spongin, are sometimes called plumose from the featherlike appearance produced by their echinating megascleres.

Some sponges in this group have megascleres with secondary rays or pseudastrose megascleres, although still classed as monaxonids. *Acarnus* GRAY has cladotylote spicules with a tylote knob at one end of a monaxial shaft and a group of recurved teeth, typically four, at the other. The teeth contain small axial filaments, which, however, are not continuous with the main axial filament. This kind of spicule seems to be a modified tylostyle with a grapnel-like group of secondary rays added at the normally oxeote end. *Cyamon* GRAY has normal echinating monaxons replaced by diactinal to pentactinal spicules, with equal or subequal rays. These spicules are outwardly similar to diactinal to pentactinal spicules of choristid sponges. In ontogeny, however, extra rays are developed from the basal (i.e., blunt, imbedded) end of an initial monactine, and in some at least their axial filaments are not continuous with the primary axial filament. By comparison with other forms, these spicules appear to be derived from a normal acanthostyle, with some basal spines enlarged as secondary rays. The spicules have hence been called pseudotetractines and other names. *Trikentrion* EHLERS is a similar sponge with echinating spicules mainly **triactinal** (or pseudotriactinal) and the other megascleres diactinal. In *Discorhabdella* DENDY, an initial monactine gives rise to pseudastrose megascleres with a massive spiny basal enlargement and a smaller one at the other end.

A variety of other monaxonids, some with microscleres, have characters that do not fit well in any of the main groups described. The most important forms here are the

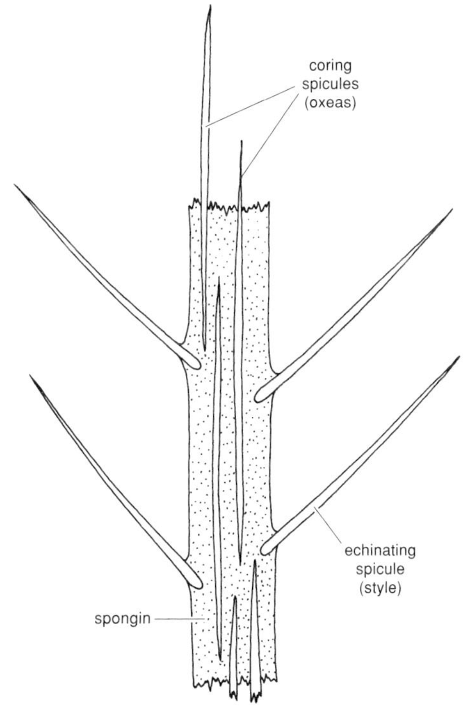

FIG. 29. Coring and echinating spicules; longitudinal section of a skeletal fiber that is cored by oxeas and echinated by styles; this type is termed ectyonid (new).

freshwater Spongillidae, with a typically confused arrangement of simple or spiny diactines. Microscleres may be absent or present in the form of amphidiscs or euastriform or other pseudasters (p. 36–37) related to spiny diactines. The amphidisc microscleres are found especially in reproductive gemmules, although not confined to these structures. These microscleres have sometimes been regarded as related to chelas (e.g., DE LAUBENFELS, 1936), but there is not sufficient evidence to establish conclusively this relationship.

MEGASPICULATION: 3. LITHISTIDS AND SUBLITHISTIDS

The lithistid Demospongea, the lithistids, are sponges in which the main choanosomal megascleres are articulated desmas, whose union produces a loosely coherent to rigid

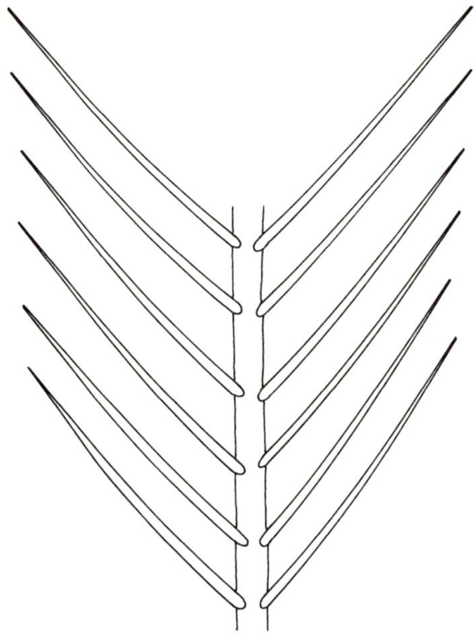

FIG. 30. Axinellid type of skeletal fiber with a plumose arrangement of monactines (styles) (new).

and stony skeletal framework. In addition to this desmal framework there may also be other types of megascleres comprising a supplemental skeleton. These may be of types also present in choristids or monaxonids or found only in lithistids. Microscleres may be present in living lithistid sponges and, when present, may be (i) microrhabds; (ii) recognizable streptoscleres; (iii) unidentified amphiasters or spirasters; or (iv) sigmaspires.

Existing lithistid sponges are considerably less abundant than choristids and monaxonids and seem generally to represent remnants of stocks that were formerly more abundant. In paleontology, however, the lithistids are considered to be the most important Demospongea. Only lithistid sponges are normally preserved as macrofossils, other types being rare in this condition; the skeleton does not collapse on decay of the soft parts, as in choristids and monaxonids, or decay as in keratose sponges. Further, many lithistid genera and some families and higher groups are known only from fossil material. This is in contrast to choristid and monaxonid sponges, which are known best from their modern representatives. Methods used in studying lithistids, as a consequence, have been mainly developed in paleontology (ZITTEL, 1878a; RAUFF, 1893, 1894, 1895; SCHRAMMEN, 1910), with zoology lagging behind.

Fossil material has, however, two major limitations by comparison with modern material. First, the most important character in the identification and classification of lithistids is usually the nature of the desmas, of which there are a number of types. A desma is identified partly by its form when fully grown but in part by the nature of the initial body or crepis from which it develops. The latter is important because crepides are of more than one type and because outwardly similar adult desmas may have different types of crepides. In modern material, the nature of the crepis can be determined by tracing the ontogeny of the desmas or recognized from axial canals (p. 19) that occur in the fully grown desmas. In fossils, however, these criteria may not be available. The articulation of desmas occurs at the end of their ontogeny in modern forms, and crepides are often of microscleric size; correspondingly, developmental stages cannot be studied often in the fossils. Canal-like spaces seen in fossil desmas may be genuine axial canals but may also be pseudaxial features produced by internal solution or diagenesis. In a modern form, the absence of axial canals marks a desma as anaxial; in a fossil, this condition may be genuine or due to diagenesis. Thus assessment may need to be based on comparison with better material or even modern material. Identification on this basis can never be certain and none may be possible if a desma is equally similar to two forms having different types of crepides. Some fossils have also desmas (**didymoclones, tricranoclones**) with no modern counterparts, although in one of these (didymoclones; p. 54–56) the nature of the crepis is known from ontogeny (REID, 1963b).

Second, the absence of loose spicules in fossils presents several types of problems. The absence of microscleres means that classification must be based on the megascleres only; but, as is usual in fossil Demospongea, forms with similar megascleres may in fact have had different types of microscleres. Modern lithistids may lack microscleres, supplemental megascleres, or both during life; but their absence in fossils may be due either to absence in life or to loss after death. This may be unimportant sometimes, so far as is known, although involving some uncertainty. For instance, if supplemental triaenes are present or absent in specimens that are otherwise apparently conspecific, and are similar in age, it is reasonably likely that those without the triaenes lost these spicules after death. In some modern forms, however, the presence or absence of triaenes in forms having comparable desmas corresponds with occurrence of different types of specialized microscleres (streptoscleres, sigmaspires). Hence supplemental triaenes may be relevant as evidence of relationships, if no microscleres are known. Their absence in fossils may be of doubtful significance, because of being possibly not genuine and especially if sponges of possible but uncertain relationship are of markedly different ages. For instance, some late Paleozoic lithistids without triaenes resemble Jurassic or Cretaceous sponges in which triaenes are present.

These limitations of fossil material must always be kept firmly in mind and are especially important in discussion of relationships between different groups of lithistids. But the absence of distinctive types of microscleres is also a problem in the study of modern forms, in which they are rarely present.

The sublithistid sponges comprise a few genera that either lack a truly articulated skeletal framework in some or all individual sponges, although desmalike spicules are present or in which either desmas or loose, desmalike spicules are less abundant than simple monaxons. In the latter instance, gradation may be seen from a simple monaxon to a desma. Though numerically few, these sublithistids provide suggestive evidence of the origin of lithistid sponges.

For further description of the spicules of these sponges, several subheaded sections are convenient.

General Features of Desmas

Despite wide variations in detail, the desmas of different types of lithistids have a number of features in common.

1. The ontogeny of any type of desmas begins with the formation of an initial body, the crepis, which lacks the form of the fully grown desma. In most types in which any crepis is known, it is usually a regular tetraxon (e.g., a calthrops) or a monaxon (e.g., a strongyle); but in variants of the tetraxon type it may be a triaene, a triactine, an angled diactine, a centrotriaene, or an amphitriaene. In one type of desma (sphaeroclones of *Vetulina* SCHMIDT; see p. 57) the crepis is instead an anaxial corpuscle. After inception, the desma then grows by formation of anaxial outgrowths, comprising articulatory structures (or zygomes) or both these and anaxial raylike arms (clones). In formation of these outgrowths, silica is secreted in successive layers, as in formation of typical rays; but no axial rod or protorhabd is present. A desma may be classified as tetraxial, triodal, diaxial, centrotriaenose, amphitriaenose, monaxial, or anaxial according to the nature of the crepis. Its status as belonging to one of these types is not affected by changes in form produced by post-inceptional formation of anaxial outgrowths.

In some publications (e.g., SOLLAS, 1888), desmas of tetraxial, triodal, monaxial, or anaxial types are called tetracrepid, tricrepid, monocrepid, and **acrepid** respectively. These terms are misleading if read literally, since a single crepis is present in each type so described. The anaxial type of desma or forms that resemble it and are supposed to be anaxial have sometimes been called cryptaxial (e.g., SCHRAMMEN, 1924a, 1936). The terms tetraxial, monaxial, etc. are, however, most convenient to convey a direct indication of the nature of the crepis.

The crepis is inevitably smaller than the fully grown desma, ranging from a little to much smaller according to how much of the desma is formed by anaxial outgrowth. The part of the desma enclosing the crepis may be termed epicrepid and is also called the epirhabd in monaxial desmas. The epicrepid part of a desma cannot normally be recognized by any external feature but can be identified if axial features are present internally. With fossils, one needs to remember that absence of the latter may be due to diagenesis as well as to their absence during life and, conversely, that pseudaxial canals may be produced by internal solution. In the living *Corallistes* SCHMIDT, this may even occur during life, so that arms of desmas become cored by a granular marrow that does not represent the crepis. It is also unsafe to identify the epicrepid region with the part of a desma from which growth seems to have started. This may be correct if the desma is tetraxial, triodal, or anaxial; but raylike clones may be emitted from one end of a monaxial crepis, producing a misleading suggestion that the crepis was a radiate (e.g., tetraxial) body.

The form of the crepis may correspond with a phylogenetic prototype. It is difficult to generalize, but probably the larger the crepis compared with the fully formed desma, the more likely is ontogeny to correspond with phylogeny. This is because transitions from normal megascleres to desmas that occur in sublithistid sponges have a progressive diminution of the crepis. But several instances are known in which, for example, a tetraxial desma intergrades with a monaxial desma; whichever direction is ascribed here to phylogeny, the ontogenetic prototype of the phylogenetic end form does not correspond with the phylogenetic prototype. Judged on this basis, it seems likely that tetraxial desmas and their variants are forms derived from simple tetraxons but that monaxial desmas may either be related to tetraxial desmas (e.g., dicranoclones and some rhizoclones) or to simple monaxons (e.g., heloclones, megaclones, or other **rhizoclones**).

Nothing is known of the origin of anaxial desmas; but reduction of the crepis in passage from normal monaxons to desmas suggests that they represent the ultimate stage in reduction (see SOLLAS, 1888).

2. The articulation of desmas is affected by features known collectively as zygomes or described as **syzygial.** The simplest zygomes are terminal or lateral notches developed at points where monaxial desmas are in contact and without formation of appreciable outgrowths (e.g., in heloclones, p. 56). More commonly, however, they are outgrowths of tonguelike, handlike, or rootlike appearance; zygosis then results from the zygomes interlocking. There is no essential difference between these modes of zygosis; both may be shown by different desmas of one sponge or different zygomes of one desma when the zygomes are rootlike. According to the form of the zygomes, the skeletal framework varies from loosely coherent to so rigid that the desmas are inseparable without breakage.

3. In addition to zygomes, many desmas have raylike arms or clones that are partly or entirely anaxial. If the crepis is radiate (e.g., tetraxial or triodal), the growth of a clone may continue in the line of a crepidal axis, as in tetraclones (p. 52); the basal part of the clone is then epicrepid in character, but the rest is anaxial. Other clones have no relationship to crepidal rays or axes, arising, for example, from the sides of an epirhabd or from an epicrepid central part in an anaxial desma.

When clones are present, the zygomes are commonly located at their ends; but rootlike zygomes or similar outgrowths may also arise laterally. No fundamental distinction can be made between zygomes and clones, although these terms are convenient. On the contrary, the two are homologous, as produced by one process, and also intergrading. For instance, some desmas of one sponge may have simple clones only, with a group of rootlike terminal zygomes; but in others the zygomes arise from two main terminal branches, or the clones themselves divide into zygome-bearing branches at any point down to their roots.

Clones are called rays, cladi, or clads by some authors; but these terms are misleading. A clone, being partly or entirely anaxial, is not a true spicular ray. Cladi or clads should refer to branches of true spicular rays and particularly the similar rays of triaenes. When authors use these terms, there is generally no implication that the spicules are related to triaenes. The description of a desma as having three, four, or more radiating rays should also identify it as triodal, tetraxial, or polyaxial; but many such desmas are monaxial, and some are anaxial.

4) It is usual for all clones of a desma to have zygosis with other desmas, irrespective of how desmas are oriented or added to the skeleton. To permit this, examples may have various distortions that vary with the orientation and spacing of adjacent individual desmas. The growing parts of young desmas thus seem to have a tendency to grow toward pre-existing desmas, ensuring construction of a continuous framework. This in turn is probably related to the anaxial character of the postcrepidal parts. Desmas may also have further variations in relation to (i) random, layered, or other constructional patterns in the skeletal framework; (ii) longitudinal or transverse orientations; or (iii) canalization of the skeleton (p. 16–17). Again, this plasticity is probably due to their partly anaxial character or to their being entirely anaxial.

NOMENCLATURE AND CLASSIFICATION

The nomenclature and classification of desmas is based generally on the form of the crepis and the fully grown desma. It is sometimes convenient, however, to use certain terms homologically, for instance in dealing with desmas in which the crepis is unknown. This applies specifically to the desmas of Paleozoic lithistids.

The nomenclature adopted here is based on that of RAUFF (1893, 1894, 1895) with additions by SCHRAMMEN (1910) and some further modifications. The following types are distinguished:

1. tetraclonar desmas (p. 52), comprising tetraclones and associated variants;
2. rhizoclonar desmas (p. 54), comprising rhizoclones, dicranoclones, and didymoclones;
3. megaclonar desmas, comprising heloclones and megaclones (p. 56);
4. sphaeroclonar desmas (p. 57), comprising sphaeroclones *s.s.* and astroclones;
5. orchocladine desmas (p. 57), comprising dendroclones, chiastoclones, and anomoclones; and
6. tricranocladine desmas (p. 58), or tricranoclones.

The two last groups comprise Paleozoic types of desmas in which the nature of the crepis is unknown or uncertain.

The division of desmas into named types and the use of different types in the taxonomic grouping of lithistids (e.g., ZITTEL, 1878b; RAUFF, 1893, 1894, 1895; SCHRAMMEN, 1910) may give an impression that each named type is sharply distinct. On the contrary, some named morphological types are completely intergrading, although others are not. For instance, no transitions are known between tetraclones and megaclones, but dicranoclones and didymoclones both pass into rhizoclones. Such intergradations have been recognized by specialists from ZITTEL (1878b) onward but are not mentioned often in textbooks. There are also grounds for thinking that a given morphological type may sometimes have had more than one origin. For instance, the living Corallistidae, with desmas passing from dicranoclones to rhizoclones, have supplemental triaenes and streptoscleres microscleres; but the Scleritodermidae, with rhizoclones only and no triaenes, have sigmaspire microscleres. A desma with the form of a megaclone may occur (1) in a true lithistid sponge, without other types of desmas; (2) as a sublithistid spicule, intergrading with a simple monaxon with or without an intermediate heloclone; or (3) as an occasional variant of a late tricranoclone.

The classification of desmas on a basis of their morphology and of lithistids on a basis of their desmas (e.g., ZITTEL, 1878b; RAUFF, 1893, 1894, 1895) is thus evidently partly artificial. This has led to taxonomic emphasis on other types of spicules by some authors (e.g., SCHRAMMEN, 1910; REID, 1963b), in particular microscleres and supplemental triaenes. But often the form of the desmas is the only available basis for classification.

TETRACLONAR DESMAS

Tetraclonar desmas comprise tetraclones s.s. and variant types of desmas found with tetraclones or sometimes replacing them in which the crepis is a calthrops, a triaene, a triactine, an angled diactine, a monaxon, a centrotriaene, or an amphitriaene. Typical examples have an outwardly tetraxial form, a tetraxial crepis, and clones bearing zygomes in the form of rootlike terminal outgrowths. The monaxial forms included are occasional tetraclone variants or special radical desmas. The tetraclonar group by definition (RAUFF, 1893, 1894, 1895) comprises desmas that are characteristic of lithistids grouped as Tetracladina (Paleozoic and Mesozoic and Cenozoic Lithistid Demosponges, *Treatise Part E (Revised),* vol. 3, in press).

The most important tetraclonar desmas are tetraclones *s.s.,* which are always tetraxial, and triders, which differ from tetraclones in showing a triaenose symmetry and may have a triactinal crepis.

A tetraclone *s.s.* is a tetraxial desma, which is not outwardly triaene-like. The crepis is commonly a calthrops but may be a short-shafted triaene. The simplest type of regular tetraclone has four equal, radiating clones that continue from the four crepidal rays and are unbranched before the terminal zygomes. In more advanced conditions the clones are branched at any point down to their origins. In irregular tetraclones the clones are unequally developed, and some may be aborted, although the desma remains tetraxial. In all types, the clones may be smooth to the zygomes, tuberculate, or beset with branching, zygomelike outgrowths.

A **trider** is a triaene-like or tripod-like desma in which the crepis is tetraxial or triactinal. When four clones are present, three are similar, comprising a **clonome** that is analogous but not homologous to the cladome of a triaene. These clones may be longer than or shorter than the fourth or may differ, for example, by being branched. The fourth may be a longer megalome or a shorter brachyome; in the latter instance it may lack a zygome or be replaced by an unmodified crepidal ray. This type passes into tripodal forms in which the fourth clone is absent. As in tetraclones, the clones may be smooth, tuberculate, or spinulated.

The triaene-like character of triders is not determined by the form of the crepis. The crepis of a desma with a **megalome** or a brachyome may be regular calthrops; while, conversely, a regular tetraclone may develop from a crepidal triaene. A trider with the brachyome aborted may still be tetraxial, with the corresponding crepidal ray rudimentary to fully developed; but a desma with four clones, bearing terminal zygomes, may have one clone entirely anaxial and the crepis triactinal. Accordingly, clonome and megalome or brachyome should not be equated with cladome and rhabdome as has been done by some authors.

Triders with a megalome or with the clonome distinguished by branching occur in forms with tetraclones, to which they are usually subordinate; but triders with a brachyome or three clones only may replace tetraclones completely in some genera (e.g., *Plinthosella* ZITTEL).

In forms with irregular tetraclones, these desmas may pass into variants in which the crepis is a triactine, an angled diactine, or sometimes a monaxon. The passage is gradational through forms in which one to three crepidal rays or axes are rudimentary. These desmas, which are minor variants only, are not given special names. Similar diactinal or monaxial forms may occur when the principal desmas are tripodal triders.

Mesotriders and **amphitriders** are rare types of desmas, in which the crepis is a mesotriaene (=centrotriaene) and an amphitriaene respectively; in the former, it may be dichomesotriaene, with dichotomizing cladi but with the rhabdal rays unbranched (in *Brachiaster* WILSON) and occur as a variant of a dichotrider with a dichotriaene crepis. In either type, the form of the desma may follow the form of the crepis or disguise it by irregular growth. The central shaft of an amphitrider, corresponding with that of the crepidal amphitriaene, is called herein a **centrome.**

Zygosis between tetraclonar desmas occurs in various ways, depending on their form and arrangement. When the desmas are regular tetraclones, they may all be united by interlocking of their terminal zygomes to form syzygial nodes. These are sometimes conspicuous spherical features. Alternatively, some or all zygomes may clasp individual clones or the central parts of other desmas with no regular pattern. If so, the zygomes are more or less clawlike. All three modes of zygosis may occur among zygomes of one desma. A mixture of modes of zygosis is usual in forms with irregular tetraclones. In spiny tetraclones, branching spinules that grow laterally from the clones may form zygomes.

When the desmas are triders with a short brachyome or no fourth clone, they often have a regular arrangement in which converging clones from three desmas are applied to the brachyome or central parts of each desma, which is clasped here by the zygomes, from the side opposite the clonome. The desmas are then typically oriented with the clonome facing inward and the brachyome outward when present. This type of orientation is recurrent in various types of desmas with clones grouped directionally on one side of a center (see also dicranoclones, p. 54; sphaeroclones, p. 57).

The more irregular types of tetraclones or tetraclone variants may vary in ways related to orientation or to canalization of the skeleton. On the other hand, desmas that are regular tetraclones or triders may be united in regular patterns or quite irregularly. The form of the desmas is thus clearly not determined solely by their arrangement; on the contrary, the form of some desmas, presumably controlled genetically, determines the modes of zygosis that are possible.

In tuberculate tetraclones and other forms, the tubercles arise as lateral spinules that are subsequently thickened. When developed fully they are often capstanlike, with a constricted neck below a buttonlike head. The head may be bifid or trifid, due to corresponding branching of the original spinule. The spinules are comparable with the outgrowths forming terminal zygomes, whose branches may develop into comparable tubercles; in some, tubercles become replaced irregularly by adventitious zygomes or zygomelike outgrowths that do not form articulating zygomes. Zygomes applied to the clones or central parts of tuberculate desmas may grasp the tubercles rather than the clones bearing them; hence the tubercles are sometimes called syzygial tubercles.

Some forms with tetraclone desmas in the body proper are supported by a stalk or by root processes containing special monaxial radical desmas. These have a fiberlike shaft, a typical zygome at each end, and a short axial canal in the middle part. They may intergrade with the typical tetraclones through triders with a long megalome and a small clonome or through forms with reduction and abortion of two clones, while the two that persist are rearranged into line.

Two possible sources of confusion are worth noting here. First, all types of tetraclonar desmas are called tetraclones by some authors (e.g., MORET, 1926b; LAGNEAU-HÉRENGER, 1962), so that a genus or species said to have tetraclones may in fact have only triders. Second, monaxial or anaxial desmas (rhizoclones, p. 54; astroclones, p. 57) may also have tetraclone-like shapes, so that a desma with the form of a tetraclone need not always be a tetraclone. This leads to problems in interpreting some fossils of uncertain character. The presence of a supplemental triaene suggests that such desmas are true tetraclones, but this may not be correct, since tetraclone-like desmas that are monaxial occur with triaenes in the living *Corallistes bowerbanki* (JOHNSON). Tuberculate desmas that are irregular tetraclones, triders, or their variants are also externally indistinguishable from monaxial dicranoclones, which again occur in forms possessing triaenes (Corallistidae). Separation of such desmas, thus, depends on recognition of the nature of the crepis, which may not be possible.

RHIZOCLONAR DESMAS

This term designates all monaxial desmas, except those grouped with tetraclones, in which the zygomes are branching, rootlike processes. Raylike anaxial clones may be present or lacking. The desmas placed in this group have probably had more than one origin.

The types of desmas included are rhizoclones, dicranoclones, and didymoclones. Both the latter types intergrade with true rhizoclones, although they are distinct when typically developed.

Rhizoclones are nontuberculate desmas, with or without distinct clones, in which the zygomes are typically branching, rootlike outgrowths. Other simple or branching spinules, which do not take part in zygosis, may also be present. These features are often present on the whole desma, although not always so. There are three main types of rhizoclones morphologically. In linear rhizoclones, the zygomes arise from a simple unbranched shaft, which may be mainly or entirely epicrepid. The shaft may be straight with the zygomes on all sides or curved with zygomes only on the convex side. In dipolar rhizoclones, the zygomes arise from the two ends of an epicrepid or partly epicrepid shaft that may be spined or smooth. The terminal zygomes are sometimes markedly bifid or carried by short clones that arise from the ends of the epirhabd. In some forms such desmas can resemble an amphitrider (p. 53) or a **dendroclone** (p. 58). In pseudoradiate rhizoclones, three or more equal or unequally developed raylike clones appear to radiate from a center, although one of the clones contains the crepis and the others are emitted from one end of it. Desmas of this type may resemble triders or tetraclones when three or four clones are present. With all three types, zygosis is usually zygome to zygome, with the apposed zygome branches having small articulatory facets or terminal expansions in some instances.

Rhizoclones may unite to form either a dense irregular structure, without interruptions unless skeletal canals are developed, or a network of composite skeletal fibers. These types of structures are called confused and fibrous respectively. The rhizoclones may all be linear forms or a mixture of different types with any type predominating. In fibrous skeletons, variation between linear and dipolar rhizoclones may depend on orientation. For instance, in *Seliscothon* ZITTEL, skeletal fibers with a longitudinal direction may contain straight linear rhizoclones; but desmas connecting these fibers transversely are typically dipolar. Skeletal fibers may also be formed largely from the zygomes or dipolar rhizoclones, arranged transversely in ladderlike series like the dendroclones of Anthaspidellidae (Paleozoic Demosponges, *Treatise Part E (Revised)*, vol. 3, in press). The curved type of linear rhizoclone is often associated with canalization of the skeleton, occurring curved around the walls of the canals with the zygomes facing outwardly.

Dicranoclones are tuberculate desmas of generally rhizoclonar type, which have tubercles and zygomes like those seen in tetraclonar desmas, and are sometimes tripodal to polypodal in form. The central type, morphologically, is a bow or arch-shaped desma with branching, clawlike zygomes at each end and the tubercles strongest on the convex side. At the center of this side of the desma there is often an especially prominent tubercle or a tuberculate brachyome-like outgrowth; or two of these features may be present. The curved shaft consists of two clones growing in opposite directions from the center. Variations on this central dipodal type occur in two main ways. First, growth of additional clones from one or both sides of the central part produces tripodal, tetrapodal, or sometimes polypodal desmas, which are the most distinctive type of dicranoclones. Second, clones may branch before the emission of terminal zygomes and may lack the directional grouping of dipodal to polypodal forms; the desmas then pass into irregularly shaped forms resembling irregular tetraclones. The tubercles originate in the same way as those seen in tuberculate tetraclones; thus immature desmas are morphologically rhizoclones, since the tubercles originate as spinules. Dicranoclones also grade into rhizoclone forms, in which spinules on the clones do not develop into tubercles, and may form zygomes.

In modern examples of these desmas, the crepis is typically minute (e.g., 0.02 mm long in a desma several millimeters in length). It may also be destroyed by internal solution, which causes the desma to be cored by a granular marrow extending into its clones and even zygomes. In fossils, there is often no trace of it, presumably due to diagenesis but perhaps due to the desmas sometimes having become wholly anaxial. In modern forms, the crepidal axis lies in the line of the two clones of a dipodal form, and in a comparable position in others; in *Schrammeniella* BREISTOFFER (=*Phalangium* SCHRAMMEN, *non* LINNÉ; *Iouea* DE LAUBENFELS); however, the tripodal type may have the crepidal canal in a brachyome-like feature, with the clones radiating from one end.

In RAUFF's original terminology (1893, 1894) of lithistid desmas, dicranoclones were included with rhizoclones *s.s.* as rhizoclones. Their present treatment as a distinct type follows SCHRAMMEN (1910), who, however, regarded the type that occurs in *Schrammeniella* as typical. The usual position of the crepis, when any is detectable, is as stated above; but this type of desma occurs also in *Schrammeniella;* and SCHRAMMEN (1910) in practice called both types dicranoclones. Accompanying desmas, which are rhizoclones morphologically, he called megarhizoclonids.

Dicranoclones are the characteristic desmas of lithistids grouped as Dicranocladina (Mesozoic and Cenozoic Lithistid Demosponges, *Treatise Part E (Revised),* vol. 3, in press), although not well developed in all genera. They appear to be related to the tetraclones of some Tetracladina. The two groups differ chiefly in that desmas are typically tetraxial in the Tetracladina but monaxial in Dicranocladina. Except when polypodal, dicranoclones are externally indistinguishable from desmas or from the tetracladines *Discodermia* DU BOCAGE, *Plinthosella* ZITTEL, and *Acrochordonia* SCHRAMMEN, in which variants of tetraclones or triders may have a diactinal or monaxial crepis. In the dicranocladine *Macandrewia* GRAY, the normally monaxial desmas intergrade with minor variants in which the crepis is a triaene with short to rudimentary cladi. There is no conclusive evidence of the direction of phylogeny, but indirect evidence is provided by the size of the crepis. In dicranocladine desmas, this is often minute; in tetracladine desmas it is usually larger and may sometimes exceed half the size of the fully formed desma. Since transition from normal megascleres to desmas is accompanied by reduction of the crepis in observed instances, the likely direction of phylogeny is from tetraclones to dicranoclones.

Didymoclones are monaxial desmas in which simple or branching clones arise from spherical swellings at the ends of a short straight **epirhabd.** They are typically directed in groups toward one side of the desma. The zygomes are formed by spinular outgrowths at the ends of the clones or by expansions with digitated margins and a central syzygial facet. Simple or branching spines often occur on the terminal swellings of the epirhabd, on the side opposite the grouped clones. Zygosis results from application of the zygomes to this part of other desmas or to their clones in its vicinity. The typical didymoclones are accompanied by desmas with various stages of passage into dipolar rhizoclones, with irregularly branching clones spreading from each end of the epirhabd and no epirhabdal swellings.

A structure formed from typical didymoclones is similar in appearance to one formed from sphaeroclones (p. 57), in which grouped clones are emitted from one side of a spherical centrum, and the opposite side may be spiny. This resemblance has led to the two types being confused sometimes. In ZITTEL's classification (1878a), his family Anomocladina was defined in terms of sphaeroclones, which occur in *Mastosia* ZITTEL; but the included genera *Cylindrophyma* ZITTEL and *Melonella* ZITTEL have didymoclones. Later (1884), he redefined the family in terms of didymoclones but still included *Mastosia,* which was removed only after two types of desmas (called didymoclones and **ennomoclones**) had been recognized by RAUFF (1893, 1894, 1895). In addition, some authors have thought that didymoclones may be bodies formed from linked pairs of sphaeroclones (e.g., SCHRAMMEN, 1910, *not* 1936). This cannot, however, be correct, because the desma develops from a crepidal strongyle (REID, 1963b) represented by the shaft between the two swollen ends in the adult form.

Didymoclones are the typical desmas of the Jurassic Cylindrophymatidae or Didymmorina, but desmas approaching didymoclones occur in the Ordovician-Silurian Anomoclonellidae. The didymoclone-like desmas of the latter are classed here as orchocladine desmas and may be of different origin.

MEGACLONAR DESMAS

This group comprises monaxial desmas not of rhizoclonar types in which the zygomes range from simple lateral facets to tonguelike or handlike terminal expansions. They are characteristic of lithistids grouped as Megamorina, but similar desmas occur in some sublithistid sponges (p. 60). The two types included are heloclones and megaclones.

A **heloclone** is a monaxial desma that normally lacks distinct clones, in which the zygomes are simple lateral facets or notches. The crepis is a more or less sinuous monaxon that in many instances is almost as long as the fully formed desma. The desma is typically sinuous with the ends markedly twisted, hooked, or sometimes digitate. The syzygial facets or notches occur mainly at the ends of the desma but may also occur at any point between the ends if two desmas cross one another. In incipient form, they are simply shallow indentations with no syzygial function. In abnormal examples, clonelike outgrowths may occur but bear no zygomes. The desmas are smooth, without spines or tubercles.

Megaclones are desmas in which distinct clones are present. They bear zygomes ranging from longitudinal facets to tonguelike or handlike expansions. The crepis is typically short and located near the center of the desma. The simplest type of megaclone is dipodal, with two opposite clones curving to one side from the ends of a short epirhabd. Other types arise by branching of the primary clones, by emission of further clones from the sides of the epirhabd, or in both ways together. The clones may lie along the side of other desmas and have zygomes in the form of longitudinal facets, or expanded terminal zygomes may clasp the epirhabds or clones of other desmas. All the clones then tend to be curved toward one side, which faces inward in relation to the skeletal surface. The desmas are normally smooth; fine tubercles may occur on the epirhabd but are never capstanlike.

Heloclones and megaclones are typical of different families (Isoraphiniidae, Pleromatidae) but are linked by intermediates in some sponges (e.g., *Nematinion* HINDE). They do not appear to be related to rhizoclonar desmas, except that some rhizoclones (megarhizoclones; SCHRAMMEN, 1910) may be forms derived from megaclones; some megaclones pass toward rhizoclone-like desmas by repeated branching of the clones. Both types are typically large desmas, reaching lengths of several millimeters in some examples. The large crepis, simple zygomes, and lack of clones point to heloclones as being relatively primitive and probably derived from a sinuous oxea (or ophirhabd) directly. The megaclone then seems to be a more advanced desma with the crepis reduced, and true clones developed. A passage from ophirhabds through heloclones to megaclones is known from a fossil sublithistid (*Archaeodoryderma* REID, 1968c; Paleozoic Demosponges, *Treatise Part E (Revised),* vol. 3, in press), which may have been ancestral to the typical megamorine lithistids.

A heloclone with digitate ends may resemble the radical desmas of some tetracladine sponges, as first noted by ZITTEL (1878a). No genetic connection seems likely, since the latter are obviously specialized; but at least one form supposed to have heloclones, *Inodia* MORET, may in fact be based on radical desmas of a tetracladine lithistid.

SPHAEROCLONAR DESMAS

This group comprises the desmas of the living genus *Vetulina* SCHMIDT, which are anaxial, and fossil desmas, which appear to be of the same type.

In sphaeroclones, a group of clones with terminal zygomes is typically emitted from one side of a globular **centrum,** which may be spined on the opposite side. The zygomes vary from branching rootlike outgrowths to cuplike expansions with digitated margins and are applied to the centra or clones of other desmas. Spines on a centrum may be few and large or small and numerous; when small, they may branch and interlock with zygomes applied to the centrum. In some forms a centrum is absent but replaced by a group of large spines. The number of clones is variable and usually three or more. The desmas are often arranged with the centrum facing outward and the clones directed inward in relation to a skeletal surface.

In *Vetulina,* the crepis is an anaxial corpuscle. During development of the clones and centrum, a depression in the surface of the developing centrum, on the side opposite the grouped clones, becomes enclosed to form an internal nucleus. The spines on the centrum are homologous with the clones, although formed later. The crepis is poorly known in fossils. In several genera, the centrum may contain a contorted tubular canal, which looks like an axial structure (*Ozotrachelus* DE LAUBENFELS, *Mastosia* ZITTEL, *Exodictydia* MORET), but the desmas do not seem to develop from a corresponding contort monaxon. Nothing is known of the crepis in Paleozoic forms.

The *Vetulina* sphaeroclone also passes into desmas in which clones are emitted from a center without directional grouping. Similar desmas accompany sphaeroclones in some fossils (e.g., *Cladodia* MORET) or may be the only type present (e.g., *Lecanella* ZITTEL). Herein these desmas are called **astroclones.** They have also been called **anomoclones** (SCHRAMMEN, 1936), but this name was proposed for desmas not certainly related to sphaeroclones. In some forms, an astroclone can resemble a tetraclone or have zygomes like those of a megaclone; it is possible that such desmas had a different origin from true sphaeroclones.

ORCHOCLADINE DESMAS

This group comprises desmas of lithistids grouped as Orchocladina (Paleozoic Demosponges, *Treatise Part E (Revised),* vol. 3, in press) in which the nature of the crepis is uncertain. These lithistids were formerly grouped as Tetracladina (RAUFF, 1895) because some of the desmas were interpreted as tetraclones or amphitriders. The desmas included are dendroclones, chiastoclones, tetraclone-like variants of these types, and anomoclones (*sensu* RAUFF, 1895; *not* SCHRAMMEN, 1936).

Dendroclones are sometimes forms resembling a tetraclonar amphitrider, except that the number of clones at the two ends may be two, three, or four (e.g., in *Dendroclonella* RAUFF). The central shaft and clones are smooth, and the zygomes are branching or nodular processes. This type may pass into tetraclone-like desmas, with clones at one end of a primary shaft. In simpler forms, however, the dendroclone resembles a dipolar rhizoclone (e.g., in *Calycocoelia* BASSLER, *Nevadocoelia* BASSLER). The shafts of the desmas extend transversely between skeletal fibers, termed **trabs** by RIGBY and BAYER (1971), formed mainly or entirely by their interlocked zygomes; the zygomes at the ends are not trifid but simply grow up and down the columns. The ends may, however, be bifid, as in dipolar rhizoclones. By shortening of the shaft, this type of desma passes into cross-shaped variants or into tetraclone-like desmas if the branches at one end are at right angles to those at the other. This can occur through the branches extending to two different skeletal columns at one end of the desma. There may also be passage into forms resembling linear or pseudoradiate rhizoclones.

Chiastoclones appear to be comparable with the amphitrider-like form of dendroclones, with the central shaft shortened so that the clones radiate from a very short shaft or from a center. They occur mixed with forms that are dendroclones or tetraclone-like desmas.

The nature of these desmas is uncertain. RAUFF (1893, 1894, 1895) considered them as tetraclonar desmas, and this view has been widely accepted until recently; but the simplest type of dendroclone, which is typical of the earliest genera (FINKS, 1960), appears to be monaxial. Tetraclone-like desmas occur, but their shape, at least in one type, seems related to skeletal structure, not the form of a crepis. It would be possible for all types to be monaxial or perhaps anaxial in some chiastoclones. For the present, it seems best to treat them as a separate group of desmas. Even if they prove to be monaxial generally, the names used will still be convenient, since location of the zygomes at the ends of clones is not typical of rhizoclones. For further discussion of the Orchocladina, see Paleozoic Demosponges (*Treatise Part E (Revised),* vol. 3, in press).

Anomoclones are irregularly shaped desmas of the Paleozoic Anomoclonellidae that may resemble chiastoclones, dicranoclones, or didymoclones in some examples. These desmas were interpreted by RAUFF (1893, 1894, 1895) as irregular forms of his ennomoclones comprising desmas here called sphaeroclones and tricranoclones. Because of more probable close relationship of anomoclonellids to Chiastoclonellidae (FINKS, 1960, 1967a), the desmas are interpreted herein as being irregular chiastoclone variants.

TRICRANOCLADINE DESMAS

This group comprises the desmas of the Paleozoic Hindiidae or Tricranocladina, called tricranoclones herein (REID, 1963b).

The hindiids are small spherical sponges in which fine, canal-like passages radiate from the center of the skeleton. The passages are enclosed by a very regular arrangement of triderlike desmas, arranged with three clones directed inward. In typical *Hindia* tricranoclones, the desma has a short brachyome-like feature on the side opposite the clones; in the skeleton this points radially outward. The outwardly facing sides of the clones may be markedly tuberculate, and the brachyome has a terminal knob or ring of tubercles. The zygomes are elongate terminal expansions of the clones and are applied to the outward facing sides of the clones of other desmas. When the latter are tuberculate, the zygomes are marginally digitated. Their ends may also grasp the shaft of the brachyome, below the terminal swelling. In addition to desmas with a brachyome, as in *Hindia* DUNCAN, the type includes desmas without this feature, as in *Scheiia* TSCHERNYSCHEV & STEPANOV. These are arranged in the same way as desmas in *Hindia.*

Nothing is known of the nature of the crepis in tricranoclones. RAUFF (1893, 1894, 1895) took the *Hindia* tricranoclone and the sphaeroclones of the Paleozoic Astylospongiidae as trider and dichotrider forms of a single type of desma, called ennomoclones; but he recognized that *Vetulina* desmas are anaxial. The relationship of these two types does not seem demonstrable. SCHRAMMEN (1910, 1936) interpreted the *Hindia* desma as tetraxial, because of its consistently triderlike shape. The most nearly similar desma in which the crepis is known is a tetraxial trider found in *Kaliapsis* BOWERBANK, but this is a special basal desma of a modern sponge with nothing else especially in common with *Hindia.* FINKS (1960) compared desmas of *Scheiia* with dicranoclones; but herein dicranoclones are related to tetraclones and the pachastrellid calthrops (REID, 1963b).

SUPPLEMENTAL MEGASCLERES

There are three main types of supplemental megascleres and several minor ones. The most common supplemental megascleres are supplemental monaxons, which occur in most modern lithistids and many fossils. They are commonly oxeas but may be strongyles, styles, or tylostyles. They are usually found within the meshes of the skeletal framework; aligned radially, at random, or more or less in the local direction of growth; or found at the surface of the skeleton and lying tangentially. Monaxons occurring on the surface may differ from those of the interior; in modern forms, such spicules are located in the ectosome. In a few genera (e.g., *Climacospongia* HINDE), simple monaxons core skeletal fibers formed by the union of desmas. Such coring spicules are common in trabs of several genera in the Paleozoic Anthaspidellidae.

Some groups of lithistids (Tetracladina; Dicranocladina; Megamorina; Didymmorina; *Treatise Part E (Revised),* vol. 3, in press) have simple or specialized ectosomal triaenes, here called triaenose dermalia. These are sometimes simple triaenes (e.g., plagiotriaenes, in *Costifer* WILSON) but usually dichotriaenes or specially modified forms (phyllotriaenes or discotriaenes). Trichotriaenes replace dichotriaenes in some genera. All types typically lie with the cladome outside the skeletal framework with the rhabdome running into its meshes (not clasped by zygomes). The simple triaenes and dichotriaenes are like those of choristid sponges, except that dichotriaenes may have the rhabdome bent to one side or the cladi spined in ways that do not occur in choristids (e.g., in *Phrissospongia* MORET). In phyllotriaenes, normal cladi are replaced by digitate leaflike expansions that lie in one plane and arise as lateral outgrowths from the cladi of an initial dichotriaene or simple triaene. This type grades into discotriaenes, in which the cladome becomes a simple or marginally indented siliceous disc; the initial cladi are usually unbranched and sometimes rudimentary. The last type may pass into monaxial **discostrongyles** without cladal axes. Both of these types may occur loose or have some degree of connection in a zygosis-like manner. In phyllotriaenes, zygomelike facets are occasionally present on the cladi. In discotriaenes, the margins of overlapping cladal discs may grow around the rhabdomes of other spicules. The rhabdome is not modified in either; in both types of spicules it is typically short and may be lacking.

Some lithistids whose primary desmas are tetraclones, dicranoclones, or megaclones have small accessory desmas that unite within the primary meshes or outside them to form an external secondary covering. The desmas, described as rhizoclonids, resemble small, finely branched rhizoclones but appear to be anaxial. The origin of these bodies is unknown; they do not intergrade with the typical desmas and cannot be identified with any other skeletal element.

Other types of ectosomal megascleres that occur in occasional genera include monaxial discs with an axis in the plane of the disc (*Neopelta* SOLLAS); scalelike anaxial plates (e.g., *Plinthosella* ZITTEL); and flat, branching, desmalike bodies (e.g., *Siphonidium* SCHMIDT, *Ozotrachelus* DE LAUBENFELS). The status of these spicules is uncertain, although the first two may be degenerate discotriaenes homologically.

SUBLITHISTID SPONGES

Sublithistid sponges are known from both fossil and modern examples. They are mainly forms having monaxial spicules only. Their resemblance to true lithistid sponges ranges from slight to almost complete.

The Cretaceous *Megarhiza* SCHRAMMEN resembles true lithistids but with partial failure of zygosis. The principal megascleres, called megarhizoclones by SCHRAMMEN, resemble large rhizoclones or take shapes intermediate between rhizoclones and megaclones. They may either articulate by small facets or syzygial expansions, where their parts are in contact, or be quite unconnected. The skeleton may then still have some coherence, although their branches may be only loosely interlocking. The extant genus *Petromica* TOPSENT has a similar condition and similar desmas.

Desmatiderma TOPSENT, *Helophloeina* TOPSENT, and *Lithochela* BURTON are modern sponges in which the skeleton consists of a mixture of simple monaxons and desmas, and these spicules are also intergrading. *Desmatiderma* is a finely ramified sponge in which the principal megascleres are styles with additional strongyles, but a layer of desmas is present at the surface. The desmas are elongate bodies with numerous small syzygial outgrowths and intergrade with the styles through irregularly nodular intermediates. The desmas also pass into spicules like those in the basal parts of *Megarhiza* and are here united firmly. *Helophloeina* is a comparable sponge, with the main spicules being oxeas, styles, and subtylostyles but with desmas in the basal parts. The desmas intergrade with the oxeas and approach the form of some megaclones or astroclones. In both genera, progression from simple monaxon to the most complex desmas corresponds with reduction in length of the crepis. In *Lithochela,* a modified ectyonid skeleton has the fibers cored by styles but connected transversely by desmalike spicules with irregular zygomelike expansions at each end (**diploclones,** REID, 1963b). The expanded ends are imbedded in the spongin of the fibers without true zygosis. At the surface, some desmalike spicules take on rhizoclonelike shapes. The desmalike spicules intergrade with styles like those that core the fibers, with their crepidal bodies again being reduced by comparison.

These sponges are of interest as evidence of the origin of desmas from normal megascleres and of lithistids from nonlithistid sponges. The principal evidence is the intergradation of normal monaxons with desmas or desmalike spicules and the correlated reduction of the crepidal bodies of the desmas. The latter reduction supports SOLLAS's (1888) idea that desmas originate by loss of control over silica secretion by normal axial structures, which are formed initially but cease growth before silica secretion ceases. This agrees with TOPSENT's (1928a) observation on *Desmatiderma* that the shorter the crepidal axis, the more complex the desma. It is, of course, possible to think that phylogeny had an opposite direction; but an origin of normal monaxons from desmas seems far less likely than the opposite. A further implication is that if two intergrading types of desmas also differ in the size of the crepis (e.g., tetraclones and dicranoclones; heloclones and megaclones) the form in which the crepis is larger is more likely to be phylogenetically primitive.

Crambe VOSMAER has a mainly monaxial spiculation but has anaxial desmas in the basal parts. These resemble astroclones or sphaeroclones morphologically but do not have the secondary nucleus of *Vetulina* sphaeroclones; instead, they have a number of small granular inclusions corresponding with the number of clones. In *C. crambe* (SCHMIDT), these inclusions are clustered at the center of the desma, suggesting the rudiments of spicular axes; but in *C. chelastra* LÉVI, they occur near the ends of the clones and thus must have been formed late in ontogeny. The zygomes are more like those of astroclones of *Lecanella* ZITTEL than sphaeroclones of *Vetulina* SOLLAS. There is no gradation from normal monaxons to desmas. The desmas arise from euaster-like bodies in ontogeny and here called asteroid by LÉVI (1960); but, as stated, in fact they are anaxial. Just possibly, this sponge might be an ally of the older Lecanellidae, which has almost lost its desmas.

Tetranthella LENDENFELD is an incompletely known modern form in which apparently radiate desmas or desmalike bodies occur on the surface of ectyonid skeletal fibers. This sponge should possibly be related to *Cyamon* and *Trikentrion,* in which echinating spicules are replaced by **pseudoradiates.**

Helminthophyllum SCHRAMMEN, of Jurassic age, is unusual among sublithistids in having tetraxons as ectosomal megascleres. The main internal megascleres are short, stout, and blunt-ended monaxons, slightly arcuate, and with semiannular transverse swellings on the convex side. The ends of these spicules rest on other kinds and may have very rudimentary syzygial expansions. The dermalia are small dichotriaenes.

The Carboniferous *Archaeodoryderma* REID is an apparently sublithistid sponge, known only from dissociated spicules. These spicules grade from blunt-ended ophirhabds through heloclone-like desmas to desmas with the form of simple megaclones. They decrease in size from the ophirhabds to the megaclone-like desmas, in a way that is comparable with what is seen in *Desmatiderma.* This sponge has been responsible for a supposed Carboniferous record of a megamorine lithistid (*Doryderma* ZITTEL; HINDE, 1884a). Supplemental triaenes, which occur in true Megamorina, are not known to have been present.

DISSOCIATED SPICULES

Dissociated megascleres of lithistid or sublithistid sponges may occur loose in a sediment due to scattering of unarticulated spicules after death or to disarticulation of desmas. The latter is most common with types that have lax articulation (e.g., heloclones and megaclones). Desmas may be taken as lithistid spicules unless transitions to a nonlithistid megasclere (e.g., an ophirhabd) imply a sublithistid. With one modern exception, phyllotriaenes, discotriaenes, and similar bodies occur only in lithistids. A simple dichotriaene with the rhabdome bent markedly to one side is likely to be from a lithistid. Otherwise, however, loose spicules from lithistid or sublithistid sponges are not distinguishable from those of nonlithistids. Thus, loose dichotriaenes, oxeas, strongyles, styles, or tylostyles may be supplemental megascleres of lithistids, not spicules of choristids or monaxonids. This should always be remembered in dealing with formations in which lithistids are known to occur.

KERATOSE SPONGES

The keratose sponges are Demospongea in which the skeleton consists of spongin only, unless foreign material is incorporated. The spongin is typically secreted in the form of skeletal fibers but may also form other structures. The fibers may be reticulate or simply dendritic. The canal system is usually eurypylous or diplodal, and the ectosome is a thin dermis.

The skeletal fibers are typically cylindrical and are formed from concentric layers of spongin secreted by cells called spongoblasts. The fibers may appear homogeneous or have a distinct granular axial medulla or pith that may form a minor or major part of the fiber. In some forms, the fibers are more or less hollow or may contain living tissue in life (*Ianthella* GRAY). The axial parts may also be packed with inclusions.

Reticulate or dictyoceratid skeletons may resemble those of monaxonids with spongin fibers. They are closely similar to the chalinid type (p. 46) in which spicules may be present or absent in different individuals of some pseudoceratose species. The fibers are then often divisible into primary fibers, which radiate to the surface of the body, and secondary fibers, which connect them transversely. The ends of the fibers may raise the dermis to form conules, producing a spiny or conulose appearance. In some forms, the axes of the primary fibers or all fibers contain foreign inclusions comprising a supplementary xenoskeleton. This type passes into forms with a skeleton composed of foreign bodies, cemented together with spongin. The inclusions may be siliceous sponge spicules, sand grains, radiolarian tests, foraminifera, or shell fragments.

In dendritic or dendroceratid skeletons, branching fibers arise from a common basal plate but are otherwise typically unconnected. Occasional transverse connections may occur, however, in some species. Foreign inclusions are commonly absent, but may be present. *Darwinella* LENDENFELD has additional loose, spicule-like bodies, composed of spongin; these pseudospicules are often hexactinal, though the number of rays is not constant.

Spongin is not resistant to decay, and keratose sponges are correspondingly rare as fossils. Recognition should be based on occurrence of one of the types of skeletons that occur in modern forms or on chemical recognition of spongin. Some alleged fossil records of keratose sponges are not based on these criteria.

OTHER DEMOSPONGEA

A few Demospongea do not fit in any of the main categories now considered. Some may be degenerate relatives of various typical demosponges.

The so-called **myxosponges** (slime-sponges) are askeletose sponges that appear to be Demospongea. *Oscarella* VOSMAER is an aphodal or diplodal sponge with amphiblastula embryos like those of the microspiculate genus *Plakina* (p. 27). *Halisarca* DUJARDIN and some similar genera are eurypylous sponges with large flagellated chambers and parenchymula embryos like those of dendroceratid sponges.

Chondrilla SCHMIDT is a diplodal and corticate sponge with spherasters but no other spicules. The body is supported chiefly by a stiff mesenchyme. *Chondrosia* NARDO is a similar sponge with no spicules. The fossil *Rhaxella* HINDE has sterraster spicules only. *Epoudenoplax* TOPSENT (=*Lepidospongia* DENDY, *non* ROEMER; =*Lepidothenea* DE LAUBENFELS) is encrusting with discotriaenes but no other spicules.

The status of these sponges is debatable. *Oscarella* and *Halisarca* have been interpreted as primitively askeletose, although related to forms (e.g., *Plakina*), in which a skeleton is present (e.g., DENDY, 1905), or to degenerate forms with the skeleton lost (e.g., MINCHIN, 1900). The close resemblance of *Chondrilla* spherasters to those of some choristids (e.g., *Aurora* ROW) and monaxonids suggests an origin by loss of megascleres in phylogeny and of *Chondrosia* by loss of the spherasters (DENDY, 1916). The spicules of *Rhaxella* correspond similarly with the sterrasters of the choristid Geodiidae, unless they are really sterrospiras like those of *Placospongia*. The spicules of *Epoudenoplax* are of a type found otherwise only as dermalia in some lithistid sponges (e.g., *Discodermia* DU BOCAGE), suggesting origin by loss of the desmal skeleton. But none of these relationships is demonstrable.

PALEOZOIC DEMOSPONGEA: MORPHOLOGY AND PHYLOGENY

ROBERT M. FINKS

[Department of Geology, Queens College (CUNY)]

INTRODUCTION

The demosponges have essentially as ancient a fossil record as the hexactinellids. The oldest fossils of the class were considered for many years to be from the Lower Cambrian (RIGBY, 1987b), but more recent discoveries have extended that record back into Ediacaran-age Precambrian (BRASIER, GREEN, & SHIELDS, 1997; LI, CHEN, & HUA, 1998). Demonstration of their existence alongside hexactinellids in the lowermost Cambrian and earlier sequences is hindered by the form of their earliest spicules, which are oxeote monaxons. These are indistinguishable in normal preservation from rhabdodiactines of hexactinellids. The minute axial cross of the rhabdodiactine axial canal is rarely preserved or at least rarely observed in older fossils, and the SEM work necessary to distinguish the square cross section of the hexactinellid axial canal from the triangular one of the demosponge (REISWIG, 1974) has not yet been done. Because no coherent skeleton of demosponge morphologic type has been found in Lower Cambrian strata, their existence at that time must still be conjectural.

SKELETAL MORPHOLOGY

The principal skeleton of living demosponges is composed most often of spongin (collagen) fibers in which variable numbers of spicules may be embedded, ranging from none at all to so many that the spongin is reduced to a mere cement between the spicules. This is termed a fibrous skeleton. Other demosponges may contain only spicules scattered without apparent order throughout the flesh, a condition often referred to as confused, although diffuse might be a better term. Combinations of the two arrangements are also known. Microscleres usually have a diffuse organization, even in sponges with a fibrous skeleton containing megascleres. Although diffusely arranged, microscleres may be concentrated in some parts of the sponge, especially in a differentiated outer zone or cortex when such exists in the soft parts. Microscleres are not properly part of the principal skeleton, however, nor are the specialized megascleres that are associated with the dermal layers. A few demosponges lack spicules altogether but are supported by variable amounts of the mesohyl (or **mesoglea**) found in all demosponges. This is a nonliving gel permeated with microscopic spongin fibers and motile amoebocytes.

FIBER TYPES

Among the sponges with a fibrous skeleton, several types of fiber may be recognized, although these are best thought of as end members of a graded series. Where each fiber is coextensive with a single monaxon, a simple isodictyal net is formed in which the spicules are united tip to tip and chiefly outline triangular interspaces. This may be developed in three dimensions, but often the isodictyal net is developed only in a plane or planes parallel to the outer surface. If more than one spicule lies side by side, yet nevertheless the side of each mesh space is but one spicule long, the net is compound **isodictyal.** Where many spicules lie along the length of a fiber in succession there is a net of spiculofibers (if the spicules dominate in volume) or of spicule tracts (if the spongin dominates). Almost always these fibers or tracts consist of several spicules side by side; it is rare to have a train of single spicules end to end. The spicules may all be parallel to the longitudinal axis of the fiber, or they may diverge outward and upward from the center of the fiber. The latter is termed a plumose fiber. An outer layer of spicules directed at a higher angle to the fiber axis may be present; these are said to be echinating, while the inner layer of more parallel spicules are said to be coring. A substantial fiber of spongin may bear only echinating spicules or may contain none at all but be cored by foreign sand grains (as in many keratose sponges). The principal spiculofibers are usually arranged as upwardly and outwardly diverging ascending fibers (or tracts) connected laterally by more slender fibers. A particular species or genus is often characterized by its fiber type, but sometimes more than one type is present in a single sponge.

SKELETAL TYPES

Among Paleozoic demosponges common as fossils there are several skeletal types that are particularly widespread: 1. hazeliid; 2. heliospongiid; 3. dystactospongiid; 4. anthaspidellid; 5. chiastoclonellid; 6. anthracosyconid; 7. astylospongiid; 8. haplistiid; 9. hindiid; and 10. *Belemnospongia* types. Each of these is characteristic of a particular family, except for the somewhat doubtful *Belemnospongia.*

Hazeliid type

Anastomosing subparallel spiculofibers of plumosely arranged oxeas characterize the main skeleton. A dermal layer of tangential oxeas may also be present. The Middle Cambrian genus *Hazelia* includes the species *H. delicatula* Walcott, 1920 (not the type species), which has spiculofibers of parallel (not plumose) oxeas arranged in a nearly isodictyal net (that is, with triangular mesh spaces one spicule-length long rather than subparallel and anastomosing). Further study of more specimens will be required to demonstrate whether the two types are to be considered end members of an intergrading sympatric series. Until such time, it is better to consider the hazeliid type as the former alone. Note that the spicules are oxeas rather than the styles that seem to characterize most plumose spiculofibers.

Heliospongiid type

Spiculofibers composed of bundles of parallel oxeas form the main skeleton. The quadrangular mesh spaces are more or less one spicule length wide, and the spiculofibers are typically radial and concentric in orientation. This skeleton approaches the form of a compound isodictyal net, but the successive overlapping of spicules and their curvature around the fiber junctions accord more with properties of a spiculofiber. A dermal layer of tangentially arranged oxeas not organized in bundles may be present. The haplistiid type is similarly organized but with the oxeas replaced by rhizoclones.

Dystactospongiid type

Spiculofibers forming the principal skeleton are composed typically of styles in plumose to parallel arrangement coated with heloclonid desmoids through which echinating styles may protrude. The desmoids may, however, form the entire spiculofiber, as in *Dystactospongia* itself; and the presence of such spicules in a spiculofibrous skeleton must be considered diagnostic of the type. The principal spiculofibers are usually subparallel and connected by lateral spiculofibers or by anastomosis. Oxeas may substitute for styles, but this is uncertain. This is a sublithistid type of skeleton and resembles a hazeliid with a coating of desmoids. The possibility that *Dystactospongia,* with desmoids alone, is actually a murrayonid calcisponge with flaky spherulites cannot be totally dismissed.

Anthaspidellid type

This type is characterized by a lithistid skeleton in which the dendroclone desmas form a simple isodictyal net with triangular interspaces parallel to the upper surface. The sponge grows by adding successively such layers, with the dendroclones occupying corresponding positions so that ladderlike, vertical series of dendroclones are formed. Terminal zygoses of the dendroclones interlock to form vertical columns or trabs that may be cored by oxeas. A few dendroclones have the external form of tetraclones. An imperforate basal layer may be aspicular.

Chiastoclonellid type

This type is characterized by a lithistid skeleton in which most spicules are chiastoclones and tetraclone-like forms with some dendroclones. The spicules interlock by terminal zygoses in a three-dimensional net of irregular but obscurely concentrically layered form. Radially oriented smooth monaxons may be present. An imperforate basal layer, like that of anthaspidellids, may be present.

Anthracosyconid type

This type is characterized by a lithistid skeleton in which dendroclones are oriented perpendicular to the upper surface in concentric layers. They may be grouped in bundles and may intergrade with rhizoclones. The bundles may be regarded as short spiculofibers more or less one spicule-length long. They are similar to those of the haplistiids, but unlike the haplistiids they are rarely throughgoing from one layer to the next. Instead they are often partition-like within the layer and form the wall between one horizontal canal and its neighbor. These partition-like spiculofibers may themselves be penetrated by pores. The spicules are rhizoclone-like, that is, with lateral zygoses, only when they occur in bundles.

Haplistiid type

This type is characterized by a lithistid skeleton in which radial spiculofibers are connected by usually thinner horizontal fibers to form quadrangular mesh spaces in the radial plane and more often quadrangular than triangular mesh spaces in the tangential plane. The mesh spaces are close to one spicule length long. The horizontal fibers occupy corresponding positions in successive layers so as to outline radial canals. The spiculofibers are porous and hollow to varying degrees in different genera. They are composed of subparallel rhizoclones and smooth monaxons (oxeas and strongyles); the smooth monaxons tend to occupy a more peripheral or coating position on the fiber. In some genera the fibers are cored by larger smooth monaxons. Dendroclones occasionally occur within the fibers and may substitute for horizontal fibers in the young parts of the sponge. A dermal layer of tangential smooth monaxons may be present, which may be organized in a finer mesh of fibers. This differs from the heliospongiid skeleton in the presence of rhizoclones and from the anthracosyconid in the dominance of rhizoclones, presence of smooth monaxons, and presence of well-defined radial and horizontal spiculofibers. A subtype in which the spiculofibers are flattened sheets is sometimes called radiate-lamellate and occurs in *Chaunactis* and *Mortieria* as well as in many post-Paleozoic rhizomorines.

Astylospongiid type

This is possibly a third way in which dendroclones can be organized, namely as a simple isodictyal net with triangular interspaces in three dimensions. This interpretation was first proposed by ZITTEL (1884). It is favored herein but is not yet firmly established. The alternative interpretation of this skeletal type, proposed by HINDE (1888) and RAUFF (1894), is that of concentric layers of six-armed anapodal desmas (dichotriders). See Volume 3 for a fuller discussion (*Treatise Part E (Revised),* vol. 3, in press). In addition to the principal skeleton of desmas, there are radially arranged, long, smooth monaxons that may have supported a dermal layer analogous if not homologous to protriaenes in other groups. The individual skeletal elements, be they dendroclones or spicule arms, resemble closely anthaspidellid dendroclones except that they are oriented in three dimensions rather than parallel to separate, successive sheets.

Fig. 31. *Multistella porosa* Finks, 1960, in vertical section, with mold of a brachiopod shell just above base, and with spicule series radiating from mold surface; brachiopod was original hard surface to which sponge attached, and was subsequently overgrown, AMNH 28089, Cherry Canyon Formation, AMNH Locality 21SW, Guadalupe Mountains, Texas, USA, ×1 (Finks, 1960).

Hindiid type

This type is characterized by a lithistid skeleton composed of concentric layers of three-armed, anapodal desmas (tricranoclones). The spicules occupy corresponding positions in alternating layers so as to outline radial canals. A dermal layer of probably diffusely arranged, tangential, smooth monaxons is present as well as larger, radial, smooth monaxons that probably supported the dermal layer. These are analogous and possibly homologous to protriaenes of the Dicranocladina. In late Paleozoic genera, four-armed anapodal desmas and **megarhizoclones** may also be present. The megarhizoclones are associated with a more irregular arrangement of the spicules in the layers and occupy spaces between the desmas.

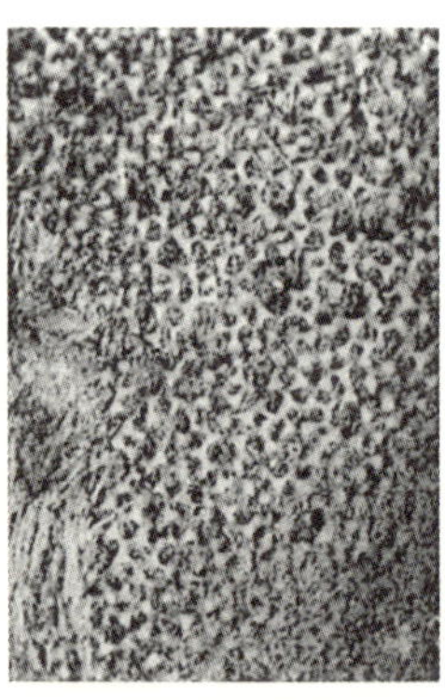

Fig. 32. Magnified view of surface of *Scheiia tuberosa* Tschernyschew & Stepanov showing monaxial spicules of possible dermal layer overlying main internal structure where spicules outline skeletal pores of hindiid skeleton, AMNH 28072, Leonard Formation, AMNH Locality 666, Guadalupe Mountains, Texas, USA, ×5 (Finks, 1960).

Belemnospongiid type

In this type, the entire skeleton consists of radial spiculofibers of long oxeas. It is possible that these are root tufts of a sponge whose principal skeleton was less coherent and more easily dispersed. The typical *Belemnospongia* with well-defined spicule bundles is late Paleozoic. In the Burgess Shale of the Cambrian there are radial masses of oxeas not grouped in bundles. Whether they are related, are root tufts, or are demosponges is equally uncertain.

BASAL ATTACHMENTS

Root tufts, that is, masses of elongate spicules that anchor the sponge in or above a soft mud bottom, are less common among demosponges of all geologic periods than among hexactinellids. Among living sponges they are best developed in such deeper water forms as *Thenea.* No Paleozoic demosponge with a well-developed root tuft is definitely known. The earliest group of demosponges to achieve widespread abundance, the anthaspidellid lithistids of the Ordovician, commonly have a short basal stalk that directly encrusts a shell or other hard object. Beginning with the Silurian *Aulocopium,* many members of this group and of the related chiastoclonellids, developed a characteristic concentrically wrinkled, imperforate, apparently aspicular but mineralized basal layer that resembles the holotheca of a compound rugose coral. This appears to have covered the part of the sponge that was sunk into soft mud. A shell at the initial ends of the ascending trabeculae and completely enclosed within the lower part of the sponge above the basal layer has been found in one specimen (Fig. 31), indicating that the initial attachment was by the usual encrusting of a hard object. The imperforate layer was a later development as the sponge sank under its own increasing weight. This is confirmed by the discovery of sponges that lack the usual imperforate layer attached to a rooted crinoid stem above the sea bottom.

Another early group, the tricranocladine Hindiidae, includes Ordovician and later *Hindia sphaeroidalis* that had no attachment whatever. These are usually perfectly spherical sponges. Since they have no structural polarity in oscules or anything else, they may have actually rolled about freely on the sea floor. A few individuals, especially in Permian species or in Australian Ordovician genera, are nonspherical and may have evidence of basal encrustation or attachment.

Although members of the genus *Astylospongia* usually live up to their name by lacking a stalk and, indeed, any signs of a basal attachment, they do have structural polarity in the presence of an osculum or exhalant pore cluster in a depression at the upper end. It is possible that the lowered center of gravity produced by the upper depression in their otherwise spheroidal bodies enabled them to right themselves automatically. Some related genera may have rolled about (*Carpospongia, Caryospongia*), while other genera had stalks and basal encrustations (*Palaeomanon*).

Stalks or basal encrustation are the rule in all other Paleozoic lithistids. *Belemnospongia,* as remarked earlier, may have been a root tuft of a nonlithistid demosponge, but it is entirely possible that it is the whole sponge.

DERMAL SPECIALIZATION

The simplest dermal specialization is a layer of small monaxons, usually smooth oxeas, that are tangential to the surface and in diffuse orientation. Such a spiculation, embedded in the exopinacoderm, is common in living demosponges. In fossils it is preserved as tangent monaxons scattered over the surface of the sponge. Where these spicules differ from those of the principal skeleton and also do not occur in the matrix the identification as a dermal layer is fairly certain. Otherwise there is a possibility that

FIG. 33. Small oxeas on surface of *Haplistion sphaericum* FINKS, 1960 probably represent remnants of a dermal layer, although such spicules also coat spiculofibers of main skeleton, USNM 127632, Magdalena Formation, USNM Locality 518q, Otero Coumnty, New Mexico, USA, ×10 (Finks, 1960).

they are foreign or are internal flesh spicules. Such a dermal skeleton is known with fair certainty in the Hindiidae (Fig. 32). It is probably also present in the Haplistiidae, for loose, small oxeas are commonly concentrated at surfaces of specimens of *Haplistion* (Fig. 33); however, because similar spicules also coat the spiculofibers, one cannot be certain there was a separate dermal layer. Nevertheless, in the related genus *Chaunactis,* there is a definite dermal layer of a more elaborate sort. There, small, smooth oxeas are organized into a quadrate mesh of

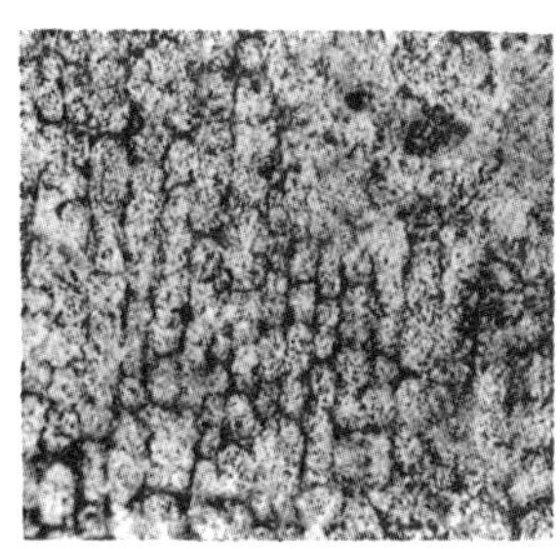

FIG. 34. Magnified view of fine, dermal mesh made of small, smooth oxeas in *Chaunactis foliata* FINKS, 1960, which has a desma-dominated, principal, internal skeleton, USNM 127640, Gaptank Formation, USNM Locality 700, Glass Mountains, Texas, USA, ×10 (Finks, 1960).

FIG. 35. Small frond of *Hazelia palmata* WALCOTT, 1920, showing obliquely crossing skeletal strands of spicules, USNM 66491, Burgess Shale, British Columbia, Canada, ×4 (Walcott, 1920).

spiculofibers finer than the desma-dominated fibers of the internal skeleton (Fig. 34). Also in *Mortieria* subparallel oxeas form a dermal layer. It is likely, therefore, that some sort of dermal layer is characteristic of other genera in the family.

Many specimens of *Hazelia* from the Burgess Shale are coated with a continuous layer of diffusely oriented, tiny, tangential monaxons. This layer is essentially continuous and the dermal pores must have been small (see Fig. 35). Thus, a dermal specialization goes back to some of the earliest known demosponges of the Cambrian. In the later Paleozoic heliospongiids a tangential dermal layer of oxeas is also developed. Here, the spicules are of the same size and form as those making up the spiculofibers of the principal skeleton, but the dermal layer is organized into a flat tangential sheet pierced by the large, circular inhalant pores (Fig. 36).

Not all Paleozoic demosponges had such a dermal skeleton. It is seemingly absent from the earliest lithistids, namely the anthaspidellids and the related chiastoclonellids. So far as can be determined, these families, however, had a nonspicular, imperforate dermal layer on the underside, as has been discussed above in the section on basal attachments.

Another form of dermal support aside from the tangential monaxons, occurs in the hindiids and in the astylospongiids. Long, smooth monaxons, much larger than desmas of the main skeleton or of the tangential dermal oxeas, are radially oriented within some of the radial skeletal canals and protrude above the outer surface. Similar spicules occur in living demosponges, for example *Spongilla* and *Ephydatia,* where they support the exopinacoderm above a subdermal space or vestibule (see also FINKS, 1971b). It is likely that the radial monaxons of the hindiids and astylospongiids had a similar function. Many living sponges have this function performed by triaenes instead of monaxons, including such common genera as *Steletta, Tetilla,* and *Geodia,* as well as many lithistids ranging from the Mesozoic to the present day. Isolated protriaenes, such as might have come from a nonlithistid sponge, are known as early as the Visean, and together with associated calthrops are the earliest record of definite tetraxonic spicules.

SPICULE FORMS

A brief review of the kinds of demosponge spicules found in Paleozoic rocks follows, for the moment without discussion of origins. The simplest and the first to appear is the oxea. The entire skeleton of the Middle Cambrian *Hazelia* is formed of smooth oxeas, larger in the main skeleton and smaller in the dermal layer. They may be diffusely arranged in the dermal layer, in an isodictyal net, or in plumose spiculofibers. The first two arrangements are typical of the occurrence of oxeas up to the Holocene; the last is more often associated with styles. From the Ordovician *Saccospongia* to the present day, styles occur characteristically in plumose spiculofibers with their blunt ends inward and their pointed ends facing outward and upward. The association of particular spicule shapes with occurrence in a specific larger organization is true of most spicule forms and points to a functional origin of spicule shapes.

The earliest lithistid desmas are the dendroclones of anthaspidellids. They appear in the Middle Cambrian. They are essentially smooth oxeas with arborescent zygoses confined to the two ends. The terminal position of the zygoses is appropriate for the isodictyal net in which they occur; the spicules are in mutual contact only at their ends. In the later (Permian) anthracosyconids, in which dendroclones often occur side by side in bundles, lateral zygoses are developed along the shaft on the side facing another spicule. The finer nature of the zygoses in such spicules distinguishes them from the similar rhizoclones of the haplistiids. It must be admitted that the distinction is sometimes difficult to make, and it is best to call all such spicules rhizoclones. In general, haplistiid rhizoclones have coarser and less finely branched zygoses. A third type of rhizoclone is found in the chiastoclonellids as well as in the anthracosyconids and haplistiids. It is a curved spicule that outlines partially a skeletal pore or canal. It is smooth on the concave side facing the pore and bears lateral zygoses on the convex side that articulates with other spicules.

In some dendroclones of Ordovician and later anthaspidellids, the smooth main shaft splits into three short, smooth branches at one end of the spicule, and these bear the terminal zygoses at their ends. Such a spicule resembles a tetraxial tetraclone, but axial canals are not demonstrable in the branches. Such spicules are more common than typical dendroclones in the anthracosyconids; the branched end uniformly faces upward or outward. Similar spicules are abundant in the chiastoclonellids and are present in smaller numbers in the haplistiids.

Another spicule that intergrades with dendroclones is the chiastoclone, characteristic of the chiastoclonellids. It has two or three branches with terminal zygoses at both ends of the spicule; the intervening shaft is very short or absent so that the spicule has an X-shaped profile. The presence of chiastoclones and tetraclone-like forms in the chiastoclonellids, rather than the more typical dendroclones of the anthaspidellids, is associated with a much more irregular skeletal net. The net may be considered isodictyal in that the spicules articulate end to end and outline mesh spaces one spicule-length wide between them. It is much denser, however, than the anthaspidellid net and is not organized into distinct layers, nor are spicules in corresponding positions in each layer to form radial series. A highly irregular or asymmetrical form of chiastoclone with many arms and no shaft occurs in the Silurian chiastoclonellids *Anomoclonella* and *Pycnopegma* and was called an anomoclone by RAUFF (1895); it is here considered a variant chiastoclone. In the Permian species assigned to the Chiastoclonellidae the tetraclone-like spicules are actually more numerous than true chiastoclones.

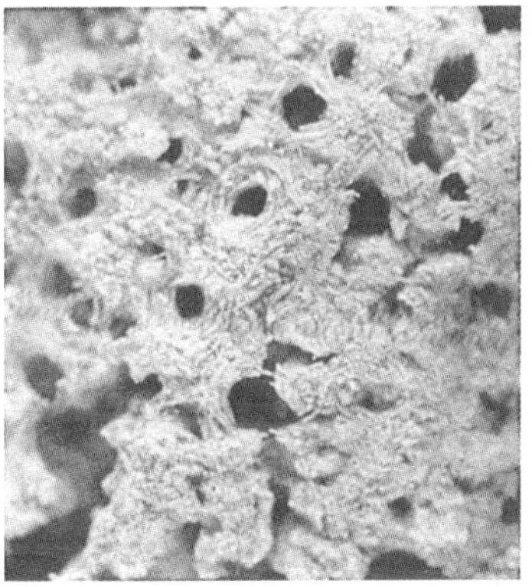

FIG. 36. Inhalant surface of *Heliospongia excavata* KING, 1933, with distinct dermal layer of spicules surrounding circular skeletal pores, USNM 127580b, Rock Hill Lithology Member of Graford Formation, USNM Locality 518g, Wise County, Texas, USA, ×10 (Finks, 1960).

It is doubtful whether true tetraclones occur in the Paleozoic. Tetraclone-like spicules of the Devonian to Permian *Jereina* superficially resemble true tetraclones of the Cretaceous *Jerea,* including the characteristically inflated junctions of the spicules. The peculiar arrangement of the canal system is also similar in the two genera. Nevertheless, axial canals are not known from the Paleozoic forms, and it is not known therefore if the branches are true rays. The presence of chiastoclones in *Jereina,* along with the tetraclone-like spicules, relates it clearly to the chiastoclonellids. On the other hand, the reported presence of chiastoclones in the Cretaceous genus suggests that the two may be related after all.

Another type of desma, superficially similar to rhizoclones but almost certainly of independent origin, are the megarhizoclones or megarhizoclonids, of the Permian hindiid *Scheiella.* These are antlerlike spicules with coarse, curving branches without the finely arborescent terminations of the dendroclone-rhizoclone type of desma. As a matter of fact, branches of the megarhizoclones do not serve primarily for articulation. They occupy spaces between tricranoclone desmas of the principal skeleton; the latter are intimately interlocked among themselves, and the megarhizoclones only loosely interlock with them, rather more like vines perhaps, twining about their supports. They are associated with the more open and irregular principal skeleton of *Scheiella,* which is in turn associated with a greater number of four-rayed tricranoclones than in its earlier relative, *Scheiia.* Before discussing tricranoclones, however, we will look at other, more clearly monaxial desmas and related simple spicules.

The strongyle, with two blunt ends, appears later in the fossil record than the oxea, although at times one may be hard pressed to distinguish between a blunt oxea and a strongyle. Like the oxea and to a lesser extent the style, strongyle-like forms also occur in the Hexactinellida. Consequently, isolated spicules cannot be identified with certainty as being of demosponge origin. The Carboniferous and later haplistiids seem to be the earliest demosponges in which strongyles are demonstrably part of the skeleton. There they comprise the spiculofibers along with oxeas and rhizoclones. Like the oxeas they often have a superficial position in the fibers. They are also sometimes concentrated on the surface of the specimen or in the immediately adjacent matrix, suggesting that they were part of a tangent dermal layer. They also occupy a corresponding dermal position and probable function in the Permian hindiid *Scheiia* and *Scheiella,* along with oxeas that apparently take their place in the early Paleozoic *Hindia.* Dermal strongyles and oxeas are also present in the Permian Anthracosyconidae.

There is a group of monaxial and possibly tetraxial desmas and desmoids that possess articular facets or cups, rather than sharp-ended and finely branched arborescent zygoses of the rhizoclone-like spicules. The group with articular facets includes heloclones and various types of anapodal spicules such as megaclones, tricranoclones, dicranoclones, sphaeroclones, and didymoclones. The simplest of these are the desmoids that coat spiculofibers of *Saccospongia* and other Ordovician dystactospongiids. They are sinuous bodies of irregular outline with one or more subcircular notches that surround partly a neighboring spicule. They are termed heloclonid desmoids because they resemble the **heloclones** of Mesozoic Helomorina.

The remaining members of this group are anapodal spicules (Finks, 1971b), that is, desmas in which the arms are all on one side of the spicule, the inward or proximal side in relation to the whole sponge. Sponges with this type of spicule build their skeletons in successive layers parallel to the sponge surface. The spicules of the latest-formed layer articulate with those of the preceding layer by means of terminal facets on their proximally directed arms to form what may be termed a simple enspicular isodictyal net, that is, one whose mesh spaces are outlined by single arms or branches of spicules.

The simplest of these are megaclones, which are clearly monaxial. Several stout, smooth arms arise from one side of a smooth monaxial central body and terminate in cuplike articular facets. The only published example from the Paleozoic of this predominantly Mesozoic group is the lower Carboniferous *Archaeodoryderma* REID, 1968b. It is known only from isolated spicules that are associated with heloclones and ophirhabds (the latter being sinuous monaxons that lack articular facets). It is assumed that all these spicules came from the same sponge, which had a relatively incoherent and perhaps sublithistid skeleton.

Anapodal spicules of the hindiids, called tricranoclones (REID, 1963b), have similar terminal articulating cups, but the distal surface of the spicule is tuberculate. In the widely occurring early Paleozoic *Hindia* and in the Australian *Belubulaspongia, Palmatospongia,* and *Arborohindia,* there are three proximal arms tetrahedrally arranged and a short fourth arm (**brachyome**) that is directed distally. The tubercles are apparently concentrated around the edge of the articular cups, encircle the brachyome, and lie along the distal side of the proximal arms. In the skeletal net the articular cups rest on the distal sides of arms of the spicule beneath, adjacent to the brachyome. The tubercles appear to interlock. The net is so arranged that spicules occupy corresponding positions in alternating layers such that each of the proximal arms of a single spicule rests upon three different spicules in the layer beneath. The corresponding spicules of alternative layers form radial series that outline radial canals between them. In the late Paleozoic *Scheiia* the brachyome is missing, the tubercles are distributed over the entire distal surface, and a few spicules have four instead of three proximal arms. This is continued in the Late Permian *Scheiella,* in which it is clearly associated with a more irregular and open skeletal net, one in which regular radial series of spicules are difficult to recognize. As noted earlier, megarhizoclones occupy interstitial positions in this irregular net. The tubercles interlock demonstrably, and their more general distribution over the distal surface, as well as the absence of the brachyome, would seem to be associated with the less regular placement of the articular facets on the underlying distal surfaces, which in turn is related to the less regular net. It is not known whether the tricranoclone is a monaxial or tetraxial form. If it is tetraxial, the earlier *Hindia* form is closer to a tetraxon in shape than the later *Scheiia-Scheiella* form.

The skeletal net of the astylospongiids was interpreted by RAUFF (1894) as similar to that of *Hindia,* and its principal spicule as similar to a tricranoclone except that it had six anapodal arms instead of three. The arms are straight rather than bowed as in the hindiids, and they are long, slender, and smooth. The center of the spicule where they meet is somewhat inflated and spheroidal, hence the name sphaeroclone. Each arm resembles in size and shape an anthaspidellid dendroclone. This was the first interpretation given to these spicules (ZITTEL, 1884). If so, the spheroidal centra would be merely inflated and possibly secondarily cemented junctions of dendroclones in a three-dimensional, simple isodictyal net. There exists no specimen or published illustration (excluding drawn reconstructions) that permits a critical rejection of one of these alternative interpretations. In some illustrations (e.g., PICKETT, 1969, pl. 10,*4*) the ends of the arms appear arborescent in the manner of dendroclones but details always seem to be at the limit of visibility or preservation.

Tricranoclones, megaclones, and sphaeroclones (if they exist) are the only anapodal spicules known so far from Paleozoic rocks. Other types are known from Mesozoic and later deposits. Of the Paleozoic forms, only the tricranoclone is possibly tetraxial.

Tetraxons of a nonlithistid type are known definitely from Paleozoic rocks beginning with the Lower Carboniferous. Isolated calthrops and various forms of triaenes are known. These include protriaenes and anatriaenes. The more elaborate dermal supporting dichotriaenes and phyllotriaenes are apparently not reported from the Paleozoic.

REID (1963c) has interpreted the supposed branching hexactines ornamented with spiral ridges of the Carboniferous *Spiractinella* as pseudohexactines of a demosponge because they intergrade with similarly ornamented but much rarer apparent tetraxons. The occurrence of similar spicules in the undoubted hexactinellid *Arakespongia,* which is roughly contemporaneous with *Spiractinella,* however, favors a hexactinellid assignment, which is followed herein. The known propensity of hexactinellid spicules for suppression of rays and variation in the angles between rays is not inconsistent with interpreting the tetraxons of *Spiractinella* as pseudotetraxons.

Finally it must be said that microscleres, of either the sigmatose or astrose variety, are so far unknown from Paleozoic rocks. Microrhabds are also unknown, but it would be hard to distinguish between a very small oxea and a microrhabd.

CANAL SYSTEMS

Herein we deal with the skeletal canals, which probably correspond only to the larger of the true canals and may be broader than these. In general, inhalant canals tend to follow the ascending tracts of the skeleton inward and downward, while the exhalant canals tend to parallel the accretionary layers of the skeleton at right angles to the inhalant passages. If a cloaca or spongocoel is present, exhalant canals run upward and inward along the accretionary layers toward the cloaca. Large exhalant canals perpendicular to the accretionary layers, however, may take the place of the cloaca. In massive sponges without a cloaca, exhalant canals may still follow the accretionary layers but converge upon numerous local exhalant centers, while inhalant canals are perpendicular to the accretionary layers as before. The simplest type of canal system is one in which both inhalant and exhalant canals are radial and perpendicular to the accretionary layers.

Although the upper depression of a cup-shaped sponge is often homologized with a cloaca, there is some evidence that this may not always be correct. The inner surface of the cup frequently bears local exhalant centers (as in *Anthaspidella*) just like the surface of a massive sponge. Truer homologues or at least structural analogues of the cloaca are exhalant canals that are perpendicular to the accretionary layers of the skeleton and therefore at right angles to the other exhalant canals. They are often grouped in the axial region of the sponge and sometimes lead into a true cloaca from below. They may also lead into local exhalant centers.

The simplest canal system or more specifically the most uniform and symmetrical is that associated with the spherical and probably unattached *Hindia* and *Scheiia* among the hindiids. All canals are radial, and the exhalant ones are distinguished, if at all, only by their greater diameter. Porelike passages between the spicule arms connect adjacent radial canals. Some of these canal-like spaces may have been occupied by choanocyte chambers, for there is no other open space in the sponge. The living flesh of the sponge may have occupied only the outer layers of the skeleton; disintegration of the interior of the skeleton has occurred in some specimens.

The spherical astylospongiid *Carpospongia* has a canal pattern like that in *Hindia* (RIGBY, 1986b). *Caryospongia,* although similar to *Hindia* in external outline, has a more complex canal system. Smaller inhalant and perhaps some exhalant canals are straight and radial. Most large exhalant canals, however, curve upward toward the outer surface and cut across the accretionary layers of the skeleton. As in all astylospongiids, the skeleton has more open space between the spicules than in the hindiids. This type of canal system is like that found in many massive sponges of less regular outline, for example the anthaspidellid *Multistella.*

Many massive chiastoclonellids (for example, *Defordia*) and anthracosyconids (for example, *Collatipora*) have this type of canal system. Often the single exhalant opening of each center (really a short cloaca) is replaced by a cluster of exhalant pores that are the termini of short radial exhalant canals (really a multiple cloaca). The species *Heliospongia excavata* also has this type of canal system, as does the cup-shaped *Anthaspidella,* although the form of the sponge is flabellate and subramose.

A variant of this form of canal system occurs in the genus *Anthracosycon.* Here the exhalant pore cluster is spread over one surface of the sponge (usually the top). The horizontal exhalant canals that converge upon the cluster are confined to the periphery of the surface, usually rising a short distance up the sides and running in a short distance toward the center of each accretionary layer. In this variant the dominant exhalant canals are now the short cloacas that run perpendicular to the accretionary layer.

Another type of canal system in which the exhalant canals are largely perpendicular to the accretionary layers is one in which these canals open into grooves on the surface of the sponge. In the spheroidal astylospongiid *Caryospongia* the grooves are meridional as they are also in the cup-shaped *Phialaspongia.* In the massive anthaspidellid *Phacellopegma* the grooves anastomose over the entire surface.

Exhalant canals perpendicular to the accretionary layers may be scattered over the sponge surface in complete isolation from one another. This is true in the haplistiid *Haplistion* and the hindiid *Scheiella.* Here we are brought back essentially to the simple condition of *Hindia* but without the spherical symmetry.

The relation of exhalant pore clusters to a cloaca occurs in a number of intermediate forms. In the astylospongiids *Astylospongia* and *Palaeomanon,* exhalant canals run parallel to the accretionary layers and open into a cup-shaped depression on the top of the sponge, which cuts down into these layers. This cup-shaped depression is homologous to a true cloaca that cuts down even deeper into the interior of a cylindrical sponge, as in *Heliospongia* and *Coelocladia* among the heliospongiids; *Exochopora, Lissocoelia, Aulocopium, Rhopalocoelia,* and *Nevadocoelia* among the anthaspidellids; *Columellaespongia* among the haplistiids; *Saccospongia* among the dystactospongiids; and *Camellaspongia, Devonoscyphia,* and *Attungaia* among the astylospongiids, in which exhalant canals follow the accretionary layers horizontally to the cloacal surface. In many of these, vertical exhalant canals perpendicular to the accretionary layers enter the cloaca from below, just as in an exhalant pore cluster. The anthaspidellid *Zittelella* is another intermediate form in which the homology between an exhalant pore cluster and a cloaca is clearly shown. Here an axial cluster of vertical exhalant canals perpendicular to the accretionary layers opens into a cup-shaped depression on the top of the sponge, where they are joined by converging exhalant canals following each accretionary layer.

For the horizontal exhalant canals to coalesce vertically to form slitlike spaces traversing several accretionary layers is a common tendency among anthaspidellids and is especially well developed in *Archaeoscyphia.* It occurs also in the astylospongiid *Devonospongia.* Indeed, it is not confined to canal systems with a cloaca, for it is also found to some extent in the anthracosyconid *Anthracosycon* and in the chiastoclonellid *Actinocoelia;* in the latter, the slitlike spaces converge upon local exhalant centers. It occurs also in the haplistiids *Chaunactis* and *Mortieria,* where it gives rise to the radiate-lamellate skeletal structure found also in many Mesozoic rhizomorines.

The axial cluster of exhalant canals that leads into the cloaca of many sponges can itself become a principal feature of the canal system, as in the chiastoclonellid *Jereina.* Each of the axial canals is essentially a separate cloaca and may even diverge from the axial region to open on the side of the sponge. This confirms the homology to a true cloaca of exhalant canals that are perpendicular to the accretionary layer.

Unlike the cup-shaped upper surfaces of *Astylospongia* and *Zittelella,* which are true homologues of a cloaca, the cup-shaped surfaces of the anthaspidellid *Anthaspidella* and the astylospongiid *Phialaspongia* are like the outer surface of a massive sponge. The former bears exhalant centers with converging canals; the latter bears meridional grooves containing exhalant pores. Nevertheless, the accretionary layers of *Anthaspidella* are transected by this cuplike surface. In this respect it is like a true cloaca and therefore may also be homologous, even though the structural correspondence with a true cloaca is not complete.

It remains to be noted that in some later Paleozoic species of anthracosyconids (*Anthracosycon auriforme*) and haplistiids (*Chaunactis*), there appears to be the first development of differentiated inhalant and exhalant surfaces on a noncloacate, sheetlike, or flabelliform sponge.

PHYLOGENY

It is apparent from the foregoing discussion that the form of the spicules and the form of the skeletal net are closely correlated, while both are relatively uncoupled from the pattern of the canal system. Families can be defined to a large extent by the form of the skeletal net. Individual spicule forms are more widely distributed, but the shape of particular spicules is often related to their spatial occurrence in the skeletal net. Thus it is likely that the evolution of spicule form is tied, in most instances if not in all, to a function of an architectural sort. Convergent evolution of spicule shape is therefore a possibility. Reconstruction of phylogeny can be based most securely on the interpenetration of spicule morphology, skeletal net morphology, and the unique sequence of these forms through time.

As is shown by their early occurrences, oxeote spicules, a spiculofibrous skeleton, and a differentiated dermal layer of tangent spicules are primitive features. These are all found in the mid-Cambrian *Hazelia,* which is one of the oldest completely preserved demosponge skeletons; several species, all sharing the enumerated features, are known. This is a nonlithistid form, preserved only because of the special conditions of the Burgess Shale. Black shales of Late Cambrian age in Quebec may give us a later glimpse of the same group in the form of the species *Lasiothrix flabellata* Dawson & Hinde, 1889, which may be a *Hazelia* or a related genus. (It is not the type species of *Lasiothrix,* which may be considered a hexactinellid.) Among the Burgess Shale specimens assigned to the various species of *Hazelia* there are different forms of the skeletal net and of the whole sponge. Perhaps some of these should be recognized as separate genera. The type species, *H. palmata,* has a skeletal net of anastomosing ascending spiculofibers; the spicules are somewhat plumosely arranged within the fibers. The entire sponge is apparently flabellate and without a cloaca. There may be a continuous brush of dermal spicules perpendicular to the surface. *H. delicatula,* on the other hand, appears to be cylindrical and branching, with a probable cloaca and distinct, circular inhalant pores. The principal net is compound isodictyal and the tangent dermal layer is simple isodictyal, at least in part.

From *Hazelia delicatula* it is possible to derive geometrically (which, of course, is not the same as saying they must of necessity have been so derived phylogenetically) several Paleozoic demosponge families by the following structural transformations. By making the compound isodictyal skeletal net more regularly rectangular, formed in accretive layers, and in general thickening both the tracts and the body wall, one can produce the structure of the late Paleozoic family Heliospongiidae. If, in addition, one converts some of the oxeas to rhizoclones by development of lateral zygoses, one arrives at the structure of some of the earliest completely preserved Haplistiidae, the Devonian genera *Columellaespongia* and *Varneycoelia,* which have a cloaca that is absent from most Paleozoic rhizomorines. Loss of the cloaca would produce their massive Devonian contemporaries (and Australian counterparts) *Crawneya* and *Oremo,* as well as the later Paleozoic haplistiids.

In another direction entirely, one can structurally derive the Anthaspidellidae from the simple isodictyal surface net of *Hazelia delicatula* by developing terminal zygoses on the oxeas, thus transforming them into dendroclones, and by building the skeleton of successive layers of these simple, isodictyal nets of dendroclones. Early anthaspidellids, possibly the Late Cambrian *Gallatinospongia* and certainly the early Ordovician *Archaeoscyphia,* also have the cloacate form of *Hazelia delicatula.*

If the skeleton of astylospongiids is composed truly of dendroclones or even of dichotriders (sphaeroclones) that are formed by the fusion of dendroclone elements, they could be derived from the earliest anthaspidellids (or directly from a *Hazelia delicatula*) by development of a three-dimensional arrangement of the simple isodictyal net. The earliest astylospongiids (*Astylospongia, Caliculospongia, Camellaspongia, Phialaspongia*) appear in the later middle Ordovician (Trentonian). They are cup shaped or cloacate with the exhalant pores arranged in vertical rows. These features are also characteristic of the earliest anthaspidellids, such as the Early Ordovician *Archaeoscyphia;* and the time relationships are such that an origin from an *Archaeoscyphia*-like sponge is quite plausible.

One can also derive the structure of the Silurian and later Chiastoclonellidae from the anthaspidellid structure. The transformation involved here is development of a three-dimensional net of rodlike elements by more elaborate branching of the dendroclone zygoses, forming chiastoclones, and the still more irregular anomoclones. This type of net (see above) is termed herein an enspicular isodictyal net, for the individual mesh elements are parts of spicules rather than entire spicules. The relationship is supported by the continued presence of dendroclones in chiastoclonellids, as well as by the characteristic concentrically wrinkled imperforate basal layer found in many chiastoclonellids and anthaspidellids. A possible connecting link is the mid-Silurian massive anthaspidellid *Dendroclonella,* which is very close in form to the contemporaneous and sympatric chiastoclonellid *Chiastoclonella* from the Niagaran of Tennessee. The latter, more than the other early chiastoclonellids, retains the radial rows of spicules that occur in the anthaspidellids. The former is unusual (although not unique) among anthaspidellids in being massive in shape. It is what one would expect for an ancestor of the uniformly massive chiastoclonellids. The greater irregularity and complexity of the late Paleozoic chiastoclonellids as compared to the Silurian forms supports the theory that the mode of origin of the chiastoclonellids was due to the decrease in the symmetry of an anthaspidellid.

Another group that contains dendroclones is the Permian Anthracosyconidae. Their structure is quite different from that of the anthaspidellid-chiastoclonellid group. Their dendroclones are arranged perpendicular to the surface in concentric shells, rather than parallel to the surface as in the anthaspidellids. In many species they are grouped into pillarlike bundles and in some species these may coalesce laterally to form wall-like structures outlining horizontal canals. The dendroclones in these bundles often develop lateral zygoses, so they become more or less rhizoclone-like. Most of them, however, are tetraclone-like. It would be possible to derive this group from the haplistiids by eliminating the transverse spiculofibers and decreasing the size, spacing, continuity, and regularity of the remaining radial spiculofibers, even to the point of rendering them single dendroclones. The fact that single dendroclones substitute for spiculofibers in juvenile parts of *Haplistion aeluroglossa* (although there for transverse fibers and not radial ones) makes this plausible. Nevertheless, the dominance of dendroclones and the absence of the coarser haplistiid-type of rhizoclone requires a reversion to a more ancestral spicule type, although perhaps this should be regarded as an example of neoteny. The absence of smooth monaxons in the anthracosyconids (except for a local patch on a specimen of *Dactylites micropora*) argue for a derivation from the haplistiids rather than to them, if one accepts the origin of the haplistiids from a hazeliid-heliospongiid lineage in which smooth monaxons are primitive. This would be consistent also with the late Paleozoic occurrence of the anthracosyconids. They could also have been developed from chiastoclonellids by reorganizing the isotropic skeleton into layers of perpendicularly oriented spicules. Forms of the spicules themselves are very similar in the two groups. The origin of the anthracosyconids must remain more problematical than some of the families discussed here, but it is clear that they belong somewhere in the complex of lineages that includes the Orchocladina and Rhizomorina and that ultimately goes back to hazeliimorph ancestors.

To return to *Hazelia,* one can derive another set of lineages from the type species, *H. palmata.* In this species ascending spiculofibers are subparallel and anastomosing, and the net is distinctly anisotropic, with mesh spaces being elongate parallel to the axis of growth. First of all, it is possible that the haplistiids were derived from this species directly rather than from *H. delicatula* via a heliospongiid-like intermediate as outlined above. One could produce the haplistiid structure by replacing the anastomoses of *H. palmata* with regularly spaced transverse spiculofibers and by developing lateral zygoses on some of the oxeas. In this connection, it may be significant that the ascending fibers in haplistiids are always thicker than the transverse fibers (unlike in the heliospongiids). It may also be significant that in the earliest haplistiids from the Devonian this distinction is more emphasized, and the mesh spaces are less quadrangular than in later forms.

A more clearly related group is *Saccospongia* and the other Ordovician dystactospongiids. All that is needed is to turn the oxeas into styles, arrange them in a more plumose fashion in the fibers, and coat the fibers with a layer of heloclone-like desmoids. Gross morphology of the net is essentially the same in both groups. Within the dystactospongiids there is a certain amount of variety. The extent of desma coating of fibers is variable in *Saccospongia* itself, a feature also of living desmacidontids with similar morphology, such as *Helophloeina* and *Desmatiderma* (see FINKS, 1967a). In *Dystactospongia* the fibers seem to be composed entirely of desmoids without the styles (see RIGBY, 1966b). In *Heterospongia* they are composed of desmoids together with seemingly nonplumose monaxons (styles or oxeas). The last two genera have a massive, subdigitate habit like *Hazelia palmata; Saccospongia* is tubular (cloacate) and branching. In *Dystactospongia* and *Heterospongia* thin transverse connecting fibers seem to dominate over true anastomosis as a means of joining the ascending fibers. This is close to the gross structure of the haplistiids, and a third possible origin of that group would be through a dystactospongiid similar to *Heterospongia;* here there is a possibility of the rhizoclones arising from the desmoids rather than from the oxeas directly.

Many living genera of sponges with chelate microscleres and stylote megascleres (and therefore belong to the natural group of the desmacidontids) have a sublithistid skeleton of often plumose spiculofibers of styles invested with monaxonic desmoids (see discussion by FINKS, 1967a). The correspondence of their structure with that of *Saccospongia* is so close that an origin from *Saccospongia* and ultimately *Hazelia palmata* of the entire closely knit group of sigma-bearing monaxonic sponges and the similar but spiculeless Keratosa is quite plausible. The axinellids, with plumose spiculofibers of styles, and even the Clavulina (hadromerids), with tylostyles, might also have had their origin from this lineage; such a hypothesis permits a one-time origin of styles, together with a functional reason for their origin, namely their participation in plumose spiculofibers where the blunt end provides an attachment surface for the spongin.

The spicule complement assigned by REID (1968b) to the Carboniferous megamorine *Archaeodoryderma* includes heloclones similar to the desmoids of *Saccospongia,* true anapodal megaclones, and ophirhabds. He considered the sponge to be sublithistid. This could conceivably have descended from a dystactospongiid, especially if still sublithistid, and thus lead ultimately from *Hazelia* to the post-Paleozoic Megamorina and Helomorina. This can have far-reaching implications, for the Megamorina share the presence of streptosclere microscleres and dermal triaenes with two other post-Paleozoic lithistid groups, the Dicranocladina and the Tetracladina, as well as with the nonlithistid theneids (pachastrellids). If these are truly related, then some tetraxon-bearing sponges could have descended from *Hazelia.*

Some light is shed on this possibility by the last remaining major lineage of Paleozoic sponges, the hindiids. At their first appearance in the Ordovician their principal spicules resemble those of no other group of sponges, not even the contemporaneous astylospongiids, even if the spicules of the latter are considered anapodal. By Late Devonian and Carboniferous time these had been transformed into the distinctive *Scheiia* spicule, which is joined in the Late Permian *Scheiella* by megarhizoclones. The combination of megarhizoclones and tuberculate anapodal spicules is characteristic of the post-Paleozoic Dicranocladina. Transformation of the *Scheiia* anapodal spicule into a dicranoclone is no greater a change than the earlier transformation of the *Hindia* anapodal spicule into the *Scheiia* one. Another transformation, however, is also necessary. This is the conversion of radial oxeas into radial dichotriaenes. Some sponge specialists find this harder to accept despite the fact that it is known to occur in ontogeny (Watanabe, 1957; Sollas, 1887). Reiswig's (1971) discovery that the axial canal of demosponges has a triangular cross section indicates how such a transformation could have taken place; the structural organization that favors the addition of three equally spaced rays is present in every oxea. It is no more mysterious to add three rays than to lose them. The possible functional morphological reasons for these transformations is discussed in greater detail elsewhere (Finks, 1971b) to provide fuller support for a more irregularly curved dermal membrane. What is important here is the likelihood, based on the *Hindia* lineage, that sponges with tetraxons can arise from ancestors that have only monaxons. It also makes more plausible the origin of tetraclones from the expansion of tripartite dendroclone zygomes and the descent of the Tetracladina from late Paleozoic Chiastoclonellidae.

True tetraxons, in the form of isolated calthrops and triaenes, are known from Carboniferous (Visean; Hinde, 1888) and later rocks. These must have originated independently from the hindiid line. They are not known to occur as part of specimens of any of the other lithistid and sublithistid lineages previously discussed. Thus they represent a separate lineage or lineages that was probably, though not necessarily, always nonlithistid. The living choristids and possibly homoscleophorids with **euastrose** microscleres may be part of this lineage.

Tracing the post-Paleozoic descendants of these Paleozoic lineages is rendered more difficult by the poor Triassic demosponge record. Nevertheless, Late Paleozoic beginnings can be seen for some later groups. The tendency to form a radial-lamellar architecture of rhizoclones in some late-Paleozoic haplistiids (*Mortieria, Chaunactis*) makes likely the descent of the Mesozoic rhizomorines, such as *Cnemidiastrum,* from them. The presence of megarhizoclones together with tuberculate anapodal spicules of variable arm number in the Late Permian hindiid *Scheiella* makes likely the descent from them of Jurassic and later Dicranocladina, such as *Dicranoclonella, Pachycothon,* and the living *Corallistes* (Finks, 1971b). The Permian chiastoclonellid *Jereina* has mostly tetraclone-like desmas with inflated junctions; the presence of similar tetraclones in the Cretaceous tetracladine *Jerea,* which also contains the chiastoclones of the Permian genus and an axial cluster of parallel apochetes, suggests a descent of at least some Tetracladina from the chiastoclonellids. The similarity of the Ordovician *Saccospongia* to living sublithistid desmacidontids (Finks, 1967a) renders almost superfluous the

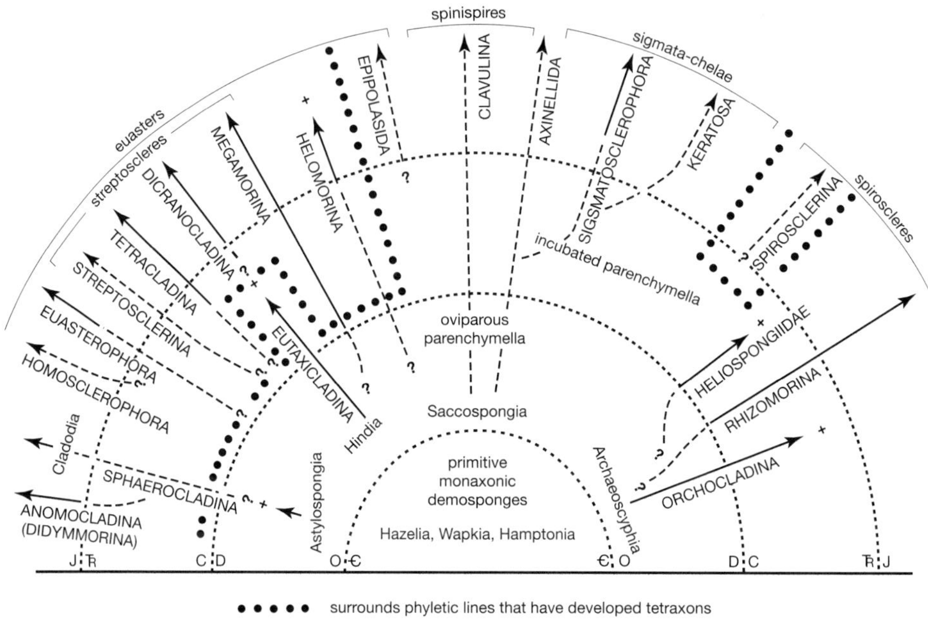

FIG. 37. Suggested phylogeny of major noncalcareous groups of the Demospongia (Finks, 1967a).

absence of an intermediate fossil record. (Here I might add that recent acceptance of the episodic nature of evolution renders less effective those objections based on the long time span involved.) If one wishes to accept a one-time origin of styles and plumose architecture one may derive the axinellids, hadromerids, and tethyids (epipolasids) from the same basic stock. If one accepts a one-time origin of sigmatose microscleres, then all the Poecilosclerida (including the desmacidontids in the narrower sense of sponges with chelae and styles) together with the related (according to many authorities) monaxonid groups Haplosclerida and Halichondrida and the similar but nonspicular Keratosa, would all go back to *Saccospongia* and its close relatives. WIEDENMAYER (1977a, 1977b), however, noted the similarity between *Heliospongia* and the living haplosclerids *Petrosia, Xestospongia,* and *Cribrochalina* with respect to radial-reticulate skeletal architecture and spiculation and suggested an origin of the sigma-bearing groups through *Heliospongia* and ultimately *Hazelia,* separate from *Saccospongia.* This is discussed further below. Finally, if one accepts a one-time origin of heloclones, as well as a connection between heloclones, **ophirhabds,** and **megaclones** as cited by REID (1968b), then *Saccospongia* may also be the source of Carboniferous and later Megamorina, the Mesozoic Helomorina, and the Mesozoic to Holocene Ophiraphiditidae (Fig. 37).

Some Paleozoic lineages seem to have become extinct. The anthaspidellids and chiastoclonellids appear not to have survived the Permian, although the Tetracladina may possibly have descended from the chiastoclonellids. The anthracosyconids appear to be a side branch without issue. The haplistiids did not survive the Paleozoic, but they lead directly into the later Rhizomorina so that the lineage did not die out. The hindiids likewise did not survive, but they seem to lead directly into the Dicranocladina. The astylospongiids appear to have died out in the Devonian. Forms with sphaeroclones, however, such as the Cretaceous *Cladodia* and the Tertiary to Holocene *Vetulina,* as well as forms with didymoclones, such as the Jurassic *Cylindrophyma* and *Melonella,* may have descended from them if the astylospongiids indeed had sphaeroclones. It is also possible that these groups descended from one of the other lineages with anapodal spicules, such as the Dicranocladina or Megamorina.

FUNCTIONAL REASONS FOR THE EVOLUTION OF PALEOZOIC DEMOSPONGES

Unlike the Hexactinellida and Calcarea, demosponges appear to have always had leuconoid architecture. This is supported by their fossil record as well as by their rhagon juvenile stage. A few of the earliest whole fossil demosponge skeletons are thin walled, but most, such as *Hazelia,* are thick bodied and suited for the support of scattered choanocyte chambers. Even the tubular, cloacate species of *Hazelia* are thick walled and bear pores that imply the presence of a well-developed canal system.

The earliest fossil demosponges were from quiet-water environments, such as *Hazelia* of the Burgess Shale. A skeleton of simple, slender monaxons held together by spongin is adequate for support, and the delicate nature of the spicules reflects the relative scarcity of dissolved silica in seawater. *Hazelia* also had two ways of achieving maximal strength with minimal material. *H. delicatula* has an isodictyal net with triangular interspaces. A triangular framework has maximum resistance to deformation and uses the least material. *H. palmata* has a skeleton of ascending fibers of plumose bundles of monaxons. The greater flexibility of such a skeleton is not disadvantageous in quiet waters.

The first lithistids appear in the Cambrian but become common in the Ordovician and are associated with shallow-water, often reefy limestones. It is hard to avoid the conclusion that the development of interlocking spicules was adaptive to higher wave-energy conditions. Among the hindiids, astylospongiids, and anthaspidellids, the skeleton consists of an isotropic (hindiids, astylospongiids) or anisotropic (anthaspidellids) triangulated net that supplements the rigidity produced by lithistid interlocking. The isotropic triangulated net (triangular in all directions) is attained by multirayed spicules; the anisotropic triangulation (triangular parallel to the surface) uses the basic monaxons. A sublithistid type is developed out of the skeleton of ascending fibers in the dystactospongiids. Here the plumose skeleton of *Hazelia palmata* is coated with desmoids that confer rigidity.

The anthaspidellids are the dominant lithistids in the reefy facies of the Ordovician and Silurian. The triangulated net is parallel to the upper and outer growing surface of the sponge, like a succession of superposed geodesic domes. Because spicules in successive layers occupy corresponding positions, the triangulation is carried downward through the entire skeleton as radial or ascending triangular compartments. The triangulated dome resists compression from above, and the longitudinal triangular compartments resist lateral compression. Such a structure is especially appropriate for the elongate tubular or conical shapes assumed by most anthaspidellids, particularly the earliest ones of the Ordovician such as *Archaeoscyphia.* It is not without significance that such anthaspidellids are important constituents of Ordovician reefs, particularly in the Arenig-Llandeilo interval.

The fibrous structure of the dystactospongiids is less strong and also less sparing of silica than that of the anthaspidellids. They are more abundant in the later Ordovician (Caradoc–Ashgill) and in somewhat deeper or at least nonreefy facies. They also tend to be smaller.

The astylospongiids, with their rigid, three-dimensionally triangulated net that is as delicate and sparing of silica as that of the anthaspidellids, reached their acme in the Silurian (Tennessee and Gotland) and Devonian (Australia). The Devonian taxa are of larger size and parallel the Ordovician anthaspidellids in shape.

In the later Paleozoic the dominant lithistids have a skeleton of bundled parallel monaxons with regular radial or ascending fibers connected by partly triangulated concentric shells of fibers. The triangulation is not carried down through the skeleton as in the anthaspidellids, and the dominant mesh space in both longitudinal and tangential view is quadrangular. Such skeletons include those of the heliospongiids, haplistiids, and anthracosyconids. Here strength is achieved by thickness of the bundled fibers, which is a more wasteful way of using silica. One may wonder whether silica may have been in better supply than in the earlier Paleozoic. This bit of speculation is not wholly unfounded, for many late Paleozoic sponges of all classes are hypersilicified and hypercalcified in the form of either heavy, excrescence-covered spicules (the siliceous hexactinellids *Stioderma, Docoderma, Carphites, Endoplegma,* and *Stereodictyum* and the calcareous heteractinids *Asteractinella* and *Wewokella*) or massive calcareous sclerosome (the sphinctozoans *Girtyocoelia, Amblysiphonella, Stylopegma* and the inozoans *Maeandrostia, Fissispongia, Catenispongia, Stratispongia*).

There is a tendency through the Paleozoic to break down or decrease the regular symmetry of these concentric and radial types of skeleton, apparently accommodating large canals and **cavaedial spaces,** and thereby shortening distances that narrower canals have to traverse between surfaces in contact with the ambient medium. The chiastoclonellids appear to be such a development out of probable anthaspidellid ancestors. Late Paleozoic genera within the hindiids (*Scheiella*) and haplistiids (*Mortieria, Chaunactis*) have similar development. In all these sponges the partial or complete triangulation of the ancestral skeleton is almost wholly lost. The more irregular skeletal net is built of such more complex spicules as chiastoclones and various rhizoclone-like forms.

POST-PALEOZOIC DEMOSPONGEA

R. E. H. REID

[formerly of Department of Geology, The Queens University of Belfast]

INTRODUCTION

The main general features of post-Paleozoic demosponge faunas may be summarized in the following ways.

1. Mesozoic demosponge genera are predominantly lithistids, a group that reached its greatest diversity in the Cretaceous period. There are no reliable Mesozoic records of modern nonlithistid genera, but some may have existed, since some lithistids (e.g., *Discodermia* BARBOZA DU BOCAGE) have survived since the Early Cretaceous.

2. Cenozoic demosponges include fewer lithistids, most of which are Miocene, but there are more supposed occurrences of modern nonlithistid genera, and some of these are probably genuine.

3. Modern demosponges are mainly nonlithistids, which greatly outnumber both all the known modern lithistids and all the known nonlithistid fossils. Cenozoic and modern lithistids appear to be mainly survivors of the large Cretaceous lithistid fauna. The large modern nonlithistid fauna may have existed at least as early as the Eocene epoch since material from the Eocene of New Zealand suggests an Indo-Pacific fauna as diverse as the modern one and includes many living genera.

4. The fossil nonlithistid Demospongea fall mainly into two categories, a) a minority comprising purely fossil genera, whose relationships to modern forms are unknown, and b) a majority comprising supposed fossil examples of modern genera recorded on the basis of isolated megascleres or microscleres that resemble those of some modern species, but which are not diagnostic of the recorded genus. A few modern genera are more or less reliably recorded from material with both megascleres and microscleres (e.g., *Ecionemia* BOWERBANK, Eocene [Plantagenet Beds], Western Australia) or from spicules found in only one modern genus (e.g., *Thrombus* SOLLAS, Eocene, New Zealand).

5. There are very few acceptable records of keratose sponges and none of askeletose genera (myxosponges).

6. Isolated microscleres include a) euasters from oxyaster to sphaeraster and sterraster, from the Upper Jurassic; b) the same forms plus plesiasters, spheres, spinispiras, discasters, sigmas, diancistras, clavidiscs, and various types of chelas, from the Upper Cretaceous; and c) the same from the Eocene (New Zealand), including many chela forms like those of various modern genera and sometimes species.

Because of its character, the nonlithistid record throws almost no useful light on the classification or phylogeny of modern forms. The record is not only sparse and unsatisfactory but probably also extremely incomplete. It is possible that the present large fauna evolved mainly in the Cretaceous period, but some of its origins appear to be much older. For instance, geodiid choristids, with long-shafted triaene megascleres and sterraster microscleres, were apparently already in existence in the Early Carboniferous (Ireland, Scotland). Other spicules of the same age resemble megascleres of some modern monaxonids.

The general characters of the modern nonlithistid demosponges have already been described, but an outline of those of characteristics of some families will be useful here.

1. Plakinidae. Microspiculate sponges, without triaenes; spicules usually mainly tetractinal but sometimes triactinal or diactinal; some with lophose tetractines (e.g., type genus *Plakina* SCHULZE); incubated amphiblastulae in *Plakina.*

2. Thrombidae. Microspiculate sponges, whose spicules are small, spiny triaenes (often trichotriaenes).

3. Pachastrellidae. Streptosclere microscleres; megascleres oxeas and calthrops, centrotriaenes, or both; aphodal canal system.

4. Poecillastridae. Streptosclere microscleres; megascleres oxeas, calthrops, and triaenes, with the last two intergrading; usually eurypylous.

5. Theneidae. Specialized deep-sea sponges, like Poecillastridae but all tetraxon megascleres long-shafted triaenes; often fixed by a root tuft.

6. Calthropellidae. Microscleres euasters or spiny microrhabds; megascleres calthrops or subtriaenes; sterrasters and aspidasters absent.

7. Ancorinidae. Microscleres euasters ranging from oxyaster to sterrosphaeraster (not sterraster or aspidaster), or with additional microrhabds or sanidasters; megascleres oxeas and triaenes, except in rare monaxonid species; aphodal.

8. Geodiidae. Similar to Ancorinidae but with sterrasters or aspidasters that pack a cortex to form a dermal armor.

9. Tetillidae. Microscleres sigmaspires when any special form is present; megascleres oxeas and triaenes, the latter almost never dichotriaenes.

10. Samidae. Microscleres sigmaspires; megascleres amphitriaenes with branched cladi; sometimes said to bore but may live in foreign borings.

Most living choristid species belong to genera of the Plakinidae, Poecillastridae, Theneidae, Ancorinidae, Geodiidae, and Tetillidae. The further nominal families Corticidae, Craniellidae, Ectyonillidae, Erylidae, Plakinastrellidae, and Halinidae also comprise choristids but are regarded here as synonyms.

The following families are monaxonid unless otherwise stated.

11. Coppatiidae. Ancorinid-like sponges without triaenes; skeleton radiate or not; megascleres diactines; microscleres euasters, sanidasters, or both.

12. Tethyidae. Sponges with a cortex and radiate skeletal structure; microscleres euasters, in forms from oxyaster to sphaeraster; megascleres typically strongyloxeas, arranged with pointed ends outward.

13. Epallacidae. Megascleres oxeas, styles, or tylostyles, arranged in columns and cemented with spongin; microscleres simple euasters.

14. Sollasellidae. Megascleres oxeas, arranged radially; no microscleres.

15. Spirastrellidae. Spinispira microscleres; predominant megascleres tylostyles, but may also have simple styles or oxeas, or these types only; monactines arranged point outward; not boring.

16. Placospongiidae. Similar, with additional sterraster-like sterrospiras.

17. Clionidae. Spirastrellid-like sponges that bore in calcareous substrata, excavating small, hemispherical lime pellets that are expelled through oscula.

18. Suberitidae, Polymastiidae, and allies. Spirastrellid-like sponges without spinispiras; microscleres are microrhabds or are absent.

19. Latrunculidae. Spirastrellid-like sponges with no microscleres but characteristic discasters that hispidate ectosome.

20. Timeidae. Spirastrellid-like sponges without typical spinispiras; characteristic microscleres sphaerasters, replaced in some by sigmasters.

21. Chondrillidae. No megascleres; only spicules typically sphaerasters, replaced in some by sigmasters.

22. Thoosidae. Megascleres styles, spiny oxeas, or lacking; microscleres are a) tuberculate microrhabds (*Alectona* CARTER) or special amphitylasters (*Thoosa* HANCOCK); b) diactinal to tetractinal oxyasters; larva may have dermal armor of tetraxial or monaxial plates, which rarely persists in adults; said to bore, and often placed in Clionidae, but lack spinispiras and tylostyles.

23. Axinellidae *sensu lato.* No microscleres in typical examples, although some may have microrhabds; predominant megascleres styles, although diactines (oxeas, strongyles, ophirhabds) may occur; the megascleres typically arranged in plumose columns, in

some forms with a distinct axial skeleton from which plumose fibers run to the surfaces; some crustose, with tangentially arranged diactines, and monactines that echinate the substratum; spongin more or less abundant.

24. Sigmaxinellidae. Axinellid-like sponges with sigma microscleres.

25. Rhabderemiidae. Axinellid-like sponges whose megascleres include hockey-sticklike rhabdostyles; microscleres toxas and contorted sigmas.

26. Raspailiidae. Axinellid-like sponges with additional spiny styles.

27. Cyamonidae. Similar forms with spiny styles replaced by diactinal to pentactinal pseudoradiates, which arise from monactines in ontogeny.

28. Chalinidae. Spicules typically diactinal and usually megascleres only, although sigmas or toxas may occur; spongin cementing to reticulate; some species with few spicules, or falsely appearing as keratose sponges; no dermal skeleton.

29. Spongillidae. Chalinid-like freshwater sponges that form gemmules; gemmule spicules typically amphidiscs.

30. Lubomirskiidae. Similar freshwater forms without gemmules or microscleres; spicules usually spiny.

31. Hamacanthidae. Megascleres oxeas or styles; characteristic microscleres diancistras; some also with sigmas.

32. Halichondriidae. Megascleres slender oxeas or styles, often matted without order; a tangential dermal skeleton usual, but no special dermal megascleres; spongin inconspicuous or lacking; no microscleres.

33. Desmacidontidae, Ectyonidae, and allies. Characteristic microscleres are chelas, often accompanied by sigmas; megascleres often of more than one type, and usually one or more of three sorts: a) principal monactines, b) echinating monactines, c) dermal diactines; spongin cementing to fibrous, with few spicules in some; fibers echinated, or cored only.

34. Astroscleridae, Ceratoporellidae, and Merliidae. Monaxonids that secrete a nonspicular aragonitic basal skeleton, resembling those of some Stromatoporoidea (Astroscleridae) or Chaetetida and favositid Tabulata (Ceratoporellidae, Merliidae); megascleres monactines; clavidisc microscleres in Merliidae; none in others.

These diagnoses should be read as outlines only, and different usages of the same names may be found elsewhere. An incorrect usage of note is use of Tethyidae for the choristid Tetillidae in some work by VON LENDENFELD (e.g., 1907), which resulted from identification of *Tethya* LAMARCK with *Tetilla* SCHMIDT. In paleontology, this error was copied in SCHRAMMEN's late monographs (1924a, 1936). Zoologists place living monaxonids into many more families than those cited above. The most useful detailed systematic accounts are in monographs by TOPSENT (1928b) and DE LAUBENFELS (1936). Of these accounts, that given by TOPSENT is in some ways the more satisfactory, but DE LAUBENFELS listed and classified almost every known modern genus.

Some nonlithistid fossils are worth notice here. *Discispongia* KOLB and *Prostolleya* LAGNEAU-HÉRENGER from the Jurassic of Europe have megascleres suggesting ancorinids or geodiids. Sterrasters may occur in *Discispongia* but are possibly foreign. The same families may be represented in the Upper Cretaceous by *Stolleya* SCHRAMMEN and *Geodiopsis* SCHRAMMEN, but smooth sterraster-like bodies ascribed to the latter could be spheres, not sterrasters. *Theneopsis* SCHRAMMEN and *Tetillopsis* SCHRAMMEN have megascleres suggesting a theneid and a tetillid.

A few fossil choristids do not correspond with any known modern forms. *Acanthastrella* SCHRAMMEN (Jurassic–Cretaceous) has small, spiny calthrops or subtriaenes, which do not have the branching of the axial enlargement of those of the living *Thrombus* SOLLAS. *Helobrachium* SCHRAMMEN (Upper Cretaceous) has triactinal megascleres with long, curved or hooked rays and a buttonlike rudiment of a fourth ray. The hooking of rays makes the skeleton loosely coherent, although no zygosis is present. In the Cephaloraphiditidae (Upper Cretaceous), the choanosomal megascleres are intertwined ophirhabds (sinuous oxeas), which are sufficiently coherent for skeletons to retain canalar features. Subtriaenes and oxeas or styles were also present. These sponges have been called Ophiraphiditidae by SCHRAMMEN (1910, 1912), because of VON ZITTEL's (1878b) identification of *Ophiraphidites* CARTER with Cretaceous sponges with subtriaenes; but the modern type species *O. tortuosus* CARTER was based on a fragment having ophirhabds only, and spicules of this type occur in various monaxonid sponges.

Helminthophyllum SCHRAMMEN is a Jurassic sublithistid with dichotriaene dermalia. The choanosomal megascleres, called **kyphorhabds**, are short, curved strongyles with transverse, weltlike swellings on the convex side and sometimes small terminal facets where the spicules were in contact.

Rhaxella HINDE is a Jurassic sponge supposed to have no spicules but sterrasters, but it may be based on partly rotted remnants of a thick geodiid cortex.

Fossil monaxonids are even more difficult to assess because of the number of modern families whose members have similar megascleres. *Opetionella* VON ZITTEL from the Jurassic and Cretaceous comprises sponges with oxea megascleres and could be coppatiids. *Stramentella* GERASIMOV (Upper Jurassic) has oxeas and styles occurring in an upwardly expanding tuft, which suggests an axinellid. The first possibly genuine examples of modern genera are supposed Upper Cretaceous species of *Axinella* SCHMIDT (Axinellidae) and *Halichondria* FLEMING (Halichondriidae). *Rhizopsis* SCHRAMMEN of the same age is a possible desmacidontid. Clavidisc and diancistra microscleres from the Upper Cretaceous suggest the existence of *Merlia* KIRKPATRICK (Merliidae) and *Hamacantha* GRAY (Hamacanthidae). Some supposed Cretaceous stromatoporoids (e.g., *Stromatoporellina* KUEHN) could be Astroscleridae.

A number of Mesozoic genera have ophirhabds as their principal megascleres, although other diactines or monactines may also be present. It is possible that sponges of this sort may be cephaloraphiditids that have lost their tetraxons before fossilization or allied forms in which the tetraxons were lost in phylogeny, but the oldest (i.e., *Euleraphe* SCHRAMMEN, *Ophiodesia* SCHRAMMEN) are Jurassic forms, considerably older than the known cephaloraphiditids. Several modern genera with ophirhabds lack tetraxons in life. The modern forms also have varied relationships. *"Jaspis" (Ophiraphidites?) serpentina* WILSON, with ophirhabds and oxeas, has euaster microscleres and is otherwise a normal coppatiid; but *Bubaris* GRAY, with styles and ophirhabds, is close to *Axinella* SCHMIDT of the family Axinellidae, and some fossil spicules ascribed to *Axinella* are ophirhabds like those of *Bubaris*.

Both Jurassic and Cretaceous sediments yield strongly annulated megascleres of the types called criccalthrops, **cricotriaenes**, cricotylotes, and cricostyles. These may represent one genus, as suggested by their ornament, or several. They are sometimes found together (in, e.g., the upper Greensand (Albian), England), and Upper Cretaceous cricotriaenes and cricostyles may be intergradational, but curved cricotylotes, as the Albian examples known as *Monilites* CARTER, occur alone in the living thoosid *Alectona higgini* CARTER. The sediments of both systems have also yielded **trachelotriaenes** with swollen rhabdomes and small dichotriaene cladomes. The form of these spicules suggests an ancorinid or a geodiid. If all fossil examples are from one genus, it existed from at least the Early Carboniferous (Visean, Ireland).

The Cenozoic records of nonlithistids are based mainly on isolated spicules ascribed to modern genera. Many are from the Eocene of New Zealand or the Miocene of Western Australia and represent Indo-Pacific faunas. Identifications of genera are often dubious, but some seem to be genuine. In the Eocene, plakinids are represented by lophose tetractines like those of living species of *Plakina* SCHULZE and *Corticium* SCHMIDT. *Thrombus* SOLLAS is represented by typical spiny trichotriaenes. These are the first sure examples of members of these families, although spicules that may represent them are known from rocks as old as the Early Carboniferous. The geodiids are represented by loose sterrasters and by such aspidasters as those of the living *Erylus* GRAY and *Triate* GRAY. An Eocene *Ecionemia* with associated megascleres and microscleres is acceptable as a genuine ancorinid. Some Eocene sphaerasters are like those of living *Tethya* species. Eocene spinispiras are presumably from spirastrellids or clionids, and the spirastrellid *Dotonella* DENDY is suggested by a special form. Some discasters correspond with those of living *Latrunculia* DU BOCAGE (Latrunculiidae). Rhabderemiids are probably represented by characteristic rhabdostyles and *Discorhabdella* DENDY by distinctive pseudastrose megascleres. There are many supposed generic records based on chelas, but none of these spicules is truly distinctive. Diancistras and clavidiscs again suggest *Hamacantha* and *Merlia*. A supposed record of *Melonanchora* CARTER (HINDE & HOLMES, 1892) was based on a clavidisc, not the typical sphaeranchora of this genus.

Spongillids and lubomirskiids, related to modern forms inhabiting Lake Baikal, are known from Miocene sedimentary rocks of that region. Spongillids are similar to the marine chalinids and point to their previous existence, if interpreted as relict derivatives of a former marine fauna. There are various supposed earlier records of spongillids (e.g., Upper Jurassic, England), but these record monaxons found in sediments with freshwater faunas.

A few apparent demosponges are known only from loose spicules that occur in both past and modern sediments. *Ditriaenella* HINDE & HOLMES is based on unusual amphimesotriaene megascleres, known from Eocene rocks (New Zealand) and the modern Indian Ocean (Seychelle Islands). An unknown sponge is represented by distinctive discs, called **pinakids**, found loose in sediments of the Upper Jurassic (Europe), the Albian and Upper Cretaceous (Europe), the Eocene (New Zealand), and the modern Indian Ocean.

Although most Cenozoic nonlithistids are poorly known, those identified with reasonable certainty are scattered through the spectrum of modern forms in a way that suggests that all the main existing groups had already existed for some time by the Eocene. As noted already, some stocks may be very much older. In particular, the choristid Geodiidae, which have the most specialized development of euaster microscleres, appear to have existed as early as the Early Carboniferous (Visean).

In dealing with the lithistids, it is helpful to begin by recalling those from Paleozoic systems (see Paleozoic Demospongea, above, p. 63). The predominant Paleozoic lithistids were the Orchocladina, with dendroclones and related forms of desmas, and the Sphaerocladina, with sphaeroclones. The remainder comprise a) the moderately diverse Tricranocladina; b) the somewhat more diverse Rhizomorina (e.g., *Haplistion* YOUNG & YOUNG); and c) some possible Tetracladina (e.g., *Jereina* FINKS), with tetraclone-like desmas but no triaenes. There are also a few sublithistid sponges, of which *Archaeodoryderma* REID (Lower Carboniferous) may be related to later forms with heloclones and megaclones.

The predominant Mesozoic lithistids are the typical Tetracladina, with both tetraclones and triaenes, and the Rhizomorina, which have rhizoclones but no tetraxons. There are several minor groups possessing triaenes: a) the Dicranocladina, whose desmas are dicranoclones or related forms grading into rhizoclones; b) the Jurassic Didymmorina, whose characteristic desmas are didymoclones; and c) the Helomorina and Megamorina, with heloclones and megaclones, which appear to be allied and to be unrelated to other forms with triaenes. The Didymmorina have been thought to lack tetraxons, but triaenes that seem to be *in situ* have been found in *Cylindrophyma milleporata* (GOLDFUSS). The name Megarhizomorina has one Cretaceous genus, *Megarhiza* SCHRAMMEN, with large rhizoclone desmas that may lack zygosis or have only loose zygosis. There are also some Sphaerocladina.

Tetracladina with triaenes appear first in the Upper Jurassic and are forms with spiny tetraclones or triders that may grade into rhizoclone-like desmas. The dermalia of these sponges (Sontheimiidae) are dichotriaenes where known. There are also some Jurassic forms with smooth desmas, which appear to be Siphoniidae, although triaenes are not recorded, and loose annulated desmas seem to represent the Phymaraphiniidae. The group became abundant and diverse in the Cretaceous Period, when it included forms with smooth and tuberculate desmas and with triaenes that range from dichotriaenes to phyllotriaenes. These spicules occur in various combinations, which are used as the basis of families (e.g., smooth desmas, dichotriaenes in Siphoniidae; smooth desmas, phyllotriaenes in Theonellidae; tuberculate desmas, phyllotriaenes, or discotriaenes in Discodermiidae). The Phymaraphiniidae, with phyllotriaenes, are especially distinguished by smooth tetraclones with prominent epicrepid annulations at the base of each clone. The peculiar family Plinthosellidae has tuberculate triders of dipodal forms as desmas and anaxial plates as dermalia. A variety of genera have more or less extensive development of anaxial supplemental rhizoclonids, which sometimes formed a supplemental cortex, the so-called **deckschicht** or **epitheca.**

Few fossils of this group are known above the Upper Cretaceous and Cenozoic and modern forms appear to be Mesozoic relicts. One of the most widespread modern forms, *Discodermia* DU BOCAGE, is reliably recorded from the Aptian (Spain) and presumably evolved earlier. Some extant genera have no microscleres but microrhabds (e.g., *Discodermia* [Discodermiidae], *Theonella* GRAY [Theonellidae]), but others have metasters and plesiasters indistinguishable from choristid streptoscleres, occurring in *Neosiphonia* SOLLAS (Siphoniidae) and *Racodiscula* VON ZITTEL (Discodermiidae).

Dicranocladine lithistids with triaenes and dicranoclones (Corallistidae) evolved in the Upper Jurassic, and one Jurassic genus (*Leiocarenus* SCHRAMMEN) is closely similar to the living *Corallistes* SCHMIDT. The desmas of this group are mainly dipodal to tetrapodal dicranoclones, but irregular forms also occur, the latter predominated in the Campanian *Procorallistes* SCHRAMMEN. Those of the two modern genera, *Corallistes* SCHMIDT and *Heterophymia* POMEL, are usually irregular, although typical dicranoclones also occur but are less massive than in the fossils. The microscleres vary from plesiasters through metasters to spirasters and are presumably streptoscleres. Some fossils and the living *Heterophymia* have supplemental rhizoclonids.

Macandrewia GRAY of the living Macandrewiidae has dermal phyllotriaenes and mainly irregular and tetraclone or rhizoclone-like desmas. The occurrence of a few tetraxial desmas among the others suggests affinity with the tetracladine family Discodermiidae. *Macandrewia* has only microrhabds as microscleres, but *Daedalopelta* SOLLAS has also streptoscleres. Another living family Neopeltidae, in which dermalia are monaxial discs, has no fossil record.

The name *Neohindia* SCHRAMMEN was based on a species of the Cretaceous corallistid *Pachinion* VON ZITTEL, now called *P. cylindratum* (SCHRAMMEN), but withdrawn by its author (SCHRAMMEN, 1910, 1912) after study of the type species of *Pachinion, P. scriptum* (F. A. ROEMER). This misleading name has led several authors to suppose a relationship between "*Neohindia*" and the Paleozoic *Hindia* DUNCAN. These nominal genera were even placed by DE LAUBENFELS (1955) in two families, with "*Neohindia*" assigned to the sphaerocladine family Astylospongiidae. *Pachinion* has no special resemblance to hindiids, and the species *P.*

scriptum and *P. cylindratum* are so similar that they could be based on individual variants of one biological species. *Pachinion* resembles the tetracladine sponge *Plinthosella* VON ZITTEL especially in the character of its desmas, which cannot be distinguished by any external feature from those of *Plinthosella.*

The Isoraphiniidae or Helomorina in which desmas are heloclones comprise a few genera scattered from the Upper Jurassic to the present. The dermalia are usually dichotriaenes in the fossils, which are Jurassic and Cretaceous, but are simple plagiotriaenes in the living *Costifer* WILSON. The Megamorina, with megaclones, have a limited Paleozoic record, but most genera are known from the Upper Cretaceous with a single modern genus (*Pleroma* SOLLAS). Most can be placed into one family, the Pleromatidae, but the Cretaceous *Heterostinia* VON ZITTEL is distinguished by having supplemental rhizoclonids. The dermalia are usually dichotriaenes but may be varied as simple triaenes. No member of these groups has phyllotriaenes, discotriaenes, or dermal discs. *Costifer* and *Pleroma* have amphiaster and spiraster microscleres of uncertain homology and additional microrhabds that are irregularly nodular in *Costifer.*

The Didymmorina comprise one small Jurassic family of uncertain affinity. The characteristic didymoclones have been thought sometimes to comprise two linked sphaeroclones (e.g., SCHRAMMEN, 1910, 1912; *not* 1936) grading morphologically into rhizoclones and developed from a crepidal strongyle. They resemble some desmas of Paleozoic Anonoclonellidae, which were also called didymoclones by RAUFF (1893, 1894, 1895), but can also be compared with dicranoclones having grouped clones at the ends of a central shaft (in e.g., *Leiocarenus* SCHRAMMEN). The dichotriaenes found in a species of *Cylindrophyma* VON ZITTEL are like those of Jurassic corallistids (REID, 1963d).

The Rhizomorina, with rhizoclone desmas and no tetraxons, have many Jurassic and Cretaceous genera. They dominate the known Jurassic fauna. There are forms possessing all types of rhizoclones and compact or fibrous skeletons. The radiate type of rhizoclone, in which three or more clones are emitted from a center, is usually subordinate or absent but occasionally predominant. The Jurassic Cnemidiastridae have predominantly bipolar desmas, analogous with simple forms of orchocladine dendroclones and also resemble some anthaspidellid Orchocladina in canalar features. Because of intergrading variations in external form, canalization, and the character of the desmas, many genera are difficult to arrange into clear-cut families.

A number of modern Rhizomorina (e.g., *Azorica* CARTER) appear to be Cretaceous relicts, and other Cretaceous genera (e.g., *Jereopsis* POMEL (*non* SCHMIDT); *Verruculina* VON ZITTEL) survived at least until the Miocene. It is difficult to relate the modern genera to nonlithistid sponges. The Scleritodermidae have sigmaspires like those of the choristid Tetillidae but are otherwise so different that relationship seems doubtful. The others have microrhabds only or no microscleres.

The Sphaerocladina have several Jurassic and Cretaceous genera and a single living genus. There is a major break in the record of the suborder between the abundant Silurian and Devonian occurrences and the Mesozoic forms (RIGBY, 1991a). No sphaerocladines are known from the Carboniferous and *Ellesmerespongia* RIGBY, 1970b, described as a Permian astylosponge from Arctic Canada, may be a didymmorine sponge instead (RIGBY, 1991a).

The fossil Mastosiidae have desmas like those of the Paleozoic Astylospongiidae but with a vermiform canal in the centrum. The centrum of sphaeroclones of the living *Vetulina* SCHMIDT (Vetulinidae) contains a granular nucleus that arises as a hilum-like pit during ontogeny. There are various similar fossils in which neither feature is known. Subordinate astroclones are commonly also present, and a Cretaceous mastosiid

(*Ozotrachelus* DE LAUBENFELS) has anaxial dermal plates. The living *Vetulina* has no microscleres. The Cretaceous mastosiids *Ozotrachelus* and *Macrobrochus* SCHRAMMEN are strikingly like Astylospongiidae, but the last surely known astylospongiid is Devonian. The long gap from Devonian to Jurassic also makes it uncertain whether the Paleozoic and later families are directly related.

The Jurassic and Cretaceous Lecanellidae comprise two genera with large desmas, which are astroclones (*Lecanella* VON ZITTEL) or forms approaching sphaeroclones (*Regnardia* MORET). No nuclear features are known, although desmas of *Regnardia* may have an internal solution cavity (moelle of LAGNEAU-HÉRENGER, 1962). The desmas of these sponges may possibly have arisen independently of typical sphaeroclones and might correspond with those of the living sublithistid *Crambe* VOSMAER.

There are various modern lithistids and sublithistids with no certain relationship to any fossils. Some of these (e.g., *Petromica* TOPSENT) may be related to the Cretaceous Megarhizidae, which they resemble in the form and loose zygosis of their desmas. These genera and some others (e.g., *Tetranthus* VON LENDENFELD, *Lophacanthus* HENTSCHEL) have features suggesting relationships to axinellid or similar sponges. Four genera with chelas appear to be related to the Desmacidontidae. *Desmatiderma* TOPSENT and *Helophloeina* TOPSENT are sublithistids with megaclone-like intergrading with normal monaxons. They do not have triaenes and do not appear to be allied to the Megamorina in which chelas are unknown. *Lithochela* BURTON has dipolar desmas set transversely to cored skeletal fibers, with a pattern like that seen in the anthaspidellid Orchocladina. *Crambe* VOSMAER has astroclone and sphaeroclone-like desmas, with multiple granular inclusions corresponding with the number of clones. These genera are not known as fossils, but some may in fact be represented by loose Cenozoic (Eocene) desmas that have been thought to belong to Megamorina or Sphaerocladina (cf. HINDE & HOLMES, 1892).

Jurassic to recent forms, in general, repeat types of canalization and skeletal growth of the Paleozoic lithistids. For instance, the type of canalization typical of Astylospongiidae was repeated in some Cretaceous Siphoniidae (Tetracladina) in which it presumably evolved independently. Its development in various Cretaceous genera was related to reduction of the paragastral cavity, with the growth plan of the skeleton showing correlated change from mainly marginal to concentric. The astylospongiid-like character of some Cretaceous Sphaerocladina (Mastosiidae) presumably evolved independently, if these forms were not direct descendants of the Astylospongiidae. The variant of this type of canal system in which the axes of elongate sponges are traversed by bundles of longitudinal aporhyses crossed by radial epirhyses recurs in various Tetracladina (e.g., *Jerea* VON ZITTEL), Megamorina (e.g., *Doryderma* VON ZITTEL), and Rhizomorina (e.g., *Jereopsis* POMEL). Ennomoclonar grouping of clones, as in desmas of the Tricranocladina (Hindiidae) and Sphaerocladina, was repeated by Tetracladina (e.g., *Plinthosella* VON ZITTEL), Dicranocladina (e.g., *Gignouxia* MORET, *Pachinion* VON ZITTEL), and Megamorina (e.g., *Heterostinia* VON ZITTEL, *Propleroma* MORET). A structure like that of the Anthaspidellidae (Orchocladina) recurs in the living *Lithochela,* which especially resembles forms in which the skeletal fibers are cored by oxeas (e.g., *Climacospongia* HINDE). The radial canalization and related skeletal structure of Tricranocladina is, however, not paralleled in any post-Paleozoic lithistids, although almost exactly reproduced in a Cretaceous minchinellid (class Calcarea: *Porosphaera* STEINMANN).

CLASSIFICATION

The classification adopted here for ordinal arrangement of the post-Paleozoic Demospongea is a hybrid of neontological and paleontological methods intended for use by both zoologists and paleontologists. In general, the nonlithistids are treated by neontological methods, but lithistids are treated by methods evolved in paleontology.

As noted already, most genera of nonlithistids recorded as post-Paleozoic fossils are represented by loose spicules, which resemble spicules found in modern species. Unless these forms are treated empirically—for instance, by methods analogous with those used for dissociated conodonts—the genera supposedly represented can be defined only in terms of modern material, which also provides the only basis for assessing their relationships. It seems best, therefore, to treat these forms by zoological methods. This means that purely fossil genera must usually be treated as forms of uncertain position, but this seems permissible because these forms are not numerous and because this is their factual status.

In contrast, the lithistids are known chiefly from articulated skeletons and greatly outnumber their living relatives. The latter have also so far yielded no useful alternative to the taxonomic methods initiated by von Zittel and very little evidence of how they are related to nonlithistids. The methods used in paleontology provide a workable classification that is also applicable to many modern forms and which seems, at least in part, to correspond with biological relationships.

A similar combination of neontological and paleontological methods was used by de Laubenfels (1955). The classification used here, however, differs widely from that of de Laubenfels in the treatment of nonlithistid sponges and in some aspects of the treatment of lithistids.

For further explanation, a number of subheadings are convenient.

STATUS OF PHYLOGENETIC CONCEPTS

The fundamental purpose of the classification presented herein is to provide a useful, orderly arrangement of the forms that are classified without further implications. The only proper primary criteria are accordingly observable characters. In practice, a phylogenetic concept is also involved when the subordinate members of any higher taxon are thought to be related biologically. In consequence, a classification can be partly or wholly a picture of inferred relationships. The closest correspondence between taxonomic and phylogenetic divisions can be thought a desirable objective, provided that the artificial nature of taxonomy is remembered. It is also true that classification can be used for the expression of a preconceived picture of phylogeny. Phylogenetic implications are properly a secondary attribute of classification, but not its primary purpose. Furthermore, nothing requires that a classification be based on an overall picture of phylogeny, or, especially, that inadequate evidence should be stretched to provide one for this purpose.

Moreover, from a practical viewpoint, it is not currently possible to give any clear picture of demosponge phylogeny. There are two major problems: a) the almost total lack of useful paleontological data on the nonlithistid sponges; and b) the ambiguous character of comparative zoological data. Reconstruction of phylogeny from stratigraphic sequences of genera is ruled out by the very sparse and plainly incomplete fossil record, by the dubious character of many of the nonlithistid records, and by the unknown relationships of all purely fossil genera. There is not even one instance in which the relationship of two modern genera can be traced through fossil species. The modern forms allow detailed study of all parts of the skeleton and also, for example, gross soft anatomy, embryology, cytology, or biochemistry, but all of them are phylogenetically end forms, and they represent only whatever stocks have survived to the present. On both these counts there is doubt regarding the status ascribed by some zoologists to a few modern genera, which are supposed to be primitive (see p. 101–102 below).

The classification adopted, accordingly, is not based on any overall picture of phylogeny nor intended to imply one. Some taxa distinguished at order or suborder level are envisaged as natural assemblages or at least as including one main natural assemblage from which possibly convergent forms cannot yet be separated. Others, however, comprise forms grouped solely in terms of morphological characters, which cannot be assumed to be distinctive of one natural assemblage, and are sometimes assessed as convergent in unrelated sponges. The choice between these methods has been based on which seemed the more appropriate in particular instances. There is no objection to the use of both methods herein, provided that what is intended is made clear in diagnosis.

CRITERIA OF CLASSIFICATION

The classification is traditional in being based chiefly on the skeleton. It is not claimed that only the skeleton can provide taxonomic criteria, but at present only skeletal data are available on a scale permitting classification of the class as a whole. In addition, only skeletal characters are apparent in the fossils.

In dealing with modern nonlithistids, there is more reliance herein on the character of the microscleres than in DE LAUBENFELS's classification (1936) but only to the level at which some families are grouped into orders. DENDY's (1921) concept of sigmatose microscleres is not accepted herein nor is the homology of all euasters assumed.

The taxonomic use of microscleres by SOLLAS (1888) was based on their successful use by SCHULZE (1887a) in dealing with the Hexactinellida. It is now clear that demosponge microscleres do not have the clearcut significance of their hexactinellidan counterparts. There are various instances in which similar microscleres, which have sometimes been used in taxonomy, occur in sponges otherwise so different that no special relationship should be assumed (e.g., the choristid Tetillidae and lithistid Scleritodermidae). There are also many forms without microscleres and instances in which morphologically similar microscleres have clearly implied different origins.

It is nonetheless a fact that some main types are characteristic microscleres of a number of groups of nonlithistids, which can be judged to be natural assemblages in terms of their overall characters. Those typical of one such group are also unusual in or absent from the others. To this extent, the microscleres seem to me to provide a real basis for the characterization of taxa, provided that their various limitations are remembered. I also think that DENDY (1921, 1924b) was well justified in distinctions that he made between asters (euasters) *sensu stricto;* dichotriacts (streptoscleres), and pseudasters. Some objections to use of the microscleres are removed by rejecting his views on how sigmatose forms are related: in particular, a) the supposed homology of tetillid sigmaspires with sigmas *sensu stricto;* and b) the alleged origin of spinispiras from sigmas through sigmodiscasters.

In dealing with lithistids, I follow SCHRAMMEN's (1910, 1912, 1924a) use of three criteria: a) the character of the desmas, in terms of the methods of VON ZITTEL (1884) and RAUFF (1893, 1894, 1895); b) the presence or absence of dermal triaenes or related types of spicules; and c) the microscleres of modern forms. The last have, however, little value, except in support of the conclusion that lithistids are polyphyletic (SCHRAMMEN, 1910, 1912; BURTON, 1929; DE LAUBENFELS, 1936).

NAMES OF TAXA

Most names proposed for divisions of the class Demospongea above family-group level have not been based historically on those of type genera, although Chalinida GRANT, Halichondrina VOSMAER, and Axinellida LÉVI are exceptions.

Names based on morphological features are open to various objections, and names based on demosponge microscleres involve special problems. A given type of microsclere may be characteristic of a taxon in the sense that this type is the main or only special form developed, but in no instance in which the names of taxa have been based on such microscleres are the characteristic forms present in all genera included. Some groups in which special microscleres occur in some genera (e.g., spinispiras in some of VOSMAER's Clavulina) contain more genera that lack them. If axinellid sponges and their allies are regarded as comprising an order, they cannot be named in terms of microscleres, which are typically absent and include no distinctive type when present. In addition, some names are misleading. The characteristic microscleres implied by the names Sigmatophora SOLLAS and Spirophorida LÉVI are sigmaspires, not sigmas *s.s.* or spires in general. The microscleres characteristic of the Streptastrosa of SOLLAS and the Streptastrosclerophora of BURTON are DENDY's dichotriacts (1924b), i.e., streptoscleres, not all forms classed as streptasters by SOLLAS (1888) and especially not the ancorinid sanidasters called streptasters by DE LAUBENFELS (1936). Names not based on microscleres (e.g., Poecilosclerina TOPSENT) may avoid this type of problem but again refer to no general feature of all forms included.

Names used here for orders of choristid and monaxonid sponges are based on those of type genera, except in one (Epipolasida) where the taxon is thought to be composite. The conceptual basis of such taxa is relationship to a fixed type genus and does not change if some included genera are later removed elsewhere. Such names can be criticized as expressing a concept of relationship—or phylogeny—that cannot be established conclusively; but their use is fixed practice in family-group nomenclature and seems an acceptable principle when relationship is what is envisaged. This method is not used, however, for taxa regarded as simply convenient assemblages of forms of uncertain or varying relationships or for any of the lithistid suborders. Some lithistid groups (e.g., Helomorina, Megamorina) are probably natural assemblages, but all consist chiefly of fossils whose relationships to even one another are strictly uncertain, and there seems to be no advantage in changing the traditional nomenclature.

RANKS OF TAXA

Taxa distinguished above family-group level are ranked as subclasses, orders, and suborders. Use of suborders is restricted to the lithistids, except for one monaxonid order (Chalinida GRANT), which unites two contrasting although apparently related groups of sponges.

SUBCLASSES

The class Demospongea SOLLAS is divided here into four subclasses: Choristida SOLLAS, Monaxonida SOLLAS, Keratosida GRANT, and Lithistida SCHMIDT. The principal members of these taxa are the sponges whose skeletons have choristid, monaxonid, keratose, and lithistid conditions, respectively; and all fossil genera are arranged on this basis. A few modern sponges that do not have the typical conditions are, however, regarded as members of the first three subclasses because of apparent relationships to typical genera. These atypical forms include sublithistids (e.g., *Crambe* VOSMAER, *Lithochela* BURTON), askeletose genera (e.g., *Oscarella* VOSMAER, *Chondrosia* NARDO, *Halisarca* DUJARDIN), some with microscleres only (e.g., *Chondrilla* SCHMIDT), and a very few monaxonids that are thought to be close allies of choristids. For instance, *Raphidotethya* BURTON is considered to be a monaxonid member of the normally choristid family Tetillidae SOLLAS, closely allied to the choristid *Amphitethya* VON LENDENFELD. One fossil sublithistid, *Helminthophyllum* SCHRAMMEN, is placed in the subclass Choristida, because the skeleton is choristid in character apart from its sublithistid features.

Each subclass includes two or more orders, whose relationships do not seem to be currently demonstrable, although grounds may exist for regarding them as probably related. The subclass that seems nearest to comprising a natural assemblage is the subclass Choristida, whose orders (Plakinida, Poecillastrida, Ancorinida, Craniellida) can be thought to have shared a common origin, although these orders, and the Ancorinida especially, may have shared the same origin as some which are classed as Monaxonida. The subclass Lithistida, in contrast, is regarded as certainly composite although convenient taxonomically. The concepts that are used in distinguishing orders and suborders are derived mainly from the methods of the following authors.

i. Choristida: SOLLAS (1888); DENDY (1905, 1924b).

ii. Monaxonida: VOSMAER (1882, 1883, 1884, 1885, 1887); TOPSENT (1928b); DE LAUBENFELS (1936); LÉVI (1955).

iii. Keratosida: MINCHIN (1900).

iv. Lithistida: SCHRAMMEN (1910, 1912, 1924a, 1936); LAGNEAU-HÉRENGER (1962).

These subclasses do not fit some major divisions of the class that have been made by zoologists and certainly not the current Tetractinomorpha and Ceractinomorpha of LÉVI (1957b). Some reasons these were not adopted are examined below (p. 101–102). For paleontology, however, it is also a matter of convenience to have at least primary divisions fit observable skeletal characters of the fossils. In addition, if most of the monaxonids have arisen independently of choristids with megascleres, which seems likely, the use of the subclasses Choristida and Monaxonida has at any rate some basis in phylogeny.

SUBCLASS CHORISTIDA

Most modern choristids can be arranged in four groups, as follows.

a. In microspiculate sponges comprising the family Plakinidae SCHULZE, the spicules are not differentiated into typical megascleres and microscleres, although meristic calthrops variants of some genera can be regarded as simple oxyasters. In the three remaining groups, there are typical megascleres and microscleres, of which the microscleres are of one of three main types.

b. In the families Pachastrellidae CARTER, Poecillastridae *nov.,* and Theneidae GRAY, the characteristic microscleres are streptoscleres (metasters *sensu* VON LENDENFELD; dichotriacts, DENDY). When others are present, they are microrhabds or simple oxyasters found as plesiaster variants.

c. In Calthropellidae VON LENDENFELD, Ancorinidae SCHMIDT, and Geodiidae GRAY the characteristic microscleres are polyactinal euasters to which sterrasters or aspidasters may be added. There may also be microrhabds or sanidasters but not streptoscleres.

d. In the Tetillidae and Samidae, the microscleres are sigmaspires or variants when any special forms are present. Some tetillids are also distinguished by occurrence of distinctive trichodal protriaenes or of incubated parenchymelloid embryos.

These groups, with some further additions, comprise the four orders Plakinida, Poecillastrida, Ancorinida, and Craniellida. The corresponding type genera are *Plakina* SCHULZE, *Poecillastra* SOLLAS, *Ancorina* SCHMIDT, and *Craniella* SCHMIDT. Each type genus is also the type of a nominal family, although Craniellidae DE LAUBENFELS is here regarded as a synonym of the older Tetillidae SOLLAS. *Tetilla* SCHMIDT was not taken as type of the order Craniellida because the type species *T. euplocamos* SCHMIDT is not known to have microscleres. Except for minor differences in the allocation of particular genera, the orders are equivalent to the following older taxa, whose names were not based on those of genera.

i. Plakinida: Carnosa CARTER (*sensu* CARTER, *not* DE LAUBENFELS); Microsclerophora SOLLAS; Megasclerophora VON LENDENFELD (*sensu* 1903); Homosclerophora DENDY.

ii. Poecillastrida: Metastrosa VON LENDENFELD; Streptosclerophora DENDY; Streptastrosclerophora BURTON; Streptastrosa SOLLAS minus Calthropellidae VON LENDENFELD (herein order Ancorinida).

iii. Ancorinida: Astrophora SOLLAS minus Poecillastrida (Metastrosa, etc.).

iv. Craniellida: Sigmatophora SOLLAS; Spirophorida LÉVI.

Inclusion of modern sponges, in addition to the typical members of the families cited, is restricted to the following instances.

a. The myxosponge *Oscarella* VOSMAER is accepted as a member of the Plakinida, closely allied to *Plakina,* which has similar amphiblastula embryos. *Thrombus* SOLLAS, the only genus of family Thrombidae SOLLAS, is placed in this order because of the small size of its spicules, although it does not seem closely related to any typical Plakinidae.

b. *Aurora* SOLLAS of the order Ancorinida, family Ancorinidae, is accepted as having apparently both choristid and monaxonid species on the basis of DENDY's (1916) demonstration of several pairs of similar species, in each of which one species is distinguished chiefly from the other by the absence of triaenes. Some purely monaxonid genera with euaster microscleres may also be monaxonid Ancorinida biologically but are excluded here for reasons given below (second paragraph).

c. The monaxonids *Raphidotethya* BURTON and *Trachygellius* TOPSENT appear to be close allies of the choristid genus *Amphitethya* VON LENDENFELD, of the otherwise choristid family Tetillidae SOLLAS (order Craniellida) and do not seem to be related to any typical Monaxonida.

The inclusion of even a few forms as Choristida that do not have choristid morphology might be criticized as inappropriate but such inclusion follows SOLLAS's (1888) usage and seems justified biologically. There appear to be genuine instances in which literal taxonomic reliance in the principal conditions of the skeleton would cut across relationships. The best known instance is that of the pseudoceratosa, in which different individuals of one species may either have or lack spicules and thus differ in a way by which genera would be placed in different subclasses (Monaxonida and Keratosida). At least one sponge normally classed as a monaxonid can sometimes have choristid characters. The megascleres of *Alectona* CARTER (family Thoosidae, order Spirastrellida) are normally spiny oxeas; but those of the type species *A. millari* CARTER, as seen in CARTER's own material, may also include a few regular triactines, and intermediates that link these with the normal diactines (oxeas) morphologically. The transition between these triactines and diactines follows the same pattern as in plakinid Choristida. In literal terms, a specimen with triactines is morphologically a choristid, although these spicules are normally absent; and the genus has often been placed in the purely monaxonid family Clionidae. Here again, a difference that is normally of high taxonomic significance can occur between species of one genus and even different individuals of one species.

The order Ancorinida is restricted here by removal of most of the monaxonids that were originally included (REID, 1968a) but which are now placed in the order Epipolasida of the subclass Monaxonida. Those comprising the family Coppatiidae TOPSENT were regarded by DENDY (1916) as monaxonid Ancorinidae (=Stellettidae, DENDY), in which triaenes or other tetraxons had been lost in phylogeny. This practice extended the concept implied by SOLLAS's (1888) family Epipolasidae and was influenced by DENDY's observations on *Aurora* (see above). It is reasonably likely that some genera grouped here as Coppatiidae are monaxonid Ancorinida biologically; but this cannot be established firmly. Some others could be allied just as well to plakinids like *Dercitopsis* DENDY, in which the largest spicules present are oxeas. In addition, monaxonids comprising the Tethyidae GRAY and Epallacidae TOPSENT are not close to any living choristid and are thought by current authors to be allied to spirastrellid and axinellid Monaxonida. It seemed best to move these forms elsewhere, which is also more convenient for paleontological arrangement.

DE LAUBENFELS (1936, 1955) placed choristids into two orders that cut across the present classification.

1. Choristida *sensu* DE LAUBENFELS (*not* SOLLAS) (=Triaenina SOLLAS): Poecillastrida, Ancorinida, and Craniellida in which tetraxon megascleres are typically long-shafted triaenes and never calthrops;

2. Carnosa *sensu* DE LAUBENFELS (*not* CARTER) (=Tetradina SOLLAS): all other choristids, including (i) Plakinida; (ii) Poecillastrida and Ancorinida having calthrops, with or without additional triaenes, or with calthrops replaced by triactines or centrotriaenes; and (iii) *Samus* GRAY, of the order Craniellida, in which the megascleres are short-shafted amphitriaenes. The arrangement used herein is regarded as clearly nearer to the probable relationships of these sponges than that used by DE LAUBENFELS. It is least satisfactory in dealing with forms that lack triaenes, which can be difficult to classify; but nearly all genera with triaenes—comprising most Poecillastrida and Ancorinida and all Craniellida—fall clearly in one of three groups that require being distinguished taxonomically. By comparison, the classification of DE LAUBENFELS both cuts across probable relationships and unites forms that need to be separated. His treatment of *Thenea* GRAY, which he dissociated from other Poecillastrida as a supposed ancorinid is especially problematic. He relied on the argument (DE LAUBENFELS, 1936, p. 167) that *Thenea* "... differs from *Ancorina* SCHMIDT only in that streptasters of the latter are much less bent than those of the former." The microscleres of *Thenea* are streptoscleres, developed mainly as plesiasters, metasters, or spirasters; but the only streptasters that occur in *Ancorina* are spinulated microrhabds, like those of *Ecionemia* BOWERBANK, that may grade toward sanidasters. The simple oxyasters of some *Thenea* species are also clearly related to the three- or four-rayed forms common as plesiaster variants in other Poecillastrida.

THE OLDEST CHORISTIDA

In the text above, which discusses the possible relationships of the choristid and monaxonid Demospongea, and elsewhere (REID, 1970), there is agreement with FINKS (1967b, 1971b) in considering the oldest known choristid spicules to be of Early Carboniferous (Visean) age. It was, nonetheless, thought likely that their Visean diversity must imply that the group is far older, unless its pre-Visean evolution was much faster than from then to the present.

In fact, REIF (1968) has recorded what seem to be older examples from the Upper Ordovician of Borehole, Estonia. The spicules occur with other types ascribed to Hexactinellida and Heteractinellida and, therefore, cannot be identified certainly as those of true choristids; but their shapes would be regarded as marking them as choristid spicules if found for example, in the Carboniferous. Morphologically they are calthrops or short-shafted triaenes or in one instance what seems to be a broken, long-shafted protriaene. In addition, a peculiar spicule (**dodecaactine**), ascribed by REIF (1968) to a hexactinellid, could also be interpreted as a tetralophose calthrops with one ray branched near its origin and compared with the candelabra spicules of the living *Corticium*.

These spicules seem to establish the existence of Ordovician choristids, which already had both calthrops and long-shafted triaenes. This in turn should imply that the group must be older than Late Ordovician, if the triaenes and calthrops are assumed to be related types of spicules. They are also almost as old as the oldest known *Hindia,* whose desmas are noted above as suggesting derivation from a pre-existing choristid.

SUBCLASS MONAXONIDA

Modern monaxonids are divided into six orders: the Epipolasida, Spirastrellida, Axinellida, Astrosclerida, Chalinida, and Desmacidontida. Fossils are referred to these orders when any arrangement is possible. The order Epipolasida is regarded as composite and is used for convenience. The remaining five orders are not known to be composite although all include genera that are nearer than others to the central type genus. The corresponding type genera are *Spirastrella* SCHMIDT, *Axinella* SCHMIDT, *Astrosclera* LISTER, *Chalina* GRANT, and *Desmacidon* BOWERBANK, each of which is also type of a family. No microscleres occur in *Axinella, Astrosclera,* or *Chalina* (also called *Haliclona* GRANT), but *Spirastrella* and *Desmacidon* have the characteristic microscleres of their orders.

The order Epipolasida comprises various monaxonid sponges that have euaster microscleres like those of the choristid Ancorinidae and some similar forms that have microrhabds, sanidasters, or no microscleres. It is also a convenient position for some of the fossils. I do not include genera with spinispira microscleres, some of which were included by DE LAUBENFELS (1936) but are here placed in the order Spirastrellida. The name Epipolasida is not meant to imply derivation from choristid Ancorinidae (="Stellettidae"), as was SOLLAS's Epipolasidae (1888), and in this sense is used as by DE LAUBENFELS. Modern forms included are mainly those comprising the families Coppatiidae TOPSENT (=Jaspidae (Jaspinae), DE LAUBENFELS), Tethyidae CARTER, Epallacidae TOPSENT, and Sollasellidae VON LENDENFELD.

The mutual relationship of the families of this order are not known with certainty, and they may be allied to members of various other orders. They are grouped together generally on the basis of having (a) true meristically varying euasters like those seen in Ancorinidae and (b) mainly oxeas as megascleres and no tylostyles except as minor variants of a style (or strongyloxea). All such sponges were regarded by DENDY (1916) as derivatives of ancorinid (stellettid) Choristida, which had lost tetraxon megascleres in phylogeny. In reality, as DE LAUBENFELS (1936) emphasized, it is not known whether their condition is secondary or primitive or whether they are allied to choristids or other monaxonids or to both in different instances. The possible relationship of coppatiids to Ancorinidae has been noted already. The Tethyidae have similar microscleres, but the megascleres are typically strongyloxeas, sometimes passing into variants developed as true styles or even tylostyles arranged radially and pointing outward as in members of the order Spirastrellida. TOPSENT (1928a) grouped these sponges with the spinispira-bearing Spirastrellida in his order Hadromerina, and this view is supported by biochemical evidence (BERGQUIST & HOGG, 1969). On the other hand, TOPSENT's view depended on the presence of apparent euasters in the family Timeidae; and DENDY (1921) seems to have been right in interpreting these microscleres as pseudoeuasters derived from spinispiras because of their replacement by sigmasters in several species of *Timea* GRAY. Typical Spirastrellida never have euastriform microscleres, and the spinispiras found in some families appear to be related to megascleres. Hence the Tethyidae seem less closely related to the forms with spinispiras than TOPSENT thought, at least in terms of their skeletal characters.

The Epallacidae (*Epallax* SOLLAS and *Hemiasterella* CARTER, sometimes thought to be identical) were placed by SOLLAS (1888) in the family Axinellidae RIDLEY and DENDY and recently in the order Axinellida by LÉVI (1955). SOLLAS (1888) thought that these forms are also close to the choristid Plakinidae from which the axinellids could thus have arisen directly; but DENDY (1922) regarded their euasters as pseudasters analogous with the pseudoradiate megascleres of *Cyamon* GRAY. Examining material that was previously studied by these authors indicates that the euasters of *Epallax callocyathus* SOLLAS are closely similar to spicules of the supposed plakinid *Astroplakina* DENDY, which was claimed to justify DENDY's own views on the origin of true euasters in choristids.

Fossils with ophirhabd megascleres have been placed into the family Ophiraphiditidae SCHRAMMEN, which should fall in the subclass Choristida because most of the genera have subtriaenes in addition to ophirhabds. A Cretaceous sponge with subtriaenes was identified by VON ZITTEL

(1878b) as a species of the modern genus *Ophiraphidites* CARTER; but the type species *O. tortuosus* CARTER was based on a macerated fragment having ophirhabds only, and similar spicules occur in the basal parts of several Axinellida, e.g., *Bubaris* GRAY. The fossils with subtriaenes were accordingly removed to a family Cephaloraphiditidae REID (1970), herein placed in the subclass Choristida. But ophirhabds, oxeas, and euasters are the only spicules present in the recent *"Jaspis" serpentina* WILSON, which is a typical coppatiid apart from having ophirhabds. This suggests a position in this family for *Ophiraphidites, sensu* CARTER and for purely monaxonid fossils such as *Euleraphe* SCHRAMMEN and *Heteroraphidites* SCHRAMMEN.

The Chondrillidae SCHMIDT normally have no spicules except euasters, which are usually sphaerasters and are often regarded (e.g., DENDY, 1916; TOPSENT, 1928a) as derived from coppatiids or tethyids by loss of megascleres. The Chondrosiidae SCHULZE appear to be allies in which all spicules have been lost. If these origins are accepted, these families can be placed into the Epipolasida. On the other hand, the sphaerasters of *Chondrilla* SCHMIDT are sometimes replaced by sigmasters, for example in *C. phyllodes* SCHMIDT. This suggests a relationship to the Timeidae TOPSENT of the order Spirastrellida unless *Chondrilla* is composite.

The order Spirastrellida comprises the Clavulina of VOSMAER (1882, 1883, 1884, 1885, 1887) or the Hadromerina *sensu* TOPSENT (1928a; not DE LAUBENFELS, 1936), which remain after removal of some families to the Epipolasida. The characteristic microscleres of the Spirastrellida are spinispiras and related forms; but these microscleres are almost restricted to the central families Spirastrellidae SCHMIDT, Clionidae D'ORBIGNY, and Placospongiidae GRAY. Other forms may have microrhabds, some of which may be homologous with spinispiras, or no microscleres. The Timeidae TOPSENT have euastriform microscleres that, however, appear to be pseudasters because sigmasters sometimes replace them (e.g. in *Timea curvistellifera* DENDY). Genera whose reproduction is known are usually oviparous, with parenchymella or parenchymella-like embryos. Spinispiras are related to megascleric oxeas, with which they sometimes share a fine spinulation or a central annulation (e.g., both in *Cliona vastifica* HANCOCK). Many genera of the order have tylostyles, but styles or oxeas may also occur and genera that lack tylostyles are not excluded herein.

The Thoosidae COCKERELL (*Thoosa* HANCOCK and *Alectona* CARTER) appear to be allied to the boring Clionidae, with which they are sometimes included, but are not typical of the order. Neither genus has spinispiras, and *Thoosa* has no megascleres. The oxea megascleres of *Alectona* are sometimes varied as triactines, and *Thoosa* may have slender triactinal or tetractinal oxyasters or dermal plates that develop from small tetractines. The inclusion of these forms in the order Spirastrellida could be doubted; but they may be archaic forms, which point to its origin (cf. below, p. 111).

A few forms with spinispira microscleres (e.g., *Trachycladus* CARTER) are axinellid according to LÉVI (1955). This could mean either than spinispira microscleres have evolved independently in axinellids or that megascleric skeletons can take on an axinellid aspect in sponges that are not Axinellida.

The order Axinellida comprises the family Axinellidae RIDLEY and DENDY and various similar sponges removed by LÉVI (1955) from the Halichondrina *sensu* TOPSENT (1928b) or the Poecillosclerina TOPSENT. These removals were made as a result of his division of the class Demospongea SOLLAS into subclasses Ceractinomorpha LÉVI and Tetractinomorpha LÉVI (1955, 1957a), which requires all forms grouped as Ceractinomorpha to be viviparous. This in turn depends on a picture of phylogeny that derives the Ceractinomorpha with spicules (orders Chalinida and Desmacidontida herein) from the keratose sponges and these from viviparous halisarcid myxosponges. Axinellida whose reproduction is known are oviparous, although with embryos of the same type as in Ceractinomorpha.

The central family Axinellidae comprises sponges without microscleres. The predominant megascleres are styles and are sometimes the only megascleres, although diactines may also occur. Plumose spicular fibers are frequent, and there may be a special axial skeleton. Spongin is more or less abundant and may form continuous fibers. Some similar sponges have microrhabds, toxas, or sigmas as microscleres, but no cheloid forms occur. Other genera have additional acanthostyle megascleres or related pseudastrose forms of megascleres or microscleres. In some genera, diactines are blunt-ended ophirhabds. Some are encrusting forms, with monactine megascleres arranged vertically so as to echinate the substratum. In addition to normal monaxonids, *Monocrepidium* TOPSENT and the fossil *Scolioraphis* VON ZITTEL have sublithistid modification of ophirhabds, which are developed as irregularly annulated scoliorhabds. A few modern lithistids with large megaclone or rhizoclone-like desmas (e.g., *Petromica* TOPSENT, *Lithobubaris* VACELET) are possibly of axinellid origin and could be placed in this order instead of the artificial subclass Lithistida. The fossil Megarhizidae SCHRAMMEN, placed here in the lithistid suborder Megarhizomorina SCHRAMMEN, have analogous desmas.

The order Astroscleridae is new and contains sponges with a monaxonid spicular skeleton and a nonspicular aragonitic basal skeleton. In the Astroscleridae LISTER, this structure resembles the skeletons of typical **astrorhiza**-bearing stromatoporoids; but in *Ceratoporella* HICKSON and *Merlia* KIRKPATRICK, sole genera of the Ceratoporellidae HICKSON and Merliidae KIRKPATRICK, it suggests those of fossil Chaetetida or favositid Tabulata. *Ceratoporella* was also mistaken for a coenothecalian octocoral before the soft parts were known (MONTANARO-GALLITELLI, 1956, p. 194). Astroscleridae and *Ceratoporella* lack microscleres and have spiny styles as megascleres, some of which have the spines arranged in whorls as in various Axinellida. On the other hand, *Merlia* has smooth megascleres only and has clavidisc microscleres that are usually regarded as related to the diancistra microscleres of hamacanthid Chalinida (e.g., DENDY, 1921; TOPSENT, 1928b). This is why the order is placed between the Axinellida and the Chalinida, although either or both of these resemblances, in fact, may be illusory.

HARTMAN and GOREAU (1970) placed these sponges into a new class Sclerospongiae and suggested that the fossil Chaetetidae and the astrorhiza-bearing Stromatoporoidea were also similar sponges. The class was so named because aragonite is commonly secreted in the form of spherulitic bodies, which they called sclerodermites; although, in general usage, the term **sclere** means spicule, as in megasclere. The few modern genera are, nonetheless, typical Demospongea apart from the special basal skeleton, and a form that lost this structure in phylogeny would appear to be a normal monaxonid. Their reference to an order of the subclass Monaxonida therefore seems more appropriate, if the modern forms only are considered. A subclass Sclerospongida would, however, be appropriate if accepted as including the fossils, and consisting for example of the orders Stromatoporida and Chaetetida. These groups are to be treated in a subsequent volume of the *Treatise.*

LECOMPTE (1956, p. 121) rejected any relationship of Stromatoporoidea to sponges without mention of *Astrosclera,* whose status as a demosponge has been known since 1910 (KIRKPATRICK, 1910b). He ruled out a relationship between **astrorhizae** and the canals of a rhagon (i.e., leuconoid) canal system on the grounds that a lamellar distribution of canals is not observed in Porifera. The radial groups of exhalant canals that occur at the surface in Astrosclerida are disposed horizontally, and corresponding astrorhiza-like channels are characteristic of *Astrosclera.* In *Ceratoporella,* the surface tissue is so thin that the canals raise the dermis, like veins under skin. The siliceous spicules lie loose in the soft tissues, and do not form part of the basal skeleton unless included incidentally. The impassable obstacle is, thus, an expected condition if the Stromatoporoidea are sponges.

A loose Campanian microsclere called a psyllium by SCHRAMMEN (1924a, pl. 4,*14*) is a clavidisc like those found in *Merlia,* which may, thus, have been a contemporary of the latest accepted stromatoporoids.

The order Chalinida comprises the Halichondrina *sensu* LÉVI (1957b; i.e., members of the Halichondrina *sensu* TOPSENT (1928b) not removed to the order Axinellida) and Haplosclerida *sensu* TOPSENT (i.e., not including chela-bearing sponges included by DE LAUBENFELS, 1936), which are placed in suborders Halichondrina and Chalinida respectively. The order Desmacidontida comprises the Poecillosclerina of TOPSENT (1928b) except for genera removed to the Axinellida by LÉVI (1955). These orders appear to have a special relationship to one another and to keratose sponges.

TOPSENT (1928a) and DE LAUBENFELS (1936) placed the orders Halichondrina and Haplosclerida on opposite sides of an order Poecillosclerina. This arrangement was based on (a) the inclusion of the present Axinellid as Halichondrina and their resemblance to some Spirastrellida (=Hadromerina, TOPSENT); and (b) resemblances between some Haplosclerida (=Chalinida herein) and dictyoceratid Keratosida. The removal of the Axinellida from the Halichondrina *sensu* TOPSENT leaves this group much restricted. Sample genera studied biochemically by BERGQUIST and HOGG (1969) are grouped as follows:

i. Halichondrina *sensu* LÉVI and Haplosclerida (=Chalinida);

ii. Poecillosclerina (=Desmacidontida) and Keratosida.

The Halichondrina *sensu* LÉVI and the marine Haplosclerida (Chalinida) are also similar in (1) the simplicity of the megascleric skeleton and (2) the frequent lack of microscleres, which never include cheloids when present (although these occur in some Haplosclerida *sensu* DE LAUBENFELS, 1936). These forms, therefore, are placed herein into one order, called Chalinida, with suborders retained for consistency with previous classifications. The name Chalinida GRANT is used as senior to Halichondrina VOSMAER. *Chalina* GRANT is currently regarded as a synonym of *Haliclona* GRANT; but herein an analogy is assumed with family-group nomenclature, which permits only a change in name when the type genus is a homonym. This analogy seems to be required by the use of a type genus.

The Chalinida and Desmacidontida are viviparous in forms whose reproduction is known, with large parenchymella embryos. They share this condition with the keratose sponges and Halisarcidae, with which they were united as Ceractinomorpha by LÉVI (1957b). The orders are distinguished by different developments of the skeleton, which is generally simpler in the Chalinida than in Desmacidontida. The Chalinida typically have megascleres of a single sort only, which are commonly diactines, and have no special dermal megascleres, echinating spicules, or cheloid microscleres. There are often no microscleres, although toxas or sigmas occur in some marine forms, and amphidiscs and other pseudasters occur in some freshwater genera (Spongillidae GRAY).

The Desmacidontida are typically forms with chelas or related forms of microscleres, although these are absent from some genera. The term lipochelous has been used by some zoologists to imply that the absence of chelas is due to their loss in phylogeny, although this is strictly an assumption. The megascleric skeleton is sometimes as simple as in Chalinida (and forms of this sort were included in TOPSENT's Haplosclerida by DE LAUBENFELS), but it typically includes two or more sorts of megascleres. In addition to the principal megascleres, which may be either diactines or monactines, there may be echinating spicules, special dermal megascleres, or both. Dermal megascleres are often diactines. Echinating spicules are typically monactines and often acanthostyles. Examples of this type may echinate skeletal fibers or be scattered through the mesenchyme but are commonly described as echinating in either instance. A few forms with chelas are sublithistids (e.g., *Helophloeina* TOPSENT, *Lithochela* BURTON), with various types of desmas (p. 88). Spongin occurs in both orders; and both include genera ranging from some without spongin or with inconspicuous amounts to others in which the main skeleton is formed by reticulate spongin fibers.

SUBCLASS KERATOSIDA

This subclass is restricted to keratose sponges and is divided into MINCHIN's orders Dictyoceratida and Dendroceratida. The halisarcid myxosponges are acceptable as askeletose Dendroceratida.

The Dictyoceratida appear to be allied certainly to the Chalinida and Desmacidontida, but how they are related is uncertain. The older view, held by VON LENDENFELD (1889a), MINCHIN (1900), and TOPSENT (1928b), is that dictyoceratids were derived from forms with spicules; but LÉVI (1957b) regarded the latter as derived from Dictyoceratida. The older view is followed herein, although solid evidence is lacking. Biochemical evidence (BERGQUIST & HOGG, 1969) related the dictyoceratids to the Desmacidontida, although based on two genera from families that VON LENDENFELD (1889a) related to the Chalinida.

SUBCLASS LITHISTIDA

Lithistids are divided here into suborders that correspond to taxa called Tribus by SCHRAMMEN in his two final monographs (1924a, 1936) but equivalent to suborders because they formed divisions of his orders. The arrangement is generally based on two characteristics: (i) the character of the desmas and (ii) the presence or absence of triaenes or related types of dermalia and their character, when present. The characters of suborders distinguished may be summarized as follows:

A. Dermal triaenes are usually present and range from dichotriaenes through phyllotriaenes to discotriaenes; some genera have monaxial or anaxial discs instead:

1. Suborder Tetracladina VON ZITTEL: desmas typically tetraclones or triders, although accessory monaxial forms may also occur (e.g., as radical desmas).

2. Suborder Dicranocladina SCHRAMMEN: desmas dicranoclones, large rhizoclone-like forms (megarhizoclonids, SCHRAMMEN) or intermediates to which a few tetraxial desmas may be added.

B. Dermal triaenes are usually dichotriaenes, rarely simple triaenes; no phyllotriaenes, discotriaenes, etc.

3. Suborder Helomorina SCHRAMMEN: desmas heloclones.

4. Suborder Megamorina VON ZITTEL: desmas megaclones.

5. Suborder Didymmorina RAUFF: desmas didymoclones, rhizoclones, and intermediates.

C. Dermal triaenes unknown and lacking in living examples.

6. Suborder Megarhizomorina SCHRAMMEN: large rhizoclone-like desmas, loosely or sometimes not articulated.

7. Suborder Rhizomorina VON ZITTEL: desmas rhizoclones only.

8. Suborder Sphaerocladina SCHRAMMEN: desmas sphaeroclones, astroclones, or intermediates.

The use of this method of classification does not imply that all named types of desmas are sharply distinct or that every named type is found in one suborder only. Some named types are completely intergrading (e.g., didymoclones and rhizoclones); others are not (e.g., didymoclones and heloclones). The type of desma cited as characteristic of a given suborder is usually a predominant type of desma and may be present in all included genera; but it may also grade in some or even all genera into some different nominal type, which may sometimes predominate or replace it. For instance, true tetraclones are present in most Tetracladina; but they may be accompanied by subordinate monaxial desmas or replaced by triders. In Dicranocladina, the desmas may be (i) dicranoclones; (ii) megarhizoclonids; (iii) both together; or (iv) part or all intermediates. The sphaeroclones of Sphaerocladina are often accompanied by astroclones and sometimes replaced by them. A given type of desma, which is characteristic in one suborder (e.g., rhizoclones in Rhizomorina) may also occur as a subordinate form in others (Dicranocladina, Didymmorina). On the other hand, some types have not been found together (e.g., tetraclones, megaclones, sphaeroclones). These facts were well known to SCHRAMMEN, but are partly disguised in his monographs (1910, 1912, 1924a, 1936) by his habit of naming desmas taxonomically (by e.g., use of rhizoclone for desmas of Rhizomorina only, irrespective of morphology).

This method provides a satisfactory arrangement of most of the post-Paleozoic fossils as well as most modern genera. The most important problematical fossils are genera with desmas like those found in various groups with triaenes, although the latter are absent. These genera are usually allocated according to the form of the desmas, although the absence of triaenes could be due to a, original absence; b, loss in phylogeny; or c, loss in fossilization.

The suborders cited are grouped into orders as follows.

1. Order Tetralithistida LAGNEAU-HÉRENGER, *sensu nov.:* Tetracladina, Dicranocladina, and (?) Didymmorina.

2. Order Megalithistida *nov.:* Helomorina and Megamorina.

3. Order Monalithistida LAGNEAU-HÉRENGER, *sensu nov.:* Megarhizomorina, Rhizomorina, and Sphaerocladina.

The Tetracladina and Dicranocladina and the Helomorina and Megamorina are interpreted herein as contrasting groups of sponges whose desmas are reasonably certain to have had different prototypes (calthrops and ophirhabds, respectively). They are placed in separate orders accordingly. The Didymmorina were classified as lithistid Monaxonia by SCHRAMMEN (1936); but there are small dichotriaenes, which do not appear to be intrusive, in a *Cylindrophyma milleporata* (GOLDFUSS) identified by SCHRAMMEN himself, and the desmas are comparable with some found in Dicranocladina. The order Monalithistida is envisaged as a composite grouping, convenient for lithistid types with monaxial megascleres only, or thought to have been of monaxonid origin. The Paleozoic Orchocladina also fall in this order. No separate order is envisaged for the Sphaerocladina, which were very probably derived from the Orchocladina and which may have normal monaxons in addition to the desmas (e.g., in the living *Vetulina* SCHMIDT).

The Paleozoic Tricranocladina (=Eutaxicladina *sensu* SCHRAMMEN: *not* RAUFF or DE LAUBENFELS) have desmas of uncertain character, regarded as tetraxons by SCHRAMMEN. If this is correct, they could be classified as Tetralithistida; but they are not related to the typical Tetracladina and Dicranocladina, whose relationships seem to lie with the choristid Poecillastrida.

ALTERNATIVE CLASSES OR SUBCLASSES

Phylogenies suggested by both DENDY (1905) and LÉVI (1957b) envisage the class Demospongea as comprising two major groups of sponges, descended independently from different myxosponge ancestors. In LÉVI's scheme, the taxa distinguished on this basis are subclasses Tetractinomorpha LÉVI and Ceractinomorpha LÉVI of the class Demospongea; but DENDY's were treated as the orders Tetraxonida VOSMAER and Euceratosa DENDY of a class non-Calcarea, which also included the Hexactinellida as an order Triaxonida SCHULZE. It might be asked why neither of these schemes is used here for division of the class Demospongea into subclasses or its replacement by two classes. This question specifically applies to the scheme of

LÉVI (1957b), which is currently widely, although not universally, accepted.

First, DENDY's views are not now acceptable because of his insistence (a) that monaxonid sponges are all forms derived from choristids by loss of tetraxons in phylogeny; and (b) that no true keratose sponges (hence Euceratosa) are related to any forms with spicules. In current perspective, a very large proportion of monaxonids have no known or likely relationship to choristid sponges, in the sense required by DENDY. The Chalinida and Desmacidontida (=Halichindrina, Poecilosclerina, and Haplosclerida LÉVI) also seem to be genuinely allied to the keratose sponges, in terms of the current biochemical and embryological evidence.

The factual basis of the subclasses proposed by LÉVI (1957b) is the embryology of their members and especially of genera grouped as Ceractinomorpha (here Chalinida, Desmacidontida, Keratosida). The latter are always viviparous with parenchymella embryos when sexual reproduction is known (although this uniformity is due partly to removal of forms that disturb it).

Sexual reproduction is uncommon in Tetractinomorpha and some forms that have it are oviparous. On the other hand, this group has no general uniformity because embryos recorded may be (a) incubated amphiblastulae (*Oscarella* VOSMAER, *Plakina* SCHULZE of Plakinida); (b) nonincubated parenchymellae (e.g., *Tethya* LAMARCK of Epipolasida); or (c) parenchymelloid types, incubated (e.g., Tetillidae, Craniellida; *Stylocordyla* THOMSON of Spirastrellida) or not (e.g., *Polymastia* BOWERBANK of Spirastrellida). It has also been found recently (BERGQUIST & HARTMAN, 1969) to be diverse biochemically, with four major patterns of amino-acid groupings, one of which can be subdivided further. Tetractinomorpha in effect means little more than Demospongea that are not Ceractinomorpha unless LÉVI's picture of phylogeny is also considered.

The critical evidence for LÉVI's picture (1957b) is derived from the early development of two modern sibling species of the myxosponge *Halisarca* DUJARDIN, which resembles dendroceratid Keratosida in the character of its soft parts. In *H. dujardini* JOHNSTON the larva developed from the embryo is an asconoid rhagon; in *H. metschnikovi* LÉVI, however, it is a syconoid rhagon like that of the dendroceratid *Aplysilla sulfurea* SCHULZE. According to LÉVI (1957b), this implies that the Dendroceratida were derived from halisarcids; the Dictyoceratida from the Dendroceratida; and the orders with spicules (Halichondrina, Poecillosclerina, Haplosclerida LÉVI; Chalinida and Desmacidontida herein) are derived from the dictyoceratids. Thus, spicules of monaxonid Ceractinomorpha are supposed to have evolved independently of those of Tetractinomorpha.

This idea may be correct but certainly can also be doubted for a number of reasons.

i. The general character of the spicules appears to be identical in both instances, and all types of megascleres found in Ceractinomorpha (except the desmas of *Crambe*) can be matched in Tetractinomorpha. The toxa and sigma types of microscleres occur also in some Axinellida, and sigmaspire-variants that are chelas morphologically occur in some Craniellida (e.g., *Chrotella amphiacantha* TOPSENT, *Tetilla sigmoanchoratum* KOLTUN). No difference in the mode of secretion of the spicules has yet been demonstrated. Last, desmas are identical in character and mode of union whether in Tetractinomorpha (e.g., *Pleroma* SOLLAS) or Ceractinomorpha (e.g., *Desmatiderma* TOPSENT). All these resemblances suggest that the spicules did not have different origins.

ii. Derivation of the spiculate Ceractinomorpha from keratose sponges implies replacement of spongin by spicules during phylogeny. On the other hand, if axinellid sponges are put into the Tetractinomorpha their spongin has presumably evolved as a replacement for spicules. Such opposite developments seem unlikely.

iii. Both DENDY (1905) and LÉVI (1953, 1957b) based their pictures of phylogeny on modern sponges only. Although they provide the only realistic basis for speculation, the modern forms are all end forms phylogenetically and have no chronological sequence. Their morphology alone, hence, provides no certain measure of phylogenetic direction. It is unknown whether the simplicity of forms like *Oscarella* and *Halisarca* is primitive or due to regression, e.g., by neotenous retention of larval characters, or which type of *Halisarca* rhagon was derived from the other. Furthermore, no modern taxon can ever be ancestral to another, although they may share common origin. Moreover, the fact that a series of modern taxa can be arranged into a sequence from simplest to most complex need not imply that this sequence represents their phylogeny. For example, such a pattern could also arise by the iterative divergence of retarded stocks from a primary progressive one, with the oldest divergent stock then retaining the most primitive characters.

In addition, the survival to the present of two groups of supposedly primitive genera (plakinids and *Oscarella;* halisarcids) does not mean that others have not existed or were the true ancestral sources of the orders with megascleres. A persistence of primitive characters, if genuine, implies in itself a nonprogressive status. In other words, the true ancestral stocks of most or all advanced modern orders may, in fact, have disappeared as completely and in the same manner as mammal-like reptiles or thecodont archosaurs.

iv. If the *Halisarca* species are siblings their status has two implications that can count against Lévi's (1957b) conclusions.

a. Their speciation must be too recent to bear directly on demosponge origins, or even nearly so.

b. If siblings can be distinguished by their embryology, this is *ipso facto* evidence that embryological characters can change cenogenetically and can be less stable phylogenetically than adult characters at even speciation level.

In addition to these problems, it should probably not be assumed that all lithistids are Tetractinomorpha. In particular, the Anthaspidellidae of the Paleozoic Orchocladina have a structure suggesting their origin from a chalinoid monaxonid, now found only among the Chalinida and Desmacidontida. It also seems likely that the Sphaerocladina are of orchocladine origin.

The classification used here makes the compromise of basing subclasses on the skeleton but adopting Lévi's order Axinellida. There is no implied rejection of his concept of a special relationship between Ceractinomorpha, and their division between the Monaxonida (as Chalinida and Desmacidontida) and Keratosida is acceptedly artificial.

SKELETAL EVOLUTION

In attempting an assessment of demosponge phylogeny, it seems best to ask first whether any general patterns can be demonstrated. One is then at once confronted by the fact that most directions of evolutionary changes cannot be firmly established, even when they can be fairly presumed.

First, no critical data are provided by the fossils, at least with respect to nonlithistids. The fossil Demospongea, unlike those now living, are predominantly lithistids, and all the nonlithistids together are considerably less numerous than the genera distinguished by zoologists. Most fossils have nothing but megascleres. Furthermore, most supposed fossil occurrences of existing genera are based on isolated megascleres or microscleres, many of which have no diagnostic value. No extinct genus or family can be classified more certainly than as choristid or monaxonid. There are very few reliable records of keratose sponges and none of myxosponges. With the evidence suggesting that the record is probably also extremely incomplete, these facts rule out normal reliance on stratigraphic sequences as a basis for establishing phylogeny.

Comparative zoological data are, therefore, the main source of evidence; but the modern forms have no chronological sequence and include only whatever stocks have survived to the present. In consequence, such data are generally of uncertain significance.

One general assumption that seems safe at present is that lithistid sponges are derivatives of nonlithistids, produced by conversion of some normal type of megasclere into a desma. This conclusion is unopposed thus far. It seems to be supported by the characters of the sublithistids at least, in which a desma intergrades with a simple monaxon, although the direction of change is not strictly demonstrable.

The most important problem is the relationship of choristid and monaxonid sponges. This can be approached by considering three contrasting possibilities.

1. The primitive demosponge spicules were tetraxons, and all monaxons are derived from them. All monaxonids have arisen from choristids by loss of tetraxons in phylogeny.

2. The primitive spicules were monaxons, and all tetraxons are derived from them. All choristid sponges have arisen from monaxonids by evolving tetraxons in phylogeny.

3. Choristid and monaxonid sponges have had separate origins.

In instances (1) and (3) tetraxons found in lithistids will have been inherited from choristids; but in instance (2) they could either be inherited or evolved from monaxons after the lithistid condition was developed.

The first of these concepts corresponds with the views held by DENDY (1905, 1916), which were based on those of SCHULZE (1887b). These authors also thought that small spicules like those of plakinids were primitive and assumed that loss of tetraxons has occurred in forms with megascleres. The principal arguments for this view are (i) the transitions from triactines or tetractines to diactines seen in plakinids and some other forms, which seem to show clearly that the diactines are produced by reduction; (ii) the apparently central status of triactines or tetractines in relation to spicules as different as oxeas, long-shafted triaenes, lithistid discotriaenes and discs, tetraxial and some monaxial desmas, sterrasters, and spiraster streptoscleres; and (iii) the close resemblance of some monaxonids to choristids, which may even be so close that both types can be regarded as species of one genus (*Aurora* SOLLAS; DENDY, 1916). The main objections are (a) the general lack of demonstrable relationships between choristids and most monaxonids and (b) the existence of evidence suggesting an opposite picture.

The second possible picture of phylogeny, deriving choristids from monaxonids, has been almost ignored except by FINKS (1967b), although hinted at by SOLLAS (1888). It is suggested by several sorts of evidence:

i. The development of long-shafted triaenes of various modern sponges from initial monaxons during ontogeny.

ii. The occurrence in *Paratetilla* DENDY (Tetillidae, Craniellida) of subtriaenes, which appear to be modified derivatives of a normally long-shafted type (as accepted by DENDY himself: DENDY, 1922).

iii. The occurrence of diactinal to pentactinal spicules that appear to be derived from monactines (echinating acanthostyles) in the axinellids *Cyamon* GRAY and *Trikentrion* WELTNER.

iv. The geological appearance of monaxonid sponges in the Cambrian Period, long before the oldest known choristids, which are Late Ordovician.

The closely similar choristids and monaxonids, used by DENDY (1905) in support of his opinions, could also be cited in this context with an opposite significance.

The objections are that none of this evidence is conclusive and some of it is certainly unreliable.

1. There is no guarantee that the ontogeny of spicules must represent their phylogeny. In the instances of (i) lithistid discs related to dermal triaenes and (ii) related tetraxial and monaxial desmas, it is clear that ontogenetic prototypes can alter cenogenetically in a way that leaves the adult form as the main indication of homology. This conclusion is independent of the direction ascribed to

phylogeny in these instances. In the instances of (a) monaxial discs related to triaenes and (b) monaxial radical desmas related to tetraclones, the change in phylogeny implied by their functional adaptations is tetraxon to monaxon, not the opposite.

In long-shafted triaenes, delay in formation of the cladi could also be cenogenetic if it is related to hypertrophy of the rhabdome. It might also lead ultimately to their suppression, with replacement of tetraxons by monaxons. This would fit DENDY's picture of phylogeny.

2. The subtriaenes of *Paratetilla* occur at or near the surface and may only be an ectosomal specialization in this genus.

3. The axinellid sponges have no certain relationship to any choristids in terms of any known evidence. The occurrence of secondary radiates in some genera has also depended presumably on their prior possession of echinating acanthostyles. There is nothing in the characters of choristids to suggest that their tetraxons were ever echinating spicules.

4. Lower Carboniferous choristids had spicules including small simple and lophose calthrops, large simple and branching calthrops, subtriaenes, long-shafted plagiotriaenes, protriaenes and anatriaenes, mesotriaene variants of protriaenes, typical dichotriaenes, trachelotriaenes, and unusual pentactinal and hexactinal megascleres. There are also rounded bodies that appear to be sterrasters. The choristids must already have existed for long enough to evolve this range of different types of spicules, which include some like those of plakinids and every major type of tetraxial megasclere found in modern forms. Unless the innovation of new types was initially much faster than since the Carboniferous Period, the time involved could be up to several times longer than that from the Early Carboniferous to the present.

5. The existence of choristids before at least the Ordovician Period is suggested by the typically tetraxial form of the desmas of *Hindia* DUNCAN and related genera and by the structural resemblance of hindiids to the minchinellid *Porosphaera* STEINMANN, in which the spicules were certainly tetractinal.

6. The pre-Carboniferous monaxonids are known only by their megascleres, and whether they are related to choristids or any later sponges is unknown. They could, in fact, represent monaxonid stocks that have descended independently of choristids or have no later relatives. If LÉVI's views are followed, such monaxonids need to be identified as Tetractinomorpha before they can be cited in evidence.

7. If choristid and monaxonid stocks have descended independently from plakinid-like microspiculate sponges, as envisaged by SOLLAS (1888), their first geological appearance will almost surely have depended on evolution of megascleres; so that which appeared first in an adequate geological record would indicate only which stock first evolved megascleres, without implication that either is derived from the other.

The concept of separate descent of most choristids and monaxonids (point 3 above) fits their general lack of evident relationship; but monaxonid stocks that were ancestral to choristids may not have living representatives. SOLLAS's (1888) concept of the descent of most choristids and monaxonids from plakinid-like sponges, by development of megascleres from tetraxons or from monaxons only, fits with the fact that the largest spicules present in plakinids may be either tetraxons (e.g., *Plakinastrella* SCHULZE) or monaxons (e.g., *Dercitopsis* DENDY). The small size of spicules in plakinids is not known to be primitive, however.

Thus, none of the three possibilities suggested above can be shown to be the truth, and some of the evidence can in fact support opposite views, according to how it is interpreted.

The most likely possibility is the third, envisaged by SOLLAS (1888), with his further conclusion that a few monaxonids only are true derivatives of choristids with megascleres. It seems clear that DENDY was badly mistaken in his concept of sigmatose microscleres, which formed the chief basis of his view that monaxonids are all derived from choristids. But the origin of choristids from monaxonids, although certainly possible, is not currently supported by any reliable evidence. In particular, it need not be implied by the development of some modern triaenes from monaxons in ontogeny. The paleontological data are also mainly of unknown significance, except for clear evidence that choristids had undergone substantial evolution before their first known appearance. On the other hand, DENDY's explanation of the skeletal evolution of choristids fits their characters sufficiently well to suggest that it is probably correct, at least for the Plakinida, Poecillastrida, and Ancorinida. There is no evidence that this applies also to the Craniellida, in which all tetraxon megascleres are triaenes; but his principal mistake was in assuming that, because a few monaxonids appear to be genuinely allied to typical choristids, this must also apply to all the others.

In attempting to reconstruct the pattern of skeletal evolution, several general points may first be considered.

1. The numerous occurrences of similar features (e.g., presence of both megascleres and microscleres or radial arrangements of megascleres) in sponges that appear to have no direct relationship implies widespread occurrence of parallel or convergent developments.

2. Assuming that the spicules were originally all of one sort, a trend toward differentiation of spicules into separate categories is implied by all instances in which two or more categories are present. On comparative grounds, size, form, or both may be affected, and there may be a correlated functional development (for instance, the arrangement of triaenes for support of the ectosome).

3. Once different categories of spicules have become differentiated, they usually evolve independently. For instance, in lithistids the megascleres developed as desmas may be accompanied by normal oxeas and triaenes; so that modification into desmas can be seen in one category of megascleres without others being affected. Once established, any category of spicules can undergo apparently further differentiation, for instance so that several sorts of triaenes are developed.

4. It is reasonably likely that spicules were initially little or no larger than the cells that secreted them. This implies that the occurrence of megascleres is a secondary development, resulting from enlargement of spicules in phylogeny. Its significance is probably functional, indicating correlation between increase in size of the spicules and their efficiency as supporting elements. Unless all the groups possessing megascleres have an orthogenetic relationship, which does not seem likely, their development has occurred independently in various lines of descent.

5. Many microscleres have shapes that do not occur in megascleres. This may be because the shapes restricted to microscleres are not suited to the supporting function of megascleres. Many shapes of microscleres have no obvious functional significance, and the microscleres themselves have often no apparent function. It was DENDY's view that most variation in the shape of sponge spicules has no primary functional significance, although types that are suited to some function may be put to it.

6. In various instances, related types of microscleres found together have an inverse relationship between size and complexity in shape. For instance, when several sorts of streptoscleres are present, it is common for the simplest (plesiasters) and most complex (spirasters) to be also the largest and smallest, respectively. This suggests that the difference in size between megascleres and microscleres can involve a reduction in the size of the microscleres in phylogeny as well as an increase in the size of the megascleres.

7. Many instances are known in which similar sponges differ only in the presence or absence of one of several categories of spicules. In addition, examples occur in which those of one category are abnormally rare, although present. This seems to represent a trend to secondary simplification of the skeleton by suppression of categories of spicules.

8. Differentiation of spicules into two or more orders of size during early phylogeny

can be pictured as the primary origin of microscleres, but they seem also sometimes to originate as secondary derivatives of megascleres. This applies especially to the spinispira series of microscleres, which seem to be related to megascleric oxeas (p. 96). Unless it is thought that a typical oxea megasclere can be derived from a spiraster, the direction of phylogeny implied is from megasclere to microsclere. The loss of some categories of megascleres (e.g., calthrops of Choristida) in phylogeny might sometimes have been due to their conversion into microscleres.

These principles apply to all types of demosponges that have spicules and do not depend on how choristids and monaxonids are related. They can, therefore, form a general background to any attempted reconstruction of demosponge phylogeny, irrespective of which of the main possibilities is envisaged.

Next, if choristids and monaxonids are related as envisaged by SOLLAS (1888), the prototypes of both would be simple, plakinid-like sponges with small spicules of a single size grade only. In some at least, the spicules would then be expected to vary from diactines to simple polyactines (e.g., five- or six-rayed); but others could have had diactines only. It does not matter what type of spicule is considered to be primitive. *A priori,* monaxons (diactines) could be thought to be the most likely prototypes, with the primitive scleroblast then forming one protorhabd only. Later, multiplication of protorhabds and their union in radial groups could have led to production of spicules with three or more rays, as seems likely to have happened in Calcarea and Heteractinellida. On the other hand, diactines of the living plakinids appear to be derived from triactines by suppression of one ray. The small size of the spicules can be pictured as related to the size of the scleroblasts or perhaps to the size of the flagellated chambers, as envisaged by SCHULZE (1887b).

If this picture is correct, the evolution of megascleres could then have the functional basis of providing more efficient support for the body, in general, and for the chambers and canal system against compression by its weight. This in turn could permit the evolution of increased size and new shapes in the body with increased physiological efficiency or new modes of function. The primary origin of microscleres would then be the retention of smaller spicules for the local support of the tissues and the circulatory system; although whether this function would account for their whole evolution is debatable. On loss of function, they might either persist as nonfunctional structures or be lost altogether. Alternatively, the development of microscleres might also have occurred as a means of controlling the number of spicules that grow to megascleric size, preventing overproduction of megascleres.

The first step in further evolution of the prototypes postulated would be differentiation of the spicules into two or more size grades, with the larger forms assuming the function of megascleres. In a stock with spicules varying meristically in the number of rays, the selective enlargement of tetractines or both tetractines and diactines would yield suitable prototypes for the typical choristid sponges. Why these shapes should be selected for enlargement is unknown; but presumably a functional adaptation would select the types best suited to the large-scale support of the body and to the movement of spicules within it. Smaller varying spicules, ranging from diactines to pentactines or hexactines for example, would be simple euasters in effect and would include the appropriate prototypes of all choristid euasters and of streptoscleres. The type of spiculation envisaged would be similar to that of the living plakinid *Plakinastrella* SCHULZE. In addition, enlargement of tetractinal to hexactinal spicules would account for the aberrant spiculation of *Spiractinella* HINDE.

On the other hand, enlargement of diactines only into megascleres would lead directly to monaxonid sponges, again with simple euaster microscleres. SOLLAS (1888) envisaged this origin for the axinellid sponges, with the root stock represented by the living Epallacidae in which simple euasters are present. The occasional occurrence of triactines as variants of *Alectona* oxeas has a similar implication if the prototype diactines are supposed to be derived from triactines. In plakinids, the largest spicules present may be oxeas in *Dercitopsis* DENDY, although this genus was supposed to establish his views (e.g., 1921) about the origin of choristid microscleres.

When euasters or related forms are absent from genera with megascleres, this could represent the loss of such microscleres in phylogeny; but this need not be postulated in monaxonids, at least. If some of the demosponge prototypes had only diactinal spicules as a primary or secondary condition, their descendants with megascleres will never have had any spicules with more than two rays, unless these have arisen from diactines as a secondary development.

Evolution on all of these lines seems to be a likely explanation of the spicular characters of the living nonlithistid Demospongea if allowance is made for some further developments. There have probably been at least some instances of monaxonids evolving from choristids by loss of tetraxial megascleres and of microscleres arising from megascleres as a secondary occurrence. No grounds exist, however, for insisting on a single unvarying pattern proceeding orthogenetically from either choristid to monaxonid or the opposite. Both extreme views are equally unlikely. There is also, then, no implication that either choristids or monaxonids should appear first stratigraphically or that whichever does so in practice must be ancestral to the other; this will simply have depended on which stocks were first to evolve spicules large enough to be preserved and recognized.

If primitive sponges were crustose, the first tetraxial megascleres could well have been short-shafted triaenes with three cladi directed to the surface. A tendency to orthotriaene shape would be likely, so that cladi could underlie the ectosome tangentially. The symmetrical calthrops, on the other hand, would be better adapted to support of a choanosome several times the depth of spicules supporting it. The triaene can be pictured also as arising as an ectosomal modification of a primary calthrops (DENDY, 1921). Either picture or both could be correct, and these spicules are certainly intergrading in the modern forms with calthrops as megascleres. The triaenes (or subtriaenes) of such sponges may be very little different from calthrops and oriented to the surface or at random. In the living *Pachastrella* SCHMIDT, a calthrops is accompanied by a separate category of oxeas; these could be envisaged as evolved independently, but diactines occur as minor variants of a calthrops in *Calthropella* SOLLAS for example.

The initial type of megascleric skeleton developed in choristids can be pictured, thus, as consisting of subtriaenes, calthrops, or both together or of these types plus oxeas. Megascleres from the choanosome would always include tetractines, and no special ectosomal skeleton would be present. The fossil *Propachastrella* SCHRAMMEN (Upper Cretaceous) has a minor advance on this condition, with choanosomal calthrops and variants accompanied by distinct ectosomal dichotriaenes.

The predominant type of modern choristid can be pictured as having developed from prototypes with diactines and tetractines as megascleres and with triaenes developed at the surface for support of the ectosome. Such triaenes are oriented with the cladi in or under the ectosome and with the rhabdome running radially inward. The triaenes could intergrade initially with a choanosomal calthrops but might also become differentiated. The prevalent condition in modern forms would then be produced by the loss of the choanosomal calthrops, leaving only the oxeas (diactines) as choanosomal megascleres. This could happen by simple suppression of a choanosomal calthrops or by its conversion into triaenes or microscleres. A more or less gradual replacement of calthrops by triaenes is suggested by the characters of the living *Poecillastra* SOLLAS. In addition, two minor trends are needed: (a) progressive replacement of short-shafted by long-shafted triaenes and (b) diversification of triaenes into more than one category, e.g., dichotriaenes and anatriaenes. A sponge of this type, without special arrangement of the oxeas, would have the felted type of choristid skeleton; but arrangement of the oxeas in parallel with the shafts of the triaenes would lead to radiate sponges. The adoption of globular habit might encourage this development.

As a final stage, sponges of this type could give rise to monaxonids, by loss of triaenes. Some modern forms have very few triaenes and have been mistaken for monaxonids before these spicules were discovered in their skeletons (e.g., "*Trachygellius*" [=*Craniella*] *cinachyra* DE LAUBENFELS). There could also be conversion of triaenes into monaxons by loss of their cladi. In *Stellettinopsis* CARTER, first thought to be monaxonid, the length of cladi may be less than the thickness of the rhabdome. If the formation of the rhabdomes of some long-shafted triaenes before their cladi is cenogenetic, as this picture of phylogeny requires, a final step could be their total suppression.

Description of evolution of sponges that are primary monaxonids needs minor further comment. Principal changes in megascleres are from oxea to style and then to tylostyle, with other forms (strongyles, amphitylotes, ophirhabds) as side products. The variously patterned types of skeletons (radiate, fibrous, reticulate) are presumably more advanced than the felted type, although secondary reversion is also possible. Spongin seems best regarded as sparsely developed, initially, and becoming fibrous later. Arrangement of spicules into fibers would encourage presumably this development. The pseudoceratosa with fibrous spongin and few spicules, or none in different specimens, should then represent sponges in the process of losing the spicules and becoming purely keratose. I prefer this view to LÉVI's (1957b) implication that Ceractinomorpha with spicules have evolved from purely keratose sponges with progressive reduction of spongin. His picture of phylogeny also does not permit this explanation for fibrous spongin in the axinellid sponges.

It is not implied, however, that all keratose sponges must be derived from monaxonids with megascleres. If the spiculate Ceractinomorpha are derived from microspiculate sponges with diactines only, it is possible that spongin or its prototype (e.g., disseminated mesenchymal fibers) was evolved before megascleres in either some or all instances; a keratose sponge could then evolve by loss of the spicules at this stage, without spongin ever having connected spicules. If spicules were evolved in some primitive forms but not others, modern keratose sponges could be forms in which spicules have never existed at any stage of phylogeny. These different postulates could represent the histories of the Dictyoceratina and Dendroceratina, which are grouped biochemically with the megasclere-bearing Desmacidontida and the askeletose Halisarcidae, respectively.

Lithistids appear to have arisen from nonlithistid sponges by conversion of one category of choanosomal megasclere into desmas. If this is correct, their primary pattern of megaspiculation must depend on the kinds of megascleres present in their nonlithistid prototypes. There are four main possibilities, assuming that the prototypes of desmas are the principal choanosomal megascleres.

1. A choristid with choanosomal calthrops and no distinct triaenes gave rise to a lithistid with tetraxial desmas but no triaenose dermalia. An additional oxea, not converted into desmas could persist as a supplemental oxea.

2. A choristid with choanosomal calthrops and ectosomal triaenes differentiated gave rise to a lithistid with tetraxial desmas and triaenes.

3. A choristid with tetraxial megascleres restricted to ectosomal triaenes and with all choanosomal forms monaxons (e.g., oxeas) yielded a lithistid with triaenes and monaxial desmas.

4. A lithistid derived from a purely monaxonid sponge had monaxial desmas and perhaps supplemental monaxons but no triaenes.

These predicted patterns correspond with those of various major groups of lithistids as well as some sublithistid sponges.

a. The Tricranocladina (or Hindiidae) could be lithistids of the first sort if their desmas were tetraxial, since no other megascleres but oxeas appear to have been present. The small size of the desmas could imply a microspiculate prototype, but this would not alter the relationship envisaged.

b. The ontogenetic development of typical tetraclones from an initial calthrops suggests the second mode of origin for the triaene-bearing Tetracladina.

c. Although ectosomal triaenes are present, the monaxial desmas of the Helomorina and Megamorina (heloclones, megaclones) have no sign of derivation from tetraxons and are probably ophirhabd derivatives. The desmalike spicules of the sublithistid *Helminthophyllum* SCHRAMMEN, which are accompanied by triaenes, are also probably of monaxon origin.

d. The Rhizomorina and Orchocladina have monaxial desmas but lack triaenes and have nothing to suggest derivation from tetraxon-bearing sponges. The Anthaspidellidae of the Orchocladina have a structure suggesting derivation from a chalinoid monaxonid, corresponding with forms now seen only among the monaxonid Chalinida and Desmacidontida. The modern *Desmatiderma* TOPSENT, *Helophloeina* TOPSENT, and *Lithochela* BURTON are sublithistid Desmacidontida with characters intermediate between monaxonids and lithistids.

On the other hand, these are not the only possibilities. In Tetracladina, the typical tetraclones may be accompanied by variants with triactinal, diactinal, or monaxial crepides found in varying abundance, of which forms with triactinal crepides may outnumber or entirely replace true tetraclones. A similar replacement of tetraclones by related monaxial desmas could result in evolution of a sponge with monaxial desmas and triaenes, not produced in the manner envisaged above, as in the Helomorina for example. This mode of origin is suggested for the Dicranocladina by their general resemblances to discodermiid and similar Tetracladina and by the presence in *Macandrewia* GRAY of tetraxial variants of the normally monaxial desmas. The two groups are not sharply separable when this genus is considered.

Anaxial desmas have no evident nonlithistid prototypes, but progressive reduction of axial structures in transitions from monaxons to desmas (e.g., in *Desmatiderma*) suggests that the anaxial type of desma has a culminating stage of this process. This could then be the origin of the Sphaerocladina. If the shortening of the axial shaft seen in transitions from dendroclones to chiastoclones of the Orchocladina is supposed to represent progressive shortening of monaxon crepis, the sphaeroclone could then be essentially a modified chiastoclone with an ennomoclonar shape correlated with growth of the skeleton in layers.

Thus, in these instances, origins of some groups of lithistids can be pictured as involving a secondary change in the desmas, whose monaxial or anaxial character does not represent that of a nonlithistid megasclere. A further possibility for secondary change is loss of triaenes, producing forms with tetraxial or monaxial desmas but no triaene dermalia. This is possibly represented in the Didymmorina, in which triaenes that appear to be intrinsic have so far only been found in one specimen of *Cylindrophyma milleporata* (GOLDFUSS).

Simple triaenes and dichotriaenes of lithistids are indistinguishable from those of choristids; but phyllotriaenes and discotriaenes have no choristid counterparts. These forms seem to represent a special lithistid modification of the ectosomal skeleton, resulting in production of an armor of overlapping megascleric scales. At the end of this sequence, discotriaenes with rudimentary cladi intergrade with monaxial discostrongyles, with an axis in a rhabdal ray only. These are not of separate origin, since their intergradation may be seen in the spicules of one specimen. This presumably represents progressive reduction of the crepis from tetraxial to monaxial unless a normal dichotriaene can be derived from a monaxial disc. The apparently anaxial dermal plates of *Plinthosella* VON ZITTEL suggest an ultimate stage in this sequence.

This suggested picture of spicular evolution can explain all the principal developments seen in demosponge megaspiculation. There is no certainty of its correctness because of its comparative basis, and other pictures are possible. In particular, some parts could require to inversion if tetraxial megascleres can originate as triaenes derived from monaxons. It is, therefore, essentially a personal impression of what seems currently most likely. On the other hand, there is no current evidence that development of tetraxons from monaxons is in any case more than simply possible, except at the most primitive level in ancestral microspiculate Demospongea.

Whether one looks at biochemical or traditional spicular evidence, by far most modern monaxonids have no evidence of special relationship to choristids, of a type implying origin of either from the other. There are especially no known choristids that have features suggesting derivation from the typically monaxonid orders (Spirastrellida, Axinellida, Chalinida, Desmacidontida; or, in other nomenclature, Hadromerina, Axinellida, Haplosclerida, Halichondrina, and Poecilosclerina). This suggests strongly evolution on the lines first suggested by SOLLAS (1888), with choristids and monaxonids arising by selection of different types of spicules for enlargement into megascleres. This could very well apply to even most supposed epipolasids (*sensu* SOLLAS), of which many coppatiids could, in fact, share no more with ancorinids than a common microspiculate ancestry.

The occurrence of forms in which triaenes are abnormally rare is also clearly comparable with instances in which various other types of spicules (e.g., monaxonid acanthostyles, chelas; spicules of pseudoceratosa) are abnormally uncommon or absent in some species, populations, or individual sponges. These instances are generally regarded as showing a process by which categories of spicules can be lost in phylogeny and cannot be interpreted differently unless the spicules concerned are all supposed to be arising *de novo.* By comparison, the choristids with rare triaenes seem almost certain to have a condition that would lead ultimately to production of monaxonid sponges and not an opposite process of production of choristids from monaxonids. In the lithistid sequence from dichotriaenes to monaxial discs, it seems almost certain, again, that the spicules have evolved from tetraxon to monaxonids and not in the opposite direction. Unless these comparisons are misleading, the development of some long-shafted triaenes from monaxons in ontogeny is probably cenogenetic and related to hypertrophy of their rhabdomes, not a feature implying that triaenes have arisen from monaxons in phylogeny.

As stated, however, this is the opinion of the author, from which others are entitled to differ. All that can strictly be insisted upon is that any assessment or opinion should always consider all the relevant possibilities and evidence and should always be adjustable, if necessary, in the light of new evidence.

As a footnote, there is some possibility that some nonlithistid sponges could be forms derived from lithistids by reduction or loss of the desmas. For example, *Crambe* VOSMAER of the Desmacidontida has a mainly monaxonid spiculation but possesses basal desmas that resemble those of some Sphaerocladina. Since Sphaerocladina may contain supplemental monaxons, *Crambe* might show a stage in reduction toward a purely monaxonid condition, although this does not seem probable.

PHYLOGENY

As emphasized already, no reliable picture of demosponge phylogeny can be given at present because of the virtual restriction of critical evidence to modern comparative data. The following suggestions are based on skeletal, biochemical, and embryological data on the general basis that the skeleton has evolved along the lines suggested above.

1. It seems likely that the primitive Demospongea were simple microspiculate sponges, some of which at least had spicules that varied meristically from diactines to simple polyactines. Others may have had diactines only, either as a primitive condition or because other forms were eliminated.

2. The orders Plakinida through Axinellida (i.e., LÉVI's Tetractinomorpha) may have descended from prototypes with spicules that varied meristically. The Plakinida, then, are to be regarded as persistently primitive in having only small spicules not differentiated into typical megascleres and microscleres but in having some differentiation, with specialized features that do not occur in typical choristids (candelabra spicules, amphiblastula embryos), and they may include forms that are not, in fact, closely related. The Poecillastrida and Ancorinida appear to be stocks of common origin, both possessing true megascleres but with different types of characteristic microscleres. The Craniellida seem to be a separate series, not related to other forms with megascleres.

3. Some monaxonids grouped here as Epipolasida (Coppatiidae) are either derived from choristid Ancorinida by loss of triaenes or other tetraxons or from allied forms with megascleres developed from monaxons only. Other Epipolasida seem to belong with the Spirastrellida (Tethyidae) or Axinellida (Epallacidae, Sollasellidae).

4. The Spirastrellida and Axinellida are two series of primary monaxonids with megascleres developed from diactines only. Euastriform microscleres are typically absent, presumably through loss in phylogeny, but have persisted in the Tethyidae and Epallacidae if these forms are included. Spirastrellida can then be pictured as including two divergent series: (a) the Tethyidae, with euasters retained and developed in parallel with those of choristid Ancorinida; and (b) other forms, with euasters lost and sometimes replaced by secondary microscleres (microrhabds, spinispiras) derived from megascleres.

5. The Chalinida and Desmacidontida are primary monaxonids, derived from primitive sponges with diactinal spicules only. According to LÉVI, their spicules were evolved independently of those of Tetractinomorpha, but this need not be correct. These forms and the keratose sponges were derived from a stock in which incubated parenchymella were an early development, and some forms may have lacked or lost spicules before any developed megascleres. The Chalinida include two main stocks with different skeletal features that are, nonetheless, similar biochemically and in having generally simple spiculation. The Desmacidontida are more nearly allied to the dictyoceratid Keratosida and are distinguished from Chalinida by more specialized spiculation and the occurrence of chelas. The reticulate structure of Dictyoceratida suggests that they once possessed spicules, although this need not be conclusive. They could also be polyphyletic, for example, if various Desmacidontida lost their spicules in different periods. The Dendroceratida and Halisarcidae may be persistently primitive or degenerate.

6. The Astrosclerida have features (verticillate spicules, clavidiscs) suggesting affinities with Axinellida or Chalinida but are too poorly known for further comment.

7. The Lithistida are polyphyletic, and their relationships are largely unknown. The following derivations are suggested.

a. Tetracladina: from pachastrellid Poecillastrida, with triaenes and choanosomal calthrops by conversion of calthrops into tetraclone desmas.

b. Dicranocladina: from Tetracladina by reduction of the crepides of desmas from tetraxons to monaxons; source stock most nearly represented by Jurassic Sontheimiidae, and probably shared with Cretaceous to recent Discodermiidae.

c. Didymmorina: unknown but perhaps as suggested for Dicranocladina.

d. Helomorina, Megamorina: from a choristid with ophirhabds and triaenes;

perhaps from a stock also leading to theneid Poecillastrida.

e. Tricranocladina (=Eutaxicladina sensu SCHRAMMEN): unknown; from an early microspiculate choristid if *Hindia* desmas were tetraxial.

f. Orchocladina: from a chalinoid monaxonid, perhaps of order Chalinida.

g. Rhizomorina: in part at least from Orchocladina by replacement of dendroclones by rhizoclones but may also include convergent forms of different (e.g., craniellid, axinellid) origin.

h. Megarhizomorina: unknown; but possibly axinellid, if regarded as including, for example, *Petromica* TOPSENT of extant forms.

i. Sphaerocladina: unknown but possibly from chiastoclonellid Orchocladina by conversion of chiastoclones into sphaeroclones and astroclones.

DEMOSPONGE PHYLOGENY

R. E. H. REID

[formerly of Department of Geology, The Queens University of Belfast]

INTRODUCTION

Two treatments of demosponge phylogeny are presented in this volume, one by FINKS (p. 63, Paleozoic Demospongea: Morphology and Phylogeny) and the present chapter. The views expressed may differ radically, but this serves to emphasize how little is really known and how subjective are the expressed views. FINKS (1967a, 1971b) has accepted the concept that tetraxial megascleres can arise from monaxial megascleres, a concept not accepted herein. In addition, FINKS's discussions (1967a, 1971b) are centered on single genera, whereas, herein the approach is to begin from the totality of all available evidence and try to work inward toward some acceptable common starting point. Moreover, while genera relied on by FINKS are fossils, herein the belief is that only modern forms, excepting some lithistids, are well enough known to be a basis for useful speculations.

In the following text, nonlithistids and lithistids are discussed separately because they present different problems.

NONLITHISTID PROBLEMS

Fossil and modern nonlithistids present different problems of interpretation. The sparse and sporadic fossil record is probably very incomplete, and nearly all fossil material is incomplete. Modern forms present problems because phylogenetic end forms from a single time plane can yield no objective indication of phylogenetic directions. These problems interact, because the fossils throw almost no light on what courses phylogeny may have followed.

Starting with the fossils, one may note first that the total of recorded fossil genera is much smaller than the total known from modern seas, while the total of purely fossil forms is smaller still. This is an unlikely picture of the true relative abundance of nonlithistids in the past and the present; and it is, at least in part, due to the spicules being loose in the tissues and hence scattered after death. Indeed, loose spicules from some deposits (e.g., cricorhabd from the Upper Jurassic and Cretaceous or trachelotriaenes from the Lower Carboniferous and Upper Jurassic) have occurrences of sponges not known from associated megascleres; and there are frustratingly rare glimpses of faunas that were probably as large as any modern ones (e.g., the lower Tertiary of Oamaru, New Zealand: HINDE & HOLMES, 1892).

A further major cause of the poor fossil record is that only some formations yield material, while in other formations fossils do not occur at all. Those that do yield material are mostly limestones, such as the Irish Glencar Limestone (Visean) and the Malm and Chalk of Germany, and some sandstones (e.g., the Upper Greensand of southern England); and, even here, remains may be limited to drifted spicules occurring in only some localities. The problem seems to be that even large siliceous megascleres are preserved only under certain physicochemical conditions and otherwise are either dissolved before fossilization can occur or rendered inextractable by later calcification. But, nonlithistid sponges almost certainly lived in many environments whose sediments have no trace of their spicules. In the Oxford Clay of England, for instance, large *Gryphaea dilatata* may be riddled with borings of the type made by *Cliona,* although spicules of the latter do not occur; and, while much of the English Chalk has no trace of spicules, the internal meal of hollow flints may be packed with them. Thus the record plainly has been subject to diagenetic deletions, and there probably have been many species that have left no trace whatever.

Because the record is restricted mostly to forms with large megascleres, this problem also goes farther, for microscleres and small megascleres are found only in rare deposits (e.g., the Oamaru sponge earth) or in protected situations (e.g., in flints). This has several consequences. First, it makes it likely that another large segment of the nonlithistid record is missing and that the apparent order in which major groups first appear in the record may depend upon when they first evolved large megascleres. The stratigraphic order of such apparent first appearances, thus, cannot be trusted as a key to phylogeny, quite apart from the possible effects of fortuitous preservation.

Second, virtual restriction of associated spicules to megascleres means that no fossil species can be referred certainly to a modern taxon, at even the generic level; and this problem is complicated further by occurrences of similar megascleres or similar skeletal architecture (e.g., radiate, plumose, or reticulate patterns) in members of different modern orders. This makes it doubtful how far back modern orders can be traced, when possible representatives might belong to more than one of them, and similar megascleres or architecture might also have evolved in extinct groups.

Third, another result is a very sparse record of microspiculate sponges, from which forms with megascleres probably derived (DENDY , 1921). Their first known occurrence in the Lower Carboniferous (REID, 1970) is probably a result of fortuitous preservation with no bearing on demosponge phylogeny.

Last, a further defect in the record is its almost complete restriction to spicular material. There are very few sure records of keratose sponges and none of dendroceratids or askeletose myxosponges. One might suspect that cementing spongin was present in the Burgess Shale monaxonids, for example, because they have well-preserved, complex, reticular arrangements of monaxial megascleres (in e.g., *Takakkawia* WALCOTT, 1920); but this excellent preservation also could have been due simply to the absence of physical disturbance and scavengers. There is also a complete lack of embryological data, used increasingly in zoology since its introduction by LÉVI (1957b), and of biochemical data of sorts used by BERGQUIST and HARTMAN (1969), for example. The first deficiency is especially relevant. For instance, were the Burgess Shale monaxonids oviparous sponges to which oviparous choristids could they be allied, or were they viviparous Ceractinomorpha with no bearing on choristid ancestry?

In summary, it is evident the fossil record is highly lacunar, recording only a tiny minority of past species and mostly forms whose relationships are either uncertain or unknown. Even the order in which they appear stratigraphically may depend upon the size of their spicules or simply on fortuitous preservation. Such a record cannot provide an adequate basis for assessing phylogeny, using any of the arguments traditional in paleontology. It could be highly misleading, for instance, to assume that some apparently younger group must have evolved from some apparently older one. In fact they may have evolved in the opposite order, without such a development being evident, or perhaps neither arose from the other. For such reasons, working backward from data provided by modern forms seems preferable, in general using the present as a key to the past. The only exception, in which the past could be the key to the present, is in the idea (e.g., VACELET, 1979, 1981) that monaxonid groups evolved polyphyletically from sclerosponges, by losing the calcareous basal skeleton. In this instance, the fossil record provides suggestive evidence (WOOD, REITNER, & WEST, 1989).

Some minor problems can also be noted here. First, some supposed fossil records of extant genera have been based on loose spicules, which may not represent the genus claimed. Some are highly dubious, as for example the records of *Thenea* GRAY, 1867, or *Geodia* LAMARCK, 1815, which have been based on triaenes, and might belong to these genera or to others. Other records have been based on apparently distinctive spicules, such as the aspidasters of *Erylus* GRAY, 1867, the discasters of *Latrunculia* BARBOZA DU BOCAGE, 1869, or the diancistra of *Hamacantha* GRAY, 1867. While these are less doubtful, there is no conclusive evidence that the sponges represented could be referred to these genera if known fully. Second, there can also be problems if microscleres appear to be preserved as well as megascleres. For instance, sterrasters occur in examples of *Discispongia* KOLB, 1910 in 1910–1911, that have associated megascleres and have been thought to belong to that genus (e.g., by SCHRAMMEN, 1936); but they may also occur within the skeletons of other Malm sponges, including hexactinellids in which they are presumably foreign or intrusive. It is therefore not certain that these spicules belong to *Discispongia.*

In modern forms, in contrast, a large amount of information on soft parts and spicules is available, due mainly to the work of such authors as VON LENDENFELD, RIDLEY, SOLLAS, HENTSCHEL, DENDY, TOPSENT, and their successors, but also in part to older authors (e.g., GRAY, BOWERBANK, CARTER). Until recently, classification has been based almost wholly on their spicular data; but in the last few decades embryology and biochemistry have made new evidence available and will continue to do so as more forms are investigated. Again, until recently, speculations on phylogeny have been centered on the spicules, most notably by DENDY (e.g., 1921); but, since 1957, embryology has assumed a major role, with LÉVI's (1957b) picture of an early division of the class into two major groups (his Tetractinomorpha and Ceractinomorpha) on this basis.

Most such data, however, are badly flawed as a basis for assessing phylogeny, because all modern species are phylogenetic end forms and because comparative study of species that are effectively all from one time can yield no objective indications of phylogenetic direction. Some examples illustrate this problem.

a. *Oscarella* VOSMAER and *Plakina* SCHULZE, 1880, are simple sponges with similar soft parts and unique amphiblastula larvae. *Oscarella* is askeletose, but *Plakina* is microspiculate. Does *Oscarella* represent a stock from which *Plakina* has arisen, or is it a plakinid that has lost its spicules?

b. "*Trachygellius cinachyra*" DE LAUBENFELS (1936) was thought to be purely monaxonid by DE LAUBENFELS but was later found to have a few triaenes by LITTLE (1963). Is this species a choristid in process of losing triaenes or an ex-monaxonid in process of becoming a choristid?

c. Various pseudokeratose monaxonids and the sclerosponge *Astrosclera* LISTER can occur in forms with or without spicules. Are they in process of losing or acquiring them?

d. The sclerosponge *Merlia normani* KIRKPATRICK can occur in forms with or without a calcareous basal skeleton. Is it a monaxonid in process of becoming a sclerosponge, or a sclerosponge becoming a monaxonid?

e. LÉVI's (1957b) concept of a subclass Ceractinomorpha was based on the presence of an asconoid larva in one of two sibling species of *Halisarca* DUJARDIN, this being assumed to be more primitive than a normal secant rhagon seen in the other. But what evidence is there that the asconoid form is not simplified rather than primitive?

f. The simplicity of such genera as *Oscarella* and *Halisarca* has led to their being seen widely as primitive; but is their simplicity a retained primitive condition or due to paedomorphic simplification?

g. In some choristids, long-shafted triaenes arise from monaxons in ontogeny, through the cladi being added at one end of an initially monaxial shaft. Does their ontogeny recapitulate phylogeny or have cenogenetic delay in formation of the cladi?

In none of these examples can the question be answered except on the basis of inference.

Different kinds of comparative evidence can also point to opposite conclusions. Most notably, LÉVI (1957b) pictured the initial evolution of his Ceractinomorpha as proceeding from askeletose *Haliscara*-like sponges through the keratose dendroceratids to dictyoceratids and, thus, in the direction of increasing morphological complexity; but the only oviparous forms among these sponges are the dictyoceratid Verongiidae, all the others including *Halisarca* having incubated parenchymellae. Assuming that oviparous sponges are unlikely to be derived from viviparous ones, embryology then points to the simplicity of *Halisarca* being secondary and not primitive. This problem cannot be solved by removing the verongiids, as suggested by BERGQUIST (1978), unless dictyoceratids have evolved from two different sources.

In any case, no modern species can have any direct bearing on the origins of major clades in demosponge phylogeny. All are phylogenetic end forms that may retain clues to their ancestry but are living long after all major divergences must be judged to have occurred. We do not know the dates of these divergences; but chela microscleres have the Desmacidontida as existing in the Late Cretaceous, for example, and sterrasters have the most specialized Ancorinida (the Geodiidae) as present by the Late Jurassic. Further, if reticulate skeletal architecture is trusted as evidence of affinity, a ceractinomorph stock must have existed by the Cambrian. Use of modern forms as keys to phylogeny, hence, needs very cautious treatment.

Moreover, if any pair of modern orders is thought to have evidence of common origin, they must still have had independent histories since divergence occurred, even if one is thought to retain primitive characters. For instance, the living plakinids may represent the type of sponges from which choristids with large megascleres were evolved, as SCHULZE (1887a) and DENDY (e.g., 1921) thought; but, if so, they have had a separate history since at least Ordovician times. Finally, it has to be realized that living forms may give a distorted picture of the overall radiation of nonlithistids as it would be seen if past faunas were known as well as modern ones. It has long been known, for instance, that modern lithistids give only a glimpse of even the Cretaceous diversity of these sponges; and there is an emerging possibility (VACELET, 1981) that monaxonid groups with no obvious evolutionary precursors may have arisen from sclerosponges by loss of the calcareous skeleton.

The extant nonlithistid fauna is, thus, also defective as a basis for assessing phylogeny, despite the fact that most speculations by zoologists are based on it. It can, of course, be hoped that cladistic analysis of biochemical data will provide clearer answers when enough forms have been studied; but even this method could be misleading. Most vertebrate specialists, for instance, do not see cladistics as proving a common origin for birds and mammals but only that extensive biochemical resemblances can arise by convergence. In the instance of birds and mammals, the fossil record is good enough to confirm their separate origins, but nonlithistids have no comparable record against which biochemical results can be checked.

In consequence, reliance still has to be placed mainly on comparative methods and on subjective assessment of which of any two alternative hypotheses is more probably correct. For example, *Merlia normani* could be in the process of either losing or gaining a calcareous accessory skeleton; but since such a complex structure, which resembles a cerioid tabulate skeleton with centrally perforated tabulae, seems an unlikely product of a single *de novo* mutation, a process of loss seems more likely. As another instance, choristids whose microscleres include euasters, streptoscleres, or both are least distinct in form with only calthrops as megascleres (pachastrellids) and most distinct in those with only long-shafted triaenes and no calthrops (theneids and geodiids). If divergence of forms with euasters and streptoscleres, respectively, from a single source is then thought more likely than convergence to pachastrellids from two sources, the direction of spicular phylogeny implied is from calthrops to triaene and not from triaene to calthrops. Furthermore, if that is correct, the development of some triaenes from monaxons in ontogeny must be cenogenetic and not reflect their phylogeny. Such judgments must also be weighted in terms of the quantity of evidence available. For these choristids, there are enough modern species available for comparative study for the conclusions just stated to be rated as probably correct. While the evidence from *M. normani* does point to loss of the calcareous skeleton, however, it does not demonstrate that all monaxonids to which sclerosponges have affinities have arisen in this manner.

HISTORICAL BACKGROUND

This section does not attempt to cover fully all the relevant literature, but as a background to the interpretation presented below outlines main lines of thought that some authors have followed.

First, in the late nineteenth and early twentieth centuries, zoological speculations on lineages were based mainly on the idea that all demosponge spicules are derived from the regular tetraxon or calthrops. This view first clearly emerged in work by SCHULZE (1880, 1887a), who saw intergradations in spicules of *Plakina* SCHULZE as showing that diactines of that genus arose from calthrops. In this and similar forms, intermediates are apparently reduced by one or two rays occurring as rudiments only, and no triaenes are present. This led SCHULZE to postulate the origin of monaxonids from choristids (his order Tetraxonia) and that of keratose sponges from monaxonids (Fig. 38). He explained the tetraxial form of the calthrops as functionally adapted to supporting tightly packed, globular, flagellated chambers arranged in a tetrahedral manner (Fig. 39). This implies that the calthrops arose as a choanosomal spicule and, hence, that triaenes are derivatives adapted to supporting the ectosome.

SOLLAS (1888), in contrast, thought that tetraxons could arise from monaxons in phylogeny, as some triaenes do in ontogeny. While following this view in interpreting

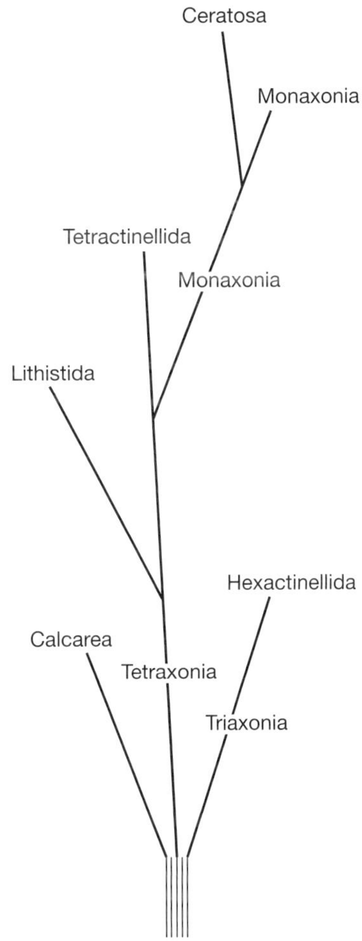

FIG. 38. Schulze's view of sponge phylogeny, with regular tetraxons assumed to be the basic type of demosponge spicule (adapted from Schulze, 1887b).

spicular phylogeny, he was unsure of its validity, however, and took a different view in interpreting demosponge phylogeny. After noting his treatment of the spicules as hypothetical, he continued: ". . . there is a good deal to be said for an opposite hypothesis which would derive the triaene from the microcalthrops . . . indeed, the simplicity which at once follows the adoption of this view is so great that nothing but the stubbornness of the ontological data prevents me from adopting it" (1888, p. lxxi–lxxii). In view of this, he suggested two alternative pictures of spicular relationships, with

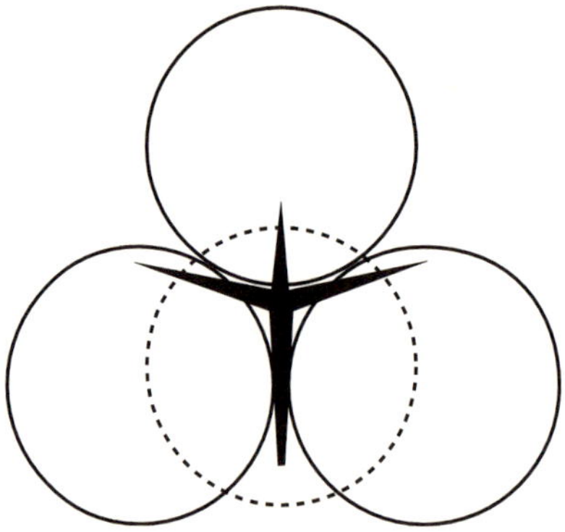

FIG. 39. Schulze's concept of the regular tetraxon as evolved to fit between closely packed, globular, flagellated chambers (adapted from Schulze, 1887b).

phylogeny proceeding in opposite directions (Fig. 40); and he also saw monaxial desmas as derived from tetraxial prototypes, despite the evidence from triaenes. Thus, while SOLLAS was influenced by the then general idea that ontogeny recapitulates phylogeny, he did not apply it rigidly and recognized contrary evidence.

SOLLAS's (1888) taxonomic groupings are more important, as having formed the basis of concepts developed by DENDY (1905, 1921, 1924b) and HENTSCHEL (1909), as well as SOLLAS's view on phylogeny. SOLLAS divided choristids as follows.

1. Microsclerophora, for microspiculate genera.

2. Astrophora, for forms with megascleres and astrose microscleres (euasters or streptoscleres).

3. Sigmatophora, with megascleres and sigmaspires but never astrose microscleres.

The Astrophora were divided into a) Streptastrosa, with streptoscleres (streptasters of SOLLAS); b) Euastrosa, with euasters not including sterrasters; and c) Sterrastrosa, with sterrasters. In addition:

4. A few monaxonids (e.g., *Asteropus* SOLLAS), called Epipolasidae, were regarded as Euastrosa in which tetraxons (triaenes) had been lost.

5. Most other monaxonids were divided into Spintharopora and Meniscophora, distinguished by having astrose and sigmatose microscleres respectively.

A few others with no microscleres were placed in a further group Asemophora.

Much in contrast to his treatment of spicular morphology, SOLLAS (1888, p. clx) based his two main pictures of demosponge phylogeny (Fig. 41) on the view that the triaene is derived from a calthrops or microcalthrops. The two schemes differ only in the pacing of monaxonids with sigmatose microscleres (Meniscophora); and he also thought that these might have been derived from both the sources suggested. The only scheme he could suggest to allow triaenes to originate from monaxons required descent from primitive sponges with only sigmaspires (Fig. 42); and he recognized this picture as improbable because of its wide separation of the Microsclerophora and Astrophora. His assessment of demosponge phylogeny, thus, did not follow the pattern of his treatment of the spicules, as one might suppose from reading it.

DENDY (1905, 1921) followed SCHULZE (1880, 1887b) in regarding tetraxons as central to demosponge spiculation, but saw only a few keratose sponges (pseudoceratosa, DENDY) as derived from monaxonids. Besides evidence relied on by SCHULZE, he emphasized meristic variation in the number of spicular rays, occurring especially in *Dercitopsis* DENDY, as a key to spicular phylogeny. This pattern (a 2, 3, 4, 5, 6 . . . sequence) is sometimes repeated in variants of a megascleric calthrops and can explain the origin of polyactinal euasters. On this basis, SOLLAS's Microsclerophora, renamed Homosclerophora, were interpreted as primitive demosponges with spicules not differentiated into megascleres and microscleres; while his Astrophora and Sigmatophora were regarded as two derived stocks with megascleres, distinguished by different types of microscleres. In addition, all monaxonids, grouped as Astromonaxonellida (=Spintharophora SOLLAS) and Sigmatomonaxonellida (=Meniscophora + Asemophora, SOLLAS), were regarded as

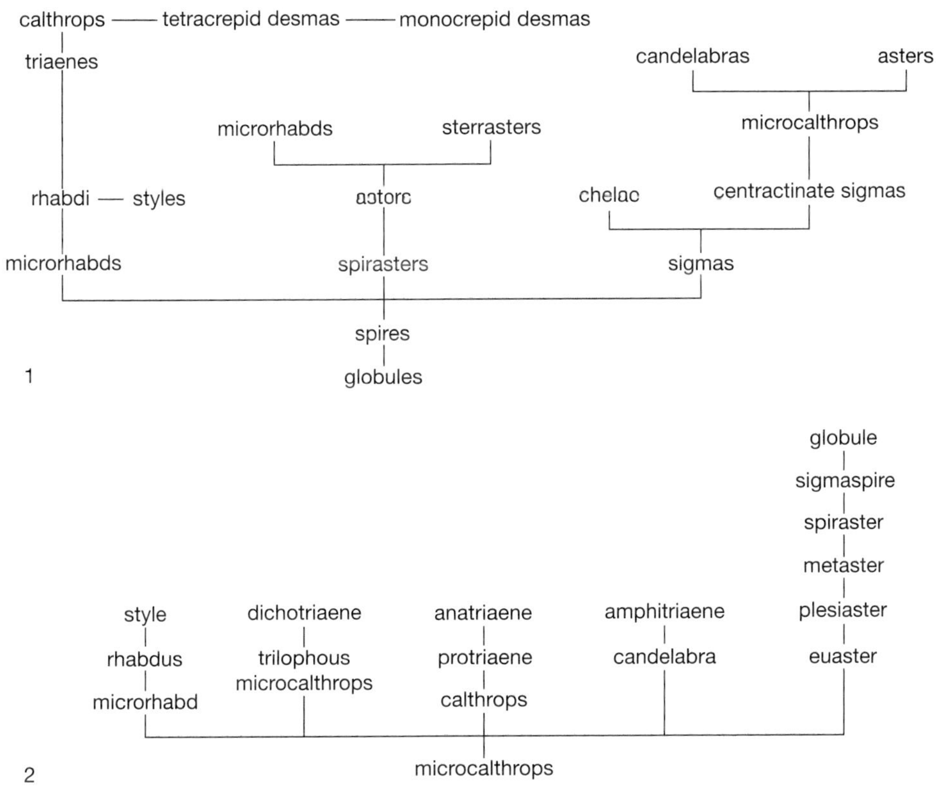

FIG. 40. Sollas's alternative interpretations of spicular phylogeny in demosponges, *1,* taking spheres as a starting point, *2,* and with microcalthrops; there are variants of modern sponges that have only microcalthrops and variants but none with only spheres (globules) (adapted from Sollas, 1888).

epipolasids, derived respectively from Astrophora and Sigmatophora (Fig. 43). DENDY thus differed radically from SOLLAS (Fig. 41), who regarded some or all monaxonids as derived from microspiculate sponges.

This picture of phylogeny led HENTSCHEL (1909) to place members of the two supposed main stocks into single taxa, named Astrotetraxonida and Sigmatotetraxonida. DENDY (1916, 1922) adopted this practice but moved forms with spinispira microscleres to the latter group on the grounds that their microscleres were related to sigmas and later (1924a) moved choristids with streptoscleres to a third group named Streptosclerophora. Figure 44 shows the pattern of skeletal evolution implied by this classification.

Reviewing the work of these authors, one may especially note SOLLAS's (1888) treatment as ambivalent and partly contradictory. Having written much of his test on the basis that ontogeny in triaenes is the key to spicular phylogeny, he first expressed doubts of this idea and then abandoned it in favor of seeing microspiculate sponges as ancestral to all forms with megascleres (Fig. 41). In these schemes, he is also to be noted as having regarded some or all monaxonids as derived directly from microspiculate sponges. DENDY, in contrast, was notable for treating all monaxonids as epipolasid derivatives of choristids with megascleres and for uncritical treatment of all diactinal (sigmatose) microscleres as the forms he took as diagnostic have had not less than four separate origins. Reaction against his views has also led

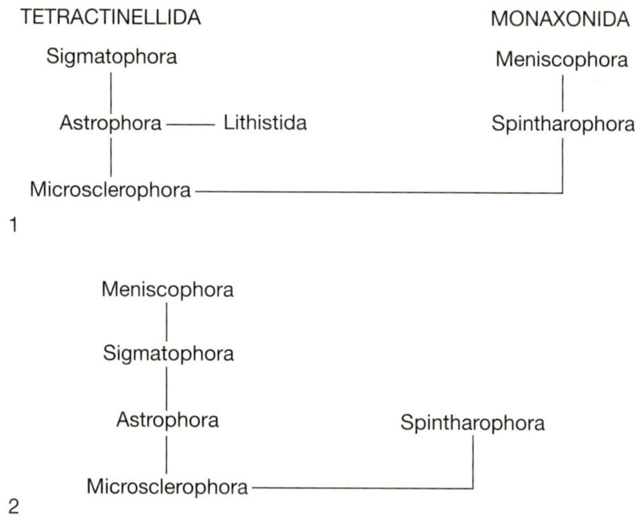

FIG. 41. Sollas's final preferred view of demosponge phylogeny, taking triaenes as derived from calthrops and not monaxons and some or all monaxonids as derived from microspiculate sponges (adapted from Sollas, 1888).

some later writers (e.g., DE LAUBENFELS, 1936; BERGQUIST, 1978) to abandon the use of microscleres in taxonomy.

BURTON (e.g., 1932) followed essentially DENDY, apart from changing names of taxa; but most later writers have used other methods. HENTSCHEL (1923–1924) himself switched to a variant of VOSMAER's (1887) classification, removing most of the sigmatose monaxonids to VOSMAER's Cornacuspongida; and this arrangement was followed by REZVOI, ZHURAVLEVA, and KOLTUN (1962) in the *Osnovy Paleontologii.* The author most followed by later writers, however, is TOPSENT (e.g., 1892, 1904, 1928b), who followed SOLLAS (1888) in grouping choristids and lithistids as Tetractinellida but placed monaxonids into a subclass Monaxonellida with orders Hadromerina, Halichondrina, Poecilosclerina, and Haplosclerina based on various features of the skeleton. These taxa, with their contents reshuffled, were adopted by DE LAUBENFELS (1936), who added a further monaxonid order Epipolasida and treated choristids as Choristida or Carnosa, based on the presence or absence of long-shafted triaenes. These changes were based on the idea that megascleres are more important than microscleres in taxonomy; and his treatment of choristids divides members of SOLLAS's Streptastrosa and Euastrosa between two orders. Regarding phylogeny, he noted various individual possibilities, but regarded most of them as unsubstantiated. The lithistids were seen as polyphyletic, as earlier by SCHRAMMEN (1910), and referred to various orders. The classifications of LÉVI (1973) and BERGQUIST (1978) are nearer to TOPSENT's, with an order Axinellida added and with subclasses based on LÉVI's picture of phylogeny (see below).

After DENDY's time, phylogenetic speculation stagnated until LÉVI (1957b) introduced new concepts based on embryological evidence (Fig. 45). After noting that various past authors had seen the Demospongea as comprising two main assemblages (e.g., VOSMAER, 1887: Spiculispongiae and Cornacuspongiae), LÉVI cited embryology as implying their separate descent from a very early stage of phylogeny. He began from the askeletose *Oscarella* and *Halisarca,* noting incubated larvae of different types. Those of *Oscarella* are hollow amphiblastulae, as in the microspiculate choristid *Plakina;* but those of *Halisarca* and all other forms whose larvae are known are solid parenchymellae. Further, those of VOSMAER's Cornacuspongiae

(=Meniscophora Sollas plus the keratose sponges) but no other forms were also incubated; and, while the rhagon of *H. dujardina* JOHNSON was asconoid, that of its sibling *H. metschnikovi* LÉVI was normally secant and like rhagons of keratose sponges. Oviparity, in contrast, was the rule in other forms whose reproduction was known. On this basis, he proposed subclasses Ceractinomorpha for *Halisarca* and conacuspongids and Tetractinomorpha for all other forms. In the former (Fig. 46, left), successive stages in evolution were represented by halisarcids, dendroceratids, dictyoceratids, and the spiculate orders. The Tetractinomorpha were divisible into three groups: a) true tetractinellids (=lithistids plus choristids with megascleres), with reproduction unknown; b) the Homosclerophora and *Oscarella,* with incubated amphiblastulae; and c) oviparous monaxonids grouped as Clavaxinellida (=Spintharophora, SOLLAS). In his figure (Fig. 46, right), the Homosclerophora appear as a central stock; but he expressed doubts of their having this status, and later (LÉVI, 1973) removed them to a separate subclass. The Clavaxinellida of this system revised SOLLAS's (1888) Spintharophora, which had the same contents (spirastrellids, axinellids). An incubated choristid embryo was figured by SOLLAS (1888, pl. 40,5), but this does not affect the main argument.

This classification has been widely accepted in zoology but still involves several problems.

1. The two *Halisarca* siblings are likely to be products of recent speciation whose divergence can have no direct bearing on that of the two main subclasses. Judged from evidence from fossils, this divergence was probably at least early Paleozoic and while there are Cambrian monaxonids that could be Ceractinomorpha, there is no way of telling whether these were viviparous or oviparous.

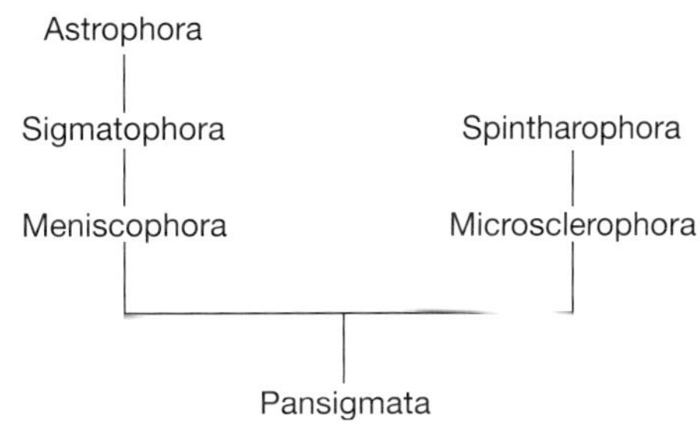

FIG. 42. Sollas's alternative to the scheme shown in Figure 41, deriving all spiculate forms from hypothetical sponges with sigmaspires only. This phylogeny was thought improbable because of wide separation of the Astrophora and Microsclerophora (adapted from Sollas, 1888).

2. The argument assumes anatomical progression from simple to complex and does not consider possible regressive evolution, for example, from dictyoceratids to halisarcids. Furthermore, while *Oscarella* and *Halisarca* appear primitive due to having simple soft parts as adults, they are less primitive than oviparous sponges in being viviparous. This could be due to their being regressive forms.

3. Although it is conventional to assume that an ascon is more primitive than a sycon, there is no objective evidence that an ascon cannot arise from a sycon. Furthermore, all that the *Halisarca* siblings show objectively is that embryological characters can be less stable than adult characters in phylogeny, even at the species level.

4. LÉVI's (1957b, 1973) picture of phylogeny requires the spicules of his subclasses to

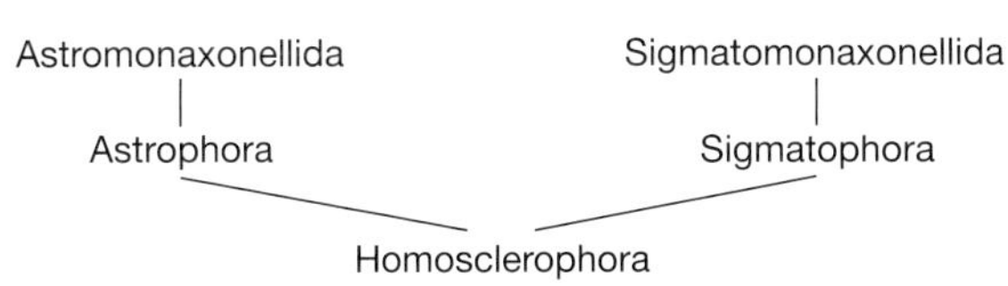

FIG. 43. Dendy's interpretation of phylogeny in spiculate demosponges, as rendered by Hentschel, 1909 (adapted from Dendy, 1905).

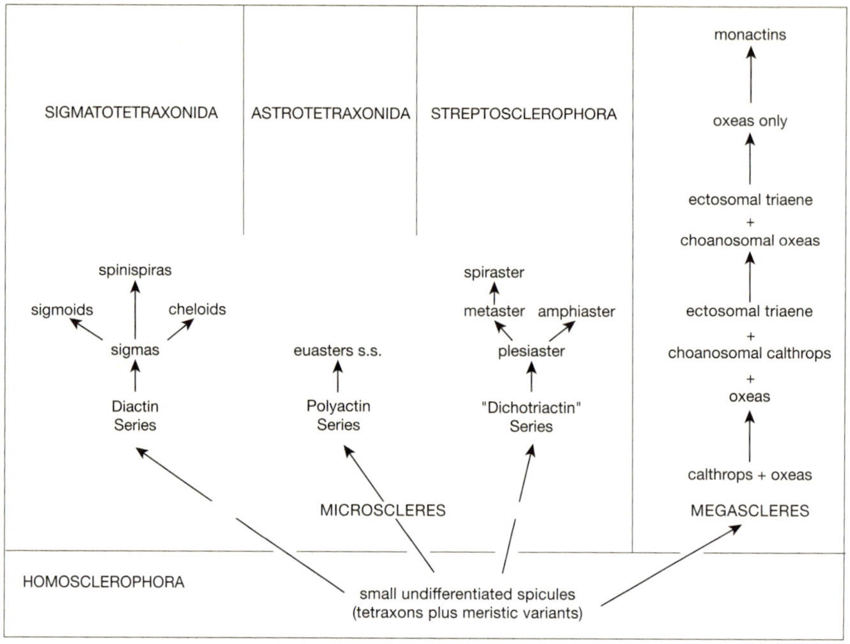

FIG. 44. Skeletal evolution in spiculate Demospongea as envisaged by Dendy in his final works (1921, 1922) (adapted from Reid, 1970; courtesy of Zoological Society of London).

have evolved independently. There is nothing in the character of the spicules themselves to suggest this, and conversion into desmas follows the same pattern in both groups.

5. The dictyoceratid verongiids are now known to be oviparous. This does not fit LÉVI's picture, unless dictyoceratids have evolved from two different sources.

6. There is no derived character by which a clade Tetractinomorpha can be defined, oviparity being a primitive character.

For reasons 1 through 4 above, REID (1968a, 1970) did not follow LÉVI in attempting a picture of phylogeny.

In the first of these papers (REID, 1968a), a critical review of the microscleres led to the conclusion that DENDY's (1921, 1924b) views were correct in some instances but mistaken in others. He was right in distinguishing his dichotriact series of microscleres (DENDY, 1924b), here called streptoscleres, from other so-called streptasters and in seeing various euaster-like spicules as

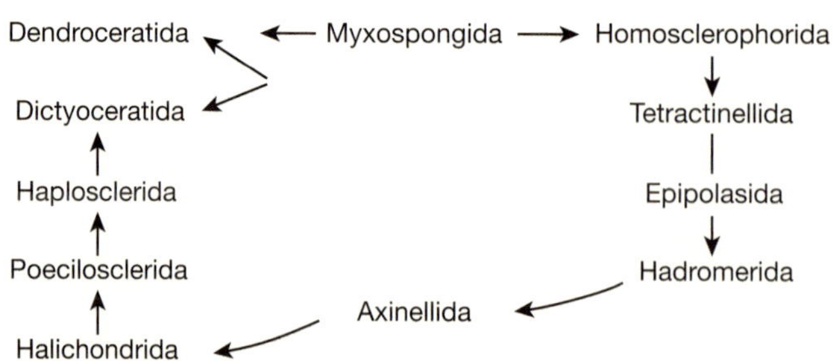

FIG. 45. Lévi's interpretation of how demosponge evolution had been viewed to that date (adapted from Lévi, 1957b).

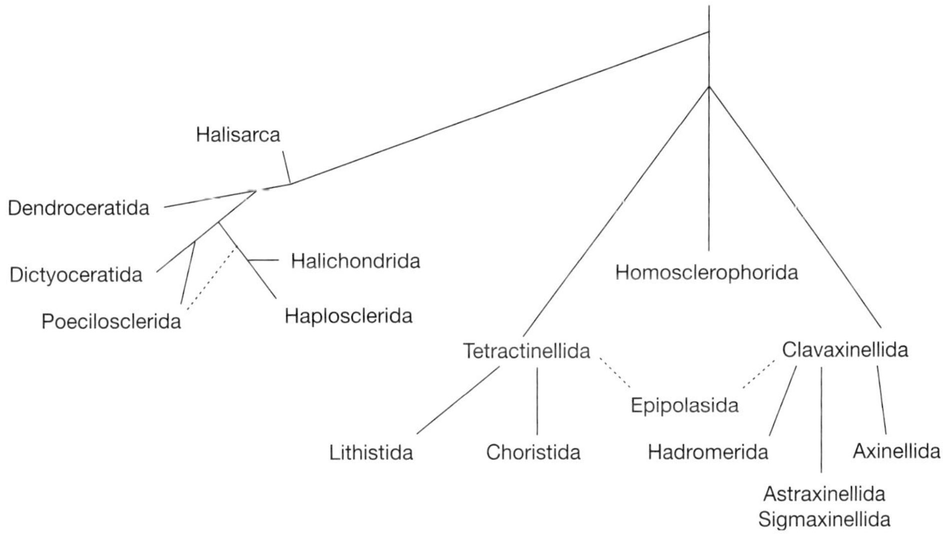

FIG. 46. Lévi's revised interpretation of demosponge phylogeny, based on embryological criteria (adapted from Lévi, 1957b).

pseudoasters; but he was badly mistaken in his concept of diactinal of sigmatose microscleres, which includes three major types with different origins (sigmaspires, spinispiras, sigmatoscleres) as well as various minor ones. Microspiculate sponges were thought the most likely prototypes of groups with megascleres, as by SOLLAS (cf. Fig. 41), with no assertion as to whether tetraxons or monaxons were primitive. DENDY's views were thought applicable to the plakinids and choristids with megascleres, in which microscleres were streptoscleres or polyactinal euasters; but choristids with sigmaspires, the Hadromerina of TOPSENT, and ceractinomorph group had no evident relationship to plakinids or astrophorous choristids. Forms with sigmaspires and the Hadromerina (Spirastrellida, REID) might be related to the euaster-bearing choristids (Ancorinida, REID); but this could not be asserted, and they could have arisen as primary monaxonids. The latter was thought to be probable for the spiculate ceractinomorphs, and keratose forms of halisarcids were thought to derive from them.

Biochemical studies by BERGQUIST and HOGG (1969) and BERGQUIST and HARTMAN (1969) on free amino acids showed further problems. These authors found a general homogeneity in the Ceractinomorpha, but not in the Tetractinomorpha, which fell into five biochemical groups and would be seen better as five separate orders (BERGQUIST & HARTMAN, 1969, p. 266). They also found no special affinity between spirastrellids and axinellids, grouped by LÉVI (1973) as Clavaxinellida and thought that even the Axinellida could be composite. In other findings, the monaxonid Epipolasida of DE LAUBENFELS (1936) were thought to belong partly with the choristids and partly with the monaxonid spirastrellids (Hadromerina in their terms). New placings of various genera were suggested, and the dictyoceratid verongiids were noted as differing from other ceractinomorphs. In a later textbook study, BERGQUIST (1978) used the term Tetractinomorpha taxonomically for a subclass but noted it as certainly polyphyletic. Her picture of ceractinomorph evolution is similar to REID's (1968a), with keratose forms and halisarcids derived from spiculate sponges. Most ceractinomorphs were seen as products of a relatively recent radiation, because of biochemical homogeneity, with verongiids

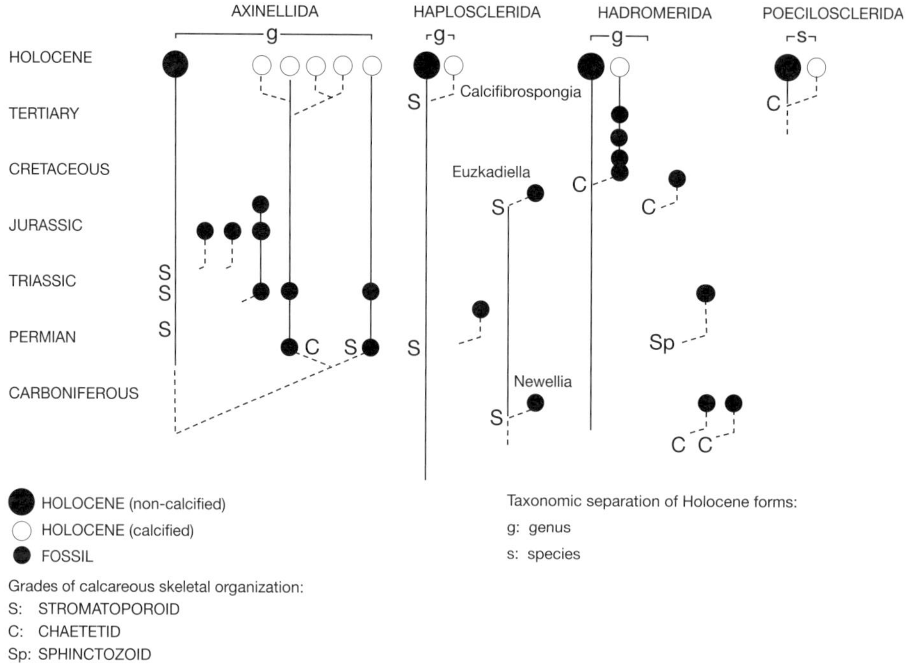

FIG. 47. Affiliation of modern and fossil sclerosponges as of 1990; for Haplosclerida, Hadromerida, Poecilosclerida, read Chalinida, Spirastrellida, and Desmacidontida (adapted from Wood, Reitner, & West, 1989).

an ancient stock with a long independent history. REID's parallel views were not mentioned, but are relevant in showing that critical study of the spicules and of microscleres especially had already led to similar suggestions. If RIGBY (1986a) is right in his placing of *Vauxia,* verongiids have existed since at least the Middle Cambrian.

In the same general period, two papers by FINKS (1967a, 1971b) revived SOLLAS's (1888) original idea that triaenes are derived from monaxons, in studies based on an Ordovician monaxonid and a Permian lithistid (*Saccospongia* ULRICH, *Scheiella* FINKS). FINKS has discussed this elsewhere in this volume (p. 77), but brief comments seem permissible here. First, *Saccospongia* could be an axinellid, an ectyonine ceractinomorph, or an unrelated convergent form with no modern relatives. Second, *Scheiella,* supposedly ancestral to dicranoclad lithistids, would need to have had monaxial desmas; but those of the related *Scheiia pumili* (HINDE), found by the hundreds in the Irish Glencar Limestone, appear to be entirely anaxial. Third, FINKS relied on no choristids being known before the Early Carboniferous; but even at that date, the diversity of their spicules (microcalthrops to long-shafted trachelotriaenes; REID, 1970) implied a much longer history, and the Ordovician record leaves only the Cambrian monaxonids as significantly older. If these are ceractinomorphs, moreover, they are not likely choristid ancestors.

The last major idea to emerge is that some or many monaxonids could have been derived from sclerosponges in phylogeny by loss of the calcareous skeleton (VACELET, 1979, 1981). This is clearly possible. The soft parts, spicules, and reproduction of known living examples are entirely like those of normal monaxonids; and *Merlia* KIRKPATRICK is known to occur in both sclerosponge and monaxonid forms (Vacelet, 1980). The modern forms are also clearly polyphyletic, having affinities to spirastrellids (hadromerids), axinellids, chalinids (haplosclerids), and desmacidontids (poecilosclerids) (Vacelet, 1985); and apparently allied fossils are known in each instance (Fig. 47), some dating as far back as Carboniferous. These are also all groups with no evident relationship to choristids. But, while related modern monaxonids could be descended from such sclerosponges, the latter could be calcified derivatives of uncalcified forms, which have not been preserved and have no modern descendants. A final verdict on these ideas is not possible at present.

Last, a final thread in demosponge phylogeny was identified by Vacelet's (1977b) discovery of a living sphinctozoid with demosponge soft parts. This line may go back to the Cambrian, where the oldest sphinctozoids are known; but their characteristic morphology is also known from Calcarea (*Sphaerocoelia, Tremacystia*), and which class most belong to is unknown.

CONCLUSIONS

In review of these various ideas, biochemistry, embryology, and microscleres are all seen as having useful data to contribute. Microscleres need critical assessment. Only streptoscleres and cheloids appear to be truly diagnostic and some (e.g., euasters) have parallel developments in different groups. Their total rejection by some authors has been as uncritical as Dendy's excessive trust in them, and some have not been understood. For instance, those of *Placospongia* Gray, called sterrasters by Bergquist (1978), make this type seem to occur in two different orders (Ancorinida, Spirastrellida). The spicules are sterrospirae, however, as stated by Dendy (1921), and indicate placing this sponge with the spirastrellids. Megascleres and skeletal architecture are unhelpful due to parallel developments in different groups, but restriction of typical dichotriaenes and large calthrops to Pachastrellida and Ancorinida fits with microscleric evidence of relationship.

Herein it is argued that only modern forms are likely to yield useful data despite their limitations and that no trust should be placed in stratigraphic sequence in fossils. There are various fossils that might be early members of modern groups, with a bearing on phylogeny; but they also might not be, and could appear in a fortuitous order. The only fossils that seem likely to be genuinely useful are the sclerosponges, but what they will yield remains to be seen.

With this background, the following suggestions are offered.

1. Judged from spicular evidence, the most likely prototypes of spiculate demosponges are oviparous microspiculate sponges, with spicules varying meristically in some but all monaxial in others. Choristids and monaxonids with megascleres should then represent various lines of descent from different parts of this spectrum, with some monaxonids never having had tetraxial spicules, as either megascleres or microscleres.

2. The Plakinida (Homosclerophora) can be seen as retaining a primitive condition in being microspiculate; but some of their features are not primitive (amphiblastula larvae, lophose calthrops), and they must have been separate from other choristids since at least the Ordovician.

3. The Pachastrellida and Ancorinida are probably related groups descended from sponges with calthrops megascleres and simple euasters, from which they have since evolved streptoscleres and polyactinal euasters, respectively.

4. The Craniellida, with sigmaspire microscleres, show no sign of relationship to other choristids and may have originated separately. In this group, the origin of triaenes from calthrops cannot be asserted since the latter are unknown.

5. The Spirastrellida have probably evolved independently of choristids from an early form with monaxon megascleres and simple euasters. In one stock leading to tethyids, polyactinal euasters were evolved;

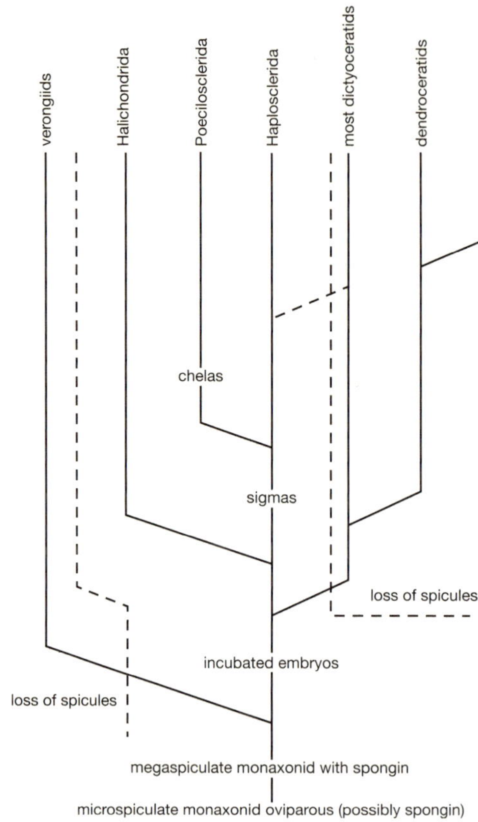

FIG. 48. Alternative interpretation of ceractinomorph evolution, using Topsent's nomenclature (1904, 1928b) (new).

but in the other, such euasters were never developed, and spinispiras arose from the megascleres.

6. The Axinellida are a second group of primary monaxonids, in some of which euasters or sigmas have evolved independently of those seen in other groups. This group may also be polyphyletic.

7. The ceratinomorph orders form a genuine clade and, except for verongiids, are likely to represent a later radiation than other groups. Their phylogeny could have followed the pattern suggested by LÉVI (1973) or some partly or wholly different one (e.g., Fig. 48).

8. The sclerosponges are polyphyletic demosponges related to the Spirastrellida, Axinellida, Chalinda, and Desmacidontida. Some fossil chaetetids and stromatoporoids are members of these taxa, but their relationships to modern forms are uncertain.

9. Some, many, or most fossil sphinctozoids may be demosponges.

Last, it might be worth noting that all groups of monaxonids include sponges with monactines, which are typically arranged with points distad, whereas, monactines are rare in choristids and have points proximad when seen. One wonders whether this could be significant.

HEXACTINELLIDA: GENERAL MORPHOLOGY AND CLASSIFICATION

R. E. H. Reid

[formerly of Department of Geology, The Queens University of Belfast]

INTRODUCTION

The Hexactinellida are sponges distinguished by siliceous spicules that are characteristically either simple orthotriaxial hexactines or variants of this type and by adult nonchoanocytal soft parts in the form of a trabecular **syncytium,** without either mesenchyme or pinacoderm. The characteristic spicules are either the only forms present or are sometimes accompanied by diactinal monaxons whose relationship to them is not always demonstrable. Tetraxial and polyaxial spicules do not occur.

Hexactinellida are recognized in paleontology by the form of the spicules or by some forms having characteristic skeletal frameworks formed by fused spicules (not united by zygosis, as in lithistid Demospongea). Skeletal remains occur certainly from the Lower Cambrian upward but also may occur in the somewhat older Ediacaran beds of Australia.

HEXACTINELLIDA AND HYALOSPONGES

The name Hexactinellida is preferred to hyalosponges, which was used by de Laubenfels (1955). Schmidt (1870) was the first to characterize the Hexactinellida as a distinct group of Porifera separated completely from Demospongea and stated the character of the spicules correctly. The taxon Hyalospongiae of Claus (1872), attributed by de Laubenfels (1955) to Vosmaer (1887), was intended to include lithistid Demospongea as well as Hexactinellida. The Hyalospongiae was based on the supposed diagnostic value of (a) the so-called fibrosiliceous skeleton thought by Bowerbank (1862) to exist in lithistid Demospongea and some Hexactinellida but now considered as imaginary and (b) the type of hexactinellidan framework known to Claus from *Euplectella aspergillum* Owen, 1841 (the Venus flower basket). Such a framework does not even occur in all species of *Euplectella.* The only other character mentioned was frequent occurrence of beardlike roots, which occur in only some Hexactinellida and demosponges. It was also claimed later by de Laubenfels (1958) that fusion of spicules, which he wrongly thought was regarded by Claus as the diagnostic character, is always significantly present at some stage in Hexactinellida. In fact, it is unknown from one entire subclass (Amphidiscophora Schulze), and not known from all members of the other.

de Laubenfels (1958) also included as hyalosponges the Heteractinida of this volume, which were unknown to Claus. These forms are clearly distinct from the Hexactinellida; their spiculation is not triaxial in basis, and their spicules were calcareous.

Finally, the spicules of Hexactinellida are no more glassy than those of Demospongea. They simply happen to be conspicuously glassy in some well-known glass sponges (e.g., *Euplectella*); equally glassy structures occur in some less well-known lithistids, e.g., *Theonella* Gray, which are also the forms that Bowerbank and Claus failed to separate from the Hexactinellida.

SOFT PARTS

Soft parts of adult Hexactinellida consist of a reticulate choanocytal membrane, which forms an internal layer of flagellated chambers, and inner and outer networks of syncytial filaments or trabeculae in which the chambers are suspended. Interspaces of the structure are filled by the external medium (i.e., seawater), which presumably circulates through them. The choanocytal membrane consists of flat choanocytes linked syncytially by lateral processes; the chambers formed by it are typically separate but joined together at their open (**apopylar**) ends by a trabecular **connecting membrane. Trabeculae** of inner and outer networks have no regular direction internally but form tangential networks at the two bounding surfaces. These are the **dermal** and **gastral membranes.** Here the terms dermal and gastral are used simply conventionally, referring to the external and paragastral sides of the wall without other homological implications. Trabecular parts immediately under the two bounding membranes, in which chambers are absent, are termed subdermal and subgastral respectively. The inner and outer networks are sometimes called the **epirhysome** and **aporhysome.**

It was argued by BIDDER (1929) that the special character of the soft parts point to the Hexactinellida originating from branching colonial Choanoflagellata, independently of sponges with a mesenchyme (Gelatinosa BIDDER). Hexactinellidan embryos are, however, much like those of Demospongea, with a cellular layer at the surface and a gelatinous interior. The post-embryonic change by which the adult type of soft parts must originate is more likely to represent their origin in phylogeny.

Hexactinellida may have a chamber system only or also possess a canal system, consisting of exhalant canals only or with additional inhalant canals. In the simplest instance, all chambers face the gastral membrane directly and are thimble shaped or modified by development of lateral diverticula. There may be no other special circulatory features, or enlarged intertrabecular lacunae may occur between the open ends (apopyles) of the chambers and the gastral bounding membrane or between their summits and under the dermal membrane. The condition of these sponges is essentially rhagonlike, although with partial transition to leuconlike if the chambers have diverticulation. More commonly a layer of simple chambers is locally protruded to form radial to branching diverticula, which may intercommunicate if their branches grow together. Occasionally, the chamber layer is convoluted to form labyrinths of chamber-lined passages, which may be open under the dermal membrane as well as on the gastral side. Interiors of structures formed by convolution of the chamber layer may form exhalant canals directly or have a trabecular lining around an intertrabecular canal, the wall of which may be formed by a canalar membrane similar to the dermal and gastral membranes. Exhalant canals may end under the gastral membrane or open directly through it because of secondary fenestration. The bounding membrane of the gastral surface may then appear to pass into the exhalant canals through becoming continuous with a lining **canalar membrane,** but the origin of this condition can be seen by comparison of young and old parts. Between diverticula of the chamber layer, there are usually corresponding inhalant canals formed as specially differentiated intertrabecular spaces. Again a lining canalar membrane may be present. These canals begin under the dermal membrane and normally never pierce it.

It needs to be noted that formation of the chamber and canal systems is not based on diverticulation or any sort of folding of the wall as a whole but on internal convolutions of the **choanocytal membrane** or the chamber layer only. There is never any outgrowth of the type seen in sycettoid Calcarea, and formation of inhalant canals involves no folding of any kind. These facts are important because various paleontologists have tried to explain features of fossil skeletons that represent the positions of chambers or canals as due to the wall being folded. This idea was first stated in the pioneer work of J. T. SMITH (1847, 1848), who attributed such features to longitudinal plications called plaits, which were supposed to fuse together laterally to produce canal-like spaces in the skeleton. Smith believed he was describing Bryozoa and had no knowledge of sponge canal systems, let alone that of the Hexactinellida, which were first described adequately by SCHULZE (1887a, 1887b). Despite this, the idea has persisted to the present, in the work of MORET (1926b), for example, who sometimes pictured the folding as running transversely. For correct interpretation, features related to the canal system must not be attributed to folding of the wall as a whole, and features related to true folding are not related to the canal system. True modes of formation of canalar interruptions of the skeleton are by progressive displacement of loose spicules as the corresponding soft features are enlarged (again demonstrable by comparison of young and old parts) or by direct interruption of developing skeletal meshwork in a growing marginal region.

Various Hexactinellida have soft parts that are secondary additions to the basic wall structure. Many that are fixed directly to the substratum have a secondary basal expansion, which may be purely trabecular or contain chamber-bearing structures. Some forms have terminal oscula covered by porous diaphragms, termed sieve plates; these are usually simply trabecular but may also contain flagellated chambers like those of the lateral wall (e.g., in *Aphrocallistes* GRAY). The living *Callicyclix zitteli* (MARSHALL & MEYER, 1879) has a body of dividing and anastomosing tubes enclosed in a peripheral capsule without flagellate chambers, and skeletal structures imply presence of a similar capsule in various fossils.

SPICULES

Spicules of Hexactinellida are formed from opaline silica with some mixture of organic matter, deposited around organic axial filaments or in various ways to form additional anaxial structures. They arise in all parts of the trabecular network and are secreted by multinucleate scleroblast-syncytia, in instances where their origin is known.

Spicules in which axial filaments extend to tips of normal rays, without raylike or branchlike anaxial prolongations, are called **holactines** and take names ending -actine. The central morphological type is the orthotriaxial hexactine, with six rays arranged as though following the axes of a cube and with their axial filaments meeting at a central axial cross. Primary variants of this type are spicules with fewer than six rays, linked with either hexactines or one another morphologically by intermediates with one ray or more short to rudimentary by comparison with the others. In many such spicules, a six-rayed axial cross is still present, with the undeveloped rays represented by axial rudiments. These forms, however, grade into spicules without axial rudiments, which may then be diaxons or monaxons. The spicules take names from pentactine to monactine according to the number of rays. Forms with four or three rays may, however, be either cross- or T-shaped stauractines (stauracts of some authors) and **tauactines,** with rays following two of the three triaxon axes only, or **orthotetractines** and orthotriactines with all three represented. Diactines may similarly be **orthodiactines,** with two rays at right angles, or rhabdodiactines with the two in line.

This group of variations is usually regarded as due to secondary reduction of a basic hexactine or similar reductions of its derivatives. It cannot be demonstrated that this view is correct, but reduction from a central type is more likely than convergence to it from multiple prototypes. MINCHIN (1905) thought that stauractines may have preceded hexactines in phylogeny because of their occurrence as the first spicules formed in ontogeny and supposed suitability for the support of thin-walled structures. This could be correct, but is not clearly demonstrable stratigraphically. The embryonic stauractines are also homologically members of one division of the adult megascleric skeleton (autodermalia: see below), not prototypes of the spicules in general, and are explicable alternatively as simply an embryonic specialization.

Holactines may be modified further by distortion of the regular triaxial pattern. Rays may be bent near their origins or curve progressively away from the ideal cubic axes. This can occur with a genetic basis or simply by growth of spicules in restricted situations, e.g., between closely spaced canals. Pentactine megascleres supporting dermal or gastral membranes often have the paired rays slightly bent toward a fifth proximal ray, directed inward. This condition is carried further in spicules that some forms protrude for attachment, in which small paired rays are bent strongly toward a large proximal ray to form grapnels that function as such. These spicules are described as anchorate. Some recent authors have called them anatetraenes by analogy with demosponge anatriaenes; but the triaenes and analogous polyaenes of Demospongea and Heteractinellida are not basically triaxial and have meristically variable cladomes in which interactinal angles depend on the number of rays. Such spicules do not occur in Hexactinellida. An anchorate pentactine with a distal axial rudiment is also not properly an anatetraene but an anisomesoanatetraene, if this nomenclature were used. In another development, specially distorted pentactines of one family (Rossellidae) have the interactinal angle between two of the paired rays more than 90°, up to more than 270°, and the rays arranged in the remaining arc at subequal intervals. These are **paratropal pentactines** or **paratropes.** Fossil examples have sometimes been mistaken for dermalia of lithistids.

Modification of holactines may occur by growth of solid (anaxial) lateral spines on the rays. These may simply grow outward but are commonly curved toward the tip of the ray. Some forms have such spines specially developed on an outwardly directed (distal) ray of pentactine or hexactine megascleres of which four paired rays support a bounding membrane; these are pinular pentactines or hexactines or simply **pinuli.** In rhabdodiactines called uncinates, thornlike spines are all directed toward one end of the spicule. In anchoring rhabdodiactines of some genera, groups of recurved spines at the end of a protruded shaft form grapnel-like **umbels,** analogous functionally with the grapnels of anchorate spicules although different in origin. Amphidisc microscleres and their variants bear similar umbels without apparent function. In microscleres called sceptrules, the cross-containing end of a monactine or a strongly unequal-rayed diactine with one ray rudimentary bears spines in forms varying from a brushlike, centrifugal cluster to an umbel-like ring of recurved teeth.

In a special group of microscleres called **rosettes,** and distinguished from holactines as astral, the true rays bear anaxial, branchlike prolongations or terminals or are replaced by anaxial outgrowths. These spicules take names ending -aster or -come, with the most common type being six-rayed hexasters. Further nomenclature of hexasters is based on the form of the terminals, which are simple, sharp spines in **oxyhexasters,** bear terminal knobs, discs or recurved teeth in **tylohexasters, discohexasters,** and onychohexasters, or are petal-like in floricomes. Some have the true rays or principals much shorter than the terminals, and some of these forms sometimes grade into triaxial **pseudopolyactines** in which the terminals radiate directly from a spherical centrum containing a six-rayed axial cross. **Octasters** are a special mutant form in which the six normal principals are replaced by eight raylike and terminal-bearing pseudoactines, which arise from the eight three-dimensional angles of an axial cross in the central part.

In anchorate pentactines or tauactines the axial filaments of the grapnel rays are sometimes rudimentary, although a solid anchor fluke is well developed. This development is not related to formation of hexaster terminals but a separate modification in the spicules concerned, which grade into normal forms through intermediates.

The rays of spicules of all sorts are normally not branched in living genera, although branching may be seen in rare abnormal examples. The fossil *Acanthocoryna* FINKS, 1960, has pentactines with a branchlike outgrowth on the outside of each paired ray; but examples from the Irish Carboniferous, at least, can have an axial canal in the typical ray but not the branch, which seems to be simply a secondary spine homologically. In *Docoderma* FINKS, 1960, massive development of tangentially radiating spines converts pentactines into spicules resembling the polyaenes of some Heteractinellida.

Some Hexactinellida have megascleres united to form rigid skeletal frameworks. The union is characteristically by fusion (not by articulation as in Demospongea), very probably because of the syncytial character of the soft parts. There are several modes of union. First, where they cross or lie side by side, rays may fuse together but still remain recognizable as individual rays. Rays close together but not in contact may then also be united by transverse siliceous bridges or **synapticula.** Second, pairs of rays apposed side by side may be enclosed in a common siliceous envelope to form beams in which individual rays are not apparent except by the presence of their axial filaments. These main modes of union may occur separately or in the same skeleton. Third, megascleres arranged with four paired rays supporting a dermal or gastral membrane (or arranged correspondingly, although the soft parts are unknown) may be connected by branching and anastomosing siliceous filaments in the meshes enclosed by the paired rays. The relationship of these filaments to the soft parts is unknown, but their pattern is typically similar to that of the trabeculae forming bounding membranes.

In many Paleozoic Hexactinellida with apparently unconnected spicules, the skeleton retains its organization in a manner that is very unusual in later comparable forms. This simply may be due to conditions of burial, but might instead indicate the presence of something like spongin.

MEGASCLERES

The lateral wall of Hexactinellida is characteristically supported by three series of megascleres. Two support the dermal and gastral bounding membranes; the third occurs in the interior and supports the wall as a whole as well as its internal choanocytal and canalar structures. The dermal and gastral megascleres or **dermalia** and **gastralia** are commonly pentactines, arranged with four paired rays in or under the bounding membrane and the fifth directed inward. The paired rays are called **tangentials** or, more strictly, **paratangentials** because their orientation is seldom truly tangential. Pentactines arranged in this manner may be replaced by stauractines, by hexactines with a projecting distal ray as well as an inwardly directed proximal ray, or by pentactines with the unpaired ray distal. Distal rays are often pinular when present. Very rarely, the gastralia are rhabdodiactines, which lie in the plane of the surface. The internal parenchymal megascleres vary from hexactines to diactines or may rarely include monactines.

There are two types of **parenchymal skeletons.** In **lyssacine** Hexactinellida, the megascleres are typically loose and of types from hexactine to rhabdodiactine, occurring in various combinations and being all rhabdodiactines in some genera. As fossils, the latter may appear falsely to be monaxonid sponges since the triaxon character of rhabdodiactines is apparent only if the axial crosses can be seen. The main supporting elements are **principalia** against which may lie smaller comitalia. The megascleres are sometimes fused together as a secondary development in the whole of the body or the older parts only. In dictyonine Hexactinellida the parenchymal megascleres or **dictyonalia** are characteristically all hexactines and are fused to form a rigid dictyonal framework as part of their normal development. In lyssacines, union between the spicules is usually more or less haphazard, but in dictyonines some or all skeletal beams are typically of the sort enclosing pairs of rays together.

Some lyssacine Hexactinellida have two series of dermalia with autodermalia whose tangential rays lie in the bounding membrane and different hypodermalia whose tangentials lie under it. The difference may be striking; *Rossella* CARTER, 1872, and related forms have small stauractine, pentactine, or hexactine **autodermalia,** contrasting with **hypodermalia** in the form of paratropal pentactines whose rays may be more than ten times longer. Where ontogeny is known, the autodermalia correspond with the embryonic stauractines, and the hypodermalia are a separate development; they seem sometimes to derive homologically from the parenchymal megascleres.

The names autodermalia and hypodermalia were originally introduced by SCHULZE (1887a, 1887b) to designate any dermalia that lie with the tangentials in the bounding membrane or under it, without necessary presence of two distinct series. Similar names were applied to the gastralia on this basis (**autogastralia, hypogastralia**). But the terms have their principal usage in instances where two systems are present, and SCHULZE himself called the outer set autodermalia even though some of its members may lie under the membrane when numerous. In paleontology, the terms are useful only if two series of dermalia are present or if apparent dermalia also seem to be special external members of a parenchymal skeleton.

Some lyssacine Hexactinellida protrude megascleres termed **prostalia** from external or marginal surfaces. The megascleres are usually either rhabdodiactines, or pentactines with a long proximal ray and may originate as hypodermalia or members of the parenchymal principalia. They are classed as **pleuralia** if protruded from lateral surfaces; marginalia, if arranged around an osculum; and basalia if protruded from the base and used for attachment. The last are typically either anchorate or provided with a terminal umbel. If a sieve plate is present, its megascleres may be special oscularia related to dermalia and gastralia but differently developed.

Last, several instances are known in which larval Hexactinellida with adult-type soft parts possess a special internal basal skeleton in the form of small, ankylosing hexactines termed **basidictyonalia.** These are distinct from the normal adult parenchymal megascleres and may have a peculiar mode of union in which fusion by their tips is preceded by zygosis-like articulation (e.g., in *Rhabdocalyptus mirabilis* SCHULZE). The ankylosing tips may also be anaxial for some distance from the end. This mode of union does not occur in adult parenchymal structures and certainly not in dictyonine frameworks, whose description as consisting of spicules fused tip-to-tip is not correct (e.g., DE LAUBENFELS, 1955).

MICROSCLERES

The spicules classed as microscleres are all those not referable to some division of the megascleres. They are typically small by comparison but sometimes reach equal or even larger sizes. Most of the main types have been noted above (see Spicules, p. 129).

The most common microscleres are small parenchymal triaxons. In the subclass Amphidiscophora these are always holactines; in the subclass Hexasterophora they are characteristically rosettes, although holactines sometimes also occur. Other microscleres are less widely distributed.

The characteristic microscleres of the subclass Amphidiscophora are amphidiscs with equal terminal umbels at the ends of a short shaft. These spicules are monaxons, but variants with four or six rays (**staurodiscs, hexadiscs**) are diaxons or triaxons. Asymmetric amphidiscs with one umbel larger than the other have been called hemidiscs (or hemiamphidiscs). The amphidiscs are typically found at right angles to bounding or canalar membranes, with one half protruded through the membrane. They are never produced by forms with rosettes.

The sceptrule group of microscleres is characteristic of one group of dictyonine Hexactinellida (suborder Hexactinosa) but unknown in another (suborder Lychniscosa) or in lyssacines. The end containing the axial cross and usually bearing spinous outgrowths is protruded through a bounding membrane with the single ray proximal, or similarly through a canalar membrane.

Uncinates occur as microscleres in dictyonines with sceptrules (Hexactinosa), in the sense of being accessory elements that take no part in formation of the main supporting skeleton. Some are small raphides, but others range up to strongly barbed forms larger than the true megascleres. These uncinates are monaxons, whose relationship to triaxons cannot be demonstrated. Uncinates present as megascleres in some forms with amphidiscs (Amphidiscophora) are, in contrast, sometimes clearly triaxial.

The Paleozoic Dictyospongiidae are reticulosids, some of which have paraclavule microscleres (umbels *sensu* HALL & CLARKE, 1899) resembling amphidiscs with one umbel missing or like a nail-like form of sceptrule called a **clavule.** The nature of these spicules is unknown, but they seem to be monaxons.

Microscleres are almost absent from fossil Hexactinellida but are sometimes found loose in sediments.

CLASSIFICATION

The classification used here above family-group level is based on the method of MORET (1926b), adopted in agreement with LAGNEAU-HÉRENGER (1962) and REZVOI, ZHURAVLEVA, and KOLTUN (1962). The reason for using this method is the absence of microscleres in the fossils and especially in lyssacines, modern representations of which may have amphidisc or rosette microscleres but similar megascleres. There is no implied rejection of the zoological use of microscleres when these are fully available. The taxa distinguished are ranked as orders (Lyssacida, Hexactinosa, and Lychniscosa).

Early attempts at classification of Hexactinellida were by KENT (1870), CARTER (1875), and MARSHALL (1876). KENT divided Hexactinellida as an order into suborders Coralliospongiae GRAY and Callicispongiae KENT, with connected spicules and loose spicules only, respectively. The first group in fact included lithistids (*Macandrewia* GRAY) as well as Hexactinellida. These were removed by CARTER, who changed the names to Vitreohexactinellida and Sarcohexactinellida, now called families, and added a third group, Sarcovitreohexactinellida, for forms with spicules united in only part of the body. His Sarcohexactinellida is of interest because of microscleric divisions Birotulifera and Rosettifera, adopted by SCHULZE (1887b) as Amphidiscophora and Hexasterophora. MARSHALL thought that some dictyonines are distinguished by a continuous system of axial canals in the beams of the skeletal framework and called these Synauloidea, as opposed to other Hexactinellida comprising Asynauloidea. In fact, the canals originate in dead specimens by secondary internal solution, enlarging the true axial canals of incorporated spicules to the point of coalescence.

ZITTEL (1877b) rejected these classifications and proposed suborders Dictyonina and Lyssakina, distinguished by the dictyonine and lyssacine types of parenchymal skeleton. These morphological terms derive from the names of his taxa, with spelling Lyssakina changed to Lyssacina by SCHULZE (1887a, 1887b).

In SCHULZE's *Challenger* system, ZITTEL's Dictyonina and Lyssacina (in this spelling) were accepted but subdivided on the basis of microscleres. The Lyssacina comprised tribes Hexasterophora and Amphidiscophora having, respectfully, hexasters (rosettes) but not amphidiscs and amphidiscs but not hexasters. The Dictyonina, with hexasters but not amphidiscs, were divided into Uncinataria and Inermia according to the presence or absence of uncinates; and the Uncinataria were divided into subtribes Clavularia and Scopularia, with sceptrules in the form of nail-like clavules and broomlike **scopules,** respectively. In later work (1899, 1904), SCHULZE gave up ZITTEL's taxa, dividing all Hexactinellida directly into Amphidiscophora and Hexasterophora. These are the main divisions currently accepted by zoologists, and they were ranked as subclasses by REID (1958a), as they are herein.

Another innovation by SCHULZE (1887a, 1887b) was the introduction of the name Triaxonia as a substitute for Hexactinellida. The diagnostic concept implied is the same in both instances, since the triaxon character of the spicules is the basis of SCHMIDT's (1870) original diagnosis; but the name Triaxonia was used by SCHULZE in connection with his theory of sponge phylogeny, in which two stocks of siliceous sponges called

Triaxonia (Hexactinellida) and Tetraxonia (choristid Demospongea) were supposed to have arisen independently from sponges without spicules. These names were used in SCHRAMMEN's work (1910, 1912, 1924a, 1936), and more recently by LAGNEAU-HÉRENGER (1962) and REZVOI, ZHURAVLEVA, and KOLTUN (1962), but the theory they depended on is doubtful; tetraxons do not occur in Hexactinellida, but triaxons occur in various demosponges, as shown long ago by SOLLAS (1888).

SCHULZE's (1899) rejection of ZITTEL's taxa left the lyssacine Hexasterophora and former Dictyonina grouped together without separation above family-group level. IJIMA (1903) suggested a division into unnamed tribes A, B, and C, consisting respectively of the lyssacines and SCHULZE's Inermia and Uncinataria. He also rejected derivation of dictyonines from the lyssacines as claimed by SCHULZE (1887a, 1887b) on grounds that parenchymal megascleres of the latter are of various types, always including rhabdodiactines, whereas those of dictyonines are normally all hexactines and never include rhabdodiactines.

SCHRAMMEN (1902) had suggested dividing the Hexactinellida (Triaxonia) into suborders Stauraktinophora, Hexaktinophora, and Lychniskophora, with the parenchymal megascleres typically stauractines in the first, including simple hexactines and their variants in the second, and specially modified hexactines called lychniscs in the third. The last group (Lychniskophora) depends on the fact that some dictyonines have the central parts of dictyonalia enclosed by an octahedral framework of 12 anaxial buttresses, which grow across the angles between the rays from points equidistant from the center. These dictyonalia are called lantern-spicules or lychniscs. No comparable structures are known in other Hexactinellida. Shortly afterward, SCHRAMMEN (1903) dropped the first group and renamed the others Hexactinosa and Lychniscosa as divisions of SCHULZE's Hexasterophora. Lyssacines, then included as Hexactinosa, were removed in 1912 so that this group then consisted of dictyonines only. This scheme was accepted by IJIMA (1927), whose last work, published six years posthumously, also characterized the lyssacine Hexasterophora as Lyssacinosa. A similar scheme given by SCHRAMMEN (1924a) used independently the new termination -aria (e.g., Hexactinaria), now used by REZVOI, ZHURAVLEVA, and KOLTUN (1962). SCHRAMMEN also divided Amphidiscophora into Amphidiscaria and Hemidiscaria, the former comprising forms with typical amphidiscs, the latter based on fossil occurrence of hemidisc microscleres, unknown in the modern forms. REID (1958) used this arrangement with the Amphidiscophora and Hexasterophora ranked as subclasses, with the orders Amphidiscosa and Hemidiscosa in the former and Hexactinosa, Lychniscosa, and Lyssacinosa in the latter. That classification scheme is followed herein, with the exception that the additional order Reticulosa is used to include many Paleozoic dictyosponges. MORET (1926b), however, preferred to retain ZITTEL's taxa in paleontology, because classification of fossils by microscleres is not normally possible, and he used the Lyssacida and Dictyonida. The Hexactinosa and Lychniscosa, however, can be recognized from their megascleres only, and they are used as subdivisions of the Hexasterophora.

DE LAUBENFELS (1955) accepted the Hexactinosa and Lychniscosa as separate orders, using the names Dictyida ZITTEL and Lychniskida SCHRAMMEN for them. While these are acceptable as separate orders on grounds of their supposed separate origin (IJIMA, 1927), equation of ZITTEL's Dictyonina (as Dictyida) with SCHRAMMEN's Hexactinosa is contrary to usual practice and is not thought desirable.

PALEOZOIC HEXACTINELLIDA: MORPHOLOGY AND PHYLOGENY

Robert M. Finks

[Department of Geology, Queens College (CUNY)]

ORGANIZATION OF THE PALEOZOIC RETICULOSID SKELETON

The simplest reticulosids, which constitute most of the early forms from the Cambrian, are conical to spherical with a very thin body wall surrounding a large interior space or cloaca. In the conical forms the wider end is open as a broad osculum; in some spheroidal forms there is seemingly no osculum at all. The body wall is most often composed of stauractines, in some sponges of pentactines or hexactines, that lie parallel to one another in the plane of the body wall. They are of several sizes and usually organized as follows: the largest are quincuncially arranged, tip to tip in vertical and horizontal series, so that **quadrules** are formed, outlined by the length of one ray. A stauractine one-half the diameter of the largest occupies each such quadrule, dividing it into four smaller quadrules. Each of the smaller quadrules is similarly occupied by a third order of stauractine, and so on to a fourth or fifth order. Spicules of each size, except the largest, have their centers quadrately arranged. Sometimes the rays overlap within each order rather than being tip to tip. Rigby (1966a, p. 554) has noted a species of *Protospongia* in which the largest spicules are quadrately arranged or, in other words, as though the largest, quincuncially arranged series were missing. This nearly monolayered sheet of stauractine spicules in parallel arrangement persists as a dermal (**autodermal**) layer in most Paleozoic reticulosids, even in those forms in which a layer of irregularly arranged parenchymal spicules forms most of the thickness of the body wall. There are some genera of thick-walled sponges in which this outer quadrate layer seems to be missing, but whether the loss is phylogenetic or merely taphonomic cannot always be ascertained.

It is worth noting that spicules of any one size in these quadrate sheets lie in linear series parallel to their rays. This seems to be a fundamental relationship, for it reappears in the Mesozoic and later dictyonine skeleton, where, as Reid (1958a, p. xxv) pointed out, the unit of construction is a fused, linear series of equally sized hexactines. There is some evidence that this may reflect a fundamental morphogenetic process, even in the unfused spicules of the Paleozoic reticulosid sponges. Some specimens of *Hydnoceras* and other dictyosponges have healed injuries in which the new growth includes long, curving lines of stauractines at variance with the original quadrate mesh that must have been formed as a linear series by addition of spicules at one end (Fig. 49). This suggests that the normal quadrate mesh described above grows by addition of spicules of like size at the upper end of each vertical series and that as the sponge expanded upward, new series were formed by branching or were intercalated between the old. It would be of interest to investigate, in specific groups of sponges, whether this was accomplished by addition of the next smaller order of size, or whether new series of several sizes are intercalated. The suggestion of Finks (1960, p. 104, 134) that in some species smaller spicules have been inserted in each quadrule as the sponge expanded, would be a completely different mode of growth involving growth over the entire body rather than at the upper or growing edge. It is possible that both kinds of growth occurred in different species or even in the same species. The ultimate cause of the rectangularity of hexactinellid spicules and spicular meshes appears to be the

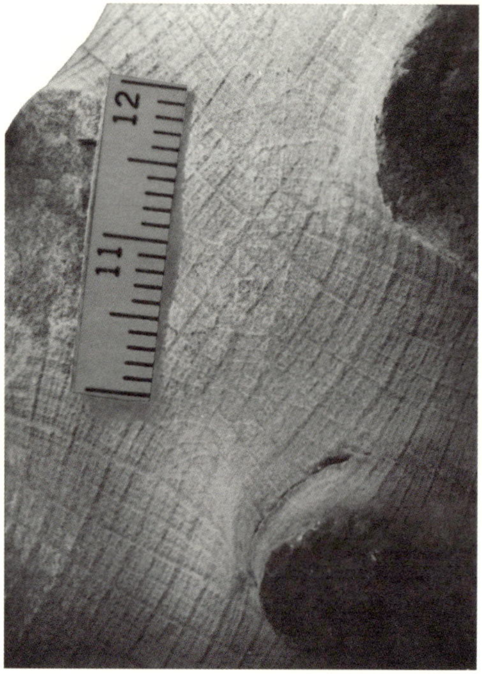

FIG. 49. *Hydnoceras tuberosum* HALL & CLARKE, 1899; healed section of skeleton with curved lines of spicules at angles to original skeletal structure to right of scale, Queens College No. 4, topotype, Devonian, New York, USA, ×2 (new).

crystallization in a cubic or tetragonal symmetry of the protein making up the axial fibers of the spicules (the fibers are square in cross section) formed within the scleroblast cells and genetically determined (REISWIG, 1971). Their formation in long lines, however, seems to require an independent cause at the cellular or syncytial level.

OKADA (1928) has shown that juvenile sponges of the living *Farrea occa* have the form of a thinly walled sphere with a skeleton of stauractines. The choanocyte membrane lines the inner cavity; that is, the sponge is asconoid. In shape and spiculation the juvenile *Farrea* resembles many Cambrian reticulosids, and it is not unreasonable to conjecture that these were also asconoid, as DE LAUBENFELS (1955) suggested. Persistence of a quadrate skeleton of stauractines as the outermost layer in thickly walled forms, in which an inner parenchymal layer of different organization may be present, suggests that the entire skeleton of the early forms is homologous to a dermal skeleton. This has been suggested by SOLLAS (1880a), REID (1958a, p. xlii), and RIGBY (1969, 1983b, 1986a).

It is possible that thinly walled forms in which the spicules are hexactines were no longer asconoid, for the proximal rays might have helped to support diverticula of the choanocyte membrane. This seems certainly to be true for species like *Multivasculatus ovatus* HOWELL & VAN HOUTEN, 1940 of the Late Cambrian (see FINKS, 1970), in which the skeleton is composed of several layers of hexactines. It is probably true of some thinly walled mid-Cambrian forms as well. Such a structure could have supported the simple or compound thimblelike diverticula of the choanocyte membrane characteristic of adult living hexactinellids. *Hintzespongia* RIGBY & GUTSCHICK, 1976 almost certainly has such diverticula, for beneath an outer layer of stauractines there is a thin layer of hexactines that are arranged irregularly about circular openings that were either exhalant canals leading from the choanocyte chambers or else contained the chambers themselves.

In most Paleozoic reticulosid hexactinellids the principal skeleton is a parenchymal one that lies beneath the dermal quadrate layer. There are so many variants that to list all the known arrangements is less than clarifying . Nevertheless some broad groupings of structure can be recognized and may indicate phylogenetic relationships. (It must be confessed, however, that we are not certain as to how much convergent evolution there has been.) In one such group the spicules of the parenchymal skeleton lack a parallel arrangement but are irregularly disposed about radial tubes, which appear to be inhalant and exhalant skeletal canals. These tubes may branch and often connect with smaller, more irregular skeletal canals; but for the most part they are straight, parallel, subequal, and penetrate completely or nearly the entire sponge wall. The simplest form of this skeletal type occurs in the mid-Cambrian *Hintzespongia* RIGBY & GUTSCHICK, 1976 (probably also in the Ordovician *Cyathophycus* WALCOTT, 1879 and *Teganium* RAUFF, 1893), in which the canalized parenchymal layer is approximately one spicule thick, beneath a similarly monolayered quadrate dermal layer. In the mid-Ordovician *Brachiospongia* MARSH, 1867 the parenchymal layer is not only many spicules thick but also is bounded by enlarged hypodermal and hypogastral spicules. In the Devonian *Pelicaspongia* RIGBY, 1970a and a number of similar genera, the outer quadrate dermal layer seems to be missing, and in the Pennsylvanian and Permian *Stioderma* FINKS, 1960 the hypodermal spicules develop sutured interlocking and fusion that forms a rigid net. The Ordovician *Pyruspongia* RIGBY, 1971 appears to be a related type that possesses hypodermalia but lacks skeletal canals and the quadrate dermal layer.

In another type, the parenchymal skeleton consists of a cubic mesh of hexactines, scarcely to be distinguished from the dermal layer. The Late Cambrian *Multivasculatus* HOWELL & VAN HOUTEN, 1940 and the Permian *Microstaura* FINKS, 1960 have this type of skeleton and are two genera that are probably not closely related.

In a third type the parenchymal skeleton consists of bundles of spicules, vertical and horizontal, in alternating layers. In many Devonian and Mississippian dictyosponges there is but one vertical and one horizontal layer, seemingly of rhabdodiactines. In the Pennsylvanian and Permian *Stereodictyum* FINKS, 1960 there are numerous such layers, and they seem to be composed of long-rayed hexactine derivatives in which two or more rays are suppressed.

In a fourth type, related to the preceding, enlarged hypodermal pentactines or hexactines are present between the autodermal quadrate mesh and the underlying layer of spicule bundles. They are parallel to the other spicules and their rays are accompanied by smaller **comitalia**, usually monaxonic. Some Devonian and Mississippian dictyosponges and possibly also some Permian genera are of this type, such as *Endoplegma* FINKS, 1960 and *Carphites* FINKS, 1960, in which the inner layer of bundles is composed of tauactines. The Permian docodermatids may be secondarily irregular derivatives of this type, with some species developing interlocking and fusion of the hypodermalia to form a rigid net.

A fifth type, represented only by the Permian *Pileolites* FINKS, 1960, has linear series of fused hexactines, the series in nonparallel arrangement, and connected by synapticulae. This may not be a reticulosid at all, but rather an aulocalycid hexactinosan.

The Ordovician *Hydriodictya* RIGBY, 1971 is a primitive, thinly walled form in which a second quadrate mesh lies beneath the dermal layer but at an angle to it. It is not clear whether this inner layer should be considered parenchymal or gastral. The Ordovician *Dierespongia* RIGBY & GUTSCHICK, 1976 is a similarly two-layered form, but the inner layer is irregular rather than quadrate.

FIG. 50. Enlarged part of dermal surface of holotype of *Docoderma papillosum* FINKS, 1960, showing small, star-shaped dermalia with numerous stubby, branching, tangential rays, between larger papillate hypodermal pentactines, PU 78885, Leonardian Bone Springs Limestone, Apache Canyon, Van Horn quadrangle, Texas, USA, ×2 (Finks, 1960).

SPECIALIZED SPICULES

Most spicules in Paleozoic reticulosids are hexactines or derivatives by reduction with simple, smooth rays. They may be accompanied by more elaborate spicules that, in many instances, appear to have been modified for special functional needs. These specialized spicules are of the following sorts: (1) enlarged hypodermal and hypogastral spicules, (2) small comitalia, (3) microscleres, (4) prostalia, including root-tuft spicules (basalia), pleuralia, and marginalia.

HYPODERMALIA AND HYPOGASTRALIA

These are usually pentactines in which the ray facing the bounding membrane (gastral or dermal) has been suppressed or reduced. They are almost always larger and stouter-rayed than the parenchymal spicules, and they frequently bear species-characteristic protuberances on the outward-facing surfaces of the tangential rays. The dermal and gastral varieties may be identical, but often the dermal are larger, and the protuberances, if any, are different on the two sorts. The function of these spicules seems to be supportive and protective, which accounts for their large size and external projections. In some late Paleozoic species they may develop special interlocking processes, sometimes accompanied by actual fusion, which results in a rigid skeleton. In the two genera in which this is known (*Stioderma, Docoderma*) the interlocking is accomplished by branching processes in the tangential plane, which are external to the regular paratangential rays. Sometimes two different forms of enlarged hypodermalia occur together as in *Docoderma papillosum* FINKS, 1960 from the Permian (Fig. 50).

COMITALIA

These smaller spicules parallel and accompany the rays of the hypodermal, hypogastral, and parenchymal spicules. They are most often monaxonic. They may be simple, smooth rhabdodiactines with pointed ends. In many members of the Dictyospongiidae described by HALL and CLARKE (1899) they include curved monaxons with blunt ends, like strongyles of the demosponges, although presumably derived from hexactines by suppression of rays. Some are so stout as to resemble beans. They have also been found in genera belonging to other families, such as the Permian *Docoderma* and *Carphites.* Tauactines also occur as comitalia in the last two genera. Another group of spicules occurring as comitalia are small or undeveloped varieties of specialized prostalia found in the same individual. These include **clemes**, normally occurring in root tufts, where they have anchorate terminations. They also include anchor-shaped anadiaenes with short shafts. Such spicules may also occur in more elongate form in root tufts, but more often they seem to be pleuralia, in which function they protrude from the sponge surface in closely packed masses. As reported from some dictyosponges by HALL and CLARKE (1899), however, they accompanied internal spicule bundles. Like microscleres, comitalia may occur throughout the sponge even though they may have originated to perform a localized special function.

MICROSCLERES

Restudy of the type specimens of dictyosponge genera of HALL and CLARKE has revealed that **paraclavule** microscleres occur in closely packed masses at the sponge surface, with the umbellate ends facing outward (Fig. 51). This is especially well illustrated in the holotype of *Cleodictya mohri* HALL, 1884 but occurs also in the original specimen of one of HALL and CLARKE's illustrations of *Physospongia dawsoni* (WHITFIELD) (HALL & CLARKE, 1899, pl. 62,*9*). It appears that these spicules formed a protective dermal armor. Amphidiscs of living hexactinellids occur in the same position (REID, 1958a, p. xxx). This discovery strengthens the inference that paraclavules are ancestral to amphidiscs and explains their original asymmetry. If they were selected originally to serve a protective function the umbel would be needed on one side only. (They may have served chiefly to discourage the settling of larvae, like other small protruding spicules, a need that has given rise to pedicellariae and avicularia in other phyla.) The oldest known paraclavules are Mississippian. By the Pennsylvanian, both **hemidiscs** (a paraclavule with a smaller umbel at the other end) and amphidiscs (equal umbels at both ends) appeared. *Microhemidiscia* KLING & REIF, 1969 has hemidiscs, *Uralonema* LIBROVICH, 1929 has amphidiscs, and *Itararella* KLING & REIF, 1969 has both (see their fig. 1*f*). Paraclavules went extinct in the Permian (see FINKS, 1960, p. 127), hemidiscs until the Cretaceous (see SCHRAMMEN, 1924a; ORTMANN, 1912). Why amphidiscs should have persisted is difficult to explain; perhaps they save the sponge energy in the process of protruding an umbellate end (the spicule would never need to be rotated through more than 90 degrees).

FIG. 51. Photomicrograph of part of dermal surface of holotype of *Cleodictya mohri* HALL, 1884, with closely spaced paraclavule microscleres at dermal surface, with umbellate ends facing outwardly, between vertical megasclere tracts, Carboniferous Keokuk Group, Crawfordsville, Indiana, USA, ×40 (new).

If paraclavules are accepted as ancestral to amphidiscs, a possibility originally suggested by REID (1958a, p. xxxii), it may aid us in reconstructing phylogenetic relationships. REID (1968b, p. 1,247) has pointed out that *Hyalostelia* ZITTEL, 1878c resembles the living amphidiscophoran family Pheronematidae in having "large pentactine hypodermalia, and parenchymal megascleres which are mainly hexactines or pentactines." He continued, ". . . but these forms have pinular autodermalia and autogastralia, and the basalia are bidentate monactines whose shafts may be developed as 'clemes'." The Dictyospongiidae are now interpreted to have paraclavules (i.e., protoamphidiscs), tripinuli (HALL & CLARKE, 1899), clemes and bidentate anchors (anadiaenes) (HALL & CLARKE, 1899), as well as large pentactine hypodermalia in some genera (*Acloeodictya, Lebedictya*). The Permian *Eudoplegma ("Carphites") diabloense* (FINKS, 1960, p. 127) has large pentactine hypodermalia, bidentate anchors, and paraclavules and, like the related cleme-bearing *Dococlerma,* has the curved, strongyle-like comitalia also found in the dictyosponges. Thus the dictyosponges, the Permian docodermatids, and perhaps such other late Paleozoic hexactinellids as *Hyalostelia* and *Stioderma* may not be so far removed from the living Amphidiscophora as once thought.

There are other complications, however; *Griphodictya* HALL & CLARKE, 1899, *Itararella* KLING & REIF, 1969, and *Endoplegma ("Carphites")* (FINKS, 1960, p. 127) have paraclavules (amphidiscs in *Itararella*) occurring with hexasters. It is possible, as suggested originally by the authors of the last two genera, that the co-occurrence is fortuitous (i.e., one or both washed in). It is also possible that the present mutual exclusion of the two microsclere types was not true of some Paleozoic groups and also that the pheronematid characters cited above were once more widespread.

The hypothesis that paraclavules are precursors to the amphidiscs is supported or at least not contradicted by the geologic history of other siliceous spicule types. Among the demosponges the monaxon is the first spicule type to appear in the geologic record even though the potentiality to form tetraxons was present from the beginning in the triangular cross section of the axial canal (REISWIG, 1971). Likewise, among the hexactinellids the first spicules to appear are stauractines, the fully developed hexactines not appearing until later. Thus there seems to be a pattern of initially not realizing fully, or perhaps suppressing the potentialities of a given spicule form.

PROSTALIA

The earliest hexactinellids preserved whole from the Cambrian bristle with protruding spicules. Thus, the presence of prostalia is a fundamental feature of the class. Because they cover the entire body as pleuralia, they seem to provide protection, probably discouraging both predation and settlement of larvae. A fringe of **marginalia** about the osculum is also present in Cambrian forms and seems to have evolved so as to prevent entrance into the cloaca by commensals. The marginalia may also enclose the stream of waste-water issuing from the cloaca, preventing its premature dispersion and recycling by the sponge. They, thus, might function as an oscular chimney in the manner analyzed by BIDDER (1923). Basalia or root-tuft spicules are also found in these early sponges. Unlike later forms, they seem to consist of only a few spicules, although they may be very long. All varieties of prostalia in these Cambrian hexactinellids (that is, pleuralia, marginalia, and basalia) appear to be simple rhabdodiactines without special terminations, spines, or other modifications. It is possible that some pleuralia may have been pentactines with an elongate distal ray. The only specialized spicules in these sponges are the stout elements twisted together like strands of rope, known as *Kiwetinokia* WALCOTT, 1920, that were probably basalia (see also *Palaeosaccus* HINDE, 1893a).

Stouter bundles of basalia are known from the Ordovician on. The oldest seem to be *"Hyalostelia" explanatum* (HICKS, 1869) from the Tremadoc (HINDE, 1888, p. 110). (*Pyritonema* M'COY, 1850, *Acestra* ROEMER, 1861, and *Hyalostelia* ZITTEL, 1878c were founded wholly or partly on isolated Paleozoic root tufts.) By the Trentonian very large root-tuft bundles of *Pattersonia* MILLER, 1882 occupy much of the sponge. Examination of similar root tufts of near topotypes with excellent preservation, which may belong to *Pattersonia,* have a few clemate spicules with a quadrianchorate termination, apparently the first appearance of specialized basalia; most of the spicules in these tufts are smooth, presumably rhabdodiactines, and have simple pointed terminations where the tips are visible. Large root tufts with accompanying hexactines are also known from a perireefal facies of the earlier Chazy Series (FINKS & TOOMEY, 1969 and unpublished). It is of interest to note that *Pattersonia* occurs in the same beds as *Brachiospongia,* which supported itself above the same soft sea floor on fingerlike, hollow, radiating protuberances of the body wall, rather resembling mangrove roots. Apparently root tufts were not the only successful hexactinellid response to identical conditions; in the later Paleozoic root tufts remained common, while the *Brachiospongia* type of adaptation did not occur again until the Mesozoic and Tertiary.

Among the diverse Devonian and Mississippian dictyosponges, surprisingly few genera have been found with root tufts, although the general shape and spiculation suggest their presence. Only *Dictyospongia* HALL & CLARKE, 1899 and *Retifungus* RIETSCHEL, 1970 have been found with root tufts attached, but the rather similar *Prismodictya* HALL & CLARKE, 1899 and *Hydnoceras* CONRAD, 1842 have not, despite their local occurrence in great numbers. Furthermore, no unattached root tufts have been found associated with them. Nevertheless, the strong inner vertical bundles of long spicules found in most dictyosponge genera could be the upper ends of root-tuft basalia as they seem to be in *Retifungus,* but we must also accept the strong possibility that in most of the dictyosponge genera they were not. One additional feature that suggests the connection of the vertical bundles with root tufts is the occurrence in some of them of clemes and short, bidentate anchors (*Physospongia, Cleodictya, Acloeodictya,* and *Lyrodictya*), which in other genera are specifically root-tuft spicules with obviously functional anchorate ends. It is possible, however, that they are vestigial in their occurrences with the dictyosponges.

Two dictyosponge genera, *Thysanodictya* HALL & CLARKE, 1899 and *Phragmodictya* HALL, 1884, had definitely flat putative bases, surrounded by an outwardly and supposedly downwardly directed frill or **periloph,** which in this interpretation would have served in place of a root tuft. This structure is not, however, composed of basalia but is an apronlike extension of the quadrate mesh of autodermalia and apparently composed of stauractines. There is a possibility that these genera have been restored upside down, and that the supposed basal disc is rather an oscular sieve surrounded by a frill as in some species of the living *Euplectella.* This is certainly a possibility in the instance of *Thysanodictya hermenia* (HALL & CLARKE, 1899, pl. 40) and *T. scyphina* (HALL & CLARKE, 1899, pl. 42,*8*) in which the supposed upper end is strongly contracted rather like a basal apex.

The short-shafted bidentate anchors (anadiaenes) in the dictyosponges could have been pleuralia. They seem to serve this function in a Permian species ("Lyssacine, species 1" of FINKS, 1960, p. 112) known only from fragments, where they occur over large areas of the sponge surface, closely packed together as a kind of dermal armor rather like the paraclavules in some dictyosponges. Their anchorate ends, which lack axial canals and are therefore not rays, are perpendicular to the sponge surface and often parallel to one another in local clusters with larger anchors in the center of the cluster. Such anchorate pleuralia are obviously homologous with anchorate basalia. It is likely that this spicule type was originally selected for its anchoring function in basalia but that its equal usefulness in a protective function caused shorter-shafted versions to be selected as pleuralia.

REID's (1958a, p. xxxiii) suggestion that the continuity of basalia with a lateral covering of pleuralia is a primitive condition seems to be borne out by the fossil record. The earliest whole hexactinellids (mid-Cambrian) have a more or less continuous covering of simple rhabdodiactine prostalia with the basalia distinguished from the pleuralia solely by their greater length. REID's further suggestion (1958a) that basalia arose by modification for attachment of some members of an original covering of pleuralia in an originally basiphytous sponge may not be correct, however. It is certainly not likely to be so in the instance of anchorate pleuralia, for their shape is obviously related to an original function as basalia. Even among the simplest rhabdodiactine prostalia, the earliest

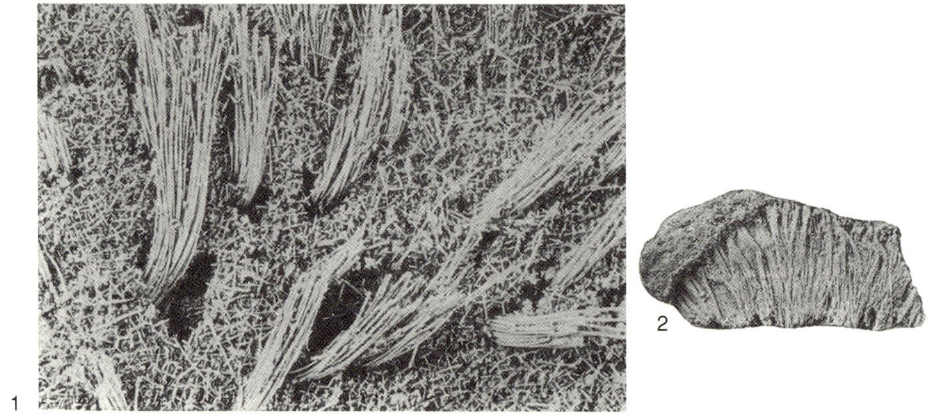

FIG. 52. *1,* Enlarged part of exterior of paratype of *Polylophidium discus* FINKS, 1960, showing dermal layer of pentactines and stauractines of various sizes pierced by tufts of monaxon pleuralia, ×5; *2,* fragmentary specimen with part of dermal layer removed to show coalesced, continuous, radiating mass of spicules in sponge interior, AMNH 28076:1, Word Formation, Word Ranch, Glass Mountains, Texas, USA, ×1 (Finks, 1960).

mid-Cambrian sponges either have longer basalia already differentiated, or the prostalia are of equally great length.

In some later forms, such as the Permian *Polylophidium* FINKS, 1960, the pleuralia are gathered into local tufts all over the sponge surface (Fig. 52.1), with a much wider, ring-like tuft providing attachment; all the tufts coalesce on the sponge interior into a continuous radiating mass of spicules (Fig. 52.2), and it is evident that the pleuralia and basalia had a common origin. The numerous tufts of the Devonian *Polylophalis* REIMANN, 1945a appear to be all basalia. HALL and CLARKE (1899) stated that many of the dictyosponges have tufts of pleuralia arising either from nodes, as in *Hydnoceras,* or from the intersections of vertical and horizontal spicule bundles, as in *Physospongia.* Examination of the type specimens indicates that in at least some of these it is not clear whether they are truly pleuralia or simply sprung-out portions of the internal horizontal spicule bundles brought about by crushing of the specimen due to burial.

The erect lamella that HALL and CLARKE (1899) interpreted as being present on the surface of such dictyosponges as *Clathrospongia* HALL, 1884 need comment. These are supposed to be extensions of the dermal quadrate mesh of stauractines perpendicular to the sponge surface along the traces of the major and minor quadrules. HALL and CLARKE believed that these, in turn, had lamella perpendicular to them, so that a three-dimensional cubic boxwork of lamella existed above the original dermal surface of the sponge (HALL & CLARKE, 1899, pl. 49,*6–7*). The evidence for these structures appears to be two-fold: (1) the surfaces of the specimens have pronounced ridges along the major and to a lesser extent along the minor quadrules; (2) at the sides of some specimens the dermal quadrate mesh seems to be prolonged as a kind of flange about one major quadrule wide that lies flat on the enclosing matrix rather than arching over the sediment filling the sponge interior, as the rest of the dermal mesh does. Close examination of many of the type specimens raises questions regarding the interpretation of HALL and CLARKE. The flanges are indeed present, but they could be interpreted as crushed edges of the specimen in which the two sides of the thin body wall were flattened together without sediment between them (or alternatively as flaps laid back along a vertical split). The ridges are also present, but they are nowhere nearly as high as the flanges they are supposed to represent. In at least one instance, the syntype of *Acloeodictya marsipus* illustrated by HALL and CLARKE (1899, pl. 55,*4*)

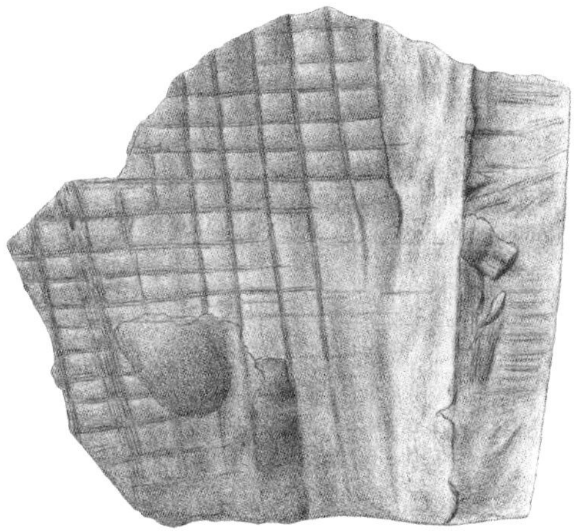

FIG. 53. Enlarged view of part of syntype of *Acleodictya marsipus* HALL & CLARKE, 1899, in which one of larger ridges is formed of vertical rays of large hypodermal pentactines or stauractines with underlying comitalia, Carboniferous Keokuk Group, Crawfordsville, Indiana, USA, ×1 (Hall & Clarke, 1899).

in which the spicules are preserved as pyrite rather than as the more usual impressions in matrix, the largest ridges are formed by the rays of large hypodermal pentactines or stauractines together with underlying bundles of comitalia (Fig. 53). It is possible that all the ridges in the genera considered by HALL and CLARKE to have erect, reticulating lamella (*Clathrospongia, Acloeodictya, Thysanodictya, Lebedictya*) are formed by the rays of large hypodermalia or internal spicule bundles. Because the usual preservation of these sponges is as impressions on the matrix, it is not always possible to be sure of the form of the spicules.

EVOLUTIONARY RELATIONSHIPS AND CLASSIFICATION

It is not unreasonable to regard the thinly walled Middle Cambrian sponges, which are the oldest whole hexactinellids preserved, as a coherent group close to the basic stock of the class. They all have in common a thin outer layer of stauractines or pentactines in parallel orientation that persists as a dermal layer in most later reticulosids. Thus, parallel orientation and lack of fully developed hexactines seem to be primitive features. Distribution of choanocytes that can be inferred from this thin skeletal wall, namely either lining the central cavity or else forming rather short choanocyte chambers between the proximal rays of the pentactines, is what one expects as a primitive feature. The nearly continuous covering of simple, rhabdodiactine prostalia is not necessarily expected but on the evidence of these sponges is also a primitive feature. Most of these early genera (*Protospongia* SALTER, 1864; *Acanthodictya* HINDE in DAWSON & HINDE, 1889; *Palaeosaccus* HINDE, 1893a; and *Diagoniella* RAUFF, 1894) are vase shaped (ranging from spheroidal to conicocylindrical) with an open osculum and with a few of the rhabdodiactine prostalia greatly elongated as basalia. Root tufts would, therefore, also seem to be primitive.

A second group, contemporary with the above, appears to represent a more advanced state. Inside the pentactine layer is an equally thin layer of hexactines not in parallel orientation but arranged irregularly about large, closely spaced circular openings that were probably exhalant skeletal canals. This is the first appearance of a parenchymal layer distinct from the dermal layer. *Hintzespongia* RIGBY & GUTSCHICK, 1976 is the only named genus, but *Ratcliffespongia* RIGBY, 1969 is probably the denuded inner layer of such a sponge. The extent to which a similar layer may have been present but not preserved in sponges of the *Protospongia* group is an open question.

All these early thinly walled hexactinellids have been found in shaly sediments. One might argue that more elaborate hexactinellids also existed at that time in less quiet-water environments but were destroyed prior to burial. It is also possible that the various features of these thinly walled forms were specifically adapted to quiet-water environments. There is some evidence for one or both of these conclusions because very similar thin-walled species not only persisted into the Ordovician Utica Shale, contemporaneous with the very elaborate, thickly walled *Brachiospongia* and *Pattersonia* of other facies but are found also as late as Pennsylvanian in a black shale (Mecca Quarry Shale) above a coal seam. No mid-Cambrian sponges of the complexity of a *Brachiospongia,* however, have been found in nonshale facies, nor have any isolated modified hypodermal spicules been found. The earliest thickly walled Cambrian sponge, *Multivasculatus* Howell & Van Houten, 1940 from nonshaly beds of the Upper Cambrian is still relatively simple and built of unmodified hexactines. Furthermore, even if some features of the thinly walled forms are specific adaptations to quiet water and muddy bottoms, the evident simplicity of their organization and spiculation accords well with their early appearance in the fossil record, in support of their being truly primitive.

Persistent descendants of *Protospongia* with monolayered body walls of stauractines or pentactines may include the late Silurian *Plectoderma* Hinde, 1884a, in which the larger spicules may be superposed to form bundles of parallel rays. Another such descendant may be *Actinodictya* Hall, 1890b from the Upper Devonian, in which the stauractines are extremely large, some more than half the sponge diameter, which itself is considerable; in this genus the stauractines lose their parallel arrangement.

Dierespongia Rigby & Gutschick, 1976 is an Ordovician form from a limy facies and has developed a separate layer of parallel hexactines. Some other poorly preserved Ordovician sponges may also be related, such as *Stephanella* Hinde, 1891; *Polyplectella* Ruedemann, 1925; *Foerstella* Ruedemann, 1925; and *Sycodictya* Ruedemann, 1925, all from the Utica Shale. Whether the Permian *Polylophidium* Finks, 1960 is an ultimate descendant of this lineage is not certain, but its general organization is very similar although more complex in its separate tufts and thicker wall.

The special structure seen in the Cambrian *Hintzespongia* Rigby & Gutschick, 1976 seems to be continued in the equally thinly walled Ordovician genera *Cyathophycus* Walcott, 1879 and *Teganium* Rauff, 1894 from the Utica Shale. There can be no doubt that topotype specimens of the type species of *Teganium* (*Cyathophycus subsphaericus* Walcott, 1879) that were illustrated by Hall and Clarke (1899, pl. 1,*14–22*) under the name of *Sphaerodictya subspherica* (Walcott) have exactly the same basic structure as *Hintzespongia,* with an inner layer of hexactines arranged irregularly around circular openings overlain by a quadrate mesh of stauractines. The outer layer of bristling prostalia seen in these specimens of *Teganium* is not known in *Hintzespongia;* but, as discussed above, it is a primitive feature of many Cambrian forms. (The actual type specimens of *Teganium* have no spicules; see Finks, 1960, p. 111.) *Cyathophycus* is closer to *Hintzespongia* in external form, being conicocylindrical rather than spherical. The spicules of the inner layer are not known, but Hall and Clarke (1899, pl. 1,*1–13*) showed beneath the outer quadrate mesh a layer of granular pyrite outlining circular openings of the same relative size and distribution as those in the inner layers of *Hintzespongia* and *Teganium.* It is worth noting that all these thinly walled forms are from shales. Thickening of the inner layer with its circular openings and nonparallel hexactines results in a wall structure like that of *Brachiospongia* and *Pattersonia.* As will be discussed below, *Hintzespongia* may have given rise by such a process to these thickly walled contemporaries of *Teganium* and *Cyathophycus,* which were perhaps adapted to a rougher-water environment.

The quadrate mesh of parallel stauractines that forms the principal skeleton of *Protospongia* is not only homologous with the outer layer of *Hintzespongia, Teganium, Cyathophycus,* and their thicker-walled descendants such as *Brachiospongia* and *Pattersonia;* but it also persists as the principal skeleton of another lineage, the great group known as the Dictyospongiidae HALL & CLARKE, 1899. In this family a different type of inner parenchymal layer is developed, namely vertical and horizontal bundles of long spicules, perhaps rhabdodiactines, that parallel the outer quadrate mesh. The beginning of this structure is perhaps seen in the Cambrian *Acanthodictya,* in which vertical bundles of spicule rays, perhaps including rhabdodiactines, run the length of the sponge and are prolonged into the root tuft. This sponge, at least as reconstructed, has a prismatic appearance because of the prominent vertical bundles. It is probably not without significance that the oldest true dictyosponge, *Prismodictya* HALL & CLARKE, 1899 from the upper Silurian (in the form of *Dictyophytra* RAUFF, 1894 and *Phormosella* HINDE, 1889b, *nomina oblita*) has a similar, prismatic, conicocylindrical form.

In the Lower Devonian *Retifungus* RIETSCHEL, 1970 the upper part of the sponge resembles *Prismodictya,* but the lower part is cylindrical and nonprismatic, passing downward into a long root tuft. This may mark the beginning of an evolutionary side branch that led to the cylindrical *Dictyospongia* HALL & CLARKE, 1899 of the Late Devonian and early Mississippian by a process of neoteny, for it is the lower, first-formed part of *Retifungus* that is cylindrical. The order of succession in time is right, and, moreover, the upper ends of large specimens of *Dictyospongia* may have incipient prism faces (HALL & CLARKE, 1899, p. 72). *Microstaura* FINKS, 1960 from the Permian may be a last descendant of this lineage. It, too, is cylindrical with a faint suggestion of prism faces in the largest specimens. The body well is thicker than in earlier dictyosponges by multiplication of the layers of parallel hexactines to form a cubic mesh. Whether this is, in turn, ancestral to Mesozoic dictyonines by fusion of the mesh is an open question. The dictyonines may have had another origin, in a nonparallel mesh, to be discussed below. It should be noted that only in this *Retifungus-Dictyospongia-Microstaura* group have root tufts been found attached to dictyosponges.

Hydnoceras CONRAD, 1842, the first of the dictyosponges to be described and probably the best known, is another likely derivative of *Prismodictya* in which the eight-sided prismatic shape is retained. *Hydnoceras* has horizontal expansions and contractions, each expansion bearing a whorl of eight nodes. The nodes are located at the interfacial angles and vary in size from low peaks to large, pendant, saccular protuberances. They tend to be consistent in form, not only within a single individual but among individuals at a given locality, suggesting either genetically differentiated local demes, a certain amount of environmental control, or both. One might expect that the more elaborately protruded nodes formed in quieter water, where mechanical strength was less necessary and where the need for more intake surface was greater (or perhaps where mechanical considerations did not forbid the packing of more sponge into a given volume of space).

Genera of related shape include *Rhabdosispongia* HALL & CLARKE, 1899, which lack the nodes but have the annular expansions and contractions as well as the prismatic sides that are initially eight in some species but generally more. This genus may have arisen, like *Hydnoceras,* directly from *Prismodictya. Ceratodictya* HALL & CLARKE, 1899 has only the annular expansions and contractions without trace of either prismatic sides or nodes. *Hydnocerina* CLARKE, 1918a is another Late Devonian member of the *Hydnoceras* group, as are the others so far discussed; it lacks prismatic sides but bears whorls of nodes that are more rounded and more closely spaced than those of

Hydnoceras. The Mississippian *Cleodictya* HALL, 1884 resembles a *Hydnocerina* with but one whorl of nodes and may have been derived from it.

Botryodictya HALL & CLARKE, 1899 is a Late Devonian to Mississippian nonprismatic sponge with large, saccular protuberances like the most extreme ones seen in *Hydnoceras* but that often split further into two or four subprotuberances. The lower part of the body of *Botryodictya* is a narrow cylinder like *Dictyospongia* (some specimens of *Dictyospongia* might be juveniles of *Botryodictya*), but it expands abruptly into the protuberance-bearing main part of the sponge. It is possible that it arose from *Dictyospongia,* acquiring its resemblances to *Hydnoceras* by convergence; but an origin from *Hydnoceras* by suppression of prism faces (a common change) and elaboration of nodes is more direct.

Interfacial ridges between prism faces in *Prismodictya* and *Hydnoceras* seem to be emphasized by extra large, internal, vertical spicule bundles. Enlargement of these bundles and the similar horizontal bundles may have led to another group of dictyosponges derived from *Prismodictya.* In the one that is closest to *Prismodictya,* namely the late Devonian *Gongylospongia* HALL & CLARKE, 1899, each prism face bears a vertical series of alternate protuberances and depressions, separated by the unbroken and rather strong interfacial angles; unlike *Hydnoceras,* the entire sponge does not have expansions and contractions, nor do the nodes occur on the interfacial angles.

Another Late Devonian genus has both vertical and horizontal spicule bundles that outline rectangular spaces. Every alternative horizontal space in alternative vertical rows was occupied by what may have been either a protuberance of the dermal mesh or a parietal gap. If it were a protuberance then the vertical rows would resemble the vertical rows of *Gongylospongia* minus the depressions. It must be admitted that the interpretation of this genus, *Uphantenia* VANUXEM, 1842, is open to doubt. That of the Mississippian *Physospongia* HALL, 1884 is not, however, and its resemblances to *Uphantenia* help in the reconstruction of the latter. In *Physospongia* strong horizontal and vertical spicule bundles outline rectangles that contain alternating protuberances and depressions in both vertical and horizontal directions so as to form a checkerboard pattern. Each vertical row resembles those of *Gongylospongia,* and the pattern of *Uphantenia* would be produced by the omission of alternating vertical rows. These alternating vertical rows that lack protuberances in *Uphantenia* are narrower than the rows that bear protuberances. If they are made narrower still so that they become an interfacial ridge, they produce the pattern of *Gongylospongia.* The sequence *Prismodictya-Gongylospongia-Uphantenia-Physospongia* forms a graded morphological series that is consistent with their occurrence in time, and it may represent an actual phylogenetic sequence.

The presence of vertical and horizontal spicule bundles outlining quadrules larger than the small squares of the stauractine mesh leads to another group of dictyosponges characterized by large quadrules bounded by coarse ridges. The genera of this group in which spicules are actually preserved reveal two kinds of structure: in the Late Devonian *Arystidictya* HALL & CLARKE, 1899 and in the early Mississippian *Thamnodictya* HALL, 1884 the large quadrules are outlined by bundles of spicules, possibly rhabdodiactines; in the early Mississippian genera *Lebedictya* HALL & CLARKE, 1899 and *Acloeodictya* HALL & CLARKE, 1899 the large quadrules are outlined by large stauractines or pentactines whose rays are underlain by bundles of comitalia, possibly rhabdodiactines. The first type of structure may represent merely the second type from which the stauractines have been lost. (Many of the specimens are internal molds from which the entire dermal layer has been lost.) Alternatively the stauractines may have been added in the course of evolution, inasmuch as the second type appears later in time. HALL and CLARKE (1899) interpreted the structural elements outlining the large quadrules in *Lebedictya* and *Acloeodictya,* as well as in the similar Late Devonian to early Mississippian genera *Thysanodictya* HALL & CLARKE, 1899 and *Clathrospongia* HALL, 1884, to have been erect lamella formed of the stauractine mesh. This has been discussed above and reasons given for doubting their interpretation. Examination of many of the type specimens shows that the large quadrules of *Arystidictya, Thamnodictya, Lebedictya,* and *Acloeodictya* are outlined by spicule bundles, whatever the further contributions of other structures may have been. Thus the whole group may be related to and perhaps descended from the *Gongylospongia-Uphantenia-Physospongia* group.

The Permian genera *Endoplegma* FINKS, 1960 and *Carphites* FINKS, 1960 have a structure of large pentactines or hexactines in parallel orientation underlain by vertical and horizontal bundles of tauactines. This structure is similar to that of *Acloeodictya* and *Lebedictya,* although the pentactines are very much coarser, as are the inner bundles. Nevertheless, the presence in *Endoplegma "Carphites" diabloense* of a paraclavule, bidentate anchor, and short, curved strongyle-like spicules (FINKS, 1960, p. 127), all of which are characteristic dictyosponge accessory spicules (bidentate anchors are known from *Acloeodictya,* paraclavules and strongyles from *Lebedictya*), suggest that *Endoplegma* and *Carphites* arose from this group of dictyosponges by a kind of gigantism. Of the two Mississippian genera cited, *Acloeodictya* is most similar to the Permian sponges, both in its larger pentactines and in its external shape, which is like that of *Endoplegma.*

Mattaspongia RIGBY, 1970a from the Late Devonian may be related to this group of dictyosponges. Its large parallel hexactines, occurring in a single layer, are similar in size and arrangement to the large hypodermal pentactines of *Lebedictya* and *Acloeodictya,* and they are similarly accompanied by comitalia of rhabdodiactines.

Another genus from the late Pennsylvanian and Early Permian, *Stereodictyum* FINKS, 1960, consists of horizontal and vertical spicule bundles in multiple layers, without any dermalia or hypodermalia. (The type species lacks them, but the early Pennsylvanian *S. proteron* RIGBY & WASHBURN, 1972 is reported to have a fine surficial mesh of hexactines on one specimen.) The sponge occurs in broad, curving sheets, and perhaps the explanate *Thamnodictya* from the Mississippian is near its ancestry. Its origin from this group of dictyosponges involves the loss of dermal and hypodermal spicules and the multiplication of spicule bundle layers. The spicules making up the bundles in *Stereodictum* appear to be largely reduced hexactines in which at least one ray is developed in each of the three axes. Thus the origin might involve the addition of rays to the bundle spicules. Inasmuch as we are not certain that the spicules in the bundles of dictyosponges are rhabdodiactines, there may not be a very great difference in spicule form, after all.

Another group of dictyosponges has a smooth surface without protuberances, ridges, or flat faces. Wherever they are known from sufficiently complete material, the body is cylindrical, often narrowest in the middle, with a wide, sometimes flat base and a wide upper end; they are sometimes vertically fluted. They may have evolved from the *Hydnoceras* group by subduing of annulations and protuberances, comparable to a progression from *Hydnocerina* through *Cleodictya* to *Calathospongia.* Late Devonian members are *Hydriodictya* HALL & CLARKE, 1899 and *Corticospongia* CASTER, 1939. Mississippian genera are *Calathospongia* HALL & CLARKE, 1899; *Lyrodictya* HALL, 1884; *Phragmodictya* HALL, 1884; and *Griphodictya* HALL & CLARKE, 1899. In addition, the Mississippian *Ectenodictya* HALL, 1884 appears to be based on a fragment of one of these sponges, according to HALL and CLARKE (1899, p. 164–165), probably *Calathospongia.* Two other supposed dictyosponges, *Hallodictya* HALL & CLARKE, 1889 and *Cryptodictya* HALL, 1890 from the Upper Devonian, may be similar fragments, but the paucity of spicules makes them somewhat dubious even as sponges. This whole group may be heterogeneous.

Griphodictya, in particular, seems to be different from the other dictyosponges in that its interior is filled with a continuous mass of rhabdodiactines or similar long spicules not organized into separate bundles, at least as preserved. *Phragmodictya* is another special instance with its supposed basal disk and frill; it could be descended from *Botryodictya* by the loss of stalk and protuberances, especially if the diaphragm between body and stalk in *Botryodictya* is homologous to the basal disk of *Phragmodictya.*

Griphodictya brings us to another problem, namely the origin of the Hexasterophora and Amphidiscophora. As mentioned above, *Griphodictya* was reported to contain numerous hexasters along with paraclavules. Oxyhexasters and small hexactine microscleres occur with clemes as early as in the middle Silurian *Corticulospongia* RIGBY & CHATTERTON (1989, p. 41) in the thin-walled euplectellid from Arctic Canada. The Late Carboniferous *Itararella* and the Permian *Endoplegma ("Carphites")* contain what may be adventitious hexasters together with amphidiscs and a paraclavule, respectively. In addition, the Early Carboniferous *Erythrospongia* HUDSON, 1929 apparently contains hexasters along with such dictyosponge-like spicules as bidentate anchors, curved strongyles and possibly clemes, although not paraclavules. If paraclavules are homologous to amphidiscs, which seems probable, then either the supposed hexasters are not true hexasters, or there were sponges in Carboniferous and possibly in Silurian times that contained two types of microscleres that today are mutually exclusive. If so, the two groups, Amphidiscophora and Hexasterophora, may not yet have differentiated one from the other. The Early Carboniferous *Uralonema* LIBROVICH, 1929 is the earliest known sponge to contain true amphidiscs. It is approximately contemporaneous with the hexaster-bearing forms *Griphodictya* and *Erythrospongia* and for that matter with the earliest paraclavule-bearing forms. *Uralonema* seems to be assignable not only to the true Amphidiscophora (that is, the Amphidiscosa) but specifically to the Pheronematidae or close to them due to its possessing dermal and gastral pinules and parenchymal hexactines. Perhaps the true Amphidiscophora arose at this time alongside a persistent earlier group in which amphidiscophoran and hexasterophoran characteristics were still combined.

As for the earliest Hexasterophora, the Early Carboniferous *Erythrospongia* is a possible candidate, but its complement of dictyosponge-like accessory spicules, as well as its somewhat obscure gross morphology, raise some doubts. The Permian *Pileolites* FINKS, 1960 may be a better candidate, but no hexasters have been found in it. Isolated hexasters have been recovered from the sediment associated with the type specimens, however; see FINKS, 1960, p. 142, and pl. 33*b*. Nevertheless, its gross morphology is hexasterophoran-like. Originally described as a euplectelloid (FINKS, 1960), it seems to have even more interesting hexasterophoran affinities. REID's emphasis (1964, p. lxxiii ff.) on the dictyonal strand as the primary unit of organization in dictyonines requires a reexamination of earlier interpretations of its structure.

Pileolites is built of horizontal layers of irregularly crisscrossing, long, spicular strands apparently joined by synapticulae. These strands can be demonstrated in a few places to be composed of more than one hexactine and fused together seriatim with overlapping rays. The strands were originally interpreted as composed of a few hexactines at most, with very greatly elongated rays in the horizontal plane. An interpretation as a fused linear series of many small hexactines with nearly equidimensional rays, however, avoids the assumed discrepant length between horizontal and vertical rays and also accounts for some of the supposed synapticulae as actually hexactinal rays. (Some suppression of rays may have to be invoked, however; for six-rayed crossings are not common.) This permits each horizontal strand to be considered as a dictyonal strand unit. The nonparallel arrangement of such strands occurs in the small, supposedly aberrant, dictyonal group of the Aulocalycidae. It is possible that this group is primitive and that *Pileolites* is the ancestor of the dictyonine Hexasterophora. If so, one should look for still earlier roots among Paleozoic lyssacinosans of irregular spicular arrangement, such as the Brachiospongioidea, and not among the parallel-spiculed protosponge-dictyosponge lineage. It may be that the dictyosponges are closer to the Amphidiscophora and the brachiospongioids to the Hexasterophora.

REID's description of the aulocalycoid skeleton (1964, p. xcii) agrees with the structure of *Pileolites* in almost every detail. In his discussion (p. xciii–xciv) he pointed out the similarities with the euplectellid lyssacinosan skeleton but noted that they lack the parenchymal rhabdodiactines that both he and IJIMA (1927) accepted as diagnostic of Lyssacinosa. By this criterion *Pileolites* could be accepted as a lyssacinosan, since it has such rhabdodiactines along with the aulocalycoid skeleton, and the aulocalycoids could be regarded as descendants that had lost the rhabdodiactines. With this point of view, the aulocalycoids would not be dictyonines, not even the degenerate dictyonines envisaged by REID (1964, p. xciv), but rather a side branch of the Lyssacinosa: degenerate euplectelloids. It may be of heuristic value, however, even if not absolutely compelled by present evidence, to regard *Pileolites* as the Permian beginning of an aulocalycoid grade of dictyonine organization, which ultimately led to the first mid-Triassic dictyonines of euretoid grade.

It is now worth returning to the earliest amphidiscosan, *Uralonema.* The hypodermal and hypogastral hexactines of the Early Carboniferous genus, with their swollen distal and paratangential rays, are nearly identical in size and shape with the corresponding spicules of the Pennsylvanian and Permian *Stioderma* FINKS, 1960, although *Stioderma* lacks the hypogastralia. Both genera possess a large, stout root tuft and possibly also the inward rolling of the dermal layer at the oscular rim (see LIBROVICH, 1929, fig. 1, p. 14, 46). *Uralonema* does not possess the large, circular, parietal gaps characteristic of *Stioderma,* but it is well to remember that specimens of *Uralonema* are only as large as juveniles of *Stioderma* and that in these small individuals of *Stioderma* the parietal gaps are quite small. The pinuli of *Uralonema* do not occur in *Stioderma,* nor do the spinose parenchymal hexactines of *Stioderma* occur in *Uralonema.* Of more interest in the present context, amphidiscs have not been found in *Stioderma.* Nevertheless, the resemblances are sufficient to suggest a close relationship, which would also include the Early Carboniferous *Hyalostelia* ZITTEL, 1878c, as interpreted by REID (1968c), who has likewise suggested such a relationship. The Late Carboniferous *Itararella* KLING & REIF, 1969 also belongs here; it has amphidiscs, but its external form is poorly known. If these four late Paleozoic genera are indeed true Amphidiscosa and, as mentioned earlier here, close to the Pheronematidae of the present day, what can be said of their origins?

Brachiospongia and *Pattersonia,* from the Trentonian (Caradoc), are the earliest known lyssacinosans with differentiated hypodermal spicules. Those of *Brachiospongia* somewhat resemble the hypodermalia of *Stioderma* and *Uralonema* in having a spherical knob in place of the distal ray and in having a distal tubercle on each tangential ray. These enlargements are much less strong than in the later genera; furthermore similar enlargements of varying degree are known from other genera, such as the Silurian *Astroconia* SOLLAS, 1881 and the Permian *Docoderma.* The hypodermal spicules of *Pattersonia* are less well known but were said by BEECHER (1889, p. 26), in describing the junior synonym *Strobilospongia,* to resemble those of *Brachiospongia. Pattersonia* is closer to the late Paleozoic Amphidiscosa in having a well-developed root tuft, which *Brachiospongia* lacks. Indeed, among the thickly walled Paleozoic lyssacinosans with nonparallel parenchymal and hypodermal spicules and with large skeletal canals or parietal gaps, one can distinguish two sorts: open, cup-shaped forms with root tufts and vaselike forms with flat bases and no root tuft. Among the former there are, including *Pattersonia;* the Devonian *Pelicaspongia* RIGBY, 1970a; *Pseudohydnoceras* REIMANN, 1945a; *Bayviewia* REIMANN, 1945a; *Polylophalis* REIMANN, 1945a; and *Calicispongia* REIMANN, 1945a (the last is not reported to have basalia but resembles the others in form); and the Pennsylvanian *Arakespongia* RIGBY, CHAMBERLAIN, & BLACK, 1970. Among the latter there are, including *Brachiospongia;* the Ordovician *Rhaeaspongia* LAMONT, 1935; *Colpospongia* LAMONT, 1935; and *Pyruspongia* RIGBY, 1971; and the Silurian *Malumispongium* RIGBY, 1967b; and *Oncosella* RAUFF, 1894.

Microscleres are not known from any of these sponges. (The oxyhexaster reported from *Arakespongia* by RIGBY, CHAMBERLAIN, & BLACK [1970, p. 829 and fig. 12U] seems too large to be a microsclere and resembles in its spiral ornamentation the spicules of *Spiractinella* HINDE, 1888.) On the basis of the cup shape and root tuft, however, the Devonian *Pelicaspongia* group and perhaps the Pennsylvanian *Arakespongia* may represent a lineage connecting *Pattersonia* with the Carboniferous and Permian Amphidiscosa (*Uralonema, Hyalostelia, Stioderma*). The quadranchorate ends of the basalia in *Pattersonia, Uralonema, Hyalostelia,* and *Arakespongia* are another common feature. If true, the roots of the Amphidiscosa, and perhaps amphidiscs themselves, go back to the Ordovician *Pattersonia.*

In this view, the dictyosponges and their possible docodermatid descendants represent a sister lineage with some amphidiscophora-like features, especially the homologous paraclavules, but would not be directly ancestral to the Amphidiscosa.

The forms not bearing root tufts also make rather a coherent group. The Ordovician *Pyruspongia,* and the Silurian *Malumispongium* and *Oncosella* are very similar in shape although the Ordovician genus lacks skeletal canals. *Brachiospongia,* also Ordovician, with its radial, rhizophytous protuberances, resembles the contemporary *Rhaeaspongia. Colpospongia,* with its strong radial flutings, is intermediate between its contemporaries, the radially protuberant *Brachiospongia* on the one hand and the slightly fluted *Pyruspongia* on the other. This lineage cannot be traced clearly beyond the Silurian. The Permian *Pileolites* resembles in external form nothing so much as the tip of a *Brachiospongia* protuberance. It is the only Late Paleozoic hexactinellid with a flattish base and no root tuft. As mentioned earlier, it is also characterized by irregularly arranged parenchymal spicules and prominent skeletal canals perpendicular to the surface. The distal ray of its dermal (or hypodermal) spicules is reduced to a spherical knob. If all these similarities to *Brachiospongia* signify a true relationship of descent, it may be that the roots of the true Hexasterophora go back to *Brachiospongia.* Only the recovery of microscleres from these genera can settle the matter. Thus the hypothesis that *Pileolites* is a protodictyonine may make it possible to trace their source to the Ordovician.

The following is a proposed phylogeny that will embody so far as possible these working hypotheses. The line of descent from protospongiids like *Acanthodictya* to the dictyosponges and from them to the stereodictyids and docodermatids of the Late Paleozoic will have to be accommodated in a single high-level taxon. The order Reticulosa REID, 1958a is available for this group and was essentially defined for them. (The two Permian families here included were not yet published at that time.) The presence of paraclavules and the principal skeleton consisting of a quadrate mesh of spicules of dermal or partly dermal origin were among the original diagnostic features. As discussed above, this group has strong amphidiscophoran affinities, not the least of which are the paraclavules, here considered homologues of the amphidiscs. For this reason it is here placed into the subclass Amphidiscophora. The possible presence of hexasters in this group necessitates a redefinition of the subclass Amphidiscophora. The order Hemidiscosa may have originated during the mid-Paleozoic from dictyosponges like *Griphodictya* through hemidiscosans like *Microhemidiscia.* (The presence of a hemidisc in the amphidisc-bearing *Itararella* KLING & REIF, 1969, if it is not adventitious from the co-occurring *Microhemidiscia,* suggests an alternative origin by reduction from the Amphidiscosa.)

The remaining order of the Amphidiscophora, the Amphidiscosa, with true amphidiscs, seems to have had a separate origin, going back to the protospongiids by a different lineage. The Late Paleozoic Stiodermatidae appear to be true Amphidiscosa, based on the inclusion therein of *Uralonema,* which has demonstrable amphidiscs. A related group of mid-Paleozoic forms, which can be accommodated in the Pelicaspongiidae RIGBY, 1970a, carry this lineage back ultimately to *Pattersonia* of the Ordovician and probably to still earlier forms known only by their stout root tufts (see HINDE, 1888, p. 110) from at least as far back as the Tremadoc. Both *Pattersonia* and *Brachiospongia* have as their most similar, simpler predecessor the mid-Cambrian *Hintzespongia,* which shares with them the characteristics of a parenchymal layer of nonparallel hexactines organized around prominent skeletal canals. *Hintzespongia,* the very similar Ordovician *Cyathophycus,* and the mid-Cambrian *Ratcliffespongia,* which has only the parenchymal layer and may be a *Hintzespongia* denuded of its dermal layer, can be accommodated in the family Hintzespongiidae FINKS, 1983b.

Where to place the Hintzespongiidae is to some extent a matter of choice. Because they seem equally related both to the pattersoniid line leading to the Amphidiscosa, and to the brachiospongiid line leading to the Hexasterophora, including them in either does not seem appropriate. Their greater similarity to the equally thinly walled Protospongiidae favors their inclusion in the order Reticulosa. The closely similar, spherical Teganiidae would be placed alongside them.

As shown above, the interpretation of *Pileolites* as having an aulocalycoid skeleton enables us to connect the hexactinosan Hexasterophora with *Brachiospongia.* The order Lychniscosa probably originated from the Hexactinosa in Triassic times as suggested by the incomplete development of lychnisc nodes in the first lychniscosan, *Triadocoelia* VINASSA DE REGNY, 1901 from the Carnian. *Brachiospongia* itself and related forms with nonparallel parenchymalia enlarged hypodermalia and no root tuft, which are the Pyruspongiidae RIGBY, 1971 and the Malumispongiidae RIGBY, 1967b, would be included in a revised Brachiospongioidea and would be the earliest members of the order Lyssacinosa. The order Hexactinosa would commence with the Permian *Pileolites.*

Several other Paleozoic forms are not yet accommodated in this scheme. One is the Late Cambrian *Multivasculatus,* which is the first hexactinellid with a cubic mesh of hexactines. Although similar to the protospongiids in its parallel spicules (but not in its encrusting habit) it does not seem to be related to the line that led to the dictyosponges, with their internal spicule bundles and thin walls. It may be ancestral to the Devonian *Titusvillia* CASTER, 1939, which, with its curiously reversed flanges on the branches, might be taken for a *Scalarituba* or other ichnofossil, as CASTER (1939, p. 7) originally did, were it not for the presence of hexactines. The connection with *Titusvillia* is based on the presence in the latter of a cubic mesh of hexactines. CASTER himself expressed reservations concerning a three-dimensional cubic mesh. It is not present in the externally similar *Annulispongia* RIGBY & MOYLE, 1959, which has only a dermal and cloacal layer. Other genera related by external shape include *Armstrongia* CLARKE, 1920; *Protoarmstrongia* CASTER, 1941; and possibly *Aglithodictya* HALL & CLARKE, 1899 and *Iowaspongia* THOMAS, 1922. On the last, I could find no spicules on the holotype, and it may be a burrow. The Multivasculatiidae de Laubenfels, 1955 and the Titusvilliidae Caster, 1939 will accommodate these genera respectively, although if *Aglithodictya* is included with the *Titusvillia* group, the family Aglithodictyiidae Hall & Clarke, 1899 (*nomen translatum*) has priority.

Another early Paleozoic form may be related to the titusvilliids in that it possesses a skeleton composed solely of a dermal and a gastral quadrate mesh, as in *Annulispongia* and possibly *Titusvillia.* This is the Ordovician *Hydnodictya* Rigby, 1971, the sole member of the Hydnodictyiidae Rigby, 1971. Unlike the titusvilliids, the paratangential spicule rays of the gastral mesh are orientated at 45° to those of the dermal mesh.

Still another group characterized by two layers of quadrate mesh, at least in some species, is that including *Dierespongia* Rigby & Gutschick, 1976 and possibly *Stephanella* Hinde, 1891; *Polyplectella* Ruedemann, 1925; *Foerstella* Ruedemann, 1925; *Sycodictya* Ruedemann, 1925; and *Polylophidium* Finks, 1960. These are all characterized by a spheroidal shape and large rhabdodiactine prostalia that extend from the center of the sponge to well past the dermal layer. Rigby and Gutschick (1976) was of the opinion that the spicules of the inner layer in *Dierespongia* were not in parallel orientation. It should be noted that original parallelism of spicules may not be preserved when a small, spheroidal surface is compressed flat upon a bedding plane. In any event, the inner layer of *Dierespongia* is not organized around large, circular openings as in *Hintzespongia, Cyathophycus,* or *Teganium.* Inasmuch as the latter type of organization points toward another line of development, it is desirable to separate such forms from those in which the inner layer is continuous, whether perfectly parallel or not. Thus the concept of the Dierespongiidae should be revised to encompass those spheroidal forms with large rhabdodiactine prostalia radiating from the center of the sponge and with the inner layer not organized about circular openings. For this reason, *Hintzespongia* is here removed from the Dierespongiidae and placed with *Ratcliffespongia* and *Cyathophycus.* The Dierespongiidae, Hydnodictyiidae, and Titusvilliidae seem somewhat related in that they tend to emphasize a dermal and a gastral layer rather than a parenchymal one. They also seem in their prevailingly quadrate arrangement of spicules to be clearly derivable from the Protospongiidae and therefore to find a place among the Reticulosa.

It remains to discuss *Vauxia* Walcott, 1920. This eminently spongelike object, whose elegant colonies grace many museum dioramas of the Burgess Shale sea floor, was originally considered a hexactinellid by Walcott (1920, p. 316). The type specimens and topotype material collected by D. H. Krinsley do not have anything resembling a hexactinellid spicule (see also Rigby, 1986a). Where best preserved, the structure consists of a hexagonal to rectangular net somewhat elongate parallel to the longitudinal axis of the branch. If the hexagons or rectangles are not elongate then the zigzag lines parallel to the longitudinal axis are more strongly impressed into the matrix than the cross connections (see also Walcott, 1920, pl. 82,*1b;* Rigby, 1986a, pl. 1,*4,* pl. 2,*3,* fig. 8–10). This structure seems to be the basis for Walcott's diagram (1920, p. 319, fig. 9) in which one zigzag and its cross connection are taken to be three paratangential rays of a reduced hexactine. In most places the regular hexagonal structure breaks down into smaller, more irregular patterns that have a more curving than polygonal outline. Some of this may be due to superimposition of two layers of skeletal net, but some appear to be the irregular wrinkling due to shrivelling or flexing of a semirigid layer. The pattern is outlined by either grooves or ridges, which have been taken to be the impressions of spicules or the spicules themselves. The entire organism has the form of a branching, tubular sponge. These organisms were interpreted to be keratose sponges by Rigby (1986a).

An outline of the classification of Paleozoic hexactinellids proposed by Finks (1983, p. 109–112) and herein in Table 1.

TABLE 1. Outline of classification of Paleozoic Hexactinellids as proposed by Finks (1983b).

TAXON NAME AND LEVEL
Class Hexactinellida Schmidt
Subclass Amphidiscophora F. E. Schulze
Order Reticulosa Reid
Superfamily Protospongioidea Hinde
Family Protospongiidae Hinde
Superfamily Dierespongioidea Rigby & Gutschick
Family Dierespongiidae Rigby & Gutschick
Family Hydnodictyidae Rigby
Family Multivasculatidae de Laubenfels
Family Titusvilliidae Caster
Family Aglithodictyiidae Hall & Clarke
Superfamily Dictyospongioidea Hall & Clarke
Family Dictyospongiidae Hall & Clarke
Subfamily Prismodictyinae de Laubenfels
Subfamily Dictyospongiinae Hall & Clarke
Subfamily Hydnoceratinae Finks
Subfamily Calathospongiinae Hall & Clarke
Subfamily Physospongiinae Hall & Clarke
Subfamily Thysanodictyinae Hall & Clarke
Family Docodermatidae Finks
Family Stereodictyidae Finks
Superfamily Hintzespongioidea Finks
Family Hintzespongiidae Finks
Family Teganiidae de Laubenfels
Order Hemidiscosa Schrammen
Family Microhemidisciidae Finks
Order Amphidiscosa Schrammen
Family Pattersoniidae Miller
Family Pelicaspongiidae Rigby
Family Stiodermatidae Finks
Subclass Hexasterophora F. E. Schulze
Order Lyssacinosa Zittel
Superfamily Brachiospongioidea Beecher
Family Brachiospongiidae Beecher
Family Pyruspongiidae Rigby
Family Malumispongiidae Rigby
Order Hexactinosa Schrammen
Family Pileolitidae Finks

This scheme of phylogeny and classification differs from that proposed by RIGBY (1976b, p. 56, fig. 5) in only two essential features. The Pyruspongiidae are here placed with the Malumispongiidae (because of their similar special shape, nonparallel spicules, and enlarged hypodermalia), while RIGBY put them with the Hydnodictyidae. He considered the lack of parietal gaps (that is, circular skeletal canals) more diagnostic of relationship and the development of enlarged hypodermalia (his armored) as a convergent feature. Also the Dierespongiidae are here divided among two families, the Hintzespongiidae, including *Hintzespongia,* and placed with the Teganiidae as RIGBY did his Dierespongiidae, and the restricted Dierespongiidae; including *Dierespongia* and *Choia,* and here placed with the Hydnodictyidae. RIGBY did not deal with the Amphidiscophora nor with the problem of the origin of the Hexasterophora, two major problems with which the present classification and phylogeny are intimately bound.

The present scheme differs from that of REID (1958a, p. xlvi, fig. 24) in placing the Reticulosa with the Amphidiscophora and in deriving the Hemidiscosa and the Amphidiscosa from two separate lineages within the Reticulosa, the second from an earlier offshoot that also gave rise to the Hexasterophora Lyssacinosa, and through the latter to the Hexactinosa and Lychniscosa.

A radically different hypothesis of hexactinellid phylogeny could be proposed if elements of the dictyosponge quadrate mesh were regarded as protodictyonine strands. It has been pointed out above (p. 135) that specimens of *Hydnoceras* and other dictyosponges with repaired injuries sometimes have curving strands of the quadrate mesh connecting horizontal with vertical elements of the regular mesh, as though the linear series of spicules were the unit of construction. In all dictyosponges where the individual spicules have been seen (usually as pyritized replacements of the original opal) they have been separate and unfused. Thus dictyosponges seem not to possess true dictyonal strands. In most Devonian forms, however, including *Hydnoceras,* no spicules have been preserved, only their impressions in the sediment. Observations of these molds, including those of the injured areas referred to above, neither confirms nor denies a fusion of the spicules into strands. Even if they were fused, the dictyosponge quadrate mesh would differ from the similarly monolayered farreoid hexactinosan mesh (see REID, 1964, p. lxxv ff.) in having strands of differing thickness outlining the several orders of quadrules.

In the Carboniferous dictyosponges, where spicules are often preserved, as well as in the Cambrian and Ordovician

Protospongiidae, the larger quadrules are formed by larger individual stauractines or pentactines. Thus, the whole basis of the dictyosponge skeleton seems to be a nested set of spicules of different size, and it stands in contrast to the uniformly fine **farreoid** mesh. On the basis of presently available evidence, considering the dictyosponge mesh to be ancestral to that of the hexasterophoran dictyonines seems unwarranted. The tendency of spicules of similar size to be produced in linear series seems best regarded as a characteristic of Hexactinellida in general.

Earlier suggestions that cubic-meshed dictyosponges of nearly uniform spicule size, such as the Permian *Microstaura,* could have given rise directly to the dictyonine sponges by fusion of spicules (e.g., FINKS, 1960, p. 102) involve the implicit hypothesis that the dictyosponges were ancestral to at least some of the Hexasterophora. REID's placement of the Reticulosa as a possible offshoot of the Hexasterophora (REID, 1958a, fig. 24, p. xlvi) does not put the dictyosponges into the direct ancestry of the dictyonines but suggests a closer relationship than that of the hypothesis proposed here. REID, however, regarded the quadrate mesh of the Reticulosa as dermal or partly dermal, while the dictyonine skeleton is parenchymal (1958a, p. xliv–xlv); thus, they would not be closely homologous, and any common tendencies to form linear series of spicules would not indicate direct ancestry.

POST-PALEOZOIC LYSSACINOSA

R. E. H. REID

[formerly of Department of Geology, The Queens University of Belfast]

Post-Paleozoic **lyssacines** are known mainly from modern examples on which this account is based. In zoological classification, they are representatives of the order Amphidiscosa SCHRAMMEN of the subclass Amphidiscophora SCHULZE and the order Lyssacinosa ZITTEL, as restricted by SCHRAMMEN, of the subclass Hexasterophora.

Living Amphidiscosa are exclusively lophophytous sponges with amphidisc microscleres, to which staurodisc or hexadisc variants are occasionally added. Parenchymal megascleres vary from hexactines to rhabdodiactines and may be all of the latter type. Both autodermalia and hypodermalia are normally present. Autodermalia are usually pentactines, with the unpaired ray distal and **pinular;** hypodermalia are larger pentactines, with the unpaired ray proximal and running into the choanosome. The gastral skeleton consists of autogastralia like the autodermalia. Canalaria corresponding with the autodermalia or gastralia may also occur. Basalia are monactinal where their character is known, with a four- or six-armed axial cross at the distal end and a varying number (usually 2 to 8) of solid anchor teeth. Recurved spines are also often present along the shafts of these spicules. Monactinal or diactinal pleuralia or marginalia may also occur.

Three living families of Amphidiscosa are currently distinguished (IJIMA, 1927). The Pheronematidae are sponges whose principal parenchymal megascleres are hexactines or pentactines, although other types may also occur, including smaller rhabdodiactines. The basalia are characteristically bidentate, with two recurved flukes only, in line with axial rudiments of a four or six-armed axial cross. Uncinates are common among minor spicules present internally, and spiny monactines called scepters occur as pleuralia or marginalia. The scepters have the axial cross distal and the spines directed distally (not recurved as in the basalia), with a special distal whorl sometimes present.

The Hyalonematidae are sponges whose parenchymal megascleres include prominent rhabdodiactines, although hexactines may also be present. The basalia have grapnel-like ends with four or more recurved teeth in most instances and are usually arranged in a compact glass rope. Uncinates are nearly always absent, and no scepters occur, although the place of the latter may be taken by spiny rhabdodiactines with a central ring of buttonlike rudiments of the four reduced rays.

The third family, the Monoraphididae, comprises *Monoraphis* SCHULZE, which is distinguished by parenchymal megascleres that are mainly tauactines and by being supported by a single, very large, basal needle, the end of which is unknown.

Fossil Amphidiscosa are known mainly from occasional finds of loose amphidiscs in Mesozoic or Cenozoic sediments. No modern genera have been identified certainly as fossils, although some have been identified by inference from isolated megascleres (e.g., by HINDE & HOLMES, 1892).

The living Lyssacinosa were divided by IJIMA (1927) into four families, the Leucopsacadidae IJIMA, Euplectellidae GRAY, Caulophacidae IJIMA, and Rossellidae GRAY. This arrangement is accepted herein, although with some reservations and with the name Caulophacidae replaced by the senior synonym Asconematidae SCHULZE. The four families form two groups, with two families in each. Both groups include both basiphytous and lophophytous genera. The Leucopsacadidae and Euplectellidae are united by having simple **dermal skeletons** only, without distinct autodermalia and hypodermalia, and by basalia that are typically umbel-bearing diactines in lophophytous genera. The Asconematidae and Rossellidae have both autodermalia and hypodermalia, and the basalia of lophophytous genera (Rossellidae only) are anchorate pentactines.

The Euplectellidae are generally thin-walled sponges of normally tubular habitus that are either lophophytous or basiphytous. The dermalia are characteristically hexactinal, with a prolonged proximal ray that gives the spicules a swordlike appearance. IJIMA (1903, 1927) distinguished the subfamilies Euplectellinae GRAY and Corbitellinae IJIMA, the latter being a junior synonym of Taegerinae SCHULZE (1887a). The Euplectellinae are lophophytous sponges in which the basalia are typically rhabdodiactines with a terminal umbel at the distal end. The distal ray of the two that lie in line to form the shaft is always shorter than the proximal and sometimes is so short that the spicule is nearly a monactine. The shaft often bears recurved lateral spinules with which the umbel teeth appear to be homologous.

In simple Euplectellinae, e.g., *Holascus* SCHULZE, the wall is imperforate, with a simple parenchymal spiculation. The principalia may be all hexactines, pentactines, or stauractines, arranged in a regular manner with four tangential rays oriented longitudinally and transversely so that these apposed rays enclose square meshes. Rhabdodiactines are present, but only as comitalia.

In more advanced genera, e.g., *Euplectella* OWEN, the wall is pierced by numerous parietal oscula and has a more complex parenchymal spiculation, in which the rays of some spicules may intersect in diagonal directions. In both advanced and simple genera, the terminal osculum may be covered by a secondary sieve plate, which is formed at the end of normal growth and contains special megascleres called oscularia. In some forms, which may have advanced spiculation, all the megascleres are permanently separate. In others they may be fused, either in basal parts only or to varying extents through the upper parts or the whole body. Fusion occurs, in part, by union of the megascleres where they cross and contact one another but, in part, also by formation of synapticula where rays are close together but not in contact. Fusion may affect oscularia, as well as the parenchymal megascleres, but never the dermalia or gastralia.

The Taegerinae (=Corbitellinae IJIMA) are basiphytous sponges, with an encrusting basal skeleton composed of small fused hexactines (basidictyonalia, IJIMA). The parenchymal megascleres are mainly hexactines and rhabdodiactines, the latter generally being predominant. Some (e.g., *Regadrella* SCHMIDT) are thin-walled sponges with parietal oscula and a sieve plate, resembling some Euplectellinae although having the parenchymal skeleton formed mainly from diagonally crossing rhabdodiactines; but others have different habits. The parenchymal megascleres of the basal part are fused to one another and to the basidictyonalia; higher up they may be free or partly or entirely fused. If they are fused up to the oscular margin, the sieve plate if present becomes rigid.

The Leucopsacadidae of IJIMA (1903, 1927) are lophophytous sponges with similar basalia and no hypodermalia, but with autodermalia and gastralia that are all simple pentactines. According to IJIMA (1927), the parenchymal megascleres should be hexactines and rhabdodiactines with the former predominant; but in some of the genera he included the megascleres are mainly or all rhabdodiactines (*Caulocalys* SCHULZE, *Placopegma* SCHULZE, *Chaunangium* SCHULZE).

The Rossellidae are mainly lophophytous, but some are basiphytous sponges with both autodermalia and hypodermalia. The autodermalia are small hexactines, pentactines, or stauractines that may be spiny but never truly pinular and that have the unpaired ray proximal when pentactinal. The hypodermalia are typically much larger pentactines that may be normal pentactines with tangential rays meeting at right angles or partly or all paratropal. The parenchymal megascleres are usually mainly rhabdodiactines, with some additional hexactines or with rhabdodiactines only. The basalia of lophophytous genera are anchorate pentactines with the anchor flukes formed by true rays (not umbel teeth) and with the unpaired ray forming a long shaft that is embedded in the body. Some genera have additional pleuralia that correspond in form with the hypodermalia and originate as hypodermalia that are later protruded through the surface. The overlapped tangential rays of these pentactinal pleuralia then form an external veil outside the dermal membrane. In lophophytes, such spicules may be mixed with the typical basalia, with which they may also intergrade morphologically; correspondingly, the basalia develop with their anchors beneath the dermal membrane and are protruded later. In basiphytous sponges, the base contains a rigid basidictyonal skeleton, with which the lowermost parenchymal megascleres are united; but rigidity does not spread through the rest of the body. Some are stalked sponges, with a rigid stalk supporting a body with loose megascleres only.

Paratropal spicules and octaster microscleres, which have sometimes been found loose in sediments, occur only in the Rossellidae, although not in all genera. The presence or absence of two forms of the rosette group of microscleres (**plumicomes**, octasters) is used by zoologists (SCHULZE, 1897, 1904; IJIMA, 1904, 1927) to distinguish the subfamilies Lanuginellinae, Rossellinae, and Acanthascinae. There are no correlated general differences in the megaspiculation, but paratropal megascleres are restricted to the Rossellinae and Acanthascinae and octasters to the Acanthascinae only.

The Asconematidae are basiphytous sponges with pentactinal hypodermalia, which are not paratropal, and with autodermalia and gastralia that are typically pinular hexactines, sometimes varied as pentactines with the unpaired ray distal and pinular. The parenchymal megascleres are rhabdodiactines and hexactines or the former only, as in Rossellidae. The encrusting base has a rigid basidictyonal network with which parenchymal megascleres of the lower parts are also united; or there may be a stalk with a rigid union of the parenchymal megascleres, although these remain loose in the body above.

An alternative arrangement (IJIMA, 1904, 1927; SCHULZE, 1904) removes the type genus, *Asconema* KENT, to the Rossellidae (-inae), leaving specialized stalked asconematids as a family Caulophacidae (*ex* Caulophacinae SCHULZE, 1887a). This can be justified on the grounds that the two groups are closely related and that the autodermalia and gastralia of *Asconema* are properly only subpinular; but those of the typical Rossellidae are nonpinular, and rossellids may also be stalked (e.g., *Crateromorpha* GRAY).

There are very few post-Paleozoic fossils of the Lyssacinosa, although the group almost certainly ranges back into the Paleozoic (see below). *Stauractinella* ZITTEL of the Upper Jurassic has been thought to be a euplectellid (SCHRAMMEN, 1936) but is known only from hexactine megascleres without diagnostic value. Sponges apparently identical with the living *Regadrella* (Euplectellidae, Taegerinae) appear in the Upper Cretaceous (Cenomanian), with *Proeuplectella* MORET, which is probably an allied form, although incompletely known. *Purisiphonia* BOWERBANK comprises thick-walled forms like *Regadrella,* but with the parietal gaps in the form of perforating radial canals. DE LAUBENFELS (1955) included this genus in the Dictyonida (Hexactinosa, Staurodermatidae), but the skeleton is very clearly that of a taegerine euplectellid. Paratropal pentactines like those of some existing Rossellidae (Rossellinae or Acanthascinae) are known from the Turonian and Senonian. Otherwise the group is known chiefly from loose Cenozoic spicules, some of which, although not diagnostic, have been referred to modern genera (e.g., by HINDE & HOLMES, 1892).

From Cretaceous sediments, there is evidence of an otherwise unknown group of Amphidiscophora. Microscleres obtained loose from sediments include so-called hemidiscs or **hemiamphidiscs** that resemble amphidiscs but have one umbel larger than the other. In zoological classification the group has been made an order, the Hemidiscosa of the Amphidiscophora (SCHRAMMEN, 1924a; REID, 1958a). How large a group is represented is unknown; there could have been only one species from the evidence available.

Because of the almost blank record, assessment of the past evolution of existing Lyssacida depends on comparative evidence. The first point to note is that division of the Hexactinellida into forms possessing amphidiscs or hexasters is, so far as is known, completely sharp. All living Amphidiscosa have amphidiscs, not hexasters or any other rosette form; all living Lyssacinosa have hexasters, but never amphidiscs. In addition, the Dictyonida agree with the Lyssacida in possession of hexasters but not amphidiscs. Hence, the division into two stocks possessing either amphidiscs and hexasters, respectively, is inferred to be the primary phylogenetic division among forms now surviving. This, however, does not imply that there may not have been other stocks.

Living Amphidiscosa are all evidently closely related, although divisible into two main families, and a third comprising *Monoraphis* only. They share occurrence of pentactinal hypodermalia and of monactinal basalia whose character implies origin from pentactines. By analogy with rossellid Lyssacinosa, the basalia probably arose from hypodermalia. The Pheronematidae can be regarded as relatively primitive, because of the greater importance of hexactines or pentactines as parenchymal microscleres. In the Hyalonematidae the rhabdodiactines are often predominant and also have specialization of basalia to form glass ropes (e.g., in *Hyalonema* GRAY). Concentration of basalia as an anchor rope is correlated with increase in size but reduction in number of spicules (REID, 1958a). The family also lacks the scepter (see the Pheronematidae, *Treatise Part E (Revised),* vol. 3, in press). *Monoraphis* carries basal specialization to the extreme but seems to be closely related to the Pheronematidae because scepters are present.

The Lyssacinosa include two groups of sponges that have no obvious point of contact. The Euplectellidae and Leucopsacadidae lack hypodermalia, and basalia of lophophytes are typically umbel-bearing rhabdodiactines. Similar spicules may occur in the parenchymal skeleton, e.g., in *Euplectella* OWEN. The Rossellidae and Asconematidae have pentactinal hypodermalia (except in some genera, e.g., *Aulochone* SCHULZE, in which their absence is supposed to be secondary), and the lophophytes have pentactinal basalia that are homologous to the hypodermalia. The two groups, thus, seem to have evolved lophophytous attachment independently, using different types of spicules. In the Euplectellidae, one can probably assume evolution from forms like the simpler Euplectellinae with imperforate walls (e.g., *Euplectella*) and Taegerinae (e.g., *Regadrella*), although with the usual reservation that the genera cited are all modern. Even the simplest Euplectellinae, however, are not really primitive morphologically because their parenchymal megascleres include sharply distinct principalia and comitalia. The Rossellidae and Asconematidae are generally specialized sponges, having most advance characters in genera with paratropal hypodermalia and octaster microscleres (e.g., *Rhabdocalyptus* SCHULZE). Beyond this, it is difficult to comment, except that occurrence of taegerine euplectellids and Rossellidae with paratropal spicules in the Upper Cretaceous suggests a much longer history.

With few exceptions, no certain relationships can be claimed between Paleozoic and later Lyssacida. Almost no diagnostic microscleres are known from the Paleozoic sponges, and some of those recorded (as hexasters) are only doubtfully genuine. This leaves only the megaspiculation as a basis for assessment, but most types of megascleres and various conditions of the skeleton occur in both living series. Both include forms with the parenchymal megascleres varying from mainly hexactines to all rhabdodiactines. Pinular autodermalia are shared by most Amphidiscosa and the asconematid Lyssacinosa, and pentactinal hypodermalia by the Amphidiscosa and the rossellid and asconematid Lyssacinosa. Only euplectellid and leucopsacadid Lyssacinosa have basalia that are umbel-bearing rhabdodiactines; but external form alone, which is all that can be seen in most fossils, does not distinguish these spicules from basalia of some Hyalonematidae, and they sometimes pass into monactines. True pentactinal basalia are now seen normally only in rossellid Lyssacinosa; but the axial crosses of hyalonematid monactines imply origin from pentactines, and the normal monactines have occasional pentactinal variants. Hence none of these types can be cited as implying the occurrence of either amphidiscs or hexasters; and no useful purpose is served by comparison with a member of the Amphidiscosa, when an equally comparable genus exists in the Lyssacinosa. This leaves little but restriction of uncinates and scepters to the Amphidiscosa (although uncinates also occur in the order Dictyonida) and restriction of paratropal hypodermalia and fusion of the parenchymal megascleres to the Lyssacinosa.

A second problem is that some Paleozoic sponges have spicules of types unknown in any modern Lyssacida. This applies especially to the paraclavule microscleres of some Dictyospongiidae (*Treatise Part E (Revised)*, vol. 3, in press). These sponges have often been thought to be allied to the living Euplectellidae, which they broadly resemble, and SCHULZE (1887b) accepted them as fossil euplectellids; but no existing euplectellid, or any other lyssacine, is known to have paraclavules. When compared with other microscleres the latter resemble most an amphidisc with one umbel missing. The megascleric basalia can also be matched in the Pheronematidae or Hyalonematidae, as well as in the Euplectellidae. REID (1957a) has suggested that these sponges are not members of any living order but of an extinct one.

Because of these problems, I do not postulate herein origins of the two groups of modern Lyssacida (Amphidiscosa and Lyssacinosa) among the Paleozoic sponges. But the Amphidiscosa and probably the Lyssacinosa existed in the Late Paleozoic. First, the Carboniferous *Uralonema* LIBROVICH has recognizable amphidiscs, and was placed in the Hyalonematidae by REZVOI, ZHURAVLEVA, and KOLTUN (1962). Second, in Permian *Pileolites* FINKS the parenchymal megascleres are united as in various Lyssacinosa. As shown by FINKS (1960), the genus comes closest to the living Euplectellidae, although as he also showed, it is not a euplectellid. In particular, the outermost fused megascleres appear to be dermalia, which are never united in modern forms. For further comments on these sponges, see chapter on Paleozoic Hexactinellida: Morphology and Phylogeny (p. 135).

WORKING KEYS TO SOME LYSSACINOSID FAMILIES

R. E. H. Reid

[formerly of Department of Geology, The Queens University of Belfast]

Because of problems dealing with material lacking diagnostic microscleres, when sponges in different families distinguished by the microscleres have similar megascleres, some guidance may be helpful.

When microscleres are present, as in modern material, most Lyssacinosa (other than some rossellids that lack hypodermalia) can be placed correctly on the following basis.

A. Amphidisc microscleres present; no hexasters or other rosettes.

1. Principal parenchymal megascleres hexactines or pentactines, although rhabdodiactines may also occur; basalia monactinal and typically bidentate; scepters often present: **Pheronematidae.**

2. Fixed by a single large needle; principal parenchymal megascleres tauactines: **Monoraphididae.**

3. Principal parenchymal megascleres mainly or all rhabdodiactines; basalia monactines that have usually four or more anchor teeth; no scepters: **Hyalonematidae.**

B. Hexaster microscleres present; no amphidiscs.

1. Dermal megascleres not differentiated into distinct autodermalia and hypodermalia.

a. Dermalia swordlike hexactines: **Euplectellida.**

b. Dermalia pentactines, with unpaired ray proximal: **Leucopsacadida.**

2. With distinct autodermalia and hypodermalia.

a. Autodermalia stauractines, pentactines with the unpaired ray proximal, or nonpinular hexactines: **Rossellidae.**

b. Autodermalia pinular hexactines, or subpinular pentactines with the unpaired ray distal: **Asconematidae.**

Most genera can also be placed correctly by using the following keys, which do not rely first on microscleres, provided that megaspiculation is complete.

A. Dermal megascleres not differentiated into distinct autodermalia and hypodermalia; lophophytous genera with basalia that are umbel-bearing rhabdodiactines; parenchymal megascleres may be fused together in lower parts or throughout the body (hexaster microscleres).

1. Dermalia hexactines; generally thin-walled sponges, with megascleres all loose or fused in the lower parts or the whole body; lophophytous or basiphytous; wall with parietal oscula in some genera: **Euplectellidae.**

2. Dermalia pentactines, with the unpaired ray proximal; thick walled lophophytes, without fusion of megascleres in most instances: **Leucopsacadidae**.

B. Dermal megascleres differentiated into distinct autodermalia and hypodermalia; lophophytous genera with basalia that are monactines or pentactines; fusion of parenchymal megascleres restricted to lower parts of basiphytes (hexaster or amphidisc microscleres).

1. Basiphytous sponges, and lophophytes whose basalia are pentactines with paired rays forming four recurved anchor flukes; hypodermalia usually large pentactines, which may be paratropal, but sometimes tangentially lying rhabdodiactines; megascleres fused in the lower parts of basiphytous genera, but not in lophophytes (hexaster microscleres).

Autodermalia hexactines, pentactines with the unpaired ray proximal, or stauractines; sometimes spiny, but hexactines not pinular;

hypodermalia may be paratropal when pentactinal; lophophytous or basiphytous: **Rossellidae.**

b. Autodermalia pinular hexactines, or pentactines with a subpinular distal ray; basiphytous: **Asconematidae.**

2. Lophophytous sponges whose basalia are monactines, with two, four, or more recurved teeth arising from distal end; or sponge fixed by a single basal needle; hypodermalia pentactines with unpaired ray proximal, never paratropal; autodermalia usually pentactines with unpaired ray distal and pinular; no fusion of megascleres in any part (amphidisc microscleres).

a. Principal parenchymal megascleres hexactines or pentactines, although rhabdodiactines may also occur; basalia typically with two recurved anchor teeth only, and not arranged to form a glass rope; sceptres often present: **Pheronematidae.**

b. Sponge fixed by a single giant needle; principal parenchymal megascleres tauactines: **Monoraphididae.**

c. Principal parenchymal megascleres mainly or all rhabdodiactines; basalia monactines, which usually have four or more recurved anchor teeth at distal end, usually emitted in single tuft only and sometimes forming a glass rope; no sceptres, although diactinal pleuralia may occur: **Hyalonematidae.**

Isolated spicules are usually not diagnostic of any single genus, but particular families or groups of families may be indicated by some types.

1. Pinular hexactines; particularly **Asconematidae**, but also occur in some Dictyonida (e.g., *Bathyxiphus* SCHULZE).

2. Pinular pentactines: characteristic autodermalia of Amphidiscosa (**Peronematidae, Monoraphidae, Hyalonematidae**); but comparably subpinular pentactines occur in *Asconema* Kent (**Asconematidae**).

3. Paratropal pentactines: **Rossellidae** (**Rossellinae** or **Acanthascinae**).

4. Unequal-rayed rhabdodiactines, with an umbel at end of shorter ray: characteristic basalia of lophophytous **Euplectellidae** and **Leucopsacadidae.**

5. Anchorate pentactines: characteristic basalia of **Rossellidae**, also sometimes occurring as pleuralia.

6. Anchorate monactines with two anchor teeth only: characteristic basalia of **Pheronematidae.**

7. Anchorate monactines with four or more recurved anchor teeth: characteristic basalia of **Hyalonematidae.**

8. Octasters: **Acanthascinae (Rossellidae).**

DICTYONINE HEXASTEROPHORA

R. E. H. Reid

[formerly of Department of Geology, The Queens University of Belfast]

INTRODUCTION

Post-Paleozoic fossil Hexactinellida are predominantly **dictyonines**, here included with the Lyssacinosa in the subclass Hexasterophora SCHULZE, 1887b as in zoological classification. Dictyonine sponges are members of the orders Hexactinosa SCHRAMMEN and Lychniscosa SCHRAMMEN. In some paleontological classifications the Hexactinosa and Lychniscosa are combined into the order Dictyonida, with Hexactinosa and Lychniscosa treated as suborders. In this volume, however, the zoological classification is followed.

Dictyonine sponges are Hexactinellida with a parenchymal skeleton that is always a rigid dictyonal framework. The component megascleres or **dictyonalia** are united as part of their normal development and are hexactines except in some species in which there is suppression of rays that would otherwise project freely from skeletal surfaces. The framework is never composed of a mixture of different kinds of megascleres or partly or entirely of diactines, as in lyssacines with rigid skeletal frameworks.

The Hexactinosa and Lychniscosa are distinguished by occurrence of different kinds of dictyonalia. In Hexactinosa, the spicular centers or nodes have no special modifications except that they may sometimes be swollen. In Lychniscosa, the centers of typical dictyonalia are enclosed within an octahedral framework of **interactinal buttresses,** which extend across the twelve interactinal angles from points equidistant from the center. The buttresses of these **nodal octahedra** originate from siliceous fibers that grow across the **interactinal angles** before union of the spicules. After this union, further thickening of skeletal beams and nodal buttresses does not affect the enclosed central part of the spicule, which remains delicate. Although Lychniscosa may also have dictyonalia that lack nodal octahedra, the latter are altogether absent in Hexactinosa.

Most modern dictyonines have dermalia and gastralia, which are usually pentactines with the unpaired ray proximal, but some are hexactines. In one genus (*Aphrocallistes* GRAY) the **gastralia** are **rhabdodiactines.** Such spicules are usually absent from fossils, presumably through being unconnected, as in most modern species. Some fossils (e.g., *Craticularia* ZITTEL, *Porospongia* D'ORBIGNY, *Cypellia* POMEL) and one living genus (*Fieldingia* KENT), however, have connected dermalia or gastralia that are pentactines or stauractines. Hexasters are present in all living species but two (*Cyrtaulon sigsbeei* (SCHMIDT); *C. solutus* SCHULZE), and most of the Hexactinosa have additional **uncinates** and **sceptrules,** although either or both may be absent. These microscleres are unknown in fossils although sceptrules have been found loose in sediments with fossil Hexactinosa.

HABITUS AND GENERAL SKELETAL MORPHOLOGY

Although the Hexactinosa and Lychniscosa appear to have evolved independently from at least an early stage of their history (see Phylogeny, p. 174), they have many common features in both habitus and skeletal morphology. The implied parallel evolution is presumably the result of possession of the same type of skeleton.

Dictyonine sponges vary widely in habitus, but many are funnel-like, branched and tubular, or of some related shape. They are mainly thinly walled sponges with a lateral wall or skeletal framework, generally about 1 to 10 mm thick. Many with walls less than 5 mm thick (e.g., species of *Farrea* BOWERBANK, *Eurete* SEMPER, *Calyptrella* SCHRAMMEN, *Brachiolites* SMITH) form branching and anastomosing tubes, which presumably are stronger mechanically than simple free branches. Genera in which the wall exceeds 10 mm thick are usually funnel-like or flabellate (e.g., *Stauroderma* ZITTEL, *Rhizoporterion* ZITTEL, *Porospongia* D'ORBIGNY) or sponges that form solid branches with no paragastral cavity (e.g., *Sclerothamnus* MARSHALL). The body may be supported by an encrusting base or by rootlike basal outgrowths.

In sponges in which the body consists partly or entirely of dividing and anastomosing tubes, it encloses two systems of passages, within and between the tubes respectively. These enclosed passages, termed **cavaedia,** are lined by the soft gastral and dermal membranes in life and by the corresponding skeletal surfaces in fossil examples. The same term has also sometimes been used for spaces enclosed between plications of the wall (IJIMA, 1927) or for enclosed spaces that more probably represent a canal system (see below and p. 174).

In the simplest dictyonines (e.g., *Farrea, Calyptrella*), the skeletal framework consists of a single layer of rectangular meshwork or of a single primary layer (in *Farrea* and related forms) on which a further layered or irregular meshwork is built up by accretion of new dictyonalia (Fig. 54). The beams of the primary meshwork each contain rays of two dictyonalia laid together side by side and enclosed in a common siliceous envelope. When additional meshwork is layered, the secondary beams are also formed in this manner, and the three-dimensional meshes are roughly cubic. Irregularly oriented secondary components are fused together haphazardly where their rays come into contact with the primary beams or with one another.

In most forms, however, the dictyonal meshwork is initially three dimensional and is not constructed in layers (Fig. 55–56). Series of dictyonalia are united by the rays of one axis to form parallel or subparallel dictyonal strands with a longitudinal to radial orientation, which are connected together laterally by beams formed from the other rays. The beams forming the dictyonal strands each contain two apposed spicular

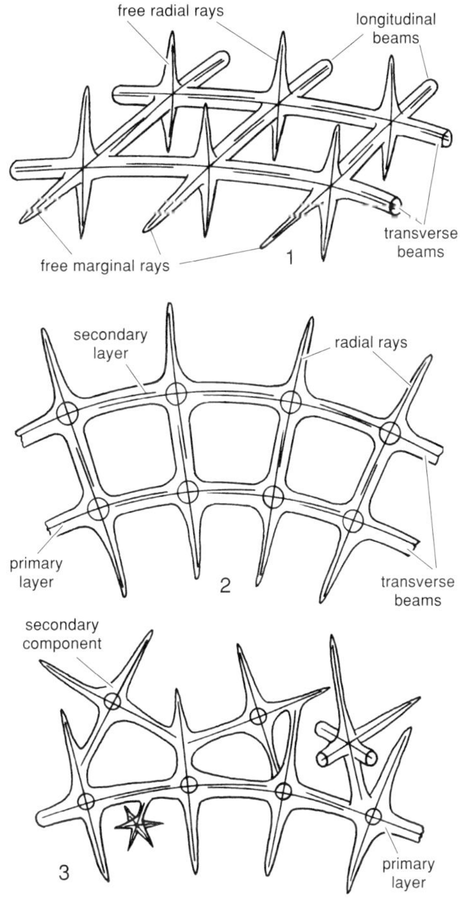

FIG. 54. Order Dictyonida, suborder Hexactinosa; farreoid structure; figures oriented dermal side upward, gastral side downward; *circles* in views *2* and *3* represent skeletal beams cut at their origins; *1,* marginal part of farreoid primary layer, with margin and free longitudinal rays toward front: each skeletal beam includes 2 oppositely directed rays of adjacent dictyonal hexactines, apposed side by side and enclosed in common siliceous envelope [not united tip-to-tip]; free rays of radially oriented spicular axes project on either side of each skeletal node; *2,* transverse section of regular three-dimensional meshwork, in which single secondary layer of meshwork is superimposed on primary layer (seen below: see view *1*); *3,* transverse section of skeleton with irregular secondary meshes, in which secondary hexactines have random orientation (new).

rays, as in the beams of the primary meshwork of the simplest type of skeleton. The lateral connecting beams may be formed in this manner, by fusion of their tips with beams or nodes in adjacent strands, or by haphazard unions where they happen to

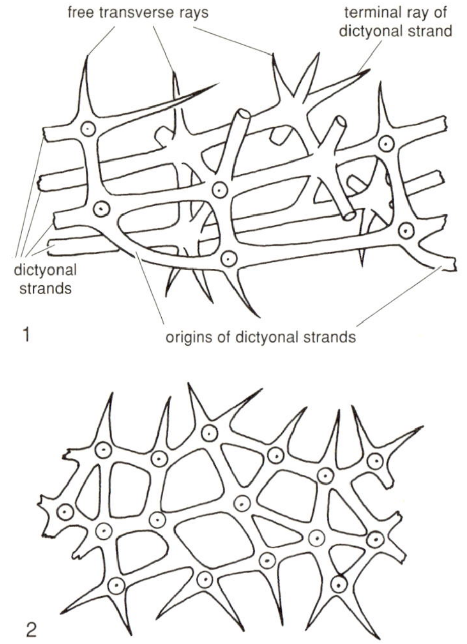

FIG. 55. Order Dictyonida, suborder Hexactinosa; sectional views of simple euretoid structure; *circles* with central spot mark beams cut at origins; *1,* simple euretoid skeleton, 1 to 2 meshes deep, in longitudinal section; dermal side upward; growth left to right; *2,* transverse section of simple euretoid skeleton, of type in which transverse (connecting) beams form regular transverse lamellae, having characteristic irregularly variable form and orientation of meshes, and lack of layered construction. Compare with Figure 54.1–54.3 (farreoid structure), and note that three-dimensional meshwork represented here corresponds with farreoid primary layer (Fig. 54.1) only; based on recent *Periphragella elisae* MARSHALL (new).

cross one another. The shape of meshes between the strands varies correspondingly from generally rectangular to mainly or all triangular or irregular.

Two genera (*Aulocalyx* SCHULZE, *Rhabdodictyum* SCHMIDT), placed in the Hexactinosa as Aulocalycidae (IJIMA, 1927), have skeletal frameworks composed of large hexactines united in an altogether haphazard manner. It is possible, however, that these are not true dictyonine sponges (see Phylogeny, p. 174).

The common type of skeleton, which is three dimensional initially, has often been wrongly assumed to have the same type of structure as the layered type that occurs only in the hexactinosan *Farrea* and some related genera. This mistake arose because skeletons of the common type are misleadingly like cubic structures if the meshes between dictyonal strands are regularly rectangular. The difference is seen easily in sections at right angles to the strands in which meshes enclosed by the connecting beams have irregularly variable shapes and orientations.

In many examples of such skeletons, the outermost meshwork at dermal or gastral surfaces differs from that of the interior and then comprises a **dictyonal cortex** (Fig. 57). This may arise by secondary accretion of dictyonalia to the primary structure or by various modifications of the outermost primary meshwork. Either type may form a thin surface layer only or be several meshes deep.

Another common development is **skeletal canalization,** here described as **intradictyonal,** representing the canal system or sometimes the flagellated chambers. As usual, this results from the skeleton growing around preexisting soft parts. The simplest development is formation of small skeletal pores (ostia, postica) or short canals (epirhyses, aporhyses) in secondary cortical meshwork, representing the positions of small inhalant canals on the dermal side and of flagellated chambers on the gastral side. This type of canalization may also occur in the secondary meshwork of *Farrea* and related genera and is the only type of canalization they exhibit. More advanced canalization affecting primary meshwork of skeletons that are initially three dimensional may be either intracortical only or affect the whole interior. In the latter, there may be two systems of skeletal canals (epirhyses, aporhyses) that typically open on the dermal and gastral sides, respectively, or a single system of canals that open on both sides. Skeletal canals of either type may be radial and separate, branching and intercommunicating from radial trunks, or irregularly labyrinthine. When two distinct systems are present, sometimes one but not the other is open through both skeletal surfaces.

Labyrinthine canalization can cause difficulties in interpretation. First, in some instances it may be difficult or almost impossible to determine whether one or two systems of skeletal canals are present. This applies especially to fossils preserved as pseudomorphs, as for example in many Cretaceous specimens. Second, in other sponges such canals are much wider than intervening skeletal partitions, and in these the skeleton may appear to consist of dividing and anastomosing tubes. Canalization can be inferred to be intradictyonal in some fossils that have this condition and have existing relatives in which the soft parts are known. But in others that have no living relatives the wall structure might be interpreted in either way.

Various canalized genera have circulatory apertures in the surfaces of the dictyonal framework spanned externally by **superficial meshwork** that are formed by union of the dermalia or gastralia, by union of adventitious dictyonal hexactines, or from siliceous filaments that grow out from the skeletal surface. Two or all of these types of superficial structures may also occur in one sponge. When dermalia or gastralia are fused together, this may happen by simple cementation of paratangential rays that lie side by side or by growth of branching and anastomosing siliceous filaments across the paratangential meshes.

Some Hexactinosa and Lychniscosa with intradictyonal epirhyses and aporhyses have longitudinal furrowing of the dermal or gastral surface, along the lines of series of ostia (e.g., *Sphenaulax* ZITTEL) or postica (e.g., *Leiostracosia* SCHRAMMEN), or irregularly reticulate furrowing of the dermal surface (*Ventriculites* MANTEL). When longitudinal, this **superficial furrowing** suggests longitudinal folding of the wall but is spanned by superficial meshwork when this is present, as in all the genera cited. It presumably represents the courses of subdermal channels on the dermal side, and subgastral channels or plications of the chamber layer on the gastral side.

Species of the living lychniscosan *Callicyclix* SCHRAMMEN (=*Aulocystis* SCHULZE, 1887b; *non Aulocystis* SCHLÜTER, 1885 [Coelenterata, Anthozoa, Tabulata, Aulo-

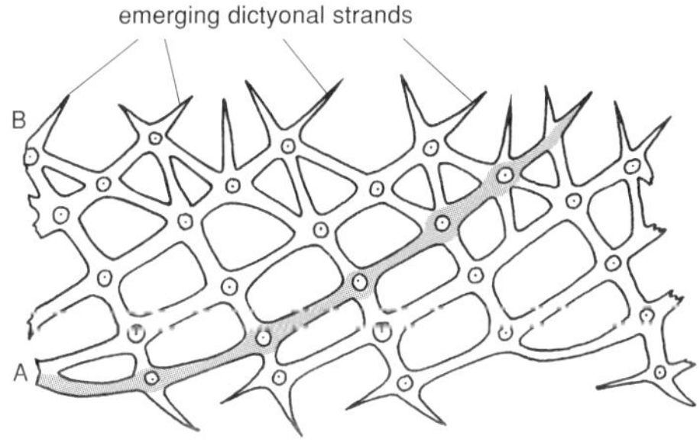

FIG. 56. Order Dictyonida, suborder Hexactinosa, euretoid structure; longitudinal section of euretoid meshwork; *A*, dermal emergence of dictyonal strands (1 *shaded*), which begin on gastral side (below) and have marked outward curvature; *B*, change in form of meshes from quadratic to triangular toward dermal side; orientation and style as in Figure 57 (new).

poridae]) have a body composed of dividing and anastomosing tubes enclosed in a **peripheral membrane** that extends across the terminal openings of the tubes and the spaces between them. This membrane may be supported by loose spicules resembling the dermalia, as in *C. zitteli sibogae* (IJIMA), or additionally by secondary dictyonal meshwork formed after the end of normal growth, as in *C. zitteli zitteli* (MARSHALL & MEYER). Some fossils have similar capsular or annular

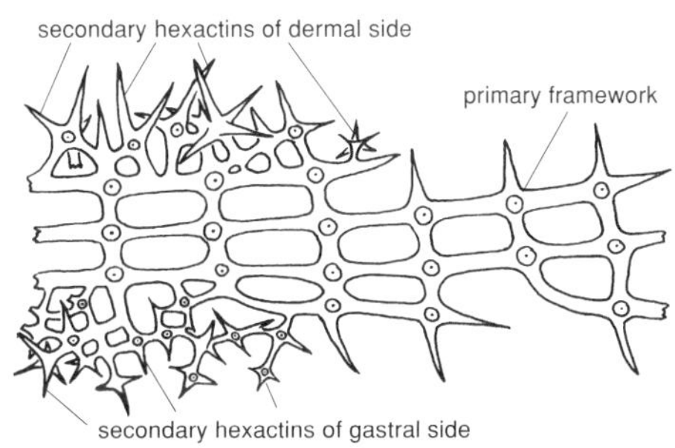

FIG. 57. Order Dictyonida, suborder Hexactinosa; simple euretoid skeleton with cortical meshwork formed by secondary accretion of dictyonal hexactines at primary surfaces; longitudinal section, dermal side upward, growth left to right; *circles with central dot* mark transverse beams cut at origins; secondary meshwork seen in older part only (left); secondary hexactines larger on dermal side; for convenience dictyonal strands are drawn as though vertically superimposed; based on Upper Cretaceous *Eurete (Aulodomus) prolatum* REID (new).

peripheral structures that imply the existence of a similar membrane in life, e.g., *Tremabolites* ZITTEL, *Coeloptychium* GOLDFUSS. Structures resembling lyssacine **sieve plates** sometimes occur (e.g., *Coeloptychium, Aphrocallistes* GRAY) but may differ in containing flagellated chambers in living sponges (*Aphrocallistes*).

The basal skeleton is always internal and rigid and consists of secondary dictyonal hexactines or of anastomosing siliceous filaments that grow downward from the dermal surface of the dictyonal framework. In some forms, e.g., *Laocoetis* POMEL, an upward extension of the basal skeleton forms a superficial meshwork far above the basal parts. Meshwork like that of the basal skeleton may also be formed on the lower parts of the paragastral surface, although it takes no part in basal fixation. Basal and analogous meshwork of the gastral side may lack canalization or have skeletal canals like those of the dictyonal framework or of a different type.

In using older literature, it is useful to note that the **Deckschichten** of ZITTEL (1877b, 1878a, 1878b) and the dermal layers of HINDE (1884a, 1887a, 1888, 1893b, 1912) may be (i) a dictyonal cortex of primary or secondary origin, (ii) superficial meshwork of any sort, or (iii) a peripheral structure. In SCHRAMMEN's work (1912, 1924a, 1936), however, a Deckschicht is nearly always a **dictyonal cortex.** In MORET's (1926b) usage, a **cortex dépendant** may be a dictyonal cortex, a superficial structure arising from the dictyonal surface, or in some sponges an upward extension of the basal skeleton. A **cortex indépendant** is the dermal or gastral skeleton or is a connected superficial meshwork developed from them. **Ectosomal** as used by IJIMA (1927) means the dermal or gastral skeleton. This was apparently not understood by DE LAUBENFELS (1955), who sometimes used ectosomal for dictyonal structures, called Deckschicht by SCHRAMMEN, e.g., in the Eubrochididae DE LAUBENFELS, 1955, p. 79).

HEXACTINOSA

The Hexactinosa are dictyonines in which **nodal octahedra,** distinctive of the Lychniscosa, are never developed at any stage of ontogeny. In most genera, the living examples are also distinguished by possession of uncinate and sceptrule microscleres.

In most Hexactinosa, the component dictyonalia of the primary dictyonal meshwork are united to form regular parallel or subparallel dictyonal strands. These occur in two main types of skeleton.

In the **farreoid** type, which occurs only in few Farreidae, the primary meshwork is formed in a single layer only (Fig. 54) apart from local irregularities, with the dictyonal strands running longitudinally (i.e., from the base to the skeletal margin). The meshes are typically rectangular, with the beams each enclosing two rays of adjacent dictyonalia. The remaining two rays of each dictyonal hexactine project at right angles on opposite sides of each intersection of the meshwork. The skeleton may retain this condition throughout life, except in the basal parts, where meshwork of the basal skeleton covers it, or have more or less extensive accretion of secondary components to form further layered or irregular meshwork. This accretion may occur in the older parts only or extend to the skeletal margin.

In **euretoid** skeletons, which occur in all other Hexactinosa except the doubtfully included Aulocalycidae (see *Treatise Part E (Revised),* vol. 3, in press), the meshwork is primarily three dimensional with a depth of one to many meshes and is not formed in layers (Fig. 55). In simple examples, the skeletal meshwork is only one to several meshes deep. The dictyonal strands then run almost longitudinally but with a gradually spreading arrangement so that many of them end at the dermal or gastral surface of the skeleton instead of at the skeletal margin (Fig. 56). This migration of strands to the surfaces is **emergence.** In more advanced skeletons that are several to many meshes thick, the strands spread out more rapidly and have more or less marked outward curvature emerging at both surfaces at angles up to 90° or begin on the gastral side and run to the dermal side. Thus, emergence of the strands may be bilateral or unilateral, according to whether they run to both surfaces or to one only. Their direction may be radial from some distance under the surface, and in some forms (*Aphrocallistes* spp., *Leptophragma pusillum* SCHRAMMEN) they are radial or subradial for most or all of their length. In all instances, the emerging dictyonal strands are continually replaced by new ones. The whole thickness of this type of three-dimensional meshwork is homologous with the single primary layer of farreoid skeletons, not with the layered or cubic type of farreoid meshwork with which it has been identified formerly.

Euretoid skeletons have four main but intergrading conditions in the form of the meshes between dictyonal strands. First, meshes between strands may be typically rectangular, apart from local irregularities. The connecting beams then form successive, platelike, transverse lamellae of irregular meshwork, through which the strands run at right angles. Because of the spreading arrangement of the strands, these lamellae are typically convex toward the skeletal margin and may overlap marginally if the spreading of the strands is pronounced. Second, meshes between the strands may have a mixture of rectangular, trapezoidal, rhomboidal, and triangular meshes in varying proportions; some connecting beams are then formed from single rays instead of pairs of rays. Third, the meshes may all be triangular and the connecting beams all formed from single rays that are attached by their tips to the centers of adjacent dictyonalia (Fig. 58). The skeletal nodes are then said to be **multiradiate** since more than six beams (typically 8–10) radiate from each of them. In this type of skeleton, dictyonal strands are often difficult to identify since the meshwork

FIG. 58. Order Dictyonida, suborder Hexactinosa, euretoid structure; skeletal meshwork with multiradiate nodes and triangular meshes regularly developed, in part also with nodes spherically enlarged; plan view, top part tilted slightly away from front (hence triangular meshes are not equilateral as in true plan view); outermost meshwork at top of figure, internal meshwork below; *A,* meshwork of interior, with 10 beams radiating from each skeletal node; *B,* node with axial canals drawn in; *arrows* show direction of rays; spicular axes follow same pattern as all other nodes except 2 surface nodes at top (views *D–E*); *C,* similar meshwork with nodes spherically swollen, although less so than at surface (views *D–E*); *D,* node at surface, with projecting tips of reduced free rays that do not follow orientations seen in interior (see view *B*); *E,* similar node with free rays suppressed (or engulfed) (new).

has a similar appearance in all directions. Last, the connecting beams may be formed in a haphazard manner, where the rays that form them happen to meet other strands or individual rays.

The types of euretoid skeleton in which meshes between dictyonal strands are mainly rectangular and mainly triangular were distinguished by MORET (1926b) as type *Craticularia* and type *Eurete* respectively, but both occur in both genera and may intergrade from one part of the skeleton to another in both genera.

In interpreting the structure of a euretoid skeleton, the orientation of dictyonal strands

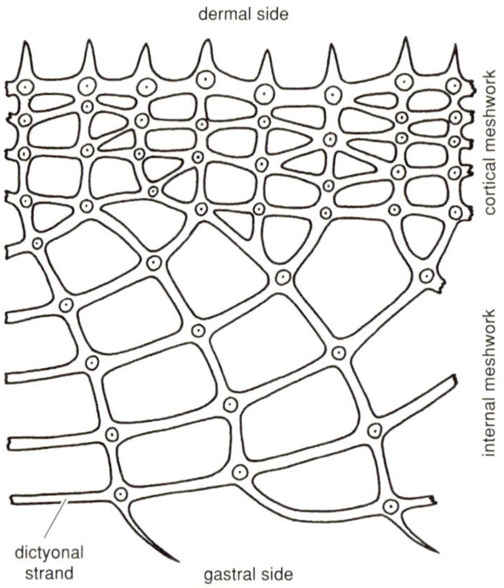

FIG. 59. Order Dictyonida, suborder Hexactinosa, euretoid structure; longitudinal section of skeleton in which dictyonal strands emerge radially on dermal side (top), and beams connect their radial parts to form roughly stratified primary cortical meshwork; also with outermost beams of dermal side differentially thickened: note that beams having stratification in cortex correspond structurally with those that connect dictyonal strands transversely in interior, and that only some radial strands of cortex are continuations of those seen internally (below cortex); orientation and style as in Figure 57; based on examples of recent *Leptophragmella choanoides* (SCHULZE & KIRKPATRICK) having minimal canalization (new).

should be established first. This is because the same type of structure may look different according to their orientation or that of sections in relation to them. In particular, a skeleton with rectangular meshes between the strands will have regular meshwork in sections roughly parallel to them but irregular meshwork in sections transverse to them. A similar difference occurs between plan views or tangential sections of skeletons in which the orientation of strands is mainly longitudinal in one instance but radial or subradial in another. Failure to understand this has led to description of imaginary differences between regular and irregular construction in various fossils.

Many euretoid skeletons have some type of dictyonal cortex (Fig. 57, 59). In the simplest, this results only from special thickening of beams or skeletal nodes at the surfaces. In others, there are structural changes as the surface is approached. Meshes between strands may change shape from rectangular in the interior to irregularly variable or triangular in the cortex (Fig. 57). Skeletons, with markedly convex transverse lamellae may have their margins overlapped and united to form irregular cortical meshwork. If strands are radial or subradial from some distance below the surface, a thickly layered cortex may result. The meshwork seen at the surface is irregular and may have diagonally intersecting series of beams running through it. Last, various developments may occur in combination.

In addition to primary developments, a cortex also may be produced by accretion of secondary dictyonalia. This type passes downward into the basal skeleton.

Farreoid and euretoid skeletons both may have a state of dictyorhysis, in which all circulation occurs through ordinary meshes. In others there is canalization of secondary meshwork or primary canalization of the euretoid type of skeleton. The simplest canalar features are small ostia or postica or short intracortical **epirhyses** or **aporhyses**, which open internally into uncanalized meshes. Fully developed intradictyonal canals may be **diplorhytic**, comprising separate series of epirhyses and aporhyses, or form single systems of **diarhyses** or **schizorhyses** that open through both skeletal surfaces. Diarhyses are oblique to radial tubular or prismatic skeletal canals, each of which contains a complex flagellated chamber in living examples. Schizorhyses are intercommunicating cleftlike features or labyrinthine tunnels that contain corresponding chamber-lined canals. Some other special developments also occur.

Some advanced euretoid Hexactinosa have a superficial meshwork that is developed outside the dictyonal skeleton proper and across canalar apertures. This is usually formed either from dictyonal hexactines, from connected dermalia and gastralia, or both together. Siliceous filaments growing out from the dictyonal skeleton also occur in some genera (e.g., *Nemarete* REID) but are less common than in Lychniscosa.

The basal skeleton is always composed of dictyonal hexactines. These have been called basidictyonalia (IJIMA, 1927) but do not appear to be homologous with lyssacine basidictyonalia. The basal meshwork may pass upward into secondary meshwork of the dictyonal framework proper in farreoids or into dictyonal superficial meshwork in euretoids (e.g., *Laocoetis* POMEL).

The **aulocalycoid** skeletons of the Aulocalycidae are composed of large, haphazardly united hexactines whose rays are interwoven diagonally (Fig. 60). True dictyonal strands seem to be absent, although strandlike series of beams may occur.

Turning to families, the Farreidae are simple Hexactinosa with farreoid skeletons and no canalization or with secondary ostia or postica only. The sceptrules of living sponges are all **clavules** or clavules and **lonchioles** or **sarules.**

The Euretidae are simple euretoid Hexactinosa that may lack canalization entirely, have intracortical features in primary or secondary meshwork, or have vague development of deeper epirhyses or aporhyses with no regular pattern. One modern genus (*Tretochone* REID) has special accessory **amararhyses** that are additional to the true canal system. These consist of longitudinal tunnels that open through slits in the gastral surface and through branch tunnels leading to the tops of papillae on the dermal side. The sceptrules of modern euretids are typically scopules but are sarules in one genus (*Sarostegia* TOPSENT) and are absent from some others (e.g., *Myliusia* GRAY).

The Craticulariidae and Cribrospongiidae are advanced euretoid Hexactinosa with deep epirhyses and aporhyses having different characteristic arrangements. In Craticulariidae, they typically occur in separate longitudinal series and represent inhalant canals and choanocytal outgrowths formed in alternating segments of the growing sponge margin. The epirhyses are often also arranged in

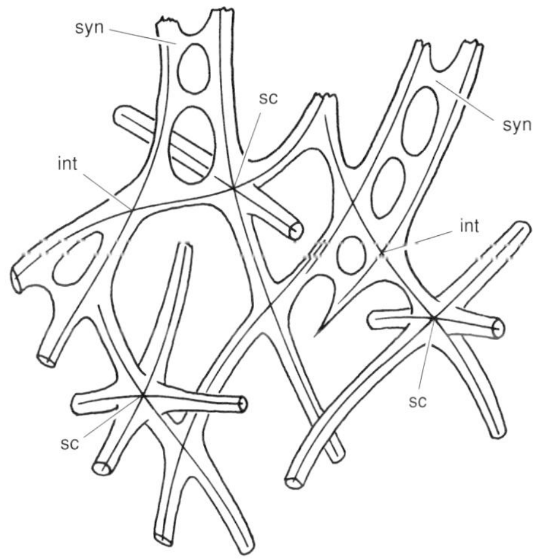

FIG. 60. Order Dictyonida, aulocalycoid structure that may occur in suborder Hexactinosa; meshwork formed by irregular union of 3 hexactines and rays from 4 others; positions of rays shown by their axial filaments; *syn,* synapticulae; *int,* intersection of rays; *sc,* spicule center (new).

transversely corresponding positions so that their ostia form a quadratic pattern. The aporhyses are then arranged similarly with each aporhysis located quincuncially in the center of a group of four epirhyses. A quincunx pattern then occurs if the wall is sectioned or eroded tangentially. In Cribrospongiidae with canals in longitudinal series, epirhyses and aporhyses occur alternately in the same series and often in alternating positions in adjacent series. The patterns of ostia and postica are consequently quincuncial, and the pattern that occurs in tangential section is quadratic. These patterns have often been confused through reliance on eroded material or on genera in which one set of canals perforates both surfaces (e.g., epirhyses in *Guettardiscyphia* DE FROMENTEL; aporhyses in *Andreaea* SCHRAMMEN).

In Craticulariidae, some genera (e.g., *Sphenaulax* ZITTEL) have longitudinal furrowing of one skeletal surface. Some Cribrospongiidae have modification by multiplication of the number of epirhyses, which may become arranged in hexagonal groups (*Andreaea*), or development of internal labyrinths from one or both systems of skeletal canals (e.g., *Polyopesia* SCHRAMMEN). Both families include genera with and without a superficial meshwork of dictyonal or dermal origin when present. These families are mainly extinct, scopule microscleres are known only from one living craticulariid (*Leptophragmella* REID).

The Aphrocallistidae and Tretodictyidae are advanced euretoids distinguished by canalization in the form of diarhyses and schizorhyses, respectively. The living forms have scopules. In Aphrocallistidae (one genus, *Aphrocallistes* GRAY), the dictyonal strands also run radially or subradially, when any are present, and dictyonalia in thin partitions between closely spaced diarhyses may be distorted so that all six rays lie roughly in one plane. In some species, oscula become covered by sieve-platelike structures, which contain flagellated chambers in a living sponge (*A. beatrix* GRAY). In Tretodictyidae, the schizorhyses vary from narrow, intercommunicating clefts to large, labyrinthine tunnels, separated by narrow partitions. A superficial meshwork composed of connected dermal pentactines occurs in one fossil and one modern genus (*Placotrema* HINDE, *Fieldingia* KENT).

The families Staurodermatidae and Cystispongiidae and the subfamily Caseariinae of the Craticulariidae comprise advanced euretoid sponges whose relationships to other forms are uncertain. All are fossils. Nothing is known of the microscleres, but the sceptrules would probably be scopules as in the previous families.

Stauroderma ZITTEL has complicated diplorhytic canalization, which may be a specialized form of the cribrospongiid type. Alternating ostia in the dermal skeletal surface lead into branching epirhyses, which unite to form a labyrinth of fine passages. Similar passages in the interspaces form an aporhytic labyrinth, arising from branching aporhyses whose openings are in pitlike depressions in the gastral skeletal surface. The epirhytic labyrinth also opens on this side through small apertures and sinuous grooves in the areas between the aporhytic depressions. This surface is usually densely coated by a superficial meshwork formed from gastral pentactines and siliceous filaments, which extends into but not over the depressions.

Casearia QUENSTEDT is a sponge with rejuvenescent growth like that of sphinctozoan demosponges, although with solid segments. A deep, narrow paragaster extends through the successive segments. The canal system, when apparent, is diplorhytic, either with no recognizable pattern or approaching the craticulariid type. The external surface of each segment is covered by paratangential meshwork, which seems to consist of fused dermalia.

Porospongia D'ORBIGNY and related genera are sponges with obscure canalization, which has been said to consist of epirhyses and aporhyses but may be schizorhytic. The dermal surface of the skeleton bears superficial meshwork formed from cemented stauractines or pentactines. On the gastral side, similar gastralia are united by siliceous filaments in the paratangential meshes to form a finely porous membrane pierced at intervals by round or ovate apertures.

Cystispongia ROEMER is a sponge with a pyriform body, composed of thinly walled lamellae enclosing labyrinthine passages that are enclosed externally by a capsulelike structure. The internal passages can be interpreted as either cavaedia or schizorhyses. The genus is a true hexactinosan, although the name *Cystispongia* has often been misapplied to similar Lychniscosa.

Last, the doubtfully included Aulocalycidae have aulocalycoid skeletal structure and either no canalization or vague ostia or postica. No uncinates or sceptrules are present.

LYCHNISCOSA

The name Lychniscosa is based on the term **lantern-spicule** or **lychnisc**, applied to dictyonalia with nodal octahedra or lanterns that are distinctive of this suborder (Fig. 61).

In typical Lychniscosa, most or all dictyonalia have nodal octahedra except sometimes in a secondary meshwork. In some forms, however, octahedra occur in only parts of the skeleton or even in only parts of some individuals. This is due either to solidification of the octahedra after formation or sometimes to simple failure to develop them in much of the skeleton. Genera of this kind (e.g., *Dactylocalyx* STUTCHBURY, *Stauronema* SOLLAS) have been mistaken for Hexactinosa although they need to be placed in this suborder.

A simple type of dictyonal framework, which occurs only in *Calyptrella* SCHRAMMEN, consists of a single layer of rectangular meshwork. This is comparable to the single primary layer of the hexactinosan Farreidae but lacks distinct **dictyonal strands.**

Other Lychniscosa have euretoid-like conditions, with dictyonal strands that are not arranged in layers and that spread and emerge comparably. The meshes between these strands are commonly predominantly rectangular, and the main variation from this condition occurs by loss of the regular arrangement of the spicules so that dictyonal strands are absent. The hexactinosan condition with regularly triangular meshes and **multiradiate nodes** is not reproduced.

Although the dictyonal strands of these forms resemble those of Hexactinosa, there seems to be a fundamental difference in the way in which their orientation is controlled. In Hexactinosa that consist of dividing and anastomosing tubes, the dictyonal strands follow the local direction of growth and pass from one tube to another at points of branching or anastomosis. An axial section of a similar lychniscosan, instead, has strands spreading out from an axis or center irrespective of the local directions of the walls of the tubes. These types of structure may be contrasted as **concordant** in Hexactinosa but **discordant** in Lychniscosa. The discordant lychniscosan type causes difficulties in interpreting some fossils because interruption of the strands by circulatory passages of doubtful character cannot be taken as a criterion of **intradictyonal** canalization (as in Hexactinosa).

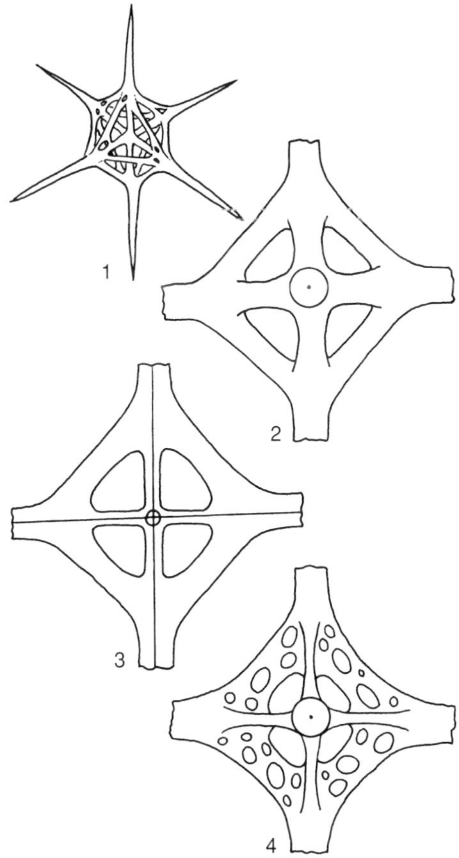

FIG. 61. Order Dictyonida, suborder Lychniscosa, lychniscs; *1,* young lychnisc, not yet incorporated into dictyonal framework, showing octahedral form of nodal lantern; at this stage 12 interactinal buttresses of lantern are more delicate than spicular rays (see view *3*); *2,* nodal part of typical adult lychnisc, drawn with 1 beam of dictyonal framework toward front, showing 8 of 12 interactinal buttresses (in this view, others are hidden behind); *circle with central dot,* cut skeletal beam facing front; *3,* section through adult lychnisc developed as in view *2* showing relatively unthickened central part; *lines* that intersect at center are axial canals of sectioned rays; *4,* nodal part of adult lychnisc, in which solid nodal buttresses are replaced by perforated plates; same view as view *2* (new).

Lychniscosa other than *Calyptrella* may have a dictyonal cortex, produced either by secondary accretion or by modification of the outermost primary meshwork. If primary meshwork is modified, this usually happens in one of three ways. First, lychnisc octahedra may be solidified at the surface although

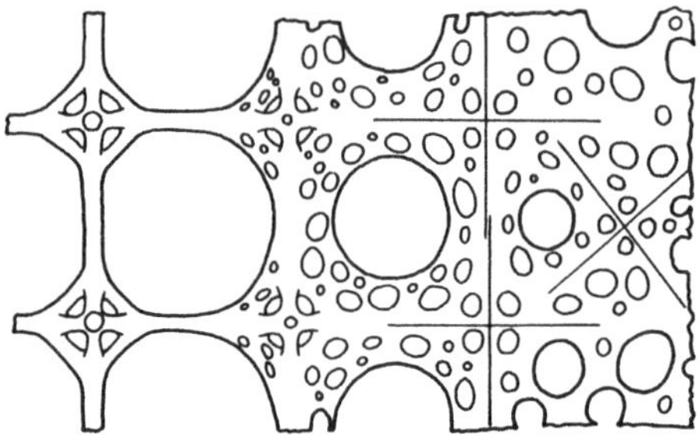

FIG. 62. Order Dictyonida, suborder Lychniscosa; cortical modification of external lychniscs to form a porous cortical membrane, seen in external view of skeletal surface; different conditions represent stages in development of this type of cortex in different sponges; *A,* unmodified lychniscs, like those of internal meshwork except that unattached rays (*central circles*) project freely from surface; *B,* lychnisc buttresses in plane of surface are replaced by perforated plates; at left, those of adjacent lychniscs are just confluent; at right, a continuous porous lamella extending between adjacent spicules; *C,* surface is formed by a porous siliceous membrane, with large pore at center of each interspicular mesh; *D,* distal rays and outer half of each nodal lantern are aborted, positions of spicules are shown by their axial canals; spicular centers indicated by axial crosses, around each of which are grouped 4 small pores that reflect underlying half-lantern structure; *E,* cortex is porous siliceous membrane with axial canals of spicules oriented irregularly (not united to form continuous meshwork, as in view *D*); positions of centers as in view *D* (new).

remaining open structures in the interior. Second, surface layers may have an irregular structure, contrasting with regular internal meshwork. Third, the meshes at the surface may be covered by a finely porous siliceous membrane that is produced by expansion of the buttresses of nodal octahedra to form perforated plates, which grow together, or by union of branching siliceous filaments that grow across the meshes (Fig. 62). Distally directed rays of dictyonalia at the surface and the corresponding halves of their nodal octahedra are then often aborted. This type of cortical membrane is pierced by small ostia or postica but often extends into larger canals if these are present.

Canalization of the skeleton may be present or absent and intracortical only or fully developed. The simplest features are small ostia or postica outlined by secondary meshwork as in *Callicyclix meandrina* (SCHRAMMEN) or by piercing a primary cortex as in *Callodictyon* ZITTEL. These presumably represent small inhalant canals and flagellated chambers like those of living *Callicyclix* species. Other fossils have diplorhytic canalization (i.e., epirhyses and aporhyses) or single systems of skeletal canals resembling diarhyses or schizorhyses. Epirhyses and aporhyses may represent inhalant and exhalant canals, respectively, although modern examples are unknown; but the nature of the soft parts is uncertain in the others. In Hexactinosa, diarhyses and schizorhyses are distinct, but skeletal canals that resemble them in fossil Lychniscosa are instead intergrading. Various fossils could have either intradictyonal canalization or cavaedial passages enclosed by anastomosing tubes or lamellae, according to how they are interpreted. The discordance between skeletal organization and local directions of growth does not permit these states to be distinguished by evidence from skeletal structure.

Superficial meshwork occurs in some genera and usually consists either of connected dermalia (in Cypelliidae) or of adventitious siliceous filaments that grow out from the surface of the dictyonal framework (Fig. 63). These filaments commonly grow from the ends of projecting dictyonal rays or from their positions when projecting rays are aborted, especially on the gastral side. They occasionally form paratangential networks that probably mark the position of the soft gastral membrane. Superficial meshwork formed from dictyonal hexactines is uncommon but occurs in some genera (e.g., *Stauronema* SOLLAS).

The basal skeleton may be formed from dictyonal hexactines, with or without nodal octahedra, or from siliceous filaments that grow downward from the skeletal surface in the lower parts of the body. The latter type occurs especially in sponges from the chalk of western Europe, in which it forms rootlike outgrowths. It seems to be an adaptation to fixation on soft sediment. Some Cretaceous

Ventriculitidae with dictyonal framework canalized by radial to labyrinthine epirhyses and aporhyses have their basal meshwork canalized by tubular longitudinal canals that run down from the positions of ostia in the dermal surface proper, as in *Rhizopoterion* ZITTEL.

The peripheral capsule of the living *Callicyclix zitteli* (MARSHALL & MEYER) may be supported entirely by loose pentactines resembling dermalia or, additionally, by a secondary dictyonal meshwork that grows into it from margins of the tubular branches at the end of normal growth. Various fossil genera (e.g., *Tremabolites* ZITTEL) have similar, rigid but usually denser structures produced by an initial growth of typical dictyonalia covered externally by a layer of smaller components and an external siliceous membrane formed from anastomosing siliceous filaments. Axial crosses of spicules that appear to be **stauractines** may occur in this membrane; but sometimes these are dictyonalia with distal rays aborted. In some forms (e.g., *Stamnia* POMEL) a similar membrane coats the marginal surface of a sponge with intradictyonal epirhyses and aporhyses.

In *Coeloptychium* GOLDFUSS the upper surface of a radially plicated funnel is covered by a sieve plate formed from siliceous filaments that grew out from the tops of upward facing plications. In *Cameroptychium* LEONHARD, a comparable sieve plate is formed from unmodified adventitious lychniscs.

Lychniscosa are difficult to classify due to our lack of knowledge of their soft parts. Consequently, the arrangement used here in some of the accepted divisions is more or less arbitrary.

The family Calyptrellidae comprises only *Calyptrella* SCHRAMMEN, with a simple netlike skeleton. The Callodictyonidae comprise all Lychniscosa with euretoid-like skeletons and no canalization or with simple ostia and postica only except for some specialized genera referred to the family Coeloptychidae. A soft peripheral capsule or a rigid **peripheral skeleton** may be present in genera with the body composed of dividing

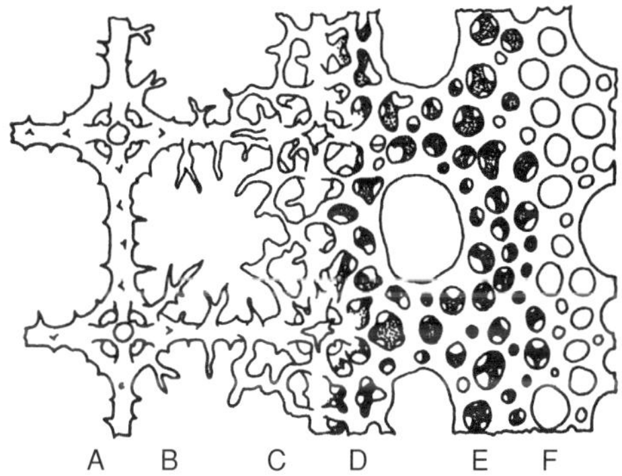

FIG. 63. Order Dictyonida, suborder Lychniscosa; formation of finely porous, cortical membrane from branched and anastomosed, siliceous filaments, which grow out from lychniscs at surface; conditions represent those of different genera or stages in development; *A,* meshwork at surface is similar to that of interior, skeletal beams ornamented with small spines; *B,* branched, siliceous filaments project into and over outermost spicular meshes; *C,* filaments are now anastomosed, although spicular meshwork is still visible under them; *D–F,* spicular meshwork covered externally by dense, three-dimensional meshwork of anastomosed, siliceous filaments; *D,* mesh spaces between filaments of irregular shapes; *E,* external filaments form a smooth, external membrane, with rounded pores; *F,* this is shown alone, with underlying structure omitted; between views *D* and *E,* an intracortical, skeletal pore opens into each underlying mesh of dictyonal framework; similar structures may occur in peripheral skeletons (new).

and anastomosing tubes (*Callicyclix, Tremabolites*); but the presence or absence of soft capsules is not determinable in fossils. The Coeloptychidae are comparable but specialized sponges in which an annular peripheral skeleton truncates either plications of a funnel-like body or radiating tubes that arise from plications of an axial funnel. The upper surface of the body or the inside of an axial funnel may be covered by a sieve plate.

The Ventriculitidae are Lychniscosa with radial to labyrinthine epirhyses and aporhyses and no special marginal structures. In some, the skeletal canals are arranged in longitudinal series, similar to the arrangement in cribrospongiid Hexactinosa. In others, this pattern is modified by multiplication of epirhyses; by furrowing of skeletal surfaces that may replace epirhyses on the dermal side; by development of internal labyrinths, although the ostia and postica retain their alternating arrangement; or in other special ways. The surfaces are usually coated by a siliceous cortical membrane, and a superficial meshwork may occur. The latter is usually formed from siliceous filaments, but sometimes from dictyonal hexactines. The body is nearly always funnel-like.

The Camerospongiidae include some sponges similar to the Ventriculitidae with a flattened oscular margin coated by an annular siliceous membrane and other sponges in which the wall resembles an axial tube with lateral outgrowths or a cavaedial labyrinth, with an annular membrane at the summit. Researchers interpret the latter types as showing the effects of coalescence of inhalant canals around radial chamber-bearing structures.

Coeloscyphia TATE (=*Polyblastidium* ZITTEL), the only genus of the Polyblastidiidae, is a sponge in which an axial tube or group of tubes emits budlike lateral outgrowths. The latter have rows of epirhyses on the outside and longitudinal aporhytic furrows on the inside. This sponge is possibly but not certainly a specialized ventriculitid.

The Pachyteichismatidae are funnel-like or toplike sponges, with a canal system of uncertain type. In some genera the wall is excavated from both sides by alternating clefts that may intercommunicate internally. These forms grade into others in which the wall is labyrinthically cavernous. Small ostia or short radial canals may occur in the walls of internal passages. The dictyonal framework is characteristically very regular, with little or no cortex. The clefts or internal passages were regarded as cavaedia by SCHRAMMEN (1936) but seem more probably intradictyonal.

Dermal surfaces of Sporadopylidae have alternating apertures of radial skeletal canals, which either pass directly through the skeleton or into a labyrinth before opening on the gastral side. The canals are, thus, comparable with both diarhyses and schizorhyses. In the Dactylocalycidae a single system of anastomosing tubular canals opens through both skeletal surfaces. Both families have extensive suppression of lychnisc octahedra in some genera (e.g., *Dactylocalyx*), which have been mistaken for hexactinosa.

The Cypelliidae have canalization that seems like that of Dactylocalycidae, but the Cypelliidae are older and have the dermal surface coated by a porous superficial membrane in which the dermalia are imbedded.

PHYLOGENY

Nothing is known of the origin of this order, which is first represented in the record by specialized Devonian *Pillaraspongia* RIGBY, 1986b from western Australia and possibly *Pseudopemmatites* FRAIPONT, 1911 and *Pachyspongia* TERMIER & TERMIER, 1981 from the Devonian of Belgium, along with figured but undescribed sponges from the Upper Devonian of Poland (RIGBY, RACKI, & WRZOLEK, 1982). Until recently the specialized hexactinosid *Cribrospongia* (=*Tremadictyon*) and *Casearia* from the Middle Triassic (Anisian) were the oldest known representatives of the order. A supposed Ordovician example (*Okulitchina* WILSON: DE LAUBENFELS, 1955) is a lithistid demosponge (FINKS, 1960). IJIMA (1927) believed that Hexactinosa and Lychniscosa had different origins because nodal octahedra of the latter are formed before the spicules unite; if this corresponds with phylogeny, the lychniscs existed in proto-Lychniscosa before the dictyonine condition was developed.

A similar problem exists in the deciphering of further phylogeny. Complex forms appear in the geologic record without apparent simpler prototypes and are also the predominant types in the earlier faunas. Most of the simple types do not appear until the Early or Late Cretaceous. These types appear primitive morphologically by comparison with earlier complex sponges but appear in the opposite stratigraphic order to what might be expected. On the other hand, forms appearing suddenly were evidently immigrants and complex forms presumably evolved from simpler ancestors. Perhaps the original dictyonines were deep-water sponges whose simple descendants were unable until at least the Devonian in some lines and later in their history in other lines to extend into depths approaching wave base, which their thicker-walled derivatives could colonize.

The theoretical prototype of Hexactinosa is a lyssacine sponge with hexactinal dermalia, principalia, and gastralia and with hexaster microscleres and some prototypes of uncinates and sceptrules. The sponge is **basiphytous** (not **lophophytous**), with a basal skeleton composed of hexactines like the **principalia.** Prototypes of sceptrules could have been small pinular hexactines that gave rise first to sarules (see IJIMA, 1927) by loss of paratangential rays and then to clavules or scopules. Uncinates do not appear to correspond with the comitalia of Lyssacinosa because the latter take part in formation of rigid frameworks and uncinates are not fused. The principalia are arranged in a single layer, with a longitudinal and transverse orientation. The prototype of Lychniscosa is a similar sponge with loose lychniscs as principalia if ontogeny follows phylogeny in modern forms. The prototypes of uncinates and sceptrules need not have been present, but the absence of these spicules in the few modern species may be due to their loss in phylogeny, as appears to be true in some Hexactinosa. This sponge is presumably derived from an older one without lychniscs.

The fossil of pre-Triassic age that comes closest to these prototypes is the Permian dictyospongiid *Microstaura* FINKS (FINKS, 1960), although nothing is known of the microscleres and most dictyospongiids are lophophytes. This middle Permian genus also seems to have evolved far too late.

If the phylogeny of Hexactinosa is assessed from the comparative evidence, the primitive stock are sponges with farreoid skeletons and sarules or some other prototypes of clavules and scopules. The Farreidae are more or less unaltered descendants of these sponges. The euretoid skeleton was presumably evolved from the farreoid type, resulting in a dictyonal framework that was three dimensional initially; this in turn allows further evolution of thicker-walled sponges and more efficient circulatory systems. Interaction of the later developments leads to canalization of the skeleton on lines determined by previous evolution of the canal or chamber systems. The simple Euretidae, in which sceptrules are normally **scopules,** represent the stock from which higher forms originated since the sceptrules of the latter are scopules in all known instances. The Craticulariidae and Cribrospongiidae are canalized sponges that retained a canal system with both inhalant and exhalant canals but diverged into forms having contrasting serial patterns. Their prototypes are forms like the living *Chonelasma* SCHULZE and *Tretochone* REID that have similar canal systems but only **intracortical** canalization. The craticulariid pattern compares with linear series of ostia and postica that occur in various euretids; the cribrospongiid type has no known euretid counterpart and, thus, seems more specialized. The Aphrocallistidae and Tretodictyidae represent independent evolutionary series of different but still euretid origin. The other families are enigmatic, but *Stauroderma* and Porospongiidae are probably derivatives of the Cribrospongiidae and Tretodictyidae, respectively.

As noted, this picture does not fit the stratigraphic sequence of first appearances, but these clearly do not correspond with any probable evolutionary sequence. Although sponges with dictyonal skeletons are known from the Devonian of Australia and Europe, the earliest known Hexactinosa are the specialized *Cribrospongia* and *Casearia* of Middle Triassic (Anisian) age. Even allowing for initially rapid evolution, the specialization and divergence of these genera imply a long previous history, perhaps from a period very much older than Permian. The most specialized genus, *Stauroderma,* appeared suddenly in the Middle Jurassic with the fully evolved Craticulariidae (*Craticularia*) and the first known euretids (*Pseudocavispongia* LAGNEAU-HÉRENGER). The first known (Oxfordian) tretodictyid, *Nitidus* de Laubenfels had already wide cavernous schizorhyses like those of the living *Psilocalyx* Ijima. The canalized Aphrocallistidae and very simple Farreidae appeared in the Cretaceous. In none of these families is there evidence of major evolutionary progress after their first appearances.

The Aulocalycidae, included doubtfully as Hexactinosa following Ijima (1927), lack the dictyonal strands, uncinates, and sceptrules that occur in farreoid and euretoid genera. They could be forms derived from simple Hexactinosa by loss of these features. They do not appear to belong with the Lyssacinosa because the parenchymal megascleres are all hexactines and all fused together as in true dictyonines. But they could mark an independent line of descent from the same source as true Dictyonida without closer relationship.

The Lychniscosa have a similar history of sudden appearance of already specialized genera, with simple forms making late appearances. The Cretaceous *Calyptrella,* with a simple netlike skeleton, is probably a late survivor of the primitive stock of the suborder. The first step in further evolution are appearances of farreid-like and euretid-like sponges, whose most direct descendants would be the Callodictyonidae. From this stock originated various canalized families. The Ventriculitidae, with diplorhytic canalization (epirhyses and aporhyses), appear to have existed by the Middle Jurassic (*Calathiscus* Sollas; although nodal octahedra are not clearly developed in most instances, and no good material is known). The Pachyteichismatidae, Sporadopylidae, and Cypelliidae, all of which flourished in the Late Jurassic, mark three further lines of specialization. The Cretaceous abundance of Callodictyondiae and Ventriculitidae should mark further evolution in these families. The Cretaceous Coeloptychidae were essentially highly specialized callodictyonids. The Dactylocalycidae, first appearing stratigraphically in the Lower Cretaceous, should represent a further line of descent, unless they are related to the earlier Cypelliidae.

GLOSSARY OF MORPHOLOGICAL TERMS

R. E. H. Reid and J. K. Rigby

[formerly of Department of Geology, The Queens University of Belfast; and Department of Geology, Brigham Young University]

INTRODUCTION

Parts of speech of terms, where not obvious, is indicated: (n), noun; (n pl), plural noun; (adj), adjective. Obsolete terms are in italics, and terms applied mainly or exclusively to members, spicules, or features of specific classes are indicated by (Cal) for Calcarea, (Dem) for Demospongea, (Hex) for Hexactinellida, and (Het) for Heteractinida.

GLOSSARY

acantho-. Ornamented with spinules.

acanthorhabd (n). Any monaxon with spinules.

acanthostrongyle (n). Strongyle with spinules (Dem).

acanthostyle (n). Style with spinules (Dem).

acanthotriaene (n). Triaene with spinules (Dem).

acanthotylostyle (n). Monaxon with swollen knob on one end and sharply pointed at the other end, with spines along the shaft.

acanthoxea (n). Oxea with spinules (Dem).

acerate (adj). Needle shaped, oxeate, referring to the extremities of megascleres.

acrepid (adj). Anaxial, refers to development from an anaxial initial body, or crepis (Dem).

actin, -e (n). True spicular ray, produced by secretion of mineral matter around an organic axial filament.

-actin, -e (n). Signifies ray in names of spicules referring to number, character, or absence of rays; in Hexactinellida, also designates holactines.

-actinal (adj). termination signifying ray.

afferent (adj). Inhalant.

agglutinating (adj). Sponge that creates its own substratum by cementing together a mixture of foreign material.

amararhysis (n). Skeletal canal in the form of a longitudinal tunnel with slitlike gastral openings, also open dermally at the ends of branch tunnels leading to the tops of papilliform external outgrowths; condition in which such skeletal canals are present (Hex).

ambiostium (n). A large exopore at the junction of two chambers that communicates with both chambers (Dem).

ambisiphonate (adj). Condition in a sphinctozoan in which an axial tube is formed by growth from both the floor and the roof of each skeletal chamber, with the two parts growing together so that a ring of perforations is left (Dem).

ambucinate (adj). Diactine with spines pointed back toward the center.

amoebocyte (n). Mesenchymal cell capable of amoeboid movement.

amphiaster (n). Microsclere with a group of raylike centrifugal spines at each end of a straight central shaft (Dem).

amphiblastula (n). Sponge embryo or free larva in which the surface is flagellate except in a small area at one pole.

amphidisc (n). Monaxial microsclere in which the ends of a straight shaft bear discoidal transverse expansions or whorls of recurved teeth (umbels) (Dem, Hex).

amphitriaene (n). Spicule similar to a triaene with a cladome at both ends of the rhabdome (Dem).

amphitrider (n). Amphitriaene-like desma, with three clones like those of a clonome at each end of a central shaft (centrome); or desma developed from an amphitriaene crepis, not retaining an obvious amphitriaene-like form (Dem).

amphityle, -ote (n). Tylote.

ana-. Designates grapnel or anchorlike spicules, with cladi recurved toward a rhabdome.

anactinal (adj). Without true cored rays.

anapolyaenes (n pl). Grapnel-like polyaenes, with the cladi recurved toward the rhabdome (Het).

anatriaene (n). Triaene with cladi recurved toward the rhabdome.

anaxial (adj). Mineral secretion in spicules not formed around an organic axial filament; see also cryptaxial.

anaxon (n). Spicule formed without organic axial structures, and hence also lacking true rays; see also cryptaxon.

anchora (n). Dentate anchor similar to chela (Dem).

anchorate (adj). With paired rays recurved toward a shaft formed by an unpaired ray, so that the spicule is anchorlike (Hex, Dem).

aniso-. Unequal; asymmetrically developed.

anisochela (n). Chela with ends developed differently (Dem).

ankylosis (n). Union of spicules by fusion.

anomoclad (n). Sphaeroclone (Dem).

anomoclone (n). Variably shaped desma intergrading with chiastoclones but not resembling an amphitrider, in some forms like a stoutly built arched rhizoclone or approaching a didymoclone; nature of the crepis unknown (Dem).

aphodal (adj). Leuconoid with aphodi but not prosodi.

aphodus (n). Fine tubular duct connecting a flagellated chamber with a main trunk or branch of an exhalant canal (apochete).

apical (adj; of choanocyte nuclei). At the end of the cell that bears the flagellum and collar (Cal).

apochete (n). Exhalant canal.

apopore (n). Exhalant pore.

apopyle (n). Open and exhalant end of a flagellated chamber.

apopylar (adj). Of or with an apopyle.
aporate (adj). Lacking pores.
aporhysis (n). Skeletal canal corresponding with a flagellated chamber or an exhalant canal of the soft parts.
aporhysome (n). Inner trabecular network (Hex).
archaeocyte (n). Reproductive cell, capable of converting into other types of cells.
ascon (n). Sponge in which choanocytes line the paragaster, without flagellated chambers or a canal system.
asconoid (adj). With the character of an ascon; as in ascons; see also homocoelous.
asiphonate (adj). Condition of thalamid sphinctozoans that lack any form of axial channel (Dem).
aspidaster (n). Discoidal microsclere with the structure of a flattened sterraster (Dem).
aster (n). Any polyactinal or seemingly polyactinal microsclere.
-aster. Designates some form of aster.
astral (adj). With anaxial terminal outgrowths of the type characteristic of rosettes (Hex).
astroclone (n). Anaxial desma with radiating clones, which are not all directed to one side as in sphaeroclones; zygomes rootlike to cuplike. Includes anomoclones *sensu* Schrammen (not Rauff) (Dem).
astrorhiza (n). traces of excurrent canal system on or within a rigid skeleton that appears as a radial or star-shaped structure.
astrose (adj). Comprising asters (Dem).
atrium (n). The spongocoel of a sponge.
aulocalycoid (adj). With diagonally interwoven or intersecting, strandlike series of skeletal beams through the whole depth of the dictyonal framework, in which typical subparallel dictyonal strands are not present (Hex).
autodermal (adj). Of or comprising autodermalia (Hex).
autodermalia (n pl). Megascleres that support the dermal membrane directly (Hex).
autogastral (adj). Of or comprising autogastralia (Hex).
autogastralia (n pl). Megascleres that support the gastral membrane directly (Hex).
axial canal (n). Intraspicular cavity left by decay of an axial filament, or the same enlarged by internal solution of mineral matter.
axial cross (n). Feature seen at the spicular center of triaxons, where axial filaments of rays arranged in opposite pairs appear to intersect (Hex).
axial filament (n). Organic coring filament, around which the mineral part of a spicular ray is deposited; see also axone.
axial rudiment (n). Rudimentary axial filament.
axial tube (n). Skeletal structure developed in the axial part of a series of skeletal chambers in some thalamid sphinctozoans, consisting of a series of discontinuous structures like the septal necks of chambered Cephalopoda, or of a single continuous tube (Dem).
-axon. General termination of names based on the number of axes of growth that rays follow.
axone (n). Axial filament.
azygose (adj). Without zygomes (Dem).
basal (adj; of choanocyte nuclei). At the end of the cell that is attached to the mesenchymal surface (Cal).
basal (adj; of body or spicules). Attached to or next to the substratum; corresponds with initial in sponges attached at one point only.
basal plate (or lamina) (n). Special skeletal layer or structure developed next the substratum in basiphytous sponges.
basal ray (n). The third ray of a sagittal triradiate, contrasted with the two similar rays (Cal).
basal skeleton (n). Any spicules or structure specially concerned with basal attachment.
basalia (n pl). Prostalia of the dermal part of the body, by which the sponge is anchored (Hex).
basidictyonalia (n). Small hexactines fused to form a special basal skeleton in some lyssacine Hexactinellida.
basiphyte (n). Sponge attached by an encrusting base.
basiphytous (adj). Attached by an encrusting base.
beam (n). Rodlike element in hexactinellid skeletons formed of merged rays of adjacent spicules.
bihamate (n). Diancistron (Dem).
bipocillate (n). Bipocillus (Dem).
bipocillus (n). Monaxial microsclere whose ends bear spoonlike or leaflike expansions, with concave sides facing together; see also diaspid (Dem).
birotulate (n). Amphidisc; an amphidisc, staurodisc, hexadisc, or hemidisc (Hex).
bispatulate (n). Palmate chela (Dem).
brachyome (n). The fourth arm of a trider when shorter than the clones of the brachyome, comprising a short clone, a clone-rudiment, or a crepidal ray from which no clone is produces (Dem).
bullipore (n). Pore within a cribribulla.
calthrops (n). Regular tetraxon with four equal rays; see also chelotrope (Dem).
canal (n). Internal passage to carry circulating water.
canal system (n). Inhalant or inhalant and exhalant canals of sycons and leucons.
canalar (adj). Related to the canal or chamber system.
canalar membrane (n). Trabecular membrane lining an inhalant or exhalant canal (Hex).
canalaria (n pl). Spicules occurring in the linings of inhalant or exhalant canals (Hex).
candelabrum (n). Tetralophose calthrops in which the branches of one ray may differ from those of the others (Dem).
capstan tubercles (n pl). Stalked tubercles, with a constricted neck and an expanded head, which may be bifid or trifid, arising by thickening of simple or branching lateral spinules (Dem).
category (of spicules) (n). One of the main types of spicules found in a sponge or a group of sponges, distinguished by form, size, location, function, or some combination of these characters.
cateniform (adj). Catenulate or moniliform (Dem).
catenulate (adj). With skeletal chambers arranged in a linear series (Dem).
cavaedia (n pl). External spaces enclosed within a body consisting of dividing and anastomosing tubes, or between folds of a plicated wall; not part of the canal system proper; see also intercanals.

cavaedial space (n). Large, deep indentations of the sponge's dermal surface.
centrifugal (adj). Directed away from the spicular center.
centrome (n). The central shaft of an amphitrider (Dem).
centrotylote (n). Monaxon with a central enlargement (Dem).
centrum (n). A differentiated central part of a spicule.
chamber (n). One of the regular juxtaposed hollow structures formed by the skeleton in sphinctozoan sponges (Dem).
chamber system (n). The flagellated chambers to which choanocytes are restricted in syconoid and leuconoid sponges.
chela (n). Microsclere developed from a C-shaped sigma, in which the ends bear inward-facing, toothlike or bladelike, lateral appendages, arranged in opposite pairs (see dentate chela, palmate chela) (Dem).
chelaster (n). Pseudaster developed from a chela (Dem).
cheloids (n). Microscleres resembling chelas and related forms.
chelotrope (n). Calthrops (Dem).
chessman spicule (n). Discaster (Dem).
chiastoclone (n). Name applied to desmas of Orchocladina having the form of an amphitrider with the central shaft shorter than the clones, or with clones emitted radially from a center; nature of the crepis uncertain (Dem).
choanocytal (adj). Lined by or composed of choanocytes.
choanocytal membrane (n). Reticulate internal membrane, diverticula of which form the flagellated chambers, composed of choanocytes connected by lateral processes (Hex).
choanocyte (n). Flagellated cell with a single flagellum, enclosed basally by a tubular protoplasmic collar; see also collar cell.
choanosomal (adj). Of the choanosome.
choanosome (n). The part of the body of a syconoid or leuconoid sponge containing flagellated chambers, when covered externally by a stratum (e.g., an ectosome) without choanocytal structures.
chondrenchyma (n). Stiff, cartilage-like mesenchyme.
chone (n). Intracortical inhalant canal, extending through a cortex from the external surface to a subdermal space underlying it (Dem).
choristid (adj). (1) With tetractinal or triactinal megascleres, to which other types may be added, but without desmas; (2) of choristids (Dem).
choristid (n). Demosponge with a choristid skeleton; member of the order Choristida.
clad (n). Any ray or axial branch in a spicule. The term is used chiefly in triaenes or tetraenes.
cladi (n pl). The similar rays of a triaene or similar spicule (e.g., diaene, tetraene).
cladocalthrops (n). Calthrops with repeatedly branching rays (Dem).
cladome (n). Group of similar rays, arranged radially in relation to a monactinal or diactinal rhabdome.
clathrate (adj). Structure resembling an open latticework.
clavidisc (n). Microsclere in the form of an ovate disc with a central perforation, produced by ingrowth and union of the ends of a diancistron-like prototype; see also psellium.
clavule (n). Pinlike or nail-like sceptrule in which the end of the spicule containing the axial cross is more or less swollen (tylote), or bears a marginally denticulate transverse disc or a ring of recurved teeth (Hex)
cleme (n). Long monactine with alternating thornlike lateral spinules arranged in two opposite rows (Hex).
cloaca (n). See spongocoel.
cloacal (adj). Of the cloaca, paragaster, or spongocoel.
clone (n). Raylike arm of a desma, partly or wholly anaxial (Dem).
clonome (n). Group of three similar clones, analogous (not homologous) with a cladome (Dem).
collar cell (n). Choanocyte.
collenchyma (n). Mainly gelatinous mesenchyme.
collencyte (n). Typically stellate mesenchymal cell, the points of which emit fine branching filaments that may be connected syncytially with those from other examples (Cal, Dem).
collosclere (n). Spicule composed of colloidal silica (Dem).
comital (adj). Comprising comitalia (Hex).
comitalia (n pl). Accessory parenchymal megascleres, occurring apposed to principalia in some lyssacine Hexactinellida (Hex).
concordant structure (n). In dictyonine Hexactinellida, condition in which the orientation of dictyonal strands is related to the local direction of growth.
connecting beams (n pl). In dictyonine Hexactinellida, skeletal beams that connect adjacent dictyonal strands laterally.
connecting membrane (n). Trabecular membrane extending between the open ends (apopyles) of flagellated chambers (Hex).
coring (adj). Running longitudinally in the core of a skeletal fiber.
coronal (adj). Fringing an osculum.
cortex (n). Thick, rindlike ectosome.
cortex (*sensu* Moret) (n). Any external part of the skeleton that differs from that of the interior, irrespective of homology.
cortex dépendant (*sensu* Moret) (n). Any specially modified external part of a skeletal framework, or superficial structure arising from the surface of such a framework.
cortex indépendant (*sensu* Moret) (n). Ectosomal triaenes of Demospongea, dermalia and gastralia of Hexactinellida.
cortical (adj). Of the cortex; of the skeletal cortex.
corynaster (n). Finely spinulate, clublike tylostrongyle (Dem).
craticula (n). A screenlike element across the outer end of an exaulos (Cal).
craticular pore (n). A pore in the craticula.
crepidal (adj). Of the crepis (Dem).
crepis (crepides, pl.) (n). Initial body in a spicule that then grows substantially by anaxial secretion (Dem).

cribribulla (n). Inward-facing, blisterlike sieve at the inner end of an exaulos.
cribripore (n). Small pores in sievelike cribribulla (Dem).
cric- (cricc-). With a typically strong annular ornament (Dem).
criccalthrops (n). Annulated calthrops (Dem).
cricophalangaster (n). Small cricorhabd with spinulate annulations (Dem).
cricorhabd (n). Strongly annulated monaxon (Dem).
cricostyle (n). Annulated style, with the annulations sometimes dying out toward the sharp end (Dem).
cricotriaene (n). Annulated triaene (Dem).
cryptaxial (adj). Anaxial (Dem).
cryptaxon (n). Anaxon (Dem).
cryptosiphonate (adj). Condition of thalamid sphinctozoans in which skeletal chambers communicate through an aperture or a group of apertures at the top of each chamber, without an axial tube (Dem, Cal).
cyathiform (adj). Cup shaped.
cystenchyma (n). Mesenchyme in which the most common cells contain conspicuous vacuoles.
Deckgespinnst (n). Superficial meshwork (Hex).
Deckschicht (n). *Sensu* Zittel, a skeletal cortex, superficial meshwork, or a peripheral skeleton (cf. dermal layer *sensu* Hinde); *sensu* Schrammen, a skeletal cortex.
dendritic (adj). Forming branching growths.
dendro-. Treelike.
dendroceratid (adj). Type of spongin skeleton consisting of dendritically branched fibers, which are unconnected except where they arise from a common basal plate (Dem).
dendroclone (n). Name applied to desmas of Orchocladina with the form of an amphitrider or a dipolar rhizoclone, in the former with a centromelike shaft roughly equalling or longer than the clones; nature of the crepis uncertain (Dem).
dentate chela (n). Chela in which the lateral appendages are toothlike, with one to several pairs at each end (Dem).
dermal (adj). (1) external or inhalant; (2) consisting of pinacocytes; (3) non-choanocytal; (4) comprising a dermis; (5) supporting a dermis or a dermal membrane; (6) comprising a special external stratum of a skeletal framework; (7) peripheral.
dermal layer (histological) (n). (1) Epidermis and mesenchyme; (2) an ectosome (Cal, Dem).
dermal layer (skeletal). Name applied to various external parts of skeletal frameworks, or to a peripheral skeleton, irrespective of whether related to a true dermal skeleton.
dermal membrane (n). Membranous trabecular network forming the dermal (outer, inhalant) surface of the body (Hex).
dermal skeleton (n). Skeleton of an ectosome or an endosome; the dermalia in Hexactinellida, or any rigid structure formed by their union.
dermal surface (n). The external and inhalant surface of the lateral wall surrounding an axial paragaster, or any equivalent surface.
dermalia (n pl). (1) Spicules supporting an ectosome or an endosome (Dem); (2) a dermal membrane (Hex); (3) (n) specialized spicule of outer or dermal part of skeleton.
dermis (n). Thin, skinlike ectosome.
desma (n). Articulating megasclere; see also desmone (Dem).
desmal framework (n). The skeletal framework of lithistid demosponges, consisting of articulated desmas (Dem).
desmoid (n). Desmalike spicule, not regarded as a true desma, or supposed to derive from a pseudaster (not a normal megasclere) (Dem).
desmone (n). Desma (Dem).
diact (n). Diactine.
diactin, -e (n). Spicule with two rays; two-rayed holactine (Hex).
diactinal (adj). Two-rayed.
diactinose (adj). Comprising diactines and modified derivatives, which are initially diactinal (Dem).
diaene (n). Triaene-like spicule with only two cladi.
diancistron (n.; pl. -a). C-shaped microsclere with bladelike lamellae developed along the inside curve, hence appearing like a partly opened penknife (Dem).
diaphragm (n). More or less rigid internal plate subdividing chambers (Cal).
diarhysis (n). Oblique to radial skeletal canal that is open at both ends, marking the position of a complex flagellated chamber, not accompanied by epirhyses; condition in which diarhyses are present (Hex).
diaspid (n). Bipocillus.
diaxon (n). Spicule with rays following two growth axes that meet at an angle or intersect.
diaxial (adj). With rays following two growth axes that meet at a point or intersect; containing two angled axial filaments.
dicho-. Signifies dichotomous branching of rays.
dichocalthrops (n). Calthrops with one or more rays branched dichotomously.
dichohexactin, -e (n). Hexactine with rays branched dichotomously.
dichotriaene (n). Triaene with the cladi branched dichotomously.
dichotetraene (n). Tuberculate monaxial desma of dipodal to polypodal or irregular form, with rootlike, terminal zygomes.
dicranoclone (n). Tuberculate monaxial desma of dipodal to polypodal or irregular form, with rootlike, terminal zygomes.
dicrepid (adj). With a diactinal crepis (Dem).
dictyoceratid (adj). Type of spongin skeleton consisting of reticulate fibers (Dem).
dictyonal (adj). Of the dictyonal framework (Hex).
dictyonal cortex (n). Specially modified external part of a dictyonal framework (Hex).
dictyonal framework (n). The rigid, parenchymal skeletal framework of dictyonine Hexactinellida.
dictyonal strand (n). Longitudinal to radial, strandlike series of skeletal beams, formed by union of a series of dictyonalia (Hex).
dictyonalia (n pl). Parenchymal megascleres that fuse to form the skeletal framework in dictyonine Hexactinellida (Hex).

dictyonine (adj). (1) With a parenchymal skeleton that is always a rigid skeletal framework, composed of fused megascleres that are always hexactines except when rays are lost at skeletal surfaces; (2) of dictyonines (Hex).

dictyonine (n). Hexactinellid sponge with a dictyonine parenchymal skeleton; member of the orders Hexactinosa or Lychniscosa.

didymoclone (n). Monaxial desma in which zygome-bearing clones are emitted from spherical swellings at the ends of an epirhabd, and are typically directed to one side of the desma; zygomes rootlike, or cuplike with denticulate margins (Dem).

dilophose (adj). With two lophose rays.

diploclone (n). Sublithistid desma with incipient to digitate expansions of the ends of an epirhabd, also grading into a normal strongyle; the zygomelike ends do not articulate as in lithistids proper (Dem).

diplodal (adj). Leuconoid with both prosodi and aphodi.

diplorhysis (n). Condition in which the skeletal framework contains both epirhyses and aporhyses.

dipolar rhizoclone (n). Rhizoclone with rootlike zygomes or distinct clones at the ends of a smooth central shaft, which is the epirhabd (Dem).

discaster (n). Microsclere with discoidal flanges or whorls of raylike spines surrounding a straight nonaxial shaft. Also called a sceptrella, discorhabd, or chessman spicule (Dem).

discohexact (n). Hexactinoid hexaster with terminal discs or umbels (Hex).

discohexaster (n). Hexaster in which the ends of the terminal outgrowths are capped by transverse discs or denticulate umbels (Hex).

discordant structure (n). In dictyonine Hexactinellida, condition in which the orientation of dictyonal strands is not related to the local direction of growth.

discorhabd (n). Discaster (Dem).

discostrongyle (n). Monaxial disc, with or without a rhabdome-like central stalk, arising from an initial strongyle (Dem).

discotriaene (n). Triaene with a specially modified cladome that consists of a siliceous disc with the axial filaments or canals of short initial cladi in the central part; see also symphyllotriaene (Dem).

distal ray (n). One of rays of a spicule, when directed outwardly at right angles to an external surface.

dodecaactine (n). Spicules with six equally spaced initial rays that diverge from a common point, but three alternating rays are shorter and the three intervening rays are longer and have trifurcate tips so that the spicule has 12 outer rays.

dragma (n). Monaxial microsclere that occurs in bundles (Dem).

dragmata (n pl). Monaxial microscleres occurring in sheaves and produced by a single scleroblast (Dem).

echinating (adj). Projecting from the surface of a skeletal fiber; used for monaxons occurring with one end imbedded in spongin fibers, which are said to be echinated by them (Dem).

ectosomal (adj). Of the ectosome.

ectosome (n). Chamberless and porous external stratum, extending across the outer ends of inhalant canals; developed as a thin dermis or a thick cortex (Cal, Dem).

efferent (adj). Exhalant.

emergence (n). In dictyonine Hexactinellida, migration of dictyonal strands to the dermal or gastral surface of a three-dimensional dictyonal framework.

endocameral (adj). Within the skeletal chambers of thalamid sphinctozoans (Dem).

endolithic (adj). Cavity-dwelling sponge occupying cavities in hard substrates.

endopore (n). Opening through the wall of a central tube, the endowall (Dem).

endopsammic (adj). Sand-dwelling sponge living in soft substrates.

endosomal (adj). Of the endosome.

endosome (n). Chamberless stratum between the choanosome and gastral or other exhalant surface. May also be misused as meaning choanosome (Cal, Dem).

endotube (n). Tube that pierces the endowall or extends into the chamber from the endowall in a sphinctozoan (Dem).

endowall (n). Wall of a central tube, cloaca, or siphon (Dem, Cal).

ennomoclone (n). Spicule with short, distal arm and three or six proximal, longer arms or rays directed symmetrically away from it; a tricranoclone or sphaeroclone (Dem).

epicrepid (adj). Formed directly around the crepis (Dem).

epidermis (n). Layer of pinacocytes or equivalent syncytial membrane, coating mesenchymal surfaces where choanocytes are absent (Cal, Dem).

ephirabd (n). The epicrepid part of a monaxial desma (Dem).

epirhysis (n). Skeletal canal corresponding with an inhalant canal in the soft parts.

epirhysome (n). Outer trabecular network (Hex).

epitheca (n). Thin, wrinkled layer different in structure from the normal skeleton.

epithelioid membrane (n). Epidermis consisting of a syncytial membrane (Dem).

epochete (n). Canal-like space that is properly an external depression, additional to true canal system.

equianchorate (n, adj). Isochela (Dem).

euaster (n). Radiate microsclere in which rays proceed from a genuine center (not from a monaxial shaft, as in pseudasters), also typically polyactinal; some with rays partly or completely enveloped by secondary silica that grows outward from the spicular center (see also sphaeraster, sterraster) (Dem).

euastrose (adj). Comprising euasters (Dem).

eulerhabd (n). Stout form of ophirhabd, often U-shaped (Dem).

euretoid (adj). With subparallel dictyonal strands spreading through three-dimensional primary meshwork (Hex).

eurypylous (adj). Leuconoid with chambers opening directly through the walls of exhalant canals (apochetes), without intervening aphodi.

eutaxiclad (n). Dicranoclone (note: this desma is not eutaxicladine, i.e., an ennomoclone, *sensu* Rauff, being based on a form of his rhizoclone) (Dem).
excurrent (adj). Flowing out.
exaulos (n). Spoutlike tube in sponge wall (Cal).
excurrent canal (n). Exhalant canal.
exhalant (adj). Through which water passes out; see also efferent.
exhalant area (n). Part of the surface of a sponge without paragaster or spongocoel, where exhalant canals open.
exhalant canal (n). Internal passage leading water from a group of flagellated chambers to the paragaster, or to an external exhalant area if a paragaster is lacking; see also apochete.
exopore (n). Pore that pierces the outer wall of a chamber (Cal).
exowall (n). External skeleton of a chamber (Cal).
farreoid (adj). With dictyonal strands restricted to a single layer or primary meshwork, to which further secondary meshwork may be added (Hex).
fiber (n). Column of spongin forming part of skeleton may or may not contain spicules or foreign materials.
filling structures (n pl). Endocameral skeletal structures of thalamid sphinctozoans (Dem).
flabellate (adj). Fanlike or tonguelike.
flagellated chamber (n). Choanocyte-lined lateral diverticulum or rounded internal cavity from which water is discharged to the paragaster directly (in sycons) or through an exhalant canal (in leucons) (Cal, Dem); diverticulum of the choanocytal membrane, which opens into meshes of the inner trabecular network (aporhysome) or into an exhalant canal (Hex).
floricome (n). Hexaster in which terminal outgrowths arranged in several whorls are S-shaped and petal-like or plumelike, with the ends facing outward (Hex).
florule (n). Scopule-like spicule with a shaft shorter than the terminal spines, which form a flowerlike ring (?Hex).
forceps (n). Monaxial microsclere shaped like a sugar tongs (Dem).
furcula (n). Spicule shaped like a wishbone (Hex).
fusiform (adj). Tapered toward both ends and in oxeate spicules.
gastral (adj). Refers to (1) choanocytes; (2) the internal surface of the sponge wall surrounding an axial paragaster, to a corresponding surface when an axial paragaster is lacking, or to spicules or any other structure specially associated with this surface.
gastral cavity (n). See spongocoel..
gastral layer (n). Choanocyte layer.
gastral membrane (n). Membranous trabecular network forming the gastral (inner, exhalant) surface of the body (Hex).
gastral skeleton (n). The gastralia in Hexactinellida or any rigid structure formed by their union.
gastral surface (n). The internal and exhalant surface of the lateral wall surrounding an axial paragaster, or an equivalent surface.
gastralia (n pl). Spicules supporting the gastral membrane (Hex).
gemmule (n). Asexually produced regenerative body.
glass sponge (n). Member of the class Hexactinellida.
globostellate (n, adj). Sphaeraster or sterraster (Dem).
glomerate (adj). With skeletal chambers arranged like the seeds in an ear of corn or like grapes in a cluster (Dem).
grantioid (adj). Syconoid, with true enclosed inhalant canals in a compact wall.
graphiocome (n). Hexaster whose rays bear brushlike clusters of fine terminals; see also graphiohexaster (Hex).
graphiohexaster (n). Graphiocome (Hex).
habitus (n). External form.
hastate (n, adj). Tornote or oxea (Dem), a spicule remaining of uniform diameter for most of its length, but which has abrupt tapering tips.
heloclone (n). Sinuous monaxial desma with zygomes in the form of lateral facets or notches, sometimes digitate at the ends but without true zygome-bearing clones; axial canal often almost as long as desma; see also megaclad, megaclone, rhabdoclone (Dem).
helotriaene (n). Tripod with three long, curved rays and a tubercle-like rudiment of a fourth (Dem).
hemiamphidisc (n). Hemidisc (Hex).
hemiaster (n). Imperfect hexaster in which some rays are holactinal (Hex).
hemidisc (n). Asymmetrical amphidisc, with one terminal umbel much larger than the other; see also hemiamphidisc (Hex).
heteractine (n). Spicule of heteractinid sponge, commonly octactine based but may be polyactine.
heteractinid (adj). Referring to heteractine sponges.
heterocoelous (adj). With choanocytes restricted to flagellated chambers, the paragaster having a pinacocytal lining (Cal).
hexact (n). Hexactine (but some so-called hexacts of Hexactinellida are hexactinoid hexasters).
hexactin, -e (n). Spicule with six rays; six-rayed holactine (Hex).
hexactinal (adj). Six-rayed.
hexactinellid (adj). Normally taxonomic, but sometimes used of spicules meaning hexactinal.
hexactinoid (adj). Hexactine-like, in the instance of hexactine-like microscleres that are properly hexasters with a single anaxial terminal continuing the line of each true ray (Hex).
hexactinose (adj). (1) of hexasters, hexactinoid; (2) of a dictyonal framework, without lychniscs (Hex).
hexadisc (n). Hexactinal amphidisc-variant (Hex).
hexaene (n). Triaene-like spicule with six cladi (Het).
hexaster (n). Six-rayed (hexactinal) and triaxial microsclere, in which the end of each ray bears a group of anaxial, branchlike, and centrifugal outgrowths, or rarely a single anaxial extension of the true cored ray (Hex).
hiloid pit (n). Hilum-like pit in the basal part of a ray (Het).

hilum (n). Smooth pit in the surface of a sterraster marking the position of the scleroblast nucleus.
hispid (adj). With a bristly appearance, due to protrusion of spicules.
hispidating (adj). Giving a bristly appearance.
holactin, -e (n). Spicule without the special terminal outgrowths of hexasters (Het).
holactinal (adj). Without terminal outgrowths of the type seen in hexasters.
holaster (n). Rosette in which all rays bear terminals (Hex).
homocoelous (adj). Asconoid (Cal).
hypercalicified sponges (n). Sponges that secrete nonspicular calcium carbonate, and may include forms that cement spicules together with coatings of calcium carbonate, as well as inozoans and related forms whose principal skeletal structure is calcium carbonate, either aragonite or calcite.
hypodermal (adj). Subdermal.
hypodermalia (n pl). Megascleres that support the dermal membrane but underlie autodermalia from which they differ (Hex).
hypogastral (adj). Subgastral (Hex).
hypogastralia (n pl). Megascleres that support the gastral membrane but underlie autogastralia, from which they differ (Hex).
hypophare (n). Basal layer of the body, apposed to the substratum (Dem).
incurrent (adj). Flowing in.
incurrent canal (n). Inhalant canal.
inequianchorate (n, adj). Anisochela (Dem).
infundibuliform (adj). Funnel shaped.
inhalant (adj). Through which water enters; see also afferent.
inhalant canal (n). Internal passage, leading water to flagellated chambers.
inner trabecular network (n). The trabeculae on the gastral side of the choanocytal and connecting membranes (Hex).
interactinal (adj). Between rays.
interactinal angle (n). Angle between two rays.
interactinal buttress (n). Anaxial strut or plate forming one of the edges of a nodal octahedron in a lychnisc (Hex).
interactinal suture (n). Joint between the bases of rays of the same spicule, which are not continuous as in normal sponge spicules (Cal, Het).
intercanals (n pl). Cavaedia.
intermedia (n pl). Smaller spicules occurring between principalia or dictyonalia, excluding lyssacine comitalia (Hex).
interpore (n). Pores that pierce the wall between chambers in sphinctozoans (Dem).
intertrabecular (adj). Between trabeculae.
interwall (n). Upwardly convex partitions between chambers in sphinctozoans (Dem).
intracortical (adj). Extending through a cortex or a skeletal cortex, but not beyond it.
intradictyonal (adj). Refers to intraskeletal canalization of a dictyonal framework (Hex).
intraskeletal (adj). Within the skeleton; designates features produced by interruption of developing skeletal meshwork, as distinct from enclosed but essentially external cavaedia.
iso-. Equal; symmetrically developed.
isochela (n). Chela with similar ends (Dem).
isodictyal (adj). With the sides of skeletal meshes formed by terminally connected monaxons, cemented together with spongin (Dem).
keratode (n). Spongin.
keratose (adj). Composed of spongin; with a spongin skeleton only (Dem).
kieselhaut (n) (*sensu* Schrammen). Dense but porous siliceous membrane, usually a peripheral structure but sometimes coating a marginal surface only (Hex).
kyphorhabd (n). Stout, curved monaxon with transverse swellings on the convex side, which may have small articulatory facets at its ends (Dem).
labid (n). Forceps.
labripore (n). Exopores surrounded by a distinct external lip (Cal).
lantern node (n). Nodal octahedron of a lychnisc (Hex).
lantern spicule (n). Lychnisc (Hex).
lateral wall (n). The part of a sponge body surrounding an axial paragaster.
leucon (n). Sponge in which flagellated chambers discharge via exhalant canals (apochetes) and inhalant canals are also present.
leuconoid (adj). With the characters of a leucon; as in leucons.
lithistid (adj). (1) With the main internal megascleres developed as articulated desmas, which form a coherent to rigid skeletal framework; (2) of lithistids (Dem).
lithistid (n). Demosponge with a lithistid skeleton; member of the order Lithistida.
lonchiole (n). Sceptrule with a single anaxial spine opposite the single ray (Hex).
long-shafted triaene (n). Triaene with the rhabdome several to many times longer than the rays of the cladome (Dem).
lophophyte (n). Sponge anchored by a root tuft.
lophophytous (adj). Anchored by a root tuft.
lophose (adj). With rays dividing into clusters of branches.
lychnisc (n). Dictyonal hexactine in which the central part is enclosed by an octahedral framework of 12 interactinal buttresses, which cross the interactinal angles from points equidistant from spicular center; see also lantern node and lantern spicule (Hex).
lychniscose (adj). Having lychniscs (Hex).
lyssacine (adj). (1) With parenchymal megascleres primarily or permanently unfused and of various sorts, often partly or all forms with less than six rays, which may be all rhabdodiactines; if secondary fusion occurs, the resultant skeletal framework retains the characteristic composition; (2) of lyssacines (Hex).
lyssacine (n). Hexactinellid sponge with a lyssacine parenchymal skeleton; member of the order Lyssacida.

marginal (adj). At the margin of an osculum, or at some other corresponding growing margin (e.g., of a funnel-like sponge).

marginalia (n pl). Prostalia of an oscular margin (Hex, Dem).

massive (adj). Forming a solid mass.

mega-, megalo-. Large.

megaclad (n). Megaclone (Dem).

megaclone (n). Monaxial desma, not of rhizoclonar type, in which distinct clones have zygomes in the form of longitudinal facets of cuplike, tonguelike, or hand-like terminal expansions; see also heloclone, rhabdoclone, megaclad (Dem).

megalome (n). The fourth clone of a trider when longer than the clones of the clonome (Dem).

megarhizoclone (n). Large rhizoclone, sometimes passing into desmas approaching a megaclone (Dem).

megarhizoclonid (n). Desma with the form of a rhizoclone, occurring in a lithistid whose typical desmas are dicranoclones (Dem).

megasclere (n). Major skeletal element, forming part of the main supporting skeleton when accompanied by accessory microscleres.

megascleric (adj). Of or comprising megascleres.

megaspiculation (n). The megascleres present in a given species, genus, etc.

membrana reticularis (n). Choanocytal membrane (Hex).

mesenchymal (adj). Of the mesenchyme (Cal, Dem).

mesenchyme (n). The gelatinous internal groundmass of Calcarea and Demospongea; see also mesoglea, parenchyma.

mesoglea (n). Mesenchyme.

mesohexaene (n). Mesopolyaene with six cladi, also called an octactine (Het).

mesopentaene (n). Mesopolyaene with five cladi (Het).

mesopolyaenes (n pl). Mesotriaene-like spicules with 4 to 6 cladi (Het).

mesotetraene (n). Mesotriaene-like spicule (or mesopolyaene) with four cladi.

mesotriaene (n). Spicule with three cladi and a diactinal rhabdome, the rays of which may be equal or unequally developed.

mesotrider (n). Tetraclonar desma in which the crepis is a mesotriaene (Dem).

metaster (n). Microsclere (streptosclere) in which raylike spines are emitted from a spiral axis of less than one turn (Dem).

micro-. May designate (1) a microsclere; (2) a small form of a particular microsclere; or (3) a spicule of microscleric size, not a microsclere homologically.

microrhabd (n). Any rodlike monaxial microsclere (Dem).

microsclere (n). Accessory skeletal element, typically but not always smaller than megascleres.

microscleric (adj). Of or comprising microscleres.

microspiculate (adj). With very small spicules, not identified as megascleres or microscleres (Dem).

microspiculation (n). The microscleres present in a given species, genus, etc.

monact (n). Monactine.

monactin, -e (n). Spicule with a single ray only; monaxon with one end sharply pointed; single-rayed holactine.

monactinal (adj). Single rayed.

monaene (n). Triaene-like spicule with only one cladus.

monaxon (n). Spicule in which rays grow along a single growth axis, in one or two directions.

monaxial (adj). With rays following a single growth axis only; containing a single axial filament.

monaxonid (adj). (1) With monaxial megascleres only, unless pseudoradiates are present, and without desmas; (2) of monaxonoids (Dem).

monaxonid (n). Demosponge with a monaxonid skeleton; member of the order Monaxonida.

moniliform (adj). Linear, threadlike growth; see also cateniform.

moniliform acuate (n). Cricostyle (Dem).

monocrepid (adj). With a monaxial crepis; see also rhabdocrepid (Dem).

monolophose (adj). With one lophose ray.

multiradiate node (n). Skeletal node from which more than six beams radiate (Hex).

myocyte (n). Contractile cell.

myxosponge (n). Demosponge with no skeleton.

neoasters (n pl). Radiate or floriform polyactines, often without distinct rhabdome and cladome but linked with true polyaenes by intermediates (Het).

nodal octahedron (n). The octahedral framework enclosing the central part of a lychnisc (Hex).

non-actinal secretion (n). Any condition in which mineral matter forming spicules or parts of spicules is not secreted concentrically around an organic axial filament.

octactin, -e (n). Spicule with eight rays; mesohexaene (Het).

octactinal (adj). Eight rayed.

octactinellid (adj). Octactinal, with the form of a mesohexaene (Het).

octaster (n). Triaxial microsclere in which eight anaxial, raylike pseudoactines are emitted from a center containing a six-rayed axial cross as though along the diagonal axes of a cube, and bear anaxial terminal outgrowths like those of a hexaster (Hex).

olynthus (n). Initial ascon formed in the early ontogeny of Calcarea, after larval fixation.

onychohexaster (n). Hexaster in which the ends of the terminal outgrowths bear umbels in the form of prominent and separate barblike teeth (Hex).

ophirhabd (n). Irregularly sinuous oxea (Dem).

orthodiactin, -e (n). Holactine with two rays growing at right angles (Hex).

orthotetractin, -e (n). Four-rayed holactine in which rays follow three growth axes, intersecting at right angles (Hex).

orthotriaene (n). Triaene in which the angle between the rhabdome and each cladus is 90 to 130°.

oscular (adj). Of an osculum.

oscule (n). Osculum.

osculum (n.; pl. -a). Aperture through which water is discharged to the exterior from a paragaster or a pseudogaster.

ostia (n pl). Inhalant skeletal pores.

ostium (n., pl. -ia). Any opening through which water enters a sponge; sometimes applied to an opening larger than a pore; used in older literature as a synonym of posticum.

outer trabecular network (n). Trabeculae on the dermal side of choanocytal and connecting membranes (Hex).

oxea (n). Monaxon sharply pointed at both ends, regarded as a diactine; see also oxeote.

oxeate (adj). Sharp ended.

oxeote (n). Oxea.

oxy-. Sharp ended.

oxyaster (n). Euaster with sharply pointed rays, the central part of which may have a small spherical mass of silica from which the rays emerge (Dem).

oxyhexaster (n). Hexaster in which the terminal outgrowths are sharply pointed (Hex).

paired rays (n pl). (1) Rays of a triaxon occurring in opposite pairs (Hex); (2) the two similar rays of a sagittal triradiate (Cal).

palm (n). One of the paired lateral processes at the ends of a chela when of bladelike form (Dem).

palmate (adj). With palms (Dem).

palmate chela (n). Chela in which the lateral appendages comprise a single pair of bladelike expansions at each end of the spicule (Dem).

paraclavule (n). Apparently monaxial microsclere with a terminal umbel at one end, resembling a short-shafted clavule or an amphidisc with one umbel missing; has also been called an umbel (Hex).

paragaster (n). See spongocoel.

paragastral (adj). Of the paragaster.

paratangential (adj). Nearly tangential; implies that the orientation of rays following the plane of a sponge surface is not strictly tangential (Hex).

paratangential (n). Tangential ray (Hex).

paratropal pentactin, -e (n). Pentactine in which the arrangement of the paired rays is distorted so that one interactinal angle is greater than the other three, which are subequal, and may exceed 180° (Hex).

paratrope (n). Paratropal pentactine (Hex).

parenchyma (n). (1) Mesenchyme; (2) mesenchyme in which cells are numerous (cf. collenchyma, with few); (3) in Hexactinellida, the trabecular network between the dermal and gastral membranes.

parenchymal (adj). Of the parenchyma.

parenchymal skeleton (n). The megascleres of the parenchyma, or any rigid structure formed by their union; hypodermalia or hypogastralia, although strictly parenchymal, are not included (Hex).

parenchymalia (n pl). Spicules of the interior, excluding the dermalia and gastralia (Hex).

parenchymella (n). Sponge embryo or free larva in which an appreciable part of the surface (up to approximately half) is formed by nonflagellate cells, the rest being flagellate.

parietal (adj). Of the lateral wall surrounding an axial paragaster.

parietal gap (n). Perforation in the lateral wall of a paragastral cavity.

parietal osculum (n). Parietal gap identified homologically as osculum.

pavement cell (n). Pinacocyte (Cal, Dem).

pentact (n). Pentactine.

pentactin, -e (n). Spicule with five rays; five-rayed holactine (Hex).

pentactinal (adj). Five rayed.

pentaene (n). Triaene-like spicule with five cladi (Het).

pentiradiate (n). Spicule with five radial rays within a single plane (Het).

periloph (n). A raised rim around an ostium or pore, a rim shorter than a tubelike exaulos (Dem).

peripheral (adj). At the periphery of a body consisting of radiating tubes, or of a radially folded disc or funnel; extending across a succession of marginal surfaces in such a body (Hex).

peripheral membrane (n). Membrane developed across and/or between a succession of marginal surfaces (Hex).

peripheral skeleton (n). Loose megascleres or rigid skeletal meshwork formed in a peripheral structure (Hex).

phalangaster (n). Very short but stout spinulate microstrongyle (Dem).

phyllopentactin, -e (n). Pentactine with bladelike lamellar expansions along the sides of the paired rays, in the plane in which they lie, so that each ray has a leaflike appearance (Hex).

phyllotriaene (n). A triaene spicule in which the three more or less equal rays are expanded into flattened, leaflike, sometimes digitate structures (Dem).

pillar (n). Rodlike skeletal structures that extend from interwall to interwall in chambers of sphinctozoan sponges (Dem).

pinacocyte (n). Cell marking limit of sponge, usually occurring in layer one cell thick.

pinacocyte layer (n). Layer of pinacocytes coating surfaces where choanocytes are lacking (Cal, Dem).

pinacoderm (n). Outer layer of a sponge.

pinakid (n). Siliceous disc with many radiating axial canals (Dem).

pinular (adj). With centripetal lateral spinules producing a resemblance to a fir tree; with a pinular ray or rays (Hex).

pinulus (n). Pinular hexactine or pentactine, with a pinular ray that is unpaired in pentactines; sometimes with less developed spinules on other rays (Hex).

plagiotriaene (n). Triaene in which the angle between the rhabdome and each cladus is about 135°.

plesiaster (n). Microsclere (streptosclere) in which raylike outgrowths are emitted from a short straight axis, sometimes grading into simple euasters with no central axis (Dem).

pleuralia (n pl). Prostalia of the sides of the body (Hex).

plumicome (n). Hexaster with S-shaped terminals, the free ends of which face outward and are arranged in several tiers (Hex).

plumose (adj). Plumelike; refers to the axinellid type of echinated skeletal fiber, without coring monaxons (Dem).

polyactin, -e (n). Spicule with many rays (in practice, any form with more than six).
polyactinal (adj). Many rayed.
polyaenes (n pl). Triaene-like spicules with 2 to 9 (typically 5 or more) rays in the cladome (Het).
polyaxon (n). Spicule with rays following more than four growth axes, arranged at subequal angles or in other ways.
polyaxial (adj). With rays following more than four growth axes.
polycladose (adj). Repeatedly branched.
porate (adj). Possessing pores.
pore (n). Any small aperture through which water passes.
pore field (n). A cluster of pores, particularly if flat and surrounded by a low rim, in an exowall.
porocyte (n). Cell perforated by an intracellular pore, through which water passes, that functions as an inhalant canal.
postica (n pl). Exhalant skeletal pores.
principalia (n pl). The main parenchymal megascleres of lyssacine Hexactinellida (Hex).
prosiphonate (adj). Condition in which the individual segments of an axial tube grow upward from the floor of each skeletal chamber, i.e. from the roof of the chamber below (Dem).
prosochete (n). Inhalant canal.
prosodus (n). Fine tubular duct leading from a main trunk of an inhalant canal (prosochete) to a flagellated chamber.
prosopore (n). Inhalant pore.
prosopyle (n). Pore through which water enters a flagellated chamber.
prostalia (n pl). Megascleres protruded from a dermal or marginal surface (Hex).
prothalmus (n). A cluster of initial few chambers of an individual sphinctozoan sponge that lacks a cloaca and is, thus, different from subsequent chambers (Dem, Cal).
protocyst (n). A chamber in a protothalus (Cal).
protriaene (n). Triaene in which the angle between the rhabdome and each cladus is more than 135°.
proximal ray (n). One of the rays of a spicule, when directed inward at right angles to an external surface.
psellium (n). Clavidisc (Dem).
pseudaster (n). Any pseudopolyactinal microsclere developed from a monaxial type (Dem).
pseudastrose (adj). Comprising pseudasters (Dem).
pseudo-. False.
pseudoactin, -e (n). Raylike structure containing no axial filament.
pseudoactinal (adj). Resembling a ray but lacking an axial filament.
pseudoaxial canal (n). Cavity resembling a true axial canal, but produced by internal solution of the axial part of an originally solid (anaxial) structure.
pseudoderm (n). Peripheral ectosome-like covering, enclosing a tubular labyrinth in some asconoid Calcarea (Cal).
pseudoeuaster (n). Oxyaster-like, sphaeraster-like, or sterraster-like microsclere, developed from a monaxial prototype (usually a spinispira) (Dem).
pseudogaster (n). Local paragaster-like cavity in a sponge with no single axial paragaster, homologically an external depression (epochete) or the trunk of an exhalant canal (apochete).
pseudohexactinose (adj). With nodal octahedra solidified so that lychniscs appear to be absent (Hex).
pseudopolyactin, -e (n). Apparently polyactinal spicule in which true cored rays are lacking (Hex).
pseudoradiate (n). Secondary diactine, triactine, tetractine, or pentactine, derived from a monactine by enlargement of spinules to the size of rays (Dem).
pseudosiphonate (adj). Condition of thalamid sphinctozoans that lack a true axial tube, but have a tubular axial passage that may or may not be outlined by an endocameral structure (Dem).
pseudospicule (n). Spicule-like body composed of spongin (Dem).
pseudotetraclone (n). Monaxial or anaxial desma with the outward form of tetraclone (Dem).
pycnaster (n). Euaster with short conical rays, which may also be a sphaeraster (Dem).
quadriradiate (n). Tetractine.
quadrule (n). Square mesh with sides formed by tangential rays of four hexactines, pentactines, or stauractines, whose centers are located at the corners, or produced by subdivision of such a square by smaller spicules within it (Hex).
radial canal (n). (1) Radially directed inhalant or exhalant canal; (2) radially directed skeletal canal; (3) a flagellated chamber of a sycon.
radiante (n). That point in the skeletal end of an anthaspidellid sponge from which the trabs radiate (Dem).
radiate (adj). (1) With radiating rays; (2) with megascleres (long-shafted triaenes or monaxons) arranged radially (Dem).
radical (adj). Rooting.
ramose (adj). Branched.
raphide (n). Hairlike monaxon.
raphidial (adj). Hairlike.
ray (n). Any portion of a spicule formed by concentric secretion of mineral matter around a centrifugal organic axial filament.
recurved (adj). Bent toward a monactinal or diactinal shaft, in the manner of spokes of a grapnel.
reticular (endocameral structure) (adj). Consisting of a three-dimensional network of skeletal beams (Cal).
reticulate (adj). Netted; in monaxonid Demospongea, refers to skeletons with spicules cemented together with spongin or imbedded in a network of spongin fibers.
retrosiphonate (adj). Condition in which an axial tube is formed by backward extensions of the wall of each skeletal chamber in sphinctozoans (Dem).
rhabd or rhabdus (n). General name for any form of monaxon; also misused in place of rhabdome.
rhabdal (adj). Of the rhabdome.
rhabdoclone (n). Megaclone or heloclone (Dem).
rhabdocrepid (adj). Monocrepid (Dem).
rhabdodiactin, -e (n). Holactine in which two opposite rays follow a single growth axis (Hex).

rhabdome (**n**). Axial (dissimilar) ray or pair of opposite rays, in spicules in which other rays form a cladome.

rhagon (**n**). Sponge with syconoid flagellated chambers but no canal system; larval demosponge having this condition; sometimes also used incorrectly as equivalent to leucon.

rhagonoid (**adj**). With syconoid flagellated chambers but no canal system.

rhax (**n**). Kidney-shaped sterraster (Dem).

rhizoclad (**n**). Rhizoclone (Dem).

rhizoclonar desmas (**n pl**). Monaxial desmas, with or without distinct clones, in which the zygomes are simple spinules or rootlike outgrowths (Dem).

rhizoclone (**n**). Generalized rhizoclonar desma, with or without distinct clones, without tubercles, and without spherical swellings at the ends of the epirhabd; see also rhyzoclad (Dem).

rhizoclonid (**n**). Anaxial supplemental desma, of rhizoclone-like form (Dem).

rhizophyte (**n**). Anaxial supplemental desma, of rhizoclone-like form (Dem).

rhizophytous (**adj**). Anchored by root processes.

rhopalostyle (**n**). Clublike or spindle-like style (Dem).

rhyses (**n.**). Canal or opening in the sponge.

root process (**n**). Rootlike basal outgrowth, by which a sponge is anchored.

root tuft (**n**). Beardlike, ropelike, or stalk-like group of protruded spicules, by which sponge is anchored.

rosettes (**n pl**). Hexasters and their variants, e.g., hemiasters, octasters (Hex).

sagittal triradiate (**n**). T- to Y-shaped triactine, with two equal interactinal angles (Cal).

sanidaster (**n**). Microsclere with raylike spines distributed along the length of a straight monaxial shaft (Dem).

sarcenchyma (**n**). Mesenchyme with many granular cells.

sarcode (n). Soft parts.

sarule (**n**). Brushlike sceptrule, in which a short, raylike rudiment opposite the single true ray bears numerous centrifugal spines, which are not arranged in regular whorls (Hex).

sceptrella (n). Discaster (Dem).

sceptrule (**n**). Monactinal but triaxial microsclere with a six-rayed axial cross at one end, which usually bears some form of anaxial outgrowth(s) externally (Hex).

schizorhyses (**n pl**). Intercommunicating cleftlike to labyrinthine skeletal canals, open at both skeletal surfaces (Hex).

schizorhysis (**n**). Condition in which schizorhyses are present (Hex).

sclere (**n**). Spicule.

scleroblast (**n**). Spicule-secreting cell.

scleroblast syncytium (**n**). Multinucleate syncytial mass within which a spicule is secreted (Hex).

sclerocyte (**n**). Scleroblast.

sclerosome (**n**). Calcareous cement uniting spicules or forming skeletal fibers in which they are imbedded (Dem, Cal).

sclerosomal trabeculae (**n**). Anastomosing skeletal elements that form the unchambered upper, exhalant, canalled layer in stratiform guadalupiid sphinctozoans (Dem).

scoliorhabd (**n**). Sinuous and irregularly annulated monaxon (Dem).

scopule (**n**). Brushlike or forklike sceptrule, with a pair or regular ring of centrifugal spines opposite the single ray (Hex).

sexiradiate (**n**). Spicule in which the six radiating rays occur in a single plane (Het).

short-shafted triaene (**n**). Triaene with the rhabdome shorter to not markedly longer than the rays of the cladome (Dem).

sieve plate (**n**). Perforated diaphragm extending across an osculum or across a paragaster below the level of its osculum.

sigma (**n; pl. -ata**). Typically C-shaped but sometimes S-shaped microsclere with sharply pointed ends, and without spinules (Dem).

sigmaspire (**n**). Spiral monaxial microsclere of about one revolution, C- to S-shaped in different views, typically blunt-ended, smooth or finely to markedly spinulate (Dem).

sigmatoscleres (**n pl**). Microscleres comprising sigmata *s.s.* and related types, including diancistra, clavidiscs, canonchelae, chelae, chelasters, chela-derivative amphidiscs, bipocilli (Dem).

sigmatose microscleres (**n pl**). General name given to diactinal demosponge microscleres, particularly sigmata, sigmaspires, and their variants.

siliceofibrous skeleton (**n**). Imaginary structure supposed to consist of a network of siliceous fibers, attributed to lithistid Demospongea and dictyonine Hexactinellida by Bowerbank (1869) and Claus (1872).

silicoblast (**n**). Silica-secreting scleroblast (Dem).

siphon (**n**). The axial tube of thalamid sphinctozoans (Dem).

skeletal canal (**n**). Canal-like internal cavity in a skeletal framework.

skeletal canalization (**n**). Interruption of skeletal meshwork, at surfaces or internally, by spaces larger than ordinary skeletal meshes, and marking the positions of canals or flagellated chambers.

skeletal cortex (**n**). Specially dense external part of a skeletal framework.

skeletal framework (**n**). Coherent internal framework, consisting of spicules united by a calcareous or spongin cement, by articulation (zygosis), or by fusion (ankylosis), or of spongin fibers with or without imbedded spicules.

skeletal node (**n**). Point at which beams of a skeletal framework come together.

skeletal pore (**n**). Canalar aperture in a skeletal surface, of inhalant, exhalant, or unknown function.

somal (**adj**). Found equally in the choanosome and ectosome (or endosome); distinguishes spicules with this distribution from other restricted to these regions (see ectosomal, choanosomal).

spathidorhabd (**n**). Spinulate oxea or strongyle, with spinules in regular whorls (Dem).

sphaeraster (**n**). Euaster in which the rays radiate from a spherical central part, the diameter of which is one third or more of their length (Dem).

sphaeroclonar (adj). Comprising sphaeroclones and astroclones (Dem).

sphaeroclone (n). Anaxial desma with clones radiating to one side of a center with the other side spinose, inflated to form globular centrum, or both; zygomes rootlike to cuplike; see also anomoclad (Dem).

sphaerohexaster (n). Tylohexaster (Hex).

sphaerule (n). See spherule.

sphere (n). Anaxial siliceous spherule produced by concentric secretion around a center (Dem).

spherical (adj) (of hexasters). With numerous terminals radiating to a spherical periphery, and arising from short actinal rays or from a central sphere containing a six-rayed axial cross (Hex).

spherule (n). Spherical body, commonly microscopic and of acicular crystals radiating from a common center, that may form part of skeletal fibers in calcareous sponges (also sphaerule, spherulite, sphaerulite) (Cal, Dem).

spherulitic (adj). Skeletal texture composed of spherulites (also sphaerulitic) (Cal, Dem).

sphinctozoan (n). Demosponge with a calcareous skeleton composed of straight, curved, branched, or glomerate series of hollow spheroidal chambers or bodies.

spicular (adj). Of spicules.

spicular center. Point within a spicule from which growth has commenced.

spiculation (n). The assemblage of spicules present in a given species, genus, etc.

spicule (n). Discrete mineral skeletal element; see also sclere.

spiculin (n). Organic substance forming axial filaments and sometimes also mixed with spicular silica.

spike (n). Unattached ray projecting freely from a surface of a dictyonal framework (Hex).

spinispira (n). Spiral monaxial microsclere of typically more than one revolution, which is finely microspinulate or developed as a spiraster, and may grade into a spinulate microstrongyle, a sigmaspire, a pseudoeuaster, or a discaster (Dem).

spinulate (adj). With spinules.

spinulate (n). Tylostyle (not spinulate in the sense of bearing spinules) (Dem).

spinule (n). Any small spine, not large enough to be called raylike.

spiraster (n). Microsclere in which raylike spines are emitted from a spiral axis making one turn or more; may be either a streptosclere or a spinispira (Dem).

spire (n). Any microsclere that is a spirally twisted monaxon (Dem).

spiroscleres (n pl). Microscleres comprising sigmaspires, toxaspires, spinispirae, and related pseudasters (Dem).

spongin (n). Horny skeletal substance, cementing spicules together or forming continuous skeletal fibers with or without imbedded spicules; sometimes also present in other forms (Dem).

spongioblast (n). Spongin-secreting cell.

spongocoel (n). Large, central exhalant opening, commonly obconical to rounded subcylindrical, without digestive function, also termed a cloaca, gastral cavity, or paragaster in zoological and paleontological literature.

spongocyte (n). Spongioblast.

spongophare (n). The chamber-containing upper part of a rhagon, which gives rise to the adult choanosome (Dem).

statoblast (n). Reproductive bud.

stauractin, -e (n). Cross-shaped tetractine (Hex).

staurodisc (n). Tetractinal amphidisc variant, with the rays arranged as in stauractines (Hex).

sterraster (n). Globular or kidney-shaped microsclere with a granular surface, formed by solidification of a many-rayed initial oxyaster (Dem).

sterrosphaeraster (n). Euaster intermediate between typical sterrasters and sphaerasters (Dem).

sterrospira (n). Sterraster-like microsclere developed from a spinispira (Dem).

stratiform (adj). Sheetlike growth form consisting of a single layer of chambers (Cal).

streptasters (n pl). (1) Originally (*sensu* Sollas) all types of pseudopolyactinal microscleres in which raylike spines are emitted from a linear axis rather than a center as in euasters, including plesiasters, metasters, spirasters, amphiasters, sanidasters, and discasters, irrespective of homology; (2) *sensu* Burton, streptoscleres only; (3) *sensu* de Laubenfels, spinulate microxeas and sanidasters (Dem).

streptoscleres (n pl). Microscleres comprising intergrading plesiasters, metasters, spirasters, and amphiasters, of which the first may pass into simple euasters from which this series appears to derive; dichotriacts of Dendy; streptasters *sensu* Burton (not de Laubenfels) (Dem).

strongylate (-ote) (adj). Round ended.

strongyle (n). Round-ended monaxon.

style (n). Monaxon with one end rounded, the other sharply pointed.

subcortical crypt (n). Subdermal space underlying a cortex (Dem).

subdermal (adj). Beneath the dermal surface, but outside the choanosome; see also hypodermal.

subdermal space (n). Internal space underlying an ectosome (Cal, Dem) or a dermal membrane (Hex) but outside the choanosome.

subgastral (adj). Beneath the gastral surface, but outside the choanosome; see also hypogastral (Hex).

subhexactin, -e (n). Hexactine with one ray markedly shorter than the rest (Hex).

sublithistid (adj). (1) With an incipient or imperfect development of the lithistid condition, or with a mixture of lithistid and nonlithistid characters; (2) of sublithistids (Dem).

sublithistid (n). Sublithistid demosponge (Dem).

subtriaene (n). Triaene in which differentiation of a cladome is limited to curvature or displacement of three rays toward or away from the fourth from the regular tetraxial positions; to shortening of the fourth ray; or to two of these developments together (Dem).

summit diaphragm (n). Sieve-platelike structure that forms the upper surface of the body in *Coeloptylum* Goldfuss and similar genera (Hex).

superficial furrowing (n). Furrowing of the surface of a skeletal framework, related to the positions of subdermal or subgastral channels in the soft parts, or to some other feature of a canal or chamber system; not related to folding of the sponge wall, as has sometimes been alleged when furrows run longitudinally.

superficial meshwork (n). Rigid skeletal meshwork, of various origins, developed outside the true surface of a dictyonal framework; distinguished by extending across ostia or postica of the skeletal surface proper (Hex).

supplemental (adj). Occurring in addition to desmas in lithistid demosponges.

sycettoid (adj). Syconoid with inhalant canals formed by interspaces between free radial diverticula of the wall of an axial paragaster.

sycon (n). Sponge with choanocytes restricted to radial flagellated chambers that discharge to the paragaster directly, with partially enclosed or distinct inhalant canals (prosochetes) but no exhalant canals (apochetes).

syconoid (adj). With the characters of a sycon; as in sycons.

symphyllotriaene (n). Discotriaene (Dem).

synapticula (n pl). Anaxial siliceous bridges uniting adjacent spicules (Hex).

synapticular filaments (n pl). Anaxial siliceous filaments growing out from spicular rays or skeletal beams (Hex).

syncytium (n). A mass of protoplasm that contains scattered nuclei but lacks distinct cells.

syzygial (adj). Taking part in or formed by zygosis (Dem).

syzygial node (n). Skeletal node formed by intergrowth of zygomes from two or more desmas (Dem).

tabulum (n). Plate or floor that divides skeletal cavity.

tangential (adj). Lying in or parallel with the plane of an external surface. This orientation is also called paratangential, as not truly tangential.

tangential rays (n pl). Rays that are oriented tangentially; used mainly of hexactinellidan dermalia and gastralia, or which four rays of two intersecting axes are typically tangential; see also paratangential ray.

tauactin, -e (n). T-shaped triactine (and holactine) (Hex).

terminal (n). Anaxial, branchlike centripetal outgrowth arising from the end of a true cored ray of a hexaster (Hex).

tetraclad (n). Tetraclone (Dem).

tetraclone (n). Tetraxial (tetracrepid) desma with four clones developed similarly or differently, without a differentiated clonome; crepis a small calthrops or short-shafted triaene; see also tetraclad (Dem).

tetraclonar desmas (n pl). Typically tetraxial desmas with zygomes in the form of anaxial rootlike outgrowths, including tetraclones, triders, and minor variants of these types (Dem).

tetracrepid (adj). With a tetraxial crepis (Dem).

tetract (n). Tetractine.

tetractin, -e (n). Spicule with four rays; four-rayed holactine.

tetractinal (adj). Four rayed.

tetraene (n). Triaene-like spicule with four cladi (Dem, Het).

tetralophose (adj). With four lophose rays.

tetraxial (adj). With rays following four growth axes, arranged tetrahedrally or in some distorted form of this arrangement.

tetraxon (n). Spicule in which rays follow four growth axes, arranged tetrahedrally or in some distorted form of this arrangement.

thalamidarium (n). Single layer of chambers in a stratiform sphinctozoan sponge in which chambers are convex toward the growing edge, commonly overlain by an unchambered, canalled trabecularium (Dem, Cal).

theca (n). Skeletal cortex in the form of a dense, sheathlike secondary covering (Dem).

tibiella (n). Tylote (Dem).

torno-. With rays abruptly pointed.

tornote (n). Monaxon with ends abruptly pointed (Dem).

toxaspire (n). Monaxial microsclere similar to a sigmaspire but making rather more than one revolution, and hence appearing bow shaped in some views (Dem).

toxon (n.; pl. -a). Bow-shaped, monaxial microsclere (Dem).

trab (n). Rodlike skeletal element in ladderlike anthaspidellid skeletons, formed by fusion of ray tips of runglike dendroclones, may be cored with monaxons (Dem).

trabeculae (n pl). Dividing and anastomosing syncytial filaments forming a network with water-filled interspaces (Hex); or any other rodlike or beamlike skeletal element other than a ray or branch of a single spicule, especially a structure of sclerosome.

trabecular network (n). The network of syncytial trabeculae in which the flagellated chambers of Hexactinellida are suspended, also forming the bounding (dermal and gastral) membranes.

trabecularium (n). Upper unchambered layer composed of anastomosing trabeculae, interrupted by groovelike canals, in the stratiform guadalupiid sphinctozoans (Dem).

trabeculae (endocameral structure) (adj). Consisting of regular pillars supporting the roofs of low chambers (Dem).

trabecular skeleton (n). Skeleton made of nonspicular, branching filaments.

trachelotriaene (n). Long-shafted dichotriaene with a very small cladome and a clublike rhabdome, with its maximum diameter just below the cladome and then tapered away from it (Dem).

triact (n). Triactine.

triactin, -e (n). Spicule with three rays; three-rayed holactine; see also triact, triradiate (Hex).

triactinal (adj). Three rayed.

triaxial (n). With rays or axial rudiments following three growth axes, which intersect at right angles.

triaxon (n). Spicule in which rays, axial rudiments, or both follow three growth axes, which intersect at right angles.

triaene (n). Regular or modified tetraxon with three similar rays (cladi) differing from the fourth.

trichotriaene (n). Triaene with cladi branched trichotomously.

tricranoclone (n). Triderlike desma of Tricranocladina, with three tripodally grouped clones bearing terminal zygomes in the form of simple or marginally denticulate expansions, and with or without a knobbed brachyome-like outgrowth on its opposite side; nature of the crepis unknown (Dem).

tricrepid (adj). With a triactinal crepis (Dem).

trider (n). Tetraclonar desma with three similar clones, comprising a clonome, and a fourth developed differently or lacking; crepis tetractinal or triactinal (Dem).

trifid spicules (n pl). Triaene (Dem).

trilophose (adj). With three lophose rays.

triod (n). Triactine with the rays arranged in one plane at 120° intervals.

triodal (adj). With three rays arranged in one plane at 120° intervals.

tripod (n). Triactine with rays arranged pyramidally.

tripodal (adj). With three rays or clones arranged pyramidally.

tripinulus (n). Spicule with pinular spines on a distal ray and the outward turned ends of two tangential rays (Hex).

triradiate (n). Triactine (Cal).

tubular (endocameral structure) (adj). Consisting of interconnected calcareous tubes, which open through pores of the external and internal walls (Dem).

tylo-. Terminally knobbed.

tylohexaster (n). Hexaster with knobbed terminals; see also sphaerohexaster (Hex).

tylostrongyle (n). Monaxon with one end knobbed, the other bluntly rounded (Dem).

tylostyle (n). Monaxon knobbed at one end, sharply pointed at the other.

tylote (n). Monaxon knobbed at both ends; see also amphityle.

tylote (adj). Terminally knobbed.

umbel (n). Whorl of recurved toothlike spicules at the end of a spicular ray or an anaxial pseudoactine.

umbel (n). Spicule with a single umbel at one end of a short shaft, now called a paraclavule (Hex).

uncinate (n). Diactinal monaxon with thornlike, lateral spinules, all directed toward one end (Hex).

unpaired ray (n). Ray of a triaxon when the opposite ray is suppressed.

vermiculate (adj). Irregularly sinuous.

verticillate (adj). With spinules in regular whorls.

vesicular (endocameral structure) (adj). Consisting of imperforate diaphragms (Dem).

wandlucken (n). Parietal gaps or oscula (Hex).

xenoskeleton (n). Foreign objects taken into the body, sometimes cemented by or imbedded in spongin (Dem).

xylotyle (n). Sceptre-like, prostal monactine (Hex).

zygome (n). Articulatory part of the desma (Dem).

zygosis (n). Coherent articulation of spicules, without fusion.

REPRODUCTION AND DEVELOPMENT

Robert M. Finks

[Department of Geology, Queens College (CUNY)]

REPRODUCTION

The normal method of reproduction by sponges is through shedding of sperm into the exhalant water, fertilization of either shed or retained ova, and dispersal through a planktonic larval stage. Asexual reproduction through budding, fragmentation, or production of **gemmules** also occurs (Bergquist, 1978).

Sponges are generally hermaphroditic but with ovarian tissue separated from the testicular tissue either spatially or temporally. In some species sexually ripe individuals seem to be either male or female, but it is not certain whether the sexes are separate or whether there is merely temporal separation of male and female phases (Brien, 1973a; Gilbert & Simpson, 1976; Kaye, 1990; Tanaka-Ichihara & Watanabe, 1990). Ova and spermatozoa are produced from cells having the form of **amoebocytes.** These in turn are produced by the transformation of **choanocytes,** at least in the Calcarea and Demospongea (Brien, 1973a; Tuzet, 1973a). Among the Hexactinellida the eggs and sperm arise from similar cells (**archaeocytes**), but these have not been seen to arise in turn from choanocytes (Tuzet, 1973b).

Among 43 temperate-water species of Demospongea, the breeding season is relatively short, generally in the summer, and averaging two months for viviparous species and one month for oviparous species (Lévi in Brien, 1973a).

Fewer data are available for tropical demosponges, but Reiswig (1973) has observed that in *Mycale* sp. from Jamaica, an r-selected opportunistic species, reproduction is extended through the six warmest months of the year, while in *Verongia gigantea* and *Tethya crypta,* both K-selected specialist species, reproduction is restricted to one and two months, respectively, during the colder season. Continuing to the temperate-water demosponges, the length of the breeding seasons are not correlated with ovoviparity versus viviparity; the species with the longest (*Mycale*) and shortest (*Verongia*) breeding seasons are both viviparous, while a third (*Tethya*) is oviparous. Perhaps among the temperate-water species the viviparous species are largely r-selected types, and this, rather than the habit of larval incubation, determines their longer breeding season (cold-water species among invertebrates in general tend toward both r-selection and larval protection). Among the Calcarea the breeding season seems more extended, and continues throughout the warmer months (Vacelet, 1965). Breeding times in the Hexactinellida are not well known, but in *Farrea sollasii* at least sexual reproduction has been observed throughout the year (Okada, 1928; Tuzet, 1973b).

Spermatozoa are discharged through the exhalant orifice of the sponge and are drawn into the inhalant stream of adjacent sponges. The spermatozoon in sponges that incubate their larvae penetrates a choanocyte or amoebocyte that becomes a carrier cell, transporting the spermatozoon to the ovum that is located in the mesoglea, whither it has previously migrated and enlarged itself through the consumption of food-bearing trophocytes. Oviparous or nonincubating sponges have direct penetration of the ovum by the sperm cell (Tuzet, 1973a, p. 15).

DEVELOPMENT

LARVAE

Free-swimming sponge larvae are at the blastula stage of embryonic development. The equivalent of gastrulation takes place at the time of fixation and metamorphosis of the larva into a sponge. There are two principal types of larvae in the Demospongea and Calcarea: (1) a **parenchymella** with a

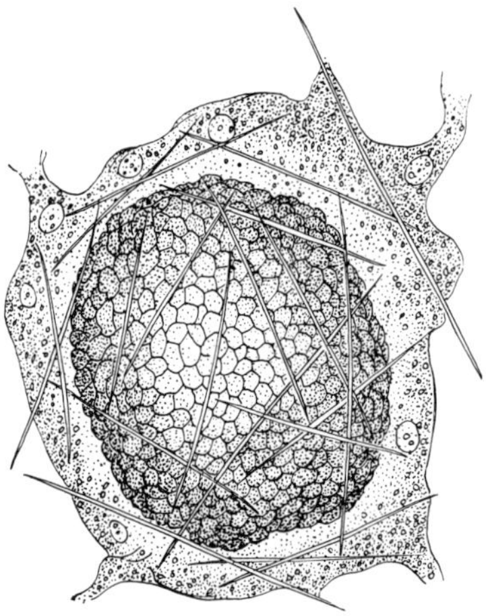

FIG. 64. Larval stage of calcareous *Sycon raphanus* with outer layer of ectoscleroblasts and irregularly oriented, diactine spicules (Tuzet, 1973a).

solid interior and a complete or nearly complete covering of flagellated cells (they may be absent at the posterior end); and (2) an **amphiblastula** with a hollow interior and a clear differentiation between the cells of the anterior half and the cells of the posterior half.

These two types, however, may not be homologous between the Demospongea and the Calcarea. In the Demospongea, both types arise from a solid nonflagellated stereoblastula, while in the Calcarea they both arise from a hollow nonflagellated blastula. The amphiblastula among the demosponges occurs only in the Homosclerophora and develops its hollow by destruction of the interior cells of the stereoblastula (BRIEN, 1973a). It differs also from the amphiblastula of the Calcarea in that cells of the posterior half are ciliated like those of the anterior half (*Octavella* is a partial exception; TUZET & PARIS, 1964). Among the Calcarea these posterior cells are nonflagellated, and there are four specialized cells in the equatorial region arranged in the form of a cross (TUZET, 1973a). It also arises from a hollow stomoblastula with a pore at one end (TUZET, 1973a). The parenchymella of the Calcarea arises from a hollow blastula that is

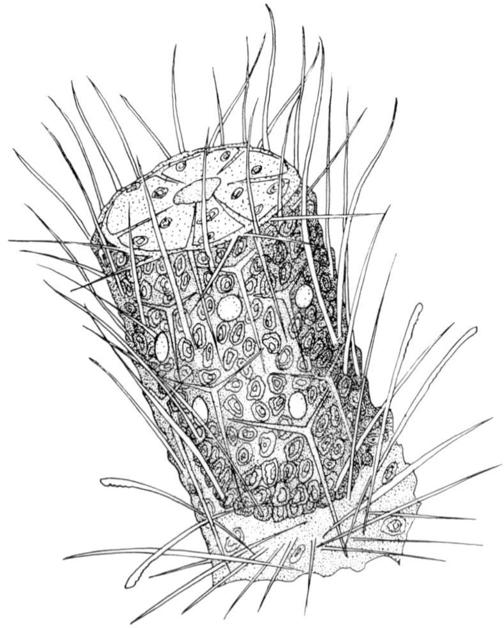

FIG. 65. Young asconoid olynthus stage of calcareous *Sycon raphanus* with regularly arranged triradiates (Tuzet, 1973a).

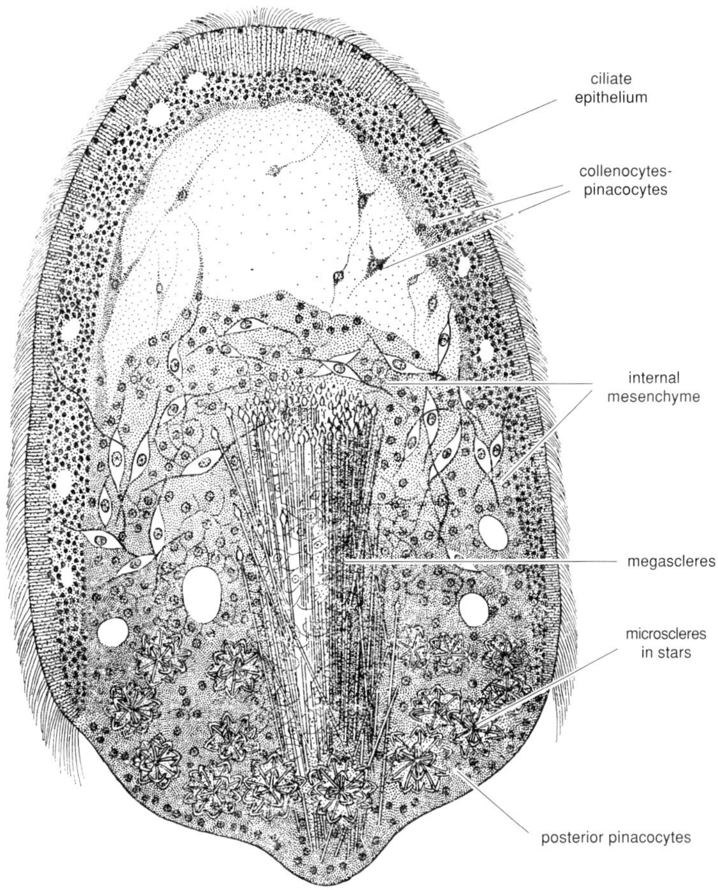

FIG. 66. Section of free-swimming paremchymella of demosponge *Esperia lorenzi* showing posterior concentration of megascleres along medial axis of sponge, and widely developed microscleres in stars; ciliate epithelium; collenocytes-pinacocytes; massive internal mesenchyme condensed in posterior polar area; and posterior pinacocytes (Brien, 1973a).

not a stomoblastula; its solid interior develops by immigration of surface cells.

There thus seems to be a fundamental distinction between the two classes that may be blurred to some extent by the use of common terms for possibly homomorphic larval types. Within the Calcarea, the two types of larvae seem to have considerable taxonomic significance in that they are correlated with the position of the nucleus in the choanocyte and with other characters (BIDDER, 1898; HARTMAN, 1958b).

The larvae of the Hexactinellida are less well known, and information comes chiefly from the study of *Farrea sollasii* by OKADA (1928). Initially the blastula is hollow with an opening to the exterior, resembling thus the stomoblastula of the Calcarea. It soon becomes solid on the interior but at or after fixation develops a choanocyte-lined central cavity with several diverticula and an osculum. At no time is the larva flagellated on the exterior; it is in this respect unlike the larvae of Demospongea and Calcarea (TUZET, 1973b).

LARVAL SPICULATION

In most sponges of all three classes the larva develops a complement of spicules before fixation. It is thus possible that a larva

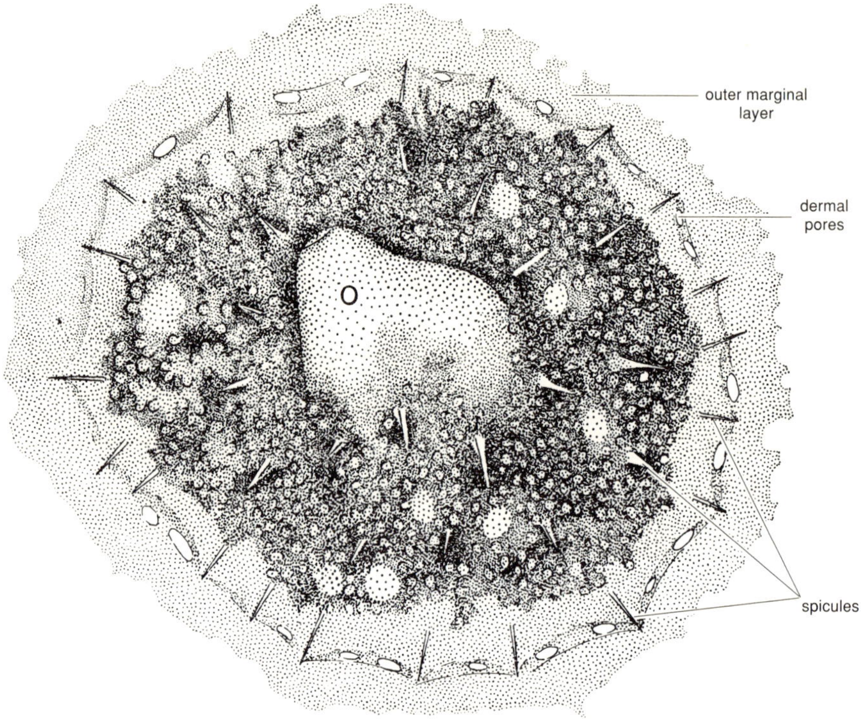

FIG. 67. Demosponge *Esperella* immediately after metamorphosis to small, leuconoid sponge with distinct osculum (*O*); outer marginal layer; dermal pores; numerous clusters of flagellae; and spicules (Brien, 1973a).

could be preserved in the fossil record, albeit under extremely favorable circumstances, inasmuch as it is less than a millimeter in maximum dimension. The spicules are not always the same as those of the adult. If ontogeny recapitulates phylogeny, the larval spiculation may give us some clues as to the evolutionary history of spicules.

Among the Calcarea, *Sycon raphanus,* which in the adult phase possesses diactines, sagittal **triradiates,** and tetraradiates, possesses only irregularly arranged diactines in

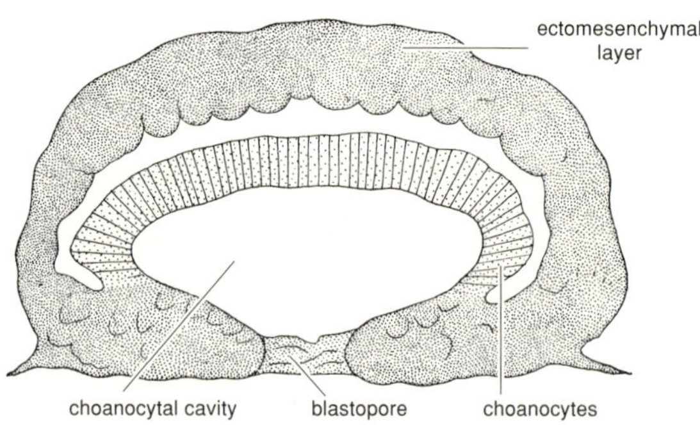

FIG. 68. Asconoid stage of metamorphosis of demosponge *Plakina monolopha* showing large, choanocytal cavity, choanocytes, blastopore, and ectomesenchymal layer (Brien, 1973a).

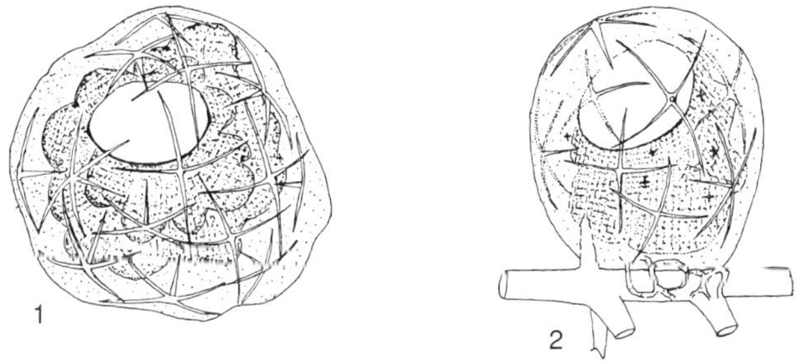

FIG. 69. Examples of young hexactinellid sponge *Farrea sollasii* SCHULZE showing *1,* stauractines of skeleton and choanosome around oscular opening; *2,* a small example (0.68 X 0.57 mm), with skeleton of stauractines and small, discohexaster microscleres, attached to skeletal fragment of larger sponge (Tuzet, 1973b).

the larval stage (Fig. 64), developing them by the time of fixation but before metamorphosis. Following metamorphosis the young sponge (**olynthus**) has the adult complement of spicules with the regular parallel arrangement of triradiates (Fig. 65) although its aquiferous system is asconoid rather than syconoid (TUZET, 1973a). This asconoid stage is characteristic of all young Calcarea.

Among the Demospongea in which the larva is incubated before being released, the free-swimming parenchymella has already

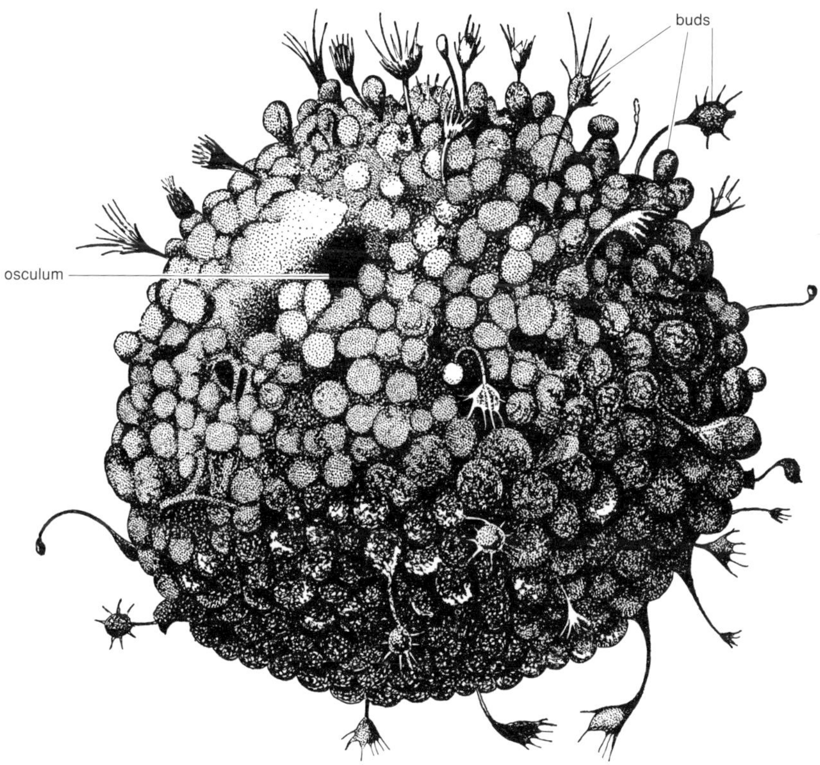

FIG. 70. Demosponge *Tethya maza* with numerous buds in various stages of development, with stalks on those nearly ready to be set free; osculum developed in upper part of parent sponge (Brien, 1973a).

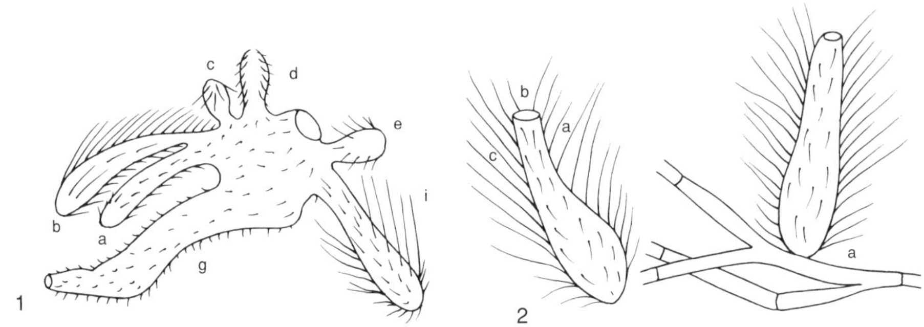

FIG. 71. The calcareous *Leucosolenia botryoides* showing *1,* several buds (*a–e*) in diverse stages of development, grading up to one (*i*), with long spicules, ready to detach from parent (*g*); *2,* free bud, on left, with a terminal osculum (*b*) and characteristic long spicules (*a, c*), and attached example, on right (Tuzet, 1973a).

developed both macroscleres and microscleres (Fig. 66). The spicules are characteristically concentrated in the posterior half of the larva, with the monaxonic megascleres arranged parallel to the anteroposterior axis of the larva (BRIEN, 1973a).

Among oviparous types the larva is less well developed and may even be benthonic rather than planktonic (e.g., *Polymastia busta;* BOROJEVIC, 1968), and the spicules may not appear until after fixation and metamorphosis (BRIEN, 1973a).

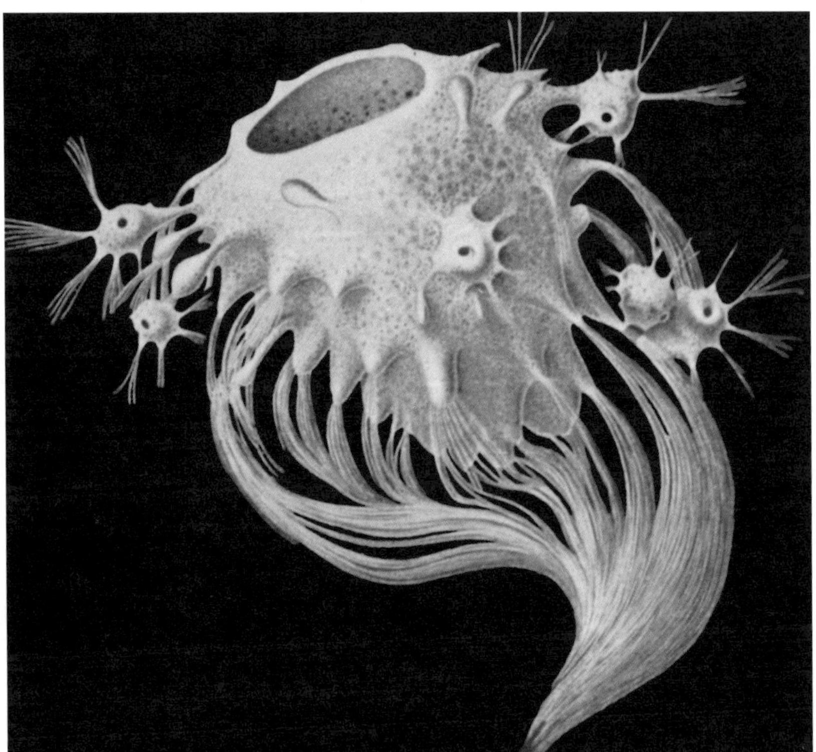

FIG. 72. *Lophocalyx (Polypophus) phillippinensis* GRAY, with several buds (Tuzet, 1973b).

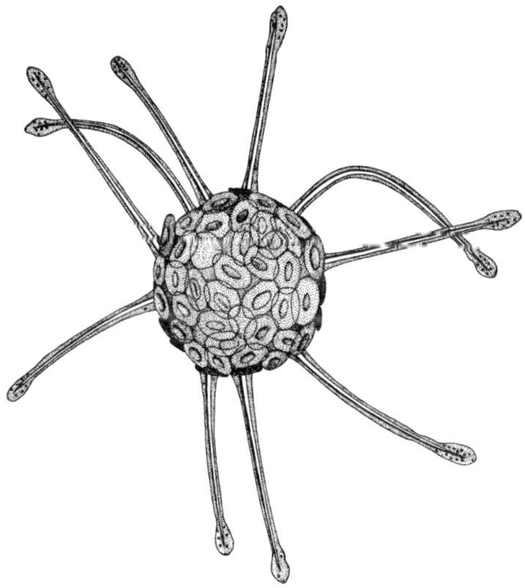

FIG. 73. Young, microscopic, pelagic propagule, approximately ×275 in diameter, coated with coccolith-like spicules and with ten projecting styles, each coated in a thin film of cytoplasm (Brien, 1973a).

Almost all demosponges develop a leuconoid aquiferous system directly at metamorphosis (Fig. 67). A very few, most notably *Halisarca dujardini* and those with amphiblastula larvae, go through brief asconoid and syconoid stages during metamorphosis (Fig. 68) before developing the leuconoid condition (BRIEN, 1973a).

The hexactinellid larva of *Farrea sollasii* develops first six stauractines oriented parallel to the anteroposterior axis; subsequently it develops six microscleres (**discohexasters**). After fixation the young sponge gradually develops the adult complement of spicules (Fig. 69), but stauractines persist for a time in the dermal layer, disappearing before the adult stage. Shortly after fixation the barrel-shaped young sponge has an asconoid structure with diverticula that are to become the pseudosyconoid choanocyte chambers of the adult (TUZET, 1973b). It is of interest that the earliest hexactinellids of the Cambrian have almost exclusively a spiculation of stauractines in a thin-walled sponge body similar to a young *Farrea* in gross shape.

WATANABE (1957), in describing the development of the demosponges *Tetilla serica*, has noted that the tetraxonic **protriaenes** first develop in the young sponge as monaxons, the **cladome** developing subsequently. Whether one sees a phylogeny reflected in this, it demonstrates that there is a developmental process whereby monaxons can become tetraxons; thus the process is not forbidden to phylogeny.

OVIPARITY VS. VIVIPARITY

Some demosponges, notably the Clavulina or Hadromerida, Epipolasida, Spirosclerida, and Axinellida, shed their eggs into the water after fertilization in the mother sponge. An even larger number of sponges, however, incubate the fertilized eggs to the larval stage before they are released into the

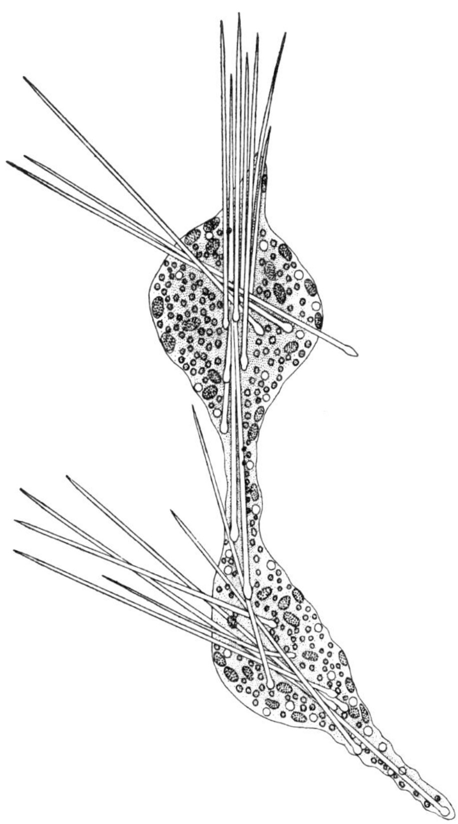

FIG. 74. Small pelagic propagule composed of two parts, each with numerous tylostyle spicules along axes of lobes and in peduncle; microscleres not present in this stage of development (Brien, 1973a).

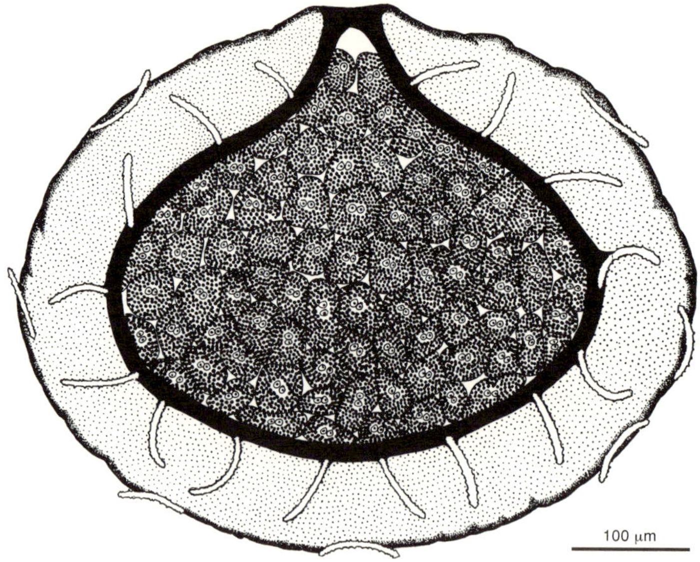

FIG. 75. Schematic drawing of vertical section through gemmule of *Spongilla lacustris* showing its general shape, with upper, micropyle opening and development of strongyles in outer, alveolar shell (Brien, 1973a).

water; these viviparous sponges include the Sigmatosclerophora, Keratosa, and Homosclerophora among the demosponges and apparently all the Calcarea and Hexactinellida that have been studied. In such sponges the developing larva is surrounded by choanocytes (in some instances an entire choanocyte chamber) that serve as food for the developing embryo. In some instances these cells penetrate the internal cavity of the embryo as well as surrounding it. The carrier cells of the sperm (a peculiar feature of vi-

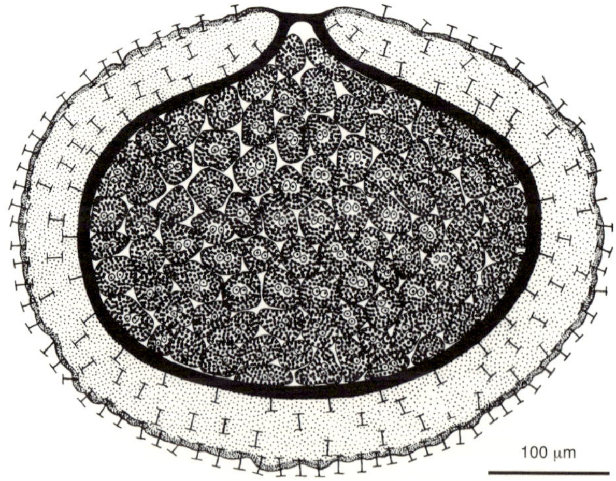

FIG. 76. Schematic drawing of vertical section through gemmule of *Ephidatia mülleri,* which has amphidiscs as spicules in outer capsule (Brien, 1973a).

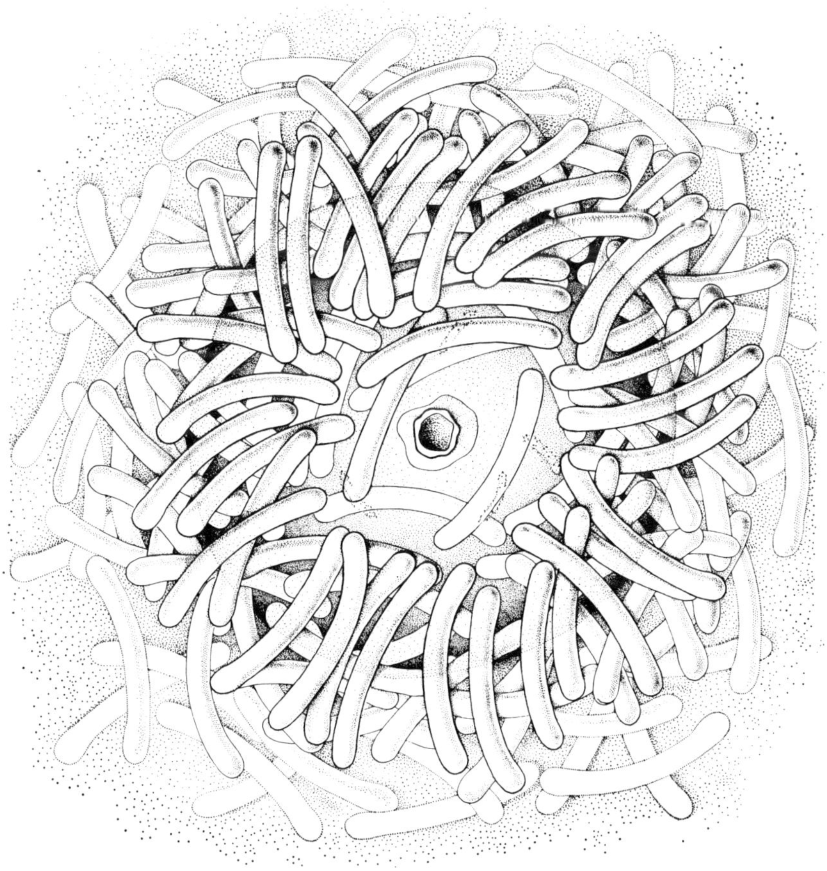

FIG. 77. Gemmule of *Corvospongylla thysi* with prominent amphistrongyles forming cuticular layer around small, central micropyle (Brien, 1973a).

viparous sponges) also serve as nourishing cells for the embryo (BRIEN, 1973a; LÉVI, 1973; TUZET, 1973a, 1973b).

Presence or absence of viviparity appears to be of fundamental taxonomic significance inasmuch as entire groups at the subordinal, ordinal, or even class level appear to be characterized by one or the other mode of reproduction. Among animals in general, incubation of larvae appears to be a more advanced character.

ASEXUAL REPRODUCTION

Reproduction by asexual means among sponges may take place through external buds, planktonic propagules, and internal statoblasts or gemmules. Such methods seem to occur most often in freshwater and shallow-marine forms in which winter conditions may kill the adults or in deep-sea forms in which, perhaps, a sparse population makes fertilization risky.

External buds, which grow from the surface of the parent and ultimately drop off to begin life as separate individuals, are known in the demosponges *Tethya maza* (Fig. 70) and *Mycale contrarenii* (BRIEN, 1973a), the calcareous sponge *Leucosolenia botryoides* (Fig. 71) (TUZET, 1973a), and the hexactinellids *Lophocalyx philippinensis* (Fig. 72), *Scyphidium longispina,* and *Aulocalyx ijimai* (TUZET, 1973b).

Propagules are microscopic bodies found in the plankton that are apparently produced

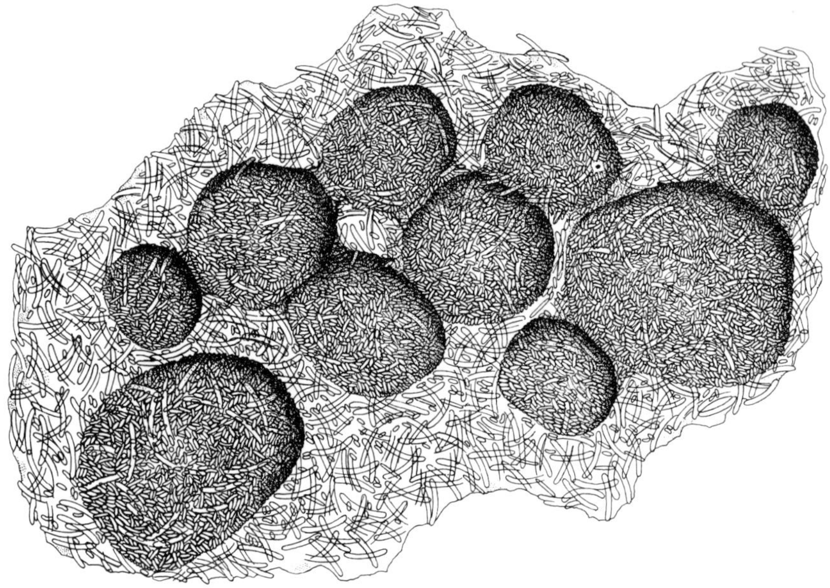

FIG. 78. Cluster of statoblasts of *Potamophloios gilberti,* up to 1 mm in diameter, that are attached to base of parent sponge; they lack a micropyle, such as is developed in gemmules, but are armored by layer of tangential strongyles (Brien, 1973a).

asexually by some demosponges. Those of *Alectona milleri* and *Thoosa armata* (both deep-water Hadromerida) have a special armor of coccolith-like spicules not found in the adult sponge, together with projecting styles (Fig. 73). The propagules have been found both in the adult sponge and free in the plankton. A different type with tylostyles and various astrose microscleres correspond with the spiculation of *Tethya aurantium,* its presumed source, although others may lack microscleres (Fig. 74) (TRÉBOUGOFF, 1942; BRIEN, 1973a).

Statoblasts, sorites, and gemmules are bodies covered with a special protective coating; they are formed within the parent sponge, generally near its base, and are capable of withstanding unfavorable conditions after being released by the death of the parent. They develop into new sponges with the return of favorable conditions.

They are best developed in the freshwater spongillids, where the body, called a gemmule, is provided with a special alveolar shell armed with spicules and bearing an opening (micropyle). The spiculation is characteristic (amphidiscs in some genera, spinose strongyles in others (Fig. 75–76). These objects are quite capable of fossilization, as are the spiculated capsules formed by the parent of *Corvospongilla thysi* around its gemmules (Fig. 77) (BRIEN, 1973a). In another freshwater family, the Potamolepidae, the genus *Potamophloios* bears statoblasts up to 1 mm in diameter that lack a micropyle as in true gemmules but are armored with tangential strongyles (Fig. 78) and are equally capable of fossilization (BRIEN, 1973a).

Several common genera of shallow-water marine demosponges (*Cliona, Chalina, Craniella, Suberites*) produce similar statoblasts, although not all are armed with spicules (BRIEN, 1973a). These statoblasts are larger than the gemmules of spongillids, reaching 1 or 2 mm in diameter and as much as 2.5 mm in the boring sponge *Cliona vastifica* (BRIEN, 1973a, p. 388).

In the spongillids the gemmule is part of the regular life cycle, the adult dying each year with the onset of the winter season or the dry season. Among the marine sponges, formation of statoblasts seems to be more sporadic (BRIEN, 1973a) and is apparently a preparation for more accidental conditions.

PHYSIOLOGY

Robert M. Finks

[Department of Geology, Queens College (CUNY)]

FEEDING MECHANISM

Sponges are filter feeders with intracellular digestion. Energy for moving water through the sponge is supplied by flagella of the choanocytes. The direction water takes seems to be determined by size of openings into and out of **flagellated chambers** (**prosopyles** and **apopyles**, respectively). The apopyle, usually single, has a larger diameter than the prosopyles. In *Ephydatia fluviatilis* the apopyle is 25 to 30 µm in diameter, as compared to 4 µm for the prosopyles (Kilian, 1952, p. 416–417). When the water within the flagellated chamber is stirred by the flagella, it tends to move through the apopyle into the exhalant canals, drawing water into the chamber through the prosopyles.

Careful observations made by Kilian (1952) on preparations of the freshwater sponge *Ephydatia fluviatilis* grown on microscope slides have confirmed this process. Kilian (1952, p. 419) further observed that collars of individual choanocytes tend to direct the water down their longitudinal axes, which are oriented toward the apopyle of the chamber. The motion of each flagellum draws water through the porous surrounding collar (see below), which acts as the ultimate filtering mechanism in the feeding system, although not the only one.

Directional or coordinated movement of the choanocyte flagella is not necessary, nor does it occur. In the asconoid calcisponge *Leucosolenia* the entire cloaca is lined with choanocytes. That opening is a space much larger than the usual flagellated chamber and has as its exit the osculum of the sponge. Even in such coarse structures the directional flow of water is maintained solely by differences in diameters between osculum and inhalant canals. Flagella of adjacent choanocytes of *Leucosolenia* beat in different planes, always, however, perpendicular to the wall, and at different frequencies (Jones, 1964).

In the small calcareous species *Leucandra aspera* studied by Bidder (1923), a 9 cm-long sponge circulated nearly a liter (936 cc) of water per hour. Sponges of this species (at 18° C) project exhalant streams up to 45 cm from the osculum at a calculated velocity of some 8.5 cm per second. Bidder (1923, p. 313) pointed out that in colder waters the combined effect of lowered metabolic rate and increased water viscosity (important at the size-level of single flagella) may reduce energy of the oscular current. This may have a bearing on distribution of sponges in cold and deep waters.

Filtration rates of marine sponges fall within the same range as those of bivalves and ascidians (Jørgensen, 1955, 1966). Expressed in terms of body weight (grams of nitrogen) for purpose of comparison, they are: sponges, 45 to 170 ml/hr/mg N; bivalves, 5 to 160 ml/hr/mg N; ascidians, 110 to 150 ml/hr/mg N.

All cells of the sponge except **scleroblasts** (Kilian, 1952) are capable of engulfing suspended particles that come into contact with the cell surface (Pourbaix, 1931, 1933; Van Weel, 1949; Kilian, 1952; Simpson, 1963). The indispensable function of the choanocytes is that of maintaining a current. In sponges with small flagellated chambers, such as the spongillids, the prosopyles are no larger than 4 µm. Many suspended particles that sponges take into their cells are larger than this (Pourbaix, 1931, cited bacteria 11 µm long), and such particles are captured from the water stream by cells lining the canals or stretching netlike across them and even by cells on the sponge surface (Kilian, 1952; Simpson, 1963).

Most observers concur, however, that the choanocytes ingest the major part of suspended particulate matter carried into the sponge and that particles are subsequently transferred to the amoebocytes of the **mesenchyme** (see KILIAN, 1952, for extensive observations on *Ephydatia* with illustrations). This has been strikingly confirmed by study of the absorption by *Sycon* of dissolved glycine tagged with C^{14} (EFREMOVA, 1965). Although the epithelial **pinacocytes** also absorbed some of the glycine, the greatest concentration of labeled carbon immediately after feeding was found in the choanocytes. Within 24 hours the concentration of C^{14} decreased in the choanocytes but increased in the adjacent mesenchyme, indicating transfer of absorbed material to mesenchymal amoebocytes.

The choanocytes are in themselves filtering mechanisms, as shown by electron-microscope investigations of the fine structure of the collar (RASMONT, 1959; FJERDINGSTAD, 1961). The collar is not solid, as once believed, but is constructed of a single layer of longitudinal fibrils (about 0.2 μm apart in Spongillidae) connected by finer transverse fibrils. Motion of the flagellum draws water through the interfibrillar spaces and filters out suspended particles that are larger than the 0.2 μm interfibrillar spaces. Uptake of food particles by choanocytes may be aided because water velocities are lowest in the flagellated chambers, as noted by BIDDER (1923).

The osculum is an essential part of the hydraulic system of many sponges. Its diameter is narrower than that of the cloaca so that the velocity of effluent water is increased and projected as a jet away from the sponge. BIDDER (1923) has shown that for the calcisponge *Leucandra,* the oscular diameter is close to the theoretical optimum, that is, the diameter that will project the longest jet for a given flagellar pressure and a given energy loss from internal friction. The longer this efferent jet the greater the separation of exhaust from intake. In all but the most highly agitated water, this increases the effective radius of unfiltered water available to the sponge.

Functioning of such a hydraulic system depends upon fluid pressure developed within the choanocyte-lined chambers. This in turn depends upon tension and elasticity of the chamber wall. Flagellated chambers are more cohesive than the rest of the sponge (KILIAN, 1952) and apparently are elastic (JONES, 1962, p. 28). Given the same properties of the chamber wall (and, of course, of the canal system and choanocytes), the smaller the chamber the larger the pressure. [Observed pressures cited by BIDDER (1923) include 0.8 mm of water in the large-chambered *Leucandra* and 4 mm of water in the small-chambered *Stylotella.*] Such a relationship provides an adaptive basis for evolution of the small flagellated chambers of the leuconoid canal system that is possessed by most sponges.

That such a canal system developed very early among demosponges is indicated by the thick-walled, three-dimensional, fine-meshed skeletal net found in some of the earliest lithistid sponges, the Lower Ordovician *Archaeoscyphia* (or probably the even earlier Upper Cambrian *Gallatinospongia* and *Wilbernicyathus*). Earlier demosponges are principally the Middle Cambrian monaxonid sponges *Hazelia, Wapkia,* and others from the Burgess Shale of British Columbia and Middle Cambrian units of Utah. These sponges appear to be relatively thin walled and could have had a less advanced form of canal system.

Among the hexactinellids there is distinct fossil evidence of increasing complexity during the Cambrian. The earliest spicules assignable to this class are stauractines of Early Cambrian age that must have supported thin-walled sponges of asconoid or near-asconoid architecture. Whole sponges (*Protospongia*) consisting of a single layer of stauractines and pentactines are well known from Middle Cambrian beds. The interpretation of an asconoid structure is supported by the ontogeny of living hexactinellids, whose embryos are asconoid sponges with a skeleton of stauractines. The first common hexactines, indicating the development of a thicker body wall, occur in upper Middle Cambrian *Bolaspis* zone. Whole sponges of a semi-encrusting habit with thick walls supported by two or three layers of hexactines (*Multivasculatus* HOWELL & VAN HOUTEN) are known from Upper Cambrian rocks. It seems probable that the semi-syconoid structure characteristic of living hexactinellids developed during the Cambrian.

Although BIDDER (1923) emphasized the adaptive value of separating influent and effluent water (partly confirmed experimentally by WARBURTON, 1960) this is by no means an absolute requirement, and other adaptive considerations must surely come into play. JØRGENSEN (1955) pointed out that a higher filtration rate can be obtained for the same expenditure of energy if internal fluid pressure is lower. Thus, if concentration of food in the ambient is low, passage of larger quantities of water per given energy expenditure may take precedence over efficient separation of water currents. Indeed, recirculation of water may insure effective removal of all available food material.

BIDDER (1923, p. 312) ascribed the open canal system of hexactinellids and their presumed lack of hydraulic evolution to constant currents in the deep-sea environment, currents that sweep through their open framework and carry away waste water, thereby obviating any need for hydraulic efficiency. Hexactinellids, however, have not always lived predominantly in deep water. It is possible that the hexactinellids have specialized in the metabolically efficient passage of large volumes of water at low pressure to extract food at low concentration. VON BRAND (1939) has shown that particulate matter rich in organic nitrogen is very rare in the deep sea as compared with surface waters. This may account for present hexactinellid abundance in the deep sea (freedom from competition of other filter feeders that may require higher food concentrations) and possibly for their earlier flourishing in shallow water at a time when ambient suspended food may have been less abundant.

FOOD SUBSTANCES

Sponge cells appear to engulf particles without regard to nutritive value. Non-nutritive substances, such as carmine or graphite particles, are subsequently excreted in normal fashion but more rapidly than such organic materials as egg-white droplets, which appear to be digested (KILIAN, 1952; JØRGENSEN, 1955). The natural food of sponges is still not known with certainty. Because digestion is exclusively intracellular, the maximum size of food particles is determined by what can be ingested by a single sponge cell. The largest cells are some 10 to 20 µm in diameter. Such a size rules out feeding on such organisms as protozoans and rotifers (KILIAN, 1952, p. 431) as well as many that are still larger. The smallest particles observed to be completely filtered out of water by sponges are 0.5 to 1.0 µm (VAN TRIGT, 1919; JØRGENSEN, 1955). Bacteria are frequently cited as a principal food of sponges. MADRI and others (1967) reported that when bacteria (*Escherichia coli*) are added to the water surrounding *Microciona prolifera* they are apparently removed from the water and concentrated in the sponge. POURBAIX (1931, 1932, 1933) has observed ingestion and apparent digestion of bacteria by archaeocytes of marine demosponges. To what extent they are a source of food is unknown. Studies of SOROKIN (1964) show that the abundance of living bacteria in ocean waters decreases sharply below 200 meters and that they are practically absent below 600 meters. The many sponges that occur below these depths must therefore feed on something else. Absence of bacteria is not a limiting factor in sponge distribution. Phytoplankton may be a food source. KILIAN (1952, p. 443) reported algae being digested apparently in cells of *Ephydatia fluviatilis,* although SIMON (1953, p. 231) was unable to maintain the same species by feeding it algae and concluded that algae could not be used as food. Suspended but nonliving, organic particles may provide a considerable source of food for most sponges. Recently it has been shown unequivocally that sponges can absorb dissolved amino acids directly (C^{14}-tagged glycine; EFREMOVA, 1965), but the extent to which this operates in nature remains undetermined, although many authors have postulated such a food source (see KILIAN, 1952, p. 430).

JØRGENSEN (1955, p. 445) estimated that sponges, bivalves, and ascidians, all of which filter about 15 liters of water for each milliliter of O_2 consumed in metabolism, can meet their food requirements for maintenance and optimal growth when 0.15 to 0.20 mg of useable organic matter is available per liter of water. JØRGENSEN gave the quantity of dissolved organic matter in seawater as 2.2 to 4.6 mg/liter and the protein fraction as one-third to one-half of this (*vis.,* 0.7 to 2.3 mg/liter). If EFREMOVA's observations are generally valid, Holocene sponges could meet all their food requirements from direct absorption of dissolved matter. Because sponges do effectively filter out particulate matter from their feeding currents down to sizes of 0.5 µm, however, it seems likely that this must provide nutrients and is not a mere exercise.

Symbiotic algae are present in many shallow-water sponges, including most freshwater species. VAN WEEL (1949) noted that starch was not present in such algae when in the amoebocytes of *Spongilla proliferans,* although starch was present when the same algae were isolated from the sponge. Upon being returned to an algae-free sponge, the starch disappeared from the algae after a week. VAN WEEL suggested that such symbionts normally supply the sponge with carbohydrates, probably directly absorbed as soluble sugars. Symbiotic algae may, thus, supply soluble foodstuffs to those sponges that possess them (KILIAN, 1952, p. 443).

DIGESTION AND EXCRETION

Such particles as have been fed to sponges under experimental conditions appear to be digested in food vacuoles over a period of 12 to 24 hours (KILIAN, 1952) or more (POURBAIX, 1931, 1933). Small, condensed masses of indigestible matter remain in the food vacuoles of amoebocytes. These cells wander to exhalant canals, where the vacuole breaks through the cell wall and releases its contents into the efferent current. The empty vacuole is shed from the cell shortly thereafter and is carried off in the excurrent stream as a bladderlike object (KILIAN, 1952, p. 439). VAN WEEL (1949) observed that if an amoebocyte protrudes a feces-laden vacuole into an inhalant canal, by mistake as it were, it is withdrawn and the cell wanders off until it encounters an exhalant canal. KILIAN (1952, p. 438) suggested that possible chemical differences between afferent and efferent water may guide the amoebocytes in this respect, and that the greater surface area of the exhalant passages reduces the probability of mistakes, even with random movement. Solid wastes from digestion by cells other than amoebocytes may be transferred to the latter for excretion (VAN WEEL, 1949), but KILIAN (1952) noted that much excretion is carried out by pinacocytes lining the exhalant canals, presumably obtained in part from amoebocytes.

The mechanism by which soluble metabolic wastes are eliminated appears to be poorly known. VAN WEEL (1949) observed that *Spongilla,* vitally stained with pyrrhol blue, eliminated this soluble dye by concentrating it in liquid droplets that were subsequently voided as food vacuoles from the amoebocytes. Excretory products are complex nitrogen bases, such as agmatine and guanidine derivatives, according to JAKOWSKA and NIGRELLI (1960).

Presence of contractile vacuoles in amoebocytes (*sensu lato,* including scleroblasts) and choanocytes of freshwater sponges (*Spongilla* and *Ephydatia*) have been demonstrated by JEPPS (1947) and GATENBY and TAHMISIAN (1959). These latter authors were unable to find such contractile vacuoles in ten genera of marine Calcarea and Demospongea, although noncontractile vacuoles were found in a similar position in the marine calcareous sponge *Grantia compressa.* Presumably contractile vacuoles of freshwater sponges eliminate excess water that enters the cells osmotically.

TRANSPORT OF METABOLITES

No circulatory system exists in a sponge. Motile amoebocytes appear to carry phagocytosed and partly digested food to other cells and to carry away solid wastes. The mechanism is either by momentary fusion of the cells involved in the transfer (VAN WEEL, 1949) or more likely by transfer of food vacuoles or solid particles through the apposed membranes of adjacent cells (KILIAN, 1952). Soluble metabolites may be transferred by diffusion across cell membranes. Scleroblasts apparently do not carry out phagocytosis (KILIAN, 1952) and presumably receive all metabolites in dissolved (i.e., molecular) form.

Specialized cells called thesocytes, possibly derived from choanocytes, have been interpreted as loci of storage of reserve metabolites. In the peculiar pharetronid calcareous sponge *Petrobiona massiliana* VACELET & LÉVI, 1958 thesocytes are concentrated in specialized areas and gradually disperse during the winter, suggesting their function as a reserve food supply for unfavorable times (VACELET, 1962). The reserves are interpreted from staining reactions as being DNA or a similar glycoprotein. Stored glycogen is present in the posterior cells of the larva of this species (VACELET, 1965). LIACI (1963) reported lipofuchsins, melanins, and sterols in thesocytes of the marine demosponge *Aaptos aaptos* that seem to lose their stored contents at the time the sponges reach the breeding season, presumably transferring the stored products to the gametes.

In the freshwater *Spongilla proliferans,* glycogen is stored in ovocytes and a concentration of protein occurs in cells making up the gemmules (VAN WEEL, 1949). LUTFY (1960) similarly found glycogen in amoebocytes and archaeocytes of *Ephydatia fluviatilis.* LÉVI (1966) noted the presence of glycogen in some cells in the cortex of the marine demosponges *Ophlitaspongia seriata, Microciona prolifera, Pachymatisma johnstoni,* and *Mycale contarenii;* their abundance shows no seasonal connection with sexual maturity, and LÉVI noted that it is not clear whether the glycogen is stockpiled for metabolism of the cells involved or is a reserve for the whole sponge. The latter would imply a greater degree of metabolic integration (Table 2).

IRRITABILITY AND BEHAVIOR

Sponges respond defensively to unfavorable stimuli by limited movements that minimize surface area and volume and close off access to interior spaces. Such responses include generalized or local contraction of the body and closure of oscules and of inhalant pores. Stimuli that evoke such responses include exposure to air, light, heat, reduced O_2, increased CO_2, stillness of ambient water, dissolved toxic substances in ambient water (e.g., alkaloids), and direct mechanical or electrical stimulation.

Few data are available for Hexactinellida. Most Calcarea respond only locally (JONES, 1957). The sponges most ready for overall contractions are Demospongea with a cortex, such as *Tethya.* BULLOCK and HORRIDGE (1965) suggested that the layer of elongate cells just beneath the surface may be responsible.

Closure of oscules is brought about by contraction of a ring of specialized amoebocytes that surrounds the orifice like a sphincter. Inhalant pores located within porocytes are closed by contraction of the latter. Generalized contraction is apparently brought about partly by contraction of the individual cells (pinacocytes, **collencytes,** and **myocytes**) and partly by collapse of internal canal spaces following cessation of choanocyte flagellar action and closure of dermal pores. The contractile cells contain

TABLE 2. Storage products in cytoplasm of sponge cells (new).

SUBSTANCE	LOCATION	SPECIES	SOURCE
glycoprotein (DNA)	thesocytes of trabecular cords	*Petrobiona massiliana*	Vacelet & Lévi, 1958
	ovocytes	*Hippospongia communis*	Tuzet & Pavans de Ceccatty, 1959
glycogen	posterior cells of larva	*Petrobiona massiliana*	Vacelet, 1965
	ovocytes	*Spongilla proliferans*	Van Weel, 1949
		Ephydatia fluviatilis	Lutfy, 1960
	archaeocytes	*Ephydatia fluviatilis*	Lutfy, 1960
		Pachymatisma johnstoni	Lévi, 1966
	pinacocytes	*Ephydatia fluviatilis*	Lutfy, 1960
	"gray cells"	*Microciona prolifera*	Simpson, 1963
		Ophlitaspongia seriata	Borojevic & Lévi, 1964
	collencytes	*Mycale contrarenii*	Lévi, 1966
melanin	thesocytes	*Aaptos aaptos*	Liaci, 1963
	archaeocytes	*Pachymatisma johnstoni*	Lévi, 1966
lipofuchsin	thesocytes	*Aaptos aaptos*	Liaci, 1963
protein	gemmules	*Spongilla proliferans*	Van Weel, 1949
sterols	thesocytes	*Aaptos aaptos*	Liaci, 1963
fats	?	?	

filaments similar to those of muscle cells of higher animals, although their chemistry is somewhat different (BAGBY, 1965).

A stimulus affects initially only the immediate area exposed to it, but the response may spread slowly from the stimulated area for a limited distance. Summation of stimuli, both temporal and spatial, has also been observed (PROSSER, 1960; PAVANS DE CECCATTY, 1960).

The means whereby excitation is spread is not clear. TUZET and PAVANS DE CECCATTY (1959) and coworkers considered some cells to be primitive nerve cells and sensory cells, but their interpretation was not accepted by reviewers of the problem (JONES, 1962; BULLOCK & HORRIDGE, 1965). Cell-to-cell conduction would seem most likely, but WINTERMANN (1951) has suggested released chemical substances carried in the water currents. This has not been tested.

LENTZ (1966) reported the presence of neurohumors (acetylcholinesterase, monoamine oxidase, epinephrine, norepinephrine, 5-hydroxytryptamine) in spindle-shaped, bipolar or multipolar cells in the mesenchyme of the calcareous sponge *Sycon*. This appears to favor the ideas of TUZET and her coworkers. REISWIG (1971) concluded that myocyte type cells occur in a network and have a pacemaker-like activity so that contractions pass quickly from cell to cell and coordinated rhythmic activity occurs.

Movements of sponges with rigid skeletons, of the sort most frequently preserved as fossils, must be extremely limited. BURTON (1948) and ARNDT (1941) have reported limited locomotion among fixed adult sponges, presumably by migration of cells over the substrate.

RESPIRATION

Gas exchange is effected by each cell, either directly with the internally circulating water of the feeding currents, by diffusion through the mesoglea for short distances, or directly with the external ambient water. Consumption of O_2 appears to be relatively low, although few measurements have been published. BERGQUIST (1978) noted that there is little consistency in rates of sponge respiration reported in the literature. She observed that the only observations to that time on respiration rates in Demospongea that need no qualification were those reported by REISWIG (1974).

POURBAIX (1939) found that in fresh slices of the marine demosponge *Tethya lyncurium,* the choanosome consumed an average of 0.237 mm^3 of O_2 per hour per mg of dry weight (less spicules). The corresponding figure for the ectosome was 0.081 mm^3/mg/hr. Choanocytes seem, therefore, to have a higher metabolic rate than other cells, a conclusion not unexpected in view of their flagellar activity and their role as principal sites of ingestion and digestion. These rates are low with respect to tissues of other invertebrates, however, which average 0.5 to 1.0 mm^3/mg (dry)/hr (POURBAIX, 1939).

HYMAN (1925) found that the O_2 consumption rate varied inversely with the size of the individual in the calcareous sponge *Sycon* and is greater in the upper half of the sponge than in the lower half.

The O_2 consumption also varies with the state of activity, as measured by the rate of water currents in the sponge. Table 3 shows several measurements of various sponges. They are not strictly comparable, as some are based on weight of dry organic matter, whereas others are fresh weights, including the nonmetabolizing spicules that may account for a significant proportion of the total weight.

Symbiotic, intracellular algae are probably a source of O_2 for many shallow-water sponges that possess them, both freshwater and marine. DE LAUBENFELS (1932) found that three species of marine demosponges have decreased O_2 consumption in sunlight as opposed to shade, and in one instance the amount of O_2 actually increased (see Table 3).

CHEMICAL COMPOSITION

Living sponges have a number of chemical characteristics that tend to emphasize their separateness from other branches of the animal kingdom. BERGMANN and his coworkers (BERGMANN & FEENEY, 1949, 1950; BERGMANN & MCTIGUE, 1949; BERGMANN & others, 1950; BERGMANN & MCALEER, 1951) have isolated a number of sterols from various sponges that either do not occur or occur but rarely in other animals. Neospongosterol and aaptostanol occur only in demosponges of the family Suberitidae (BERGMANN & others, 1950). Chondrillasterol and haliclonasterol occur elsewhere only in green algae (ALTMAN & DITTMER, 1964). Clionasterol and poriferasterol occur elsewhere only in molluscs (ALTMAN & DITTMER, 1964).

24-Methylenecholesterol occurs elsewhere only in molluscs and in the honeybee. All have 28 or 29 carbon atoms rather than the 27 found in most of the common sterols of animals. Most of the other known 28- or 29-carbon atom sterols have been recovered from plants.

Another peculiar feature of sponges is the high concentration of protein-bound halogens (iodine and bromine) present in spongin (in the form of 3, 5-diiodotyrosine and dibromotyrosine). Iodine may constitute as much as 10 percent or more of spongin (VINOGRADOV, 1953).

Sponge pigments are likewise unusual among animals in that carotenes tend to dominate over xanthophylls (NICOL, 1967). Individual sponges have revealed peculiarities of composition that may be of more general distribution in the phylum. The demosponge *Cryptotethya crypta* has yielded three unique nucleic acids: spongothymidine (2-D-arabofuranoside of thymine), spongouridine (2-d-arabofuranoside of uracil) and spongosine (2-D-ribofuranoside of 2-methoxyadenine) (STEMPIEN, 1960). The demosponge *Microciona prolifera* yields a substance or substances extractable with organic solvents and as yet undetermined chemically (but named *ectyonin*) that has antibiotic properties against *Escherichia coli,* tuberculosis bacilli, *Pseudomonas pyocyanea, Staphylococcus aureus,* and *Candida albicans* (JAKOWSKA & NIGRELLI, 1960).

The sterols studied by BERGMANN are of interest in revealing a pattern of distribution within the class Demospongea that is somewhat tied to taxonomic subdivisions erected on morphologic grounds. Clionasterol and poriferasterol have been obtained only from

TABLE 3. Respiratory rates of sponges (new).

SPECIES	RATE	SOURCE
Tethya lyncurium		Pourbaix, 1939
choanosome	0.237mm^3 O_2/hr/mg dry wt. (less spicules)	
ectosome	0.081 mm^3/hr/mg dry wt. (less spicules)	
Iotrochota birotulata		de Laubenfels, 1932b
in shade	0.154 cm^3 O_2/hr/cm^3 sponge (wet incl. spicules)	
in sun	0.067 units as above	
Haliclona rubens		de Laubenfels, 1932b
in shade	0.150 units as above	
in sun	0.055 units as above	
Haliclona longleyi		de Laubenfels, 1932b
in shade	0.053 units as above	
in sun	0.002 units as above (produced by the sponge)	
Grantia compressa	0.05 ml O_2/hr/gm (wet wt.) incl. spicules) at 21°–23° C in an individual weighing about 0.15 gm	Hyman, 1925
Sycon sp.	0.04–0.16 cm^3 O_2/hr/gm (wet wt. incl. spicules) varying inversely with size of sponge	Hyman, 1925
Suberites sp. *Aplysina* sp.	0.0117 ml O_2/hr/gm (wet wt. incl. spicules) at 20°–22° C in 20–25 gm individuals	van Budden-Brock, 1939
Suberites massa	0.34 ml O_2/hr/gm (dry wt. less spicules) at 22° C	Putter, 1914
Mycale sp.	0.126 ml O_2/hr/gm (wet wt.)	Reiswig, 1974
Verongia gigantea	0.1004 ml O_2/hr/gm (wet wt.)	Reiswig, 1974
Tethya crypta	0.0329 ml O_2/hr/gm (wet wt.)	Reiswig, 1974

species belonging to the Sigmatosclerophora, Clavulina, and Epipolasida (*vis., Spongilla,* 1 sp.; *Haliclona,* 3 spp.; *Callyspongia,* 1 sp.; *Tedania,* 1 sp.; *Spheciospongia,* 2 spp.; *Anthosigmella,* 1 sp.; *Cliona,* 2 spp.; and *Cryptotethya,* 1 sp.). Cholesterol is confined to the Sigmatosclerophora (*Haliclona,* 2 spp.; *Microciona,* 1 sp.; and *Halichondria,* 2 spp.), as is haliclonasterol (*Haliclona longleyi*) (BERGMANN & MCTIGUE, 1949; BERGMANN & others, 1950; BERGMANN & FEENEY, 1950).

The Suberitidae are the only sponges to have yielded neospongosterol (*Suberites,* 3 spp.) and aaptostanol (*Aaptos* sp., *Radiella sol, Weberella bursa, Polymastia infrapilosa*) and share cholestanol with the Sigmatosclerophora (suberitids: *Suberites,* 3 spp.; *Terpios,* 2 spp.; *Aaptos,* 1 sp.; *Weberella,* 1 sp.; *Polymastia,* 1 sp.; Sigmatosclerophora: *Microciona,* 1 sp.; *Halichondria,* 2 spp.; *Hymeniacidon,* 1 sp.) (BERGMANN & FEENEY, 1949; BERGMANN & others, 1950).

The homosclerophoran *Chondrilla nucula* is the only sponge to yield chondrillasterol. The sigmatosclerophoran *Haliclona oculata* and the spirosclerophoran *Craniella crania* are the only sponges to have yielded 24-methylenecholesterol (chalinasterol or ostreasterol) (BERGMANN & FEENEY, 1949; ALTMAN & DITTMER, 1964).

All the Keratosa (8 spp.), Axinellida (2 spp.), Euasterophora (1 sp.), Hexactinellida (1 sp.), and Calcarea (1 sp). studied by BERGMANN, as well as five species of Sigmatosclerophora and two of Epipolasida, have yielded as yet only poorly defined sterols that are not clearly any of the foregoing.

It is probably too early to draw phylogenetic conclusions from the promising study of the distribution of sterols, but in a preliminary way a relationship of the monaxonid groups Sigmatosclerophora, Clavulina, and Epipolasida is suggested through the common occurrence of clionasterol-poriferasterol, as well as their separation from the monaxonid family Suberitidae, which uniquely possesses aaptostanol-neospongosterol and lacks clionasterol-poriferasterol. Since the remaining groups of demosponges are set off from the foregoing only by the fact that their sterols have not been clearly determined, however, some sterols may turn out to have a wider distribution. Also, needless to say, we do not yet know to what extent sterol chemistry is a conservative character.

Other noteworthy substances that occur in sponges include unusually large amounts of histamine (100 mg/kg) in the tissues of the demosponge *Geodia gigas* (DUNER & PERNOW, 1963), and true chitin in the walls of gemmules of freshwater spongillids (JEUNIAUX, 1963).

The skeletal material of the sponges varies somewhat compositionally from taxon to taxon. Few analyses of scleroprotein spongin, which is confined to the class Demospongea are available. One analysis (SAPER & WHITE, 1958) of the keratose sponge *Hippospongia equina* revealed the following amino acids: alanine, Y-aminobutyric acid, arginine, aspartic acid, cystine, glutamic acid, glycine, histidine, hydroxyproline, leucin-isoleucine, lysine, ornithine, phenylalanine, proline, serine, threonine, tryptophan, tyrosine, 3,5-diiodotyrosine (plus 3-monoiodotyrosine considered an artifact of breakdown), and valine. LOW (1951) found in *Spongia officinalis obliqua* the additional amino acids: Y-aminobutyric acid, dibromotyrosine, and methionine. Elemental analyses of spongin of Keratosa must be interpreted with caution, for foreign particles are frequently incorporated into the spongin fibers.

Silica of siliceous spicules of the classes Demosponges and Hexactinellida is hydrated. The proportion of water is variable from one species to another; demosponge silica is less hydrated than that of the hexactinellids (5.97 to 7.34% and 7.16 to 13.18% respectively) (VINOGRADOV, 1953). Variability within a single spicule was reported by VOSMAER and WIJSMAN (1904) in the demosponge *Tethya aurantia*. They observed that the axial portion was more soluble in HF than the peripheral part, probably because of a greater degree of hydration of the axial part (this may account for the enlarged axial canal in many fossil spicules). A possible instance of an actual gel phase of silicic acid may obtain in the peculiar saclike bodies of gel in the aberrant demosponge *Collosclerophora arenacea* DENDY, 1917. At the other extreme, VERNADSKY (1934) reported birefringence, and therefore a crystalline state, in some hexactinellid spicules, although this has not been confirmed by later investigators. Small amounts of alkali and alkaline earth elements may be present in siliceous spicules, greater in demosponges than in hexactinellids (VINOGRADOV, 1953).

Calcareous spicules of the Calcarea are composed of magnesian calcite. The proportion of magnesium varies from about 4 percent to 14 percent (VINOGRADOV, 1953). Strontium and traces of lithium (0.005%) have been reported (FOX & RAMAGE, 1931) from *Clathrina*. The very peculiar sponges *Astrosclera willeyana* and the probably closely related *Ceratoporella nicholsonii* have massive exoskeletons of aragonite in the form of closely packed spheroids. The soft parts contain siliceous acanthostyles and are

TABLE 4. Life span of sponges (new).

SPECIES	LIFE SPAN	REFERENCE
Demospongea		
Hippospongia sp.	50 years	Altman & Dittmer, 1962
Dysidea spinifera	2–3 years (aquarium)	Arndt, 1941
Adocia "alba"	1.5 years (aquarium)	Arndt, 1941
Gellius angulatus	2 years (aquarium)	Arndt, 1941
Suberites carnosus	1.83 years (aquarium)	Arndt, 1941
Axinella sp.	4 years	Altman & Dittmer, 1962
Hymeniacidon perlevis	1.5 years, at least	Burton, 1948
Halichondria panicea	1.5 years, at least	Burton, 1948
Pachymatisma johnstoni	1.5 years, at least	Burton, 1948
Calcarea		
Grantia capillosa	3 months	Altman & Dittmer, 1962

anatomically demosponge-like. The calcareous skeleton in its gross morphology is like that of the calcitic pharetrone Calcarea (VACELET, 1965), as well as like that of the extinct Stromatoporoidea (HARTMAN & GOREAU, 1966).

Despite the wide variety of substances produced by sponges, the quantity of DNA in sponge nuclei is considerably less than that in more complex animals. Sponges give values of about 0.1 picograms per cell, bacteria (*E. coli*) about 0.01, coelenterates about 0.3, and mammals about 3.0 (RENDEL, 1965).

LIFE SPAN

The natural life span of a sponge varies from less than a year for forms that live in freshwater that freezes over during the winter and for small marine forms to more than 50 years for a large, marine, keratose sponge. Some observed life spans are given in the table above (Table 4).

An attempt has been made (FINKS, 1955) to determine the longevity of a Permian sphinctozoan calcisponge (*Guadalupia*). Two generations of empty brachiopod valves grew on the sponge, one valve overgrowing the other after its death, and the sponge overgrew both. Thus the sponge was older than two successive brachiopod lifetimes. This was estimated as a minimum of two years if the brachiopods had a limited annual breeding season, although that assumption is open to question.

FUNCTIONAL MORPHOLOGY AND ADAPTATION

ROBERT M. FINKS

[Department of Geology, Queens College (CUNY)]

INTRODUCTION

To the extent that fossilizable structures of sponges can be linked to specific modes of functioning of the organism, they can be used to identify the paleoenvironment and to interpret sponge evolution in terms of the natural selection of particular functions. The principal functional requirements of sponges include 1. hydraulic efficiency of the water system; 2. maximization of surface area in contact with the ambient water to permit gas exchange; 3. mechanical support of the tissues; 4. stabilization of the organism against displacement; and 5. protection against predation.

These requirements must be reconciled with each other as well as with environmental conditions. Moreover, these needs must be reconciled with the growth (i.e., volume increase) of the individual in the course of its lifetime. Many of these adaptations are reflected, directly or indirectly, in the preservable skeletal structure of the sponge.

HYDRAULIC SYSTEM

The central activity of the sponge is related to production of its water current, which brings food and oxygen to the sponge and removes metabolic wastes. Thus hydraulic efficiency of the water system is of prime importance. Fundamental needs of the system have been discussed in the chapter on Physiology (see p. 201).

The external form of a sponge, a readily observable feature of fossils, is related to aspects of the hydraulic function. BIDDER (1923) has analyzed this. The dispersion of metabolic wastes and depleted water from the immediate vicinity of the sponge in all but the most agitated water requires the separation of effluent water from intake and maximization of the distance traveled by effluent before any of it can be recirculated into the sponge. Diameter of the returning eddy was termed the diameter of supply by BIDDER. Length of the effluent jet is increased by concentrating its flow through one or a few large oscules, which reduces the energy loss due to friction with surrounding water for a given volume of effluent. The maximum angle between the effluent stream and the inflowing streams of water was termed the angle of supply by BIDDER. For a sponge resting on the sea floor with an oscule located on the upper surface, the angle of supply is 90°. It may be increased beyond 90° by elevating the sponge on a stalk so that water may be drawn from below the sponge. In such cases the efficiency of separation of inflow from exhaust may be enough to reduce the need for an osculum, and the sponge may open out into a cup, the interior of the cup being homologous with the cloaca.

Another method of separation is development of a **flabellate** form in which one side, homologous with the cloacal lining, contains all the exhalant pores and the opposite side all the inhalant pores. For such a structure the angle of supply would be 180°. These examples of BIDDER do not exhaust the possibilities of sponge shape. Massive and encrusting forms with multiple oscules may be regarded as a series of closely spaced hydraulic systems with a common 90° angle of supply. The more spread out such a sponge is or the more individual sponges are crowded together, the more important is it for the effluent jet to be projected for a greater distance under given wave-energy conditions. One expects such sponges to have greater development of oscular chimneys, for example, in quiet water. The common cylindrical sponge shape may be regarded as approaching the advantages of a stalked form in having a wide angle of supply.

Some forms of sponges seem to depart radically from the ideal system analyzed by BIDDER. Sponges that lack large oscules entirely seem to be incapable of projecting the effluent water very far above the sponge surface and have a diameter of supply close to zero. Spheroidal sponges, on the other hand, in which the oscules are indifferently scattered over the surface and facing in all directions, seem to dispense entirely with any concern over angle of supply. The spherical Paleozoic hindiids, indeed, have no signs of attachment and may have rolled about freely on the sea floor. Some of these forms live or may have lived in agitated water that quickly dispersed the wastes.

It would be rash, therefore, to interpret paleoenergetics of ambient water from the form of fossil sponges. The crowded conditions in which sponges sometimes live, even in the deep sea below the photic zone where antipollutant ministrations of symbiotic algae cannot be invoked, suggest a certain tolerance to recirculation of used water.

In addition to separation of inflow and exhaust outside the sponge, functional efficiency of the hydraulic system requires maintenance of pressure relationships within the sponge. The inhalant canal system has lower water pressure than the exhalant system, owing to activity of the choanocytes and the directional effect imposed by the larger diameter of the apopyles. Existence of this pressure differential can be demonstrated by distension of the oscular tube under exhalant pressure and by collapse of the dermal membrane onto the subdermal spaces of the inhalant system if its support is removed (ANKEL, WINTERMANN-KILIAN, & KILIAN, 1955). Complete separation of inhalant and exhalant canal systems except through the flagellated chambers is required. Likewise, resistance to expansion on the part of the flagellated chamber walls, and to a lesser extent of the exhalant canal walls, is necessary to maintain pressure in the system, as is resistance to collapse on the part of inhalant canals.

Smaller flagellated chambers resist expansion better than larger ones (see chapter on Physiology, p. 201), which provides one of the adaptive reasons for evolution of the leuconoid canal system (for another see below). The skeleton appears to play a principal role in keeping the inhalant canals open, at least in the case of large subdermal spaces (ANKEL, WINTERMANN-KILIAN, & KILIAN, 1955). The **mesoglea** may play a role in holding smaller inhalant canals open. Pressure relations have not been studied in Hexactinellida in the laboratory, but from the relative openness of the entire canal system they are assumed to have a lower pressure differential than other classes and consequently to filter at a lower rate. Absence of mesoglea with its possible supportive function for the canal walls may be related to the presumed lower pressures.

A third aspect of hydraulic efficiency that may have affected evolution of the canal systems is reduction of internal friction through centralization. Development of relatively few large oscules and a leuconoid canal system, as pointed out by BIDDER (1923), by increasing the volume of an individual **efferent** jet reduces friction per unit volume with the external water. Even when a cloaca with a single osculum is not developed, efferent channels tend to be collected into a few exhalant openings. Development of **chones** or large subdermal spaces in the inhalant system may also be related to reduction of friction and in some sponges, at least, appears to be functionally connected with development of a centralized exhalant system. In *Ephydatia fluviatilis* LINNÉ, 1759, grown on microscope slides, a single osculum normally serves the entire set of flagellated chambers of the small sponge, and a single subdermal space communicates almost directly with prosopyles of the flagellated chambers. If the sponge is grown in silica-free water, spicules that normally support the dermal layer above the subdermal space do not develop. The subdermal space collapses, and each flagellated chamber draws its water only through the dermal pores immediately adjacent to it. The single large osculum does not develop under such circumstances, but instead many small oscules grow, each serving only a few flagellated chambers (ANKEL, WINTERMANN-

KILIAN, & KILIAN, 1955). The apparent restriction on free inflow leads to a corresponding reduction in outflow. Incidentally, this illustrates the fact that the number of oscules is determined by purely functional considerations. Therefore, the concept that each oscule represents a so-called sponge individual, as suggested by HYMAN (1940) and followed by many authors, seems to be misleading.

SURFACE AREA

An important functional consideration in sponges and one that affects gross morphology and external shape is the lack of a circulatory system for internal transport of metabolites. The ability of most individual cells to ingest and digest food renders such a system less necessary than in higher animals, and the wandering amoebocytes serve as a primitive internal transport system. Nevertheless, gaseous metabolites involved in respiration and probably soluble metabolic wastes as well must be exchanged directly with the ambient medium by each cell.

It is the canal system, primarily, that brings the ambient water within reach of diffusion of each cell. Nevertheless, as narrow canals are increased in length, energy loss due to friction with the walls is increased and inefficiency of circulation results. These relationships set a size limit on sponges with a spheroidal or massive shape unless some means can be found to bring ambient water into the interior in the form of broad spaces. The **cavaedia** of demosponges and the various **rhyses** of hexactinellids represent such tunnels through the sponge body. A possible effect of the absence of such large spaces leading to the interior of a massive sponge occurs in the Paleozoic spherical hindiids. A central hollow space filled with loose spicules may develop in larger individuals, although not in small ones, and possibly represents a moribund area due to lack of oxygen.

Another way of solving this problem lies in development of shapes that inherently have a greater surface area per unit volume. Such forms include **clathrate** cylinders, open cups, plicate cups, expanded sheets, either stalked or encrusted forms, and various branching straplike forms. All these forms represent various ways of bending or subdividing a thin sheet, in which one side of the sheet bears the inhalant openings and the other side the exhalant ones.

The form taken by a particular sponge results from reconciliation of many functional needs and is also determined by growth possibilities of its skeletal system. A very broad expanded sheet held above the sea bottom on a stalk fulfills the requirement for maximum contact of each cell with the seawater. Such a form, however, is impossible to support without a relatively rigid spicular skeleton. Given such a skeleton it would be still more disadvantageous if the species lived in rough water. Growth would be limited in even the most quiet water by the weight that the stalk could support.

Deep-sea sponges often have elaborate lateral outgrowths of the body, particularly in stalked forms. This growth form is permitted not only by the quietness of the water but also has probably a functional advantage in enabling the sponge to draw food from as large a volume of water as possible.

Some sponges have a very limited repertory of shapes. Others have a great range of potentialities. It is the latter that may be useful in interpreting paleoenvironment, for the shape represents a response to the environment rather than reflecting genetic limitations. An example of a very plastic genus, the Permian sphinctozoan calcisponge *Guadalupia,* is shown in Figure 79.

A possible example of the inherent limitations imposed on shape by growth possibilities of its skeletal system is the group of hindiids referred to earlier. Their tripodal spicules are of such form that they can be put together most efficiently as concentric shells or layers of hexagonally packed spicules. The whole sponge, thus, is commonly limited to spherical or hemispherical shapes, although other shapes have developed in some members of the family (RIGBY & WEBBY, 1988).

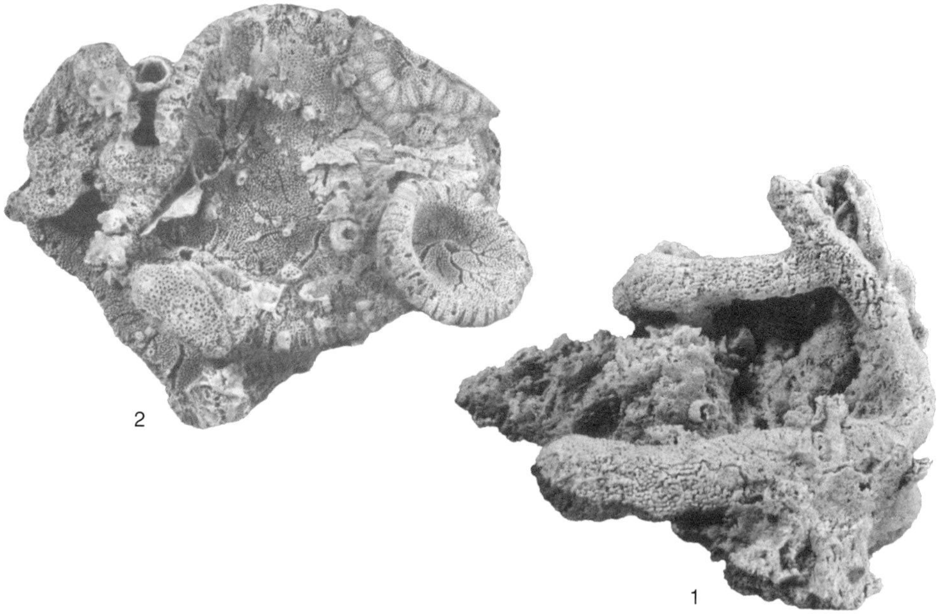

FIG. 79. *Guadalupia zitteliana* GIRTY, 1908; *1,* aberrant, irregularly branched form of species, AMNH 44652, ×1; *2,* characteristic, irregular to regular, bowl-shaped specimens demonstrating plasticity in form of species, USNM 133160, Cherry Canyon Formation, Guadalupe Mountains, Texas, USA, ×1 (new).

MECHANICAL SUPPORT

The supportive function of sponge structures is carried out by the following materials, either alone or in combination: 1. mesoglea; 2. collagen fibers; 3. spongin fibers; 4. foreign particles (sand grains, etc.); 5. spicules; 6. massive mineral deposits; and 7. chitin. Fixed structures thus maintained include the canal system, other internal spaces (parietal gaps, cavaedia, rhyses), the dermal membrane with its **porocytes,** specialized protective cortexes, stalks, branches and other lateral expansions, various forms of attachment and stabilization structures, and reproductive bodies (gemmules).

Smaller canals and flagellated chambers are supported in part by mesoglea in the Demospongea and Calcarea. In the Hexactinellida, where no mesoglea is present, the viscosity of the protoplasm apparently maintains the syncytial network of cells in position. Larger canals of the hydraulic system, as well as other spaces within the sponge, such as parietal gaps, cavaedia, and the cloaca are outlined by the spicular network or the network of spongin.

The dermal membrane may be supported above a subdermal space by spicules arranged at high angles to the membrane in the manner of tent poles. If the subdermal space is part of the inhalant system, the dermal membrane may collapse if the spicular supports are removed (ANKEL, WINTERMANN-KILIAN, & KILIAN, 1955). The dermal membrane is frequently stiffened by spicules, with some or all of their rays in the plane of the membrane. In sponges with a rigid principal skeleton, spicules of the dermal membrane are frequently not attached to the main skeleton.

External **cortexes** are often stiffened by dense concentrations of spicules and sometimes microscleres. *Geodia* has a cortex packed with sterrasters; *Cladorhiza* and *Asbestopluma* have a surface layer of chelae;

and *Desmatiderma* has one of monocrepid desmoids. In each of these cases, maintenance of characteristic form of the sponge body as a whole is due largely to the stiff outer layer, which has a supportive function as well as, probably, a protective one.

In most sponges, however, the skeleton may be regarded as performing two separate and to some extent contrasting functions: one, the provision of an open, three-dimensional scaffolding for the hydraulic system and the other, support of an outer protective and regulatory surface membrane. Because mechanical requirements are somewhat different, the two functions are often performed by different and physically separate skeletal systems, differences reflected in the forms of spicules. Those of the dermal layer have often most of their rays in one plane and are less commonly fused together. Spicules of the interior mesh tend to have several rays equally developed in different planes and are more likely to be fused together or interconnected with spongin. Monaxonic spicules may function in either situation, in the dermal layer being oriented in the plane of the membrane as stiffeners or perpendicular to it as supports and in the interior mesh being organized into strands, either singly or in bundles.

The two functions tend to converge in thin-walled asconoid sponges, such as small Calcarea, embryonic Hexactinellida, and probably some adult Paleozoic hexactinellids and heteractinids (*Eiffelia*). In such sponges a three-dimensional mesh is unnecessary to support the hydraulic system, and the spicules are planar types (triradiates, stauractines, sexiradiates).

In general, when the main interior supporting skeleton develops rigidity through interlocking or fusion of spicules, the organization is uniform throughout. In some late Paleozoic hexactinellids (*Stioderma, Docoderma*), however, the major burden of mechanical support through spicular fusion is assumed by the large, outermost spicules, probably homologous with the hypodermal spicules of the Lyssacinosa. Remaining spicules of the principal skeleton are unfused.

Gemmules of freshwater spongillids are strengthened and protected by an outer layer of specialized spicules along with chitin, apparently the only occurrence of this substance in the phylum.

PROTECTION

Protection against predation and mechanical injury due to external agents is difficult to separate from adaptations for general mechanical support, so far as morphological manifestations go. Cortical specializations mentioned above, as well as the general development of specialized dermal and hypodermal spicule types, not only support the outer layer of the sponge but also offer mechanical protection to the sponge as a whole. Nevertheless, we recognize spicule arrangements involving protrusion of sharp-pointed rays from the sponge surface, a possible specific adaptation for discouragement of predation or of settling of larvae of sessile organisms. Such arrangements may involve simple monaxons or more elaborate dermal spicules, such as pinules and scopules of some hexactinellids. Such defensive adaptations, if that is what they are, can be traced back into the early Paleozoic: pinulelike spicules occur in Mississippian dictyosponges, protruding rhabdodiactines in the Ordovician lyssacine *Cyathophycus,* hispid tufts of sharp-ended styles in the Ordovician demosponge *Saccospongia,* and oxeas in the Middle Cambrian *Hazelia.*

Some living sponges are irritating to the human skin (e.g., *Fibulia uolitaupere, Tedania ignis*), although the chemical substance responsible has not been identified. It is possible that the prevailing bright colors of many sponges (including *Tedania*) are warning colors to potential predators. Paleobiochemical methods may permit the recognition of pigment substances associated with sponge fossils, such as carotenes and malignance, although so far no such studies have been reported.

The boring habit of the clavuline demosponge *Cliona* and its relatives may be considered an adaptation for protection. Borings resembling those of *Cliona* occur in the geologic record back to the Cambrian.

An antibacterial and antifungal substance produced by *Microciona* (JAKOWSKA & NIGRELLI, 1960) may protect the sponge from infection.

ADAPTATIONS TO ROUGH OR QUIET WATER

Unfortunately, little can be said with any certainty on this subject, which is potentially very useful for environmental reconstruction. By analogy with experiments on scleractinian corals (see VAUGHAN & WELLS, 1943) one might expect encrusting forms to be characteristic of very rough water and more delicately branching forms of quiet water. To be sure, sponges growing in the surf zone are frequently encrusting forms, but the encrusting habit is found in waters below 366 m (200 fathoms), and one would be rash to use such forms as indicators of rough-water environments.

Nevertheless BURTON (1928) pointed out that in the species *Halichondria panicea,* forms from the surf zone are encrusting or irregular in shape, while those from deeper and quieter water are more symmetrical, being either cylindrical or spherical. He also noted that deep-water sponges tend toward greater symmetry as a general rule.

Another way in which a sponge may become modified in rough water is to allow for freer movement of the water through and around it. For example, BURTON (1928) cited WHITELEGGE's observations (1901) on Australian *Pachychalina communis,* which in quiet but shallow water has a flabellate or lamellose form, whereas in rougher water it assumes a digitate form.

In general, one is struck by the prevalence of forms with long, delicate branches (*Asbestopluma, Cladorhiza, Chondrocladia, Desmatiderma*) or long stalks (*Hyalonema, Stylocordyla*) among sponges from bathyal or abyssal depths. It is tempting to see in such shapes, when occurring among fossils, an indication of quiet, although not necessarily deep water. *A priori* one might expect that forms with slender bases and top-heavy shapes, such as the mushroomlike *Coeloptychium,* would indicate quiet water, as would such delicate, thin-walled, nonrigid sponges as many Paleozoic reticulosids.

Strong forms on the other hand, although capable of surviving in rough water, need not be confined to it. The Permian *Stioderma,* which has a rather rigid and heavy skeleton, is almost the only Permian lyssacine to occur in shell-bank deposits, yet it is also found in adjacent deep-basin deposits (FINKS, 1960). Those, however, may be transported occurrences.

As a preliminary test of the relationship between sponge form and water agitation, FINKS plotted the bathymetric frequency distribution of sponges by shape (Fig. 80), as reported in a broad faunal study (BURTON, 1956). One may assume that mechanical considerations would be largely operative in matters of sponge shape and that, therefore, of all the parameters that vary with depth, that of water agitation will be the only one to affect sponge shape strongly since, in a general way water agitation decreases with increasing depth.

It is apparent that there is some difference in the distribution of the different shapes. Two types, *cylindrical* and *spheroidal,* appear to be distinctly deeper-water forms than the others, not occurring shallower than 55 m (30 fathoms). The spheroidal forms have a distinct peak of abundance around 91 m (50 fathoms). The spheroidal forms have a distinct peak of abundance around 91 m (50 fathoms), and cylindrical forms are less clearly concentrated at a particular depth, but share the same overall range of 55 to 165 m (30–90 fathoms). Massive forms, on the other hand, appear to be limited to shallower water, do not occur below 91 m (50 fathoms), have a distinct peak at 55 m (30 fathoms) and range up into depths of only 27 m (15 fathoms), the shallowest depth collected. Other form categories occurring at shallow

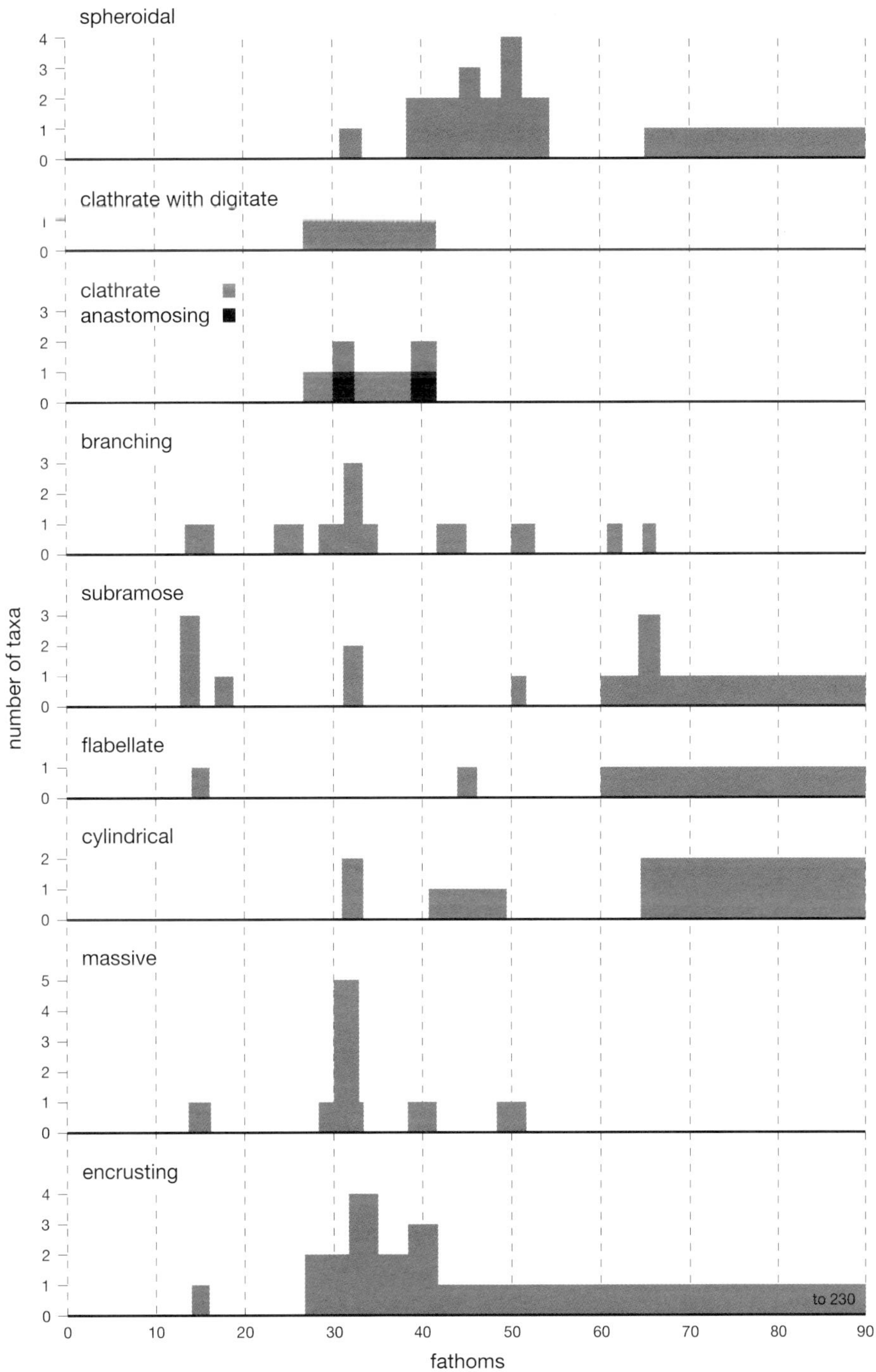

FIG. 80. Bathymetric distribution of sponges by shape, as reported in broad faunal study of sponges of West Africa by Burton (1956) (new).

depths have a wider range: branching forms and flabellate forms both range down to 165 m (90 fathoms), the former with a distinct peak at 55 m (30 fathoms). Encrusting forms have the widest range of all, extending from shallow water down to the deepest sample, from 428 m (234 fathoms). Clathrate-anastomosing forms have the most restricted range, occurring between about 55 to 73 m (30–40 fathoms).

The results tend to confirm the hypothesis that symmetrical shapes are characteristic of quiet water and do not develop in agitated water. The results also indicate that both branching and encrusting forms are indifferent to the degree of water agitation.

ADAPTATIONS TO SUSPENDED SEDIMENT

Another potentially useful indicator of paleoenvironmental conditions is morphological adaptations of sponges to excessive suspended mineral matter in the water. That such suspended matter is likely to be harmful to sponges by way of clogging their pores seems to be a safe assumption, yet observations by DE LAUBENFELS (1953a) and WIEDENMAYER (1977a) that some sponges appear to survive while growing partially buried in the mud, make even this somewhat doubtful. Nevertheless, a heavy accumulation of sediment on the upper surface of a sponge seems to be harmful, and sponges with broad horizontal expansions, such as open cups or mushroom-shaped forms, probably did not live in turbid waters.

Long stalks of deep-sea forms may represent an adaptation to elevating the sponge above an oozy bottom, from which quantities of fine, suspended matter are likely to be raised by passing vagrant benthos. Very long-stalked fossil forms may have lived under similar conditions, although one must be careful to discriminate between a root tuft buried in mud and a stalk raised above it.

STABILIZATION

One of the most obvious morphologic adaptations of most sponges is the provision for maintenance of position on the substrate. This includes keeping the sponge from becoming buried in the substrate, if the substrate happens to be soft sediment, as well as maintaining exhalant openings in a constant position, generally facing away from the substrate.

Stabilization of position may be achieved in a variety of ways. One of the simplest is development of a broad base, without other means of attachment. Such a shape not only resists overturning but also reduces the possibility of sinking into the sediment by distributing the weight over a wide area. Examples include the Permian hexactinellid *Pileolites,* which grew in flat-bottomed, cake-shaped forms, and the Ordovician hexactinellid *Brachiospongia,* whose flat base is expanded radially in lobate extensions. The lobes of *Brachiospongia* extend downward as well as outward (Fig. 81), apparently raising the main body of the sponge above the sea floor. In both these forms oscules are located opposite the base.

Encrusting forms may be considered the ultimate development of this type of stabilization, in which lateral extension is many times greater than vertical. The Cambrian hexactinellid *Multivasculatus* seems to be a form with more or less indefinite and irregular lateral extension over the sediment, with evenly spaced, low cups developed on the upper surface. More typical encrusting forms, such as many living monaxonid demosponges, are extremely thin sheets that grow on solid objects.

Boring sponges that excavate galleries in shells, corals, and limestone undoubtedly have the most intimate and fixed contact with the substrate. Many, perhaps all, species of *Cliona* and other boring sponges grow ultimately above the riddled substrate, the embedded parts of the sponge serving as a means of attachment.

Physical adherence of sponge skeletal material to the substrate is a means of stabilization in many forms. Encrusting shapes are often so attached, but so are more narrowly based forms, such as those with stalked or

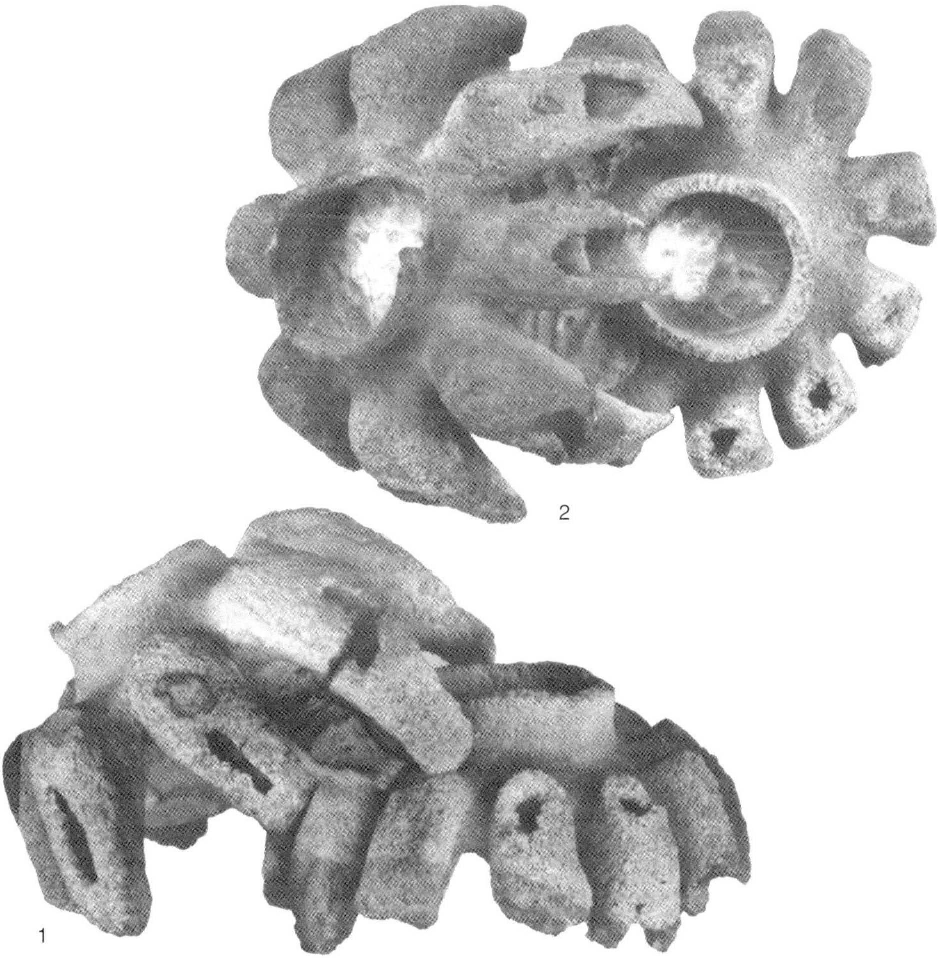

FIG. 81. Two associated specimens of *Brachiospongia digitata* (OWEN) showing lobate margins of sponge that apparently raised sponge above sea floor; *1,* from the side, and *2,* from above, FMNH 10851, Trenton, Frankfort, Kentucky, USA, ×0.5 (new).

obconical shapes. This type of attachment is common in pharetronid calcisponges, where massive deposits of calcium carbonate form an important part of the skeleton (*Stellispongia*). It is also present in many siliceous sponges, in which secondary deposits of silica in the outermost layer of the skeleton make direct contact with the substrate. Such attachment requires a solid substrate, most often the shells of other organisms.

In narrowly based, stalked, obconical, or cylindrical sponges that have direct contact with the substrate, a basal encrusting expansion (the hexactinellid *Myliusia,* the heteractinid *Wewokella,* or the pharetrone *Eusiphonella*) or branching, rootlike extensions of the sponge body (the lithistid *Siphonia,* or the hexactinellids, *Coscinopora, Camerospongia,* and *Verruculina*) may be developed. Such bases both resist overturning and distribute the weight so as to minimize sinking into the bottom. In this way stability may be achieved while enjoying the advantages of a wider angle of supply or a greater elevation above a muddy bottom than is possible with a broad-based body resting directly on the substrate. A related adaptation is seen in the hexactinellid *Becksia,* in which many slender, stiltlike processes raise the main body of the sponge above the substrate.

Many siliceous sponges have developed a root tuft of long spicules. These tufts commonly both anchor the sponge in a soft bottom and raise it above the sediment. The tufts take many forms. In the spheroidal demosponge *Radiella sol* an equatorial fringe prevents overturning. In the related *Radiella tissieri* tuft spicules are distributed over the entire lower hemisphere and fix the sponge to the bottom (VACELET, 1961). In *Tetilla grandis* the root spicules form a cushionlike mat beneath the sponge and appear to function both as a fixing device and as ballast to prevent overturning because it is probably more dense than the main body of the sponge. In *Thenea wyvilli* a similar cushionlike mat is surmounted by several stalklike tufts that hold the sponge body well above the sea floor (BURTON, 1928); here fixing, ballasting, and supporting functions seem to be combined. In the long, single root tufts of the hexactinellid *Hyalonema,* the sponge is elevated well above the sea floor, as demonstrated both by bottom photographs and by the occurrence of symbiotic anemones covering much of its length. Here the uplifting function is emphasized at some expense to stability, a situation probably permitted by quietness of the deep sea environment where these sponges live.

In Paleozoic demosponges of the order Orchocladina, the upper hemispherical body of the sponge often overhangs a conical basal portion that is covered with a dense and nearly imperforate surface layer. That this basal portion may have been embedded in sediment and to some extent served as a stabilizing device is suggested by its reduction or absence where the sponge has grown upon and surrounded a shell or crinoid stem that presumably supported it above the bottom (FINKS, 1960).

Some sponges appear actually to grow within sediment to some extent. WIEDENMAYER (1977a) cited forms that are partly buried in mud, with circulation carried on in those parts that protrude above the sediment. A possible example from the Permian is presented by a species of the pharetronid *Virgola* recovered through acid-etching of a block of sediment (Fig. 82). The very irregular sponge has incorporated quantities of shells in its lower half, whose size and spacing are identical to shells in the immediately adjacent sediment. Such relationships indicate that the sponge ramified through the loose shell hash on the sea floor, at least partly below the sediment-water interface. A possible ballasting function performed by incorporation of a mass of mud in the base of the demosponge *Radiella tissieri* has been described by VACELET (1961).

Last we must call attention to those sponges that seem to dispense entirely with the maintenance of stability. BURTON (1932) has described several demosponges (species of *Tedania, Thenea, Cinachyra, Polymastia, Monosyringa, Disyringa*) that have no signs of attachment and which he believed may have hovered above the sea floor by virtue of a density close to that of sea water, being gently wafted about by currents. He noted that specimens of *Polymastia invaginata* incorporate a small pebble or shell in the base opposite the osculum and suggested that this pebble functioned as ballast, sufficient to keep the osculum directed upward but not heavy enough to anchor the sponge in one place. BURTON mentioned that pores on the subspherical *Cinachyra antarctica* are uniformly developed over the entire sponge surface, which one would not expect if part of it continually rested on the sea floor. He suggested that it rolled about freely. The same arguments surely apply to the spherical Paleozoic lithistids such as *Hindia, Scheiia, Caryospongia,* and *Carpospongia,* which also have pores equally developed on all sides and have no sign of attachment. These were certainly too dense to have floated, with their closely packed net of siliceous spicules, but they could well have rolled on the bottom. The similarly spheroidal *Astylospongia* raises

FIG. 82. Side view of *Cooperaria getawayensis* FINKS and associated, skeletal debris exposed by etching, that suggests sponge may have lived partly buried in substrate and ramified through adjacent debris during growth, AMNH 44654, Getaway Limestone Member, Cherry Canyon Formation, Guadalupe Mountains, Texas, USA, ×1 (new).

an unsolved problem. It has no sign of attachment but it has a well-differentiated area of exhalant pores at one end, such that it could not have been a matter of functional indifference if the sponge was turned upside down. One cannot easily invoke differences in ambient wave energy, for *Astylospongia* appears to occur together with *Caryospongia* and *Carpospongia.*

MODES OF GROWTH

Sponges must reconcile the needs for a supportive and protective skeletal system with the needs of growth. Where spicules are not attached to one another they can and do change their relative positions and thereby permit internal expansion and rearrangement in connection with growth. In many hexactinellids, especially Paleozoic forms, spicules appear to pull apart from one another and increase in size as the sponge body expands. New spicules of smaller size are intercalated between them. In this manner, size may increase proportionally in all parts, including interspicular spaces, parietal gaps, and canals.

Where spicules are held together by spongin, as in most demosponges, the resulting net has a fibrous structure, with only limited capabilities for internal expansion. Enlargement of the sponge body requires lengthening of preexisting fibers, or the laying down of new fibers more or less parallel to the old. Such a mode of growth may lead to a radial structure: either symmetrical about a central point and producing a spherical or discoidal shape or radiating asymmetrically from an eccentric point or points and producing flabellate, cylindrical, or branching cylindrical shapes, with or without a central cloaca.

One of the persistent trends in all groups of sponges has been development of rigidly fused spicular skeletons. For obvious reasons, most fossil sponges are in this group. In some forms the state of rigidity is attained only after the sponge reaches advanced size, as in the Paleozoic hexactinellids *Docoderma* and *Stioderma,* in which only the dermal spicules fused (FINKS, 1960). These forms appear to have grown by expansion of the entire body and intercalation of new spicules that were continually enlarged. Fusion of the dermal layer stopped the growth process.

Among hexactinosan and lychniscosan hexactinellids the skeleton is rigid from the beginning. Growth, thus, must proceed peripherally without expansion of the already formed parts. This may explain the frequent occurrence of tubular and sheetlike structures in these groups, often forming a body of considerable internal complexity.

Lithistid demosponges, likewise, are rigid from the outset. They seem to have grown in three ways. Some have tended to grow by adding shells parallel to the surface to produce a massive sponge (Hindiidae, Astylospongiidae, Chiastoclonellidae, Anthracosyconidae). Others have tended to produce radial rows of spicules that were added to at the upper or outer end to form more or less expanded cups (Anthaspidellidae). A third group produced fiber tracts of more or less irregularity and anastomosis and grew peripherally in a variety of directions, permitting greater freedom of shape (Dystactospongiidae, Rhizomorina, most other Mesozoic and Cenozoic lithistids).

The pharetronid Calcarea developed more massive mineral deposits that sometimes exceed the spicules in volume if they do not completely substitute for them. Some forms (*Petrobiona*) have such a massive skeleton that the flesh is limited to a thin surface layer (VACELET, 1965). Most of the pharetronids have an irregular, fibrous skeleton that grows peripherally to form rather massive sponges.

VARIABILITY AND VARIATION

Robert M. Finks

[Department of Geology, Queen's College (CUNY)]

INTRODUCTION

The phenomenon of variability is a central problem to be dealt with in practical recognition of species. In sponges, both living and fossil, limits of individual variation within a species are not well known. It is apparent from studies of both living and fossil collections, however, that sponges have a wider variability in external form among members of a local population of a species than more complex animals. In this they resemble plants; and, like plants, the repertory of shapes shown by a species, although relatively broad, is not unlimited and can be very useful for species recognition (Burton, 1932, p. 376).

Sponges are peculiar in that most morphological features other than external form relate to small repetitive parts, such as spicules, pores, and canals. This introduces another aspect of variability, namely, variation within an individual organism. As with leaves of a tree, this variability is not unlimited and may characterize species [analogy courtesy of Dr. J. W. Wells of Cornell University, who introduced it during a discussion of coloniality in corals]. In sponges, however, this sort of variability has not been extensively investigated.

Study of variation is useful in another context in addition to discrimination of species. It can be used in reconstruction of ecological and environmental conditions, when the environmental factors that cause particular variations are known. Not only gross form but also the shape and dimensions of pores, pore clusters, and canals may be determined by local environmental conditions. Such local factors may operate not only between individuals but also within a single individual. Indeed, knowledge of intra-individual effects may be easier to come by and may aid us in interpreting variation between individuals. There are some kinds of variability, however, particularly the size and form of spicules, that cannot be related always to external conditions or functional needs.

Temporal variations within individuals may reflect seasonal changes in the environment; or cyclic changes in physiology, such as breeding periods; or unidirectional ontogenetic change. Among sponges with rigid skeletons, such as are most frequently found as fossils, these temporal changes may be preserved in the skeleton as intra-individual variation, but the same temporal changes will result in interindividual variation among individuals that die young, or for sponges that can reorganize their skeletons through resorption, discarding, and regrowth of skeletal elements.

SKELETAL TREATMENT

Statistical analysis has two important uses in taxonomic studies; one is descriptive

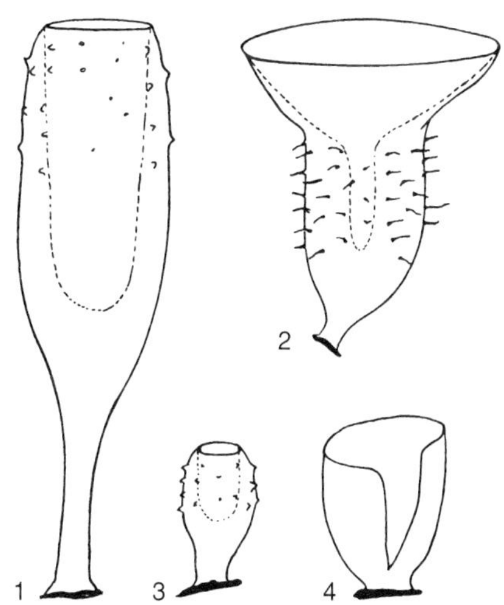

Fig. 83. Series showing external form in smaller specimens of *Haliclona bilamellata* Burton; *1–4,* diagram; *dotted lines* indicate shape and position of cloaca (Burton, 1932; courtesy of Cambridge University Press).

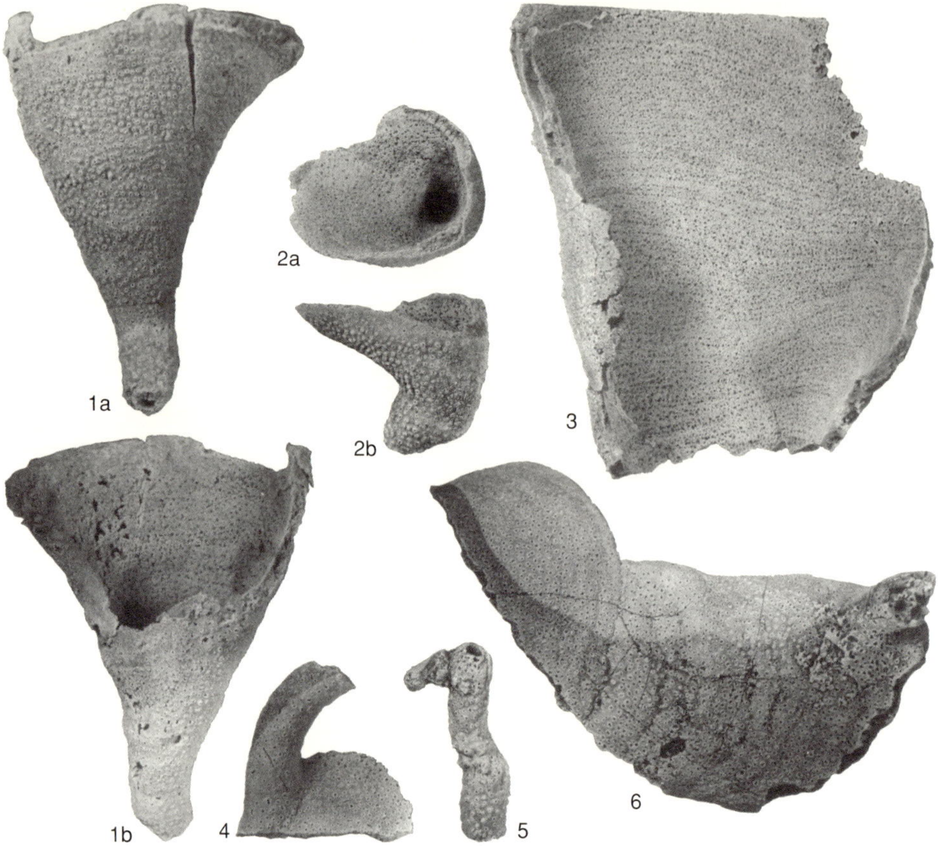

FIG. 84. Individual variation in external form in *Coelocladia spinosa* GIRTY among specimens from a single locality, Pennsylvanian Rock Hill Limestone, Bridgeport, Texas, USA; *1–2,* two views of single, funnel-shaped specimens with basal stalks and upper, thin walls; *3,* inner or gastral view of large, funnel-shaped fragment showing aligned pores; *4,* tubular branch growing from relatively flat, outer surface of wall; *5,* cylindrical specimen with osculum at top, with attached fragment of another individual, both of which are near basal parts of species; *6,* part of large frond with unbroken, upper surface, but broken left and right ends, sponge grew from left to right, ×0.7 (Finks, 1960).

characterization of a given population; the other is assessment of the probability that observed differences arise from sampling two different populations rather than a single one. The usual parametric methods of characterization of populations by the mean, the standard deviation, and so forth and the usual tests of significance of difference, such as chi-square of Student's *t,* involve the assumption that the frequencies in the population follow the normal or Gaussian distribution. FRY (1970) pointed out that characters most often measured on sponges, such as spicule sizes, are not normally distributed. Thus, nonparametric methods are to be preferred. FRY recommended the use of simple histograms for descriptions of size-frequency distributions and the comparison of these through nonparametric tests. He demonstrated the use of one such method, the Kolmogorov-Smirnov test, by analysis of the generated probability values to determine the degree of similarity between sponges from four different localities.

EXTERNAL FORM

It will suffice to point out some examples of the range of external shape to be found within a single species. The living

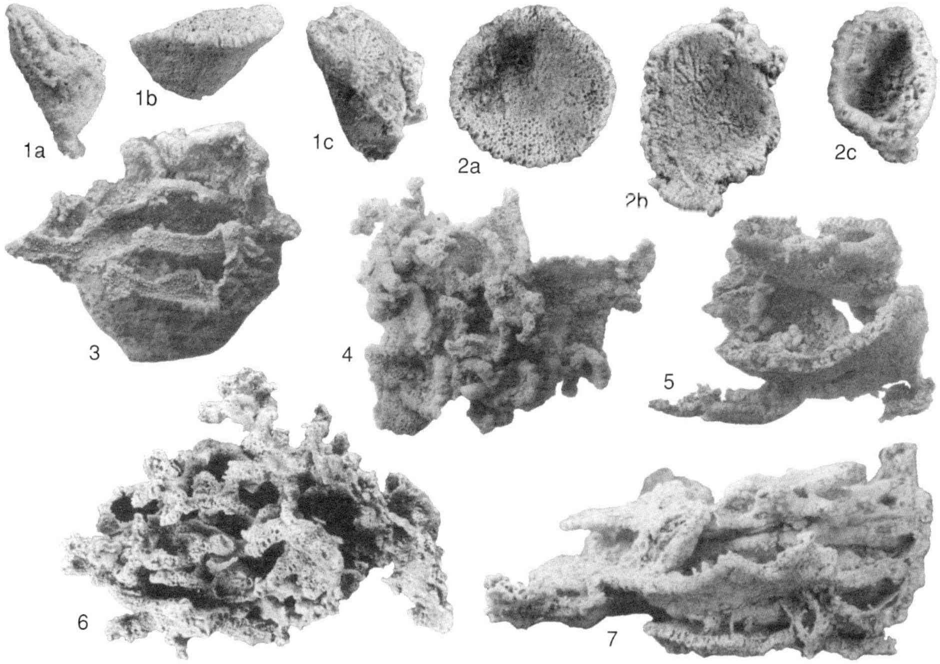

FIG. 85. Individual variation in external form in species of *Guadalupia* GIRTY from single locality, basal bioherms of Permian Road Canyon Formation, near Old Word Ranch House, Glass Mountains, Texas, USNM 703a; *1,* approximately ×0.67; *2–5,* ×1; *6,* ×0.77; *7,* approximately ×0.67 (new).

demosponge *Haliclona bilamellata* BURTON, 1932, is a stalked cylindrical form with a deep cloaca and external protuberances. Besides variation in proportionate length of the stalk, the upper end may be expanded to form a broad funnel or the sponge opened on one side in a nearly flabellate shape; likewise, the external protuberances may be reduced or absent (BURTON, 1932, p. 268, fig. 6; Fig. 83 herein).

Similar variability was reported by FINKS (1960) for the Pennsylvanian *Coelocladia*

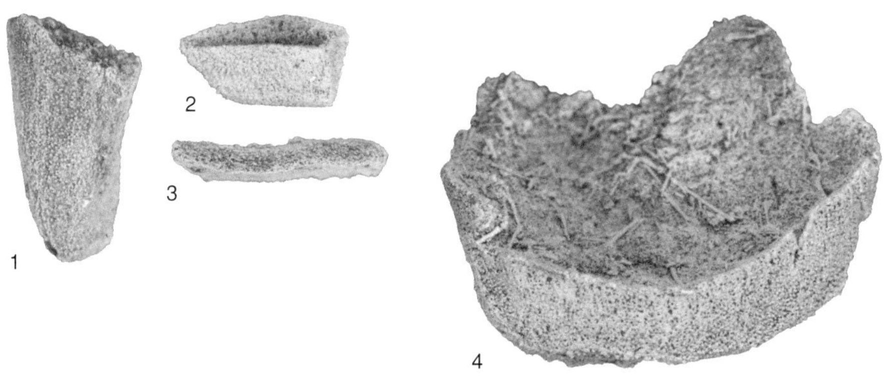

FIG. 86. Profile views of four individuals of *Pileolites baccatus* FINKS, 1960, from single block of limestone, showing individual variation; Permian Skinner Ranch Formation, Glass Mountains, Texas, USNM 707ha, ×2 (Finks, 1960).

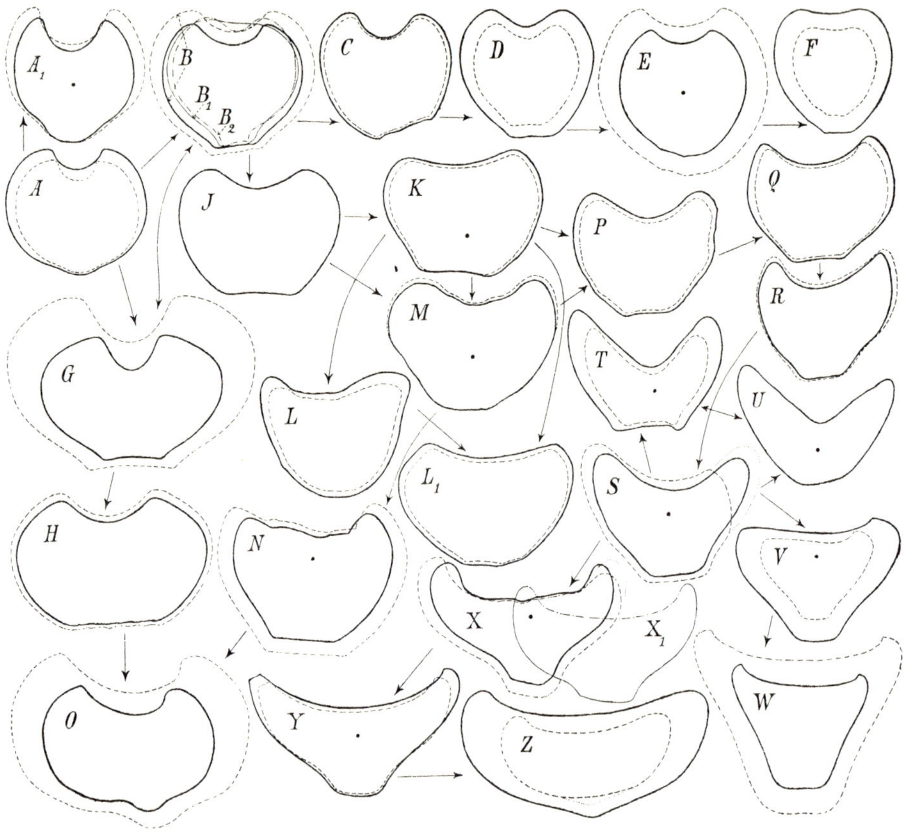

FIG. 87. Profiles of various growth forms of *Palaeomanon cratera* (ROEMER, 1848), middle Silurian (Niagaran), western Tennessee; *letters* indicate different subspecies; *arrows* indicate direction of change in morphology; *solid lines* indicate reduced size for comparison purposes; *dotted lines* indicate actual size (Rauff, 1893).

spinosa GIRTY, 1908, which varied at a single locality from simple small cylinders, through tall, narrow funnels, to broad, highly asymmetrical funnels that may be nearly laminar or tongue shaped (FINKS, 1960, pl. 5–6; Fig. 84 herein). A species of the Permian calcisponge *Guadalupia* GIRTY also varies at a single locality from circular open cups, through multitiered asymmetric cups, to similar forms with long, subparallel, tongue-like extensions on one side (Fig. 85). The Permian hexactinellid *Pileolites baccatus* FINKS, 1960, among specimens from a single block of limestone, varied from thimble-shaped, through wedge-shaped, to pancake-shaped forms (FINKS, 1960, pl. 50; Fig. 86 herein). RAUFF (1894, fig. 64, pl. 13,*1–5;* Fig. 87–88 herein) has recorded the range of form of the lithistid *Palaeomanon cratera* (ROEMER, 1848) from the middle Silurian of western Tennessee. This species has a limited range of shapes, but there is considerable variation in proportion and in depths of the bowl-like exhalant surface. It will be apparent from these examples, which from their continuous intergradation at a single locality appear to be members of a single species, that many separate species and even genera reported in the literature may be merely individual variants. Nevertheless, each of the cited species has a limited repertory of form, and it should be noted that some genera, such as the toadstool-like Cretaceous hexactinellid *Coeloptychium,* are nearly invariant in external form.

The ecologic significance of external form is briefly discussed in the chapter on Ecology and Paleoecology (p. 243) but the

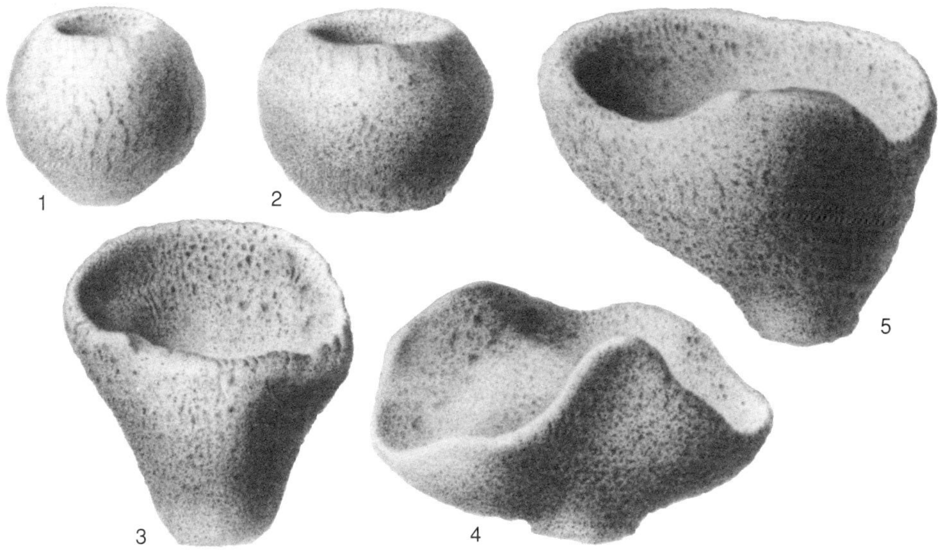

FIG. 88. *Palaeomanon cratera* (ROEMER, 1848) showing variant growth forms, Niagaran, Silurian, Decatur County, Tennessee, USA, ×1 (Rauff, 1893).

ecologic significance of the above-mentioned examples can be conjectured only. It seems reasonable to suppose, however, that at least some and perhaps all are responses to environmental circumstances rather than being reflections of genetic differences. Next to nothing is known of the genetics of living sponges and to what extent individual

FIG. 89. Shape variation (ontogenetic) with size in *Microstaura doliolum* FINKS, 1960, from single locality in Permian Road Canyon Formation, Glass Mountains, Texas, USNM 703c, ×2 (Finks, 1960).

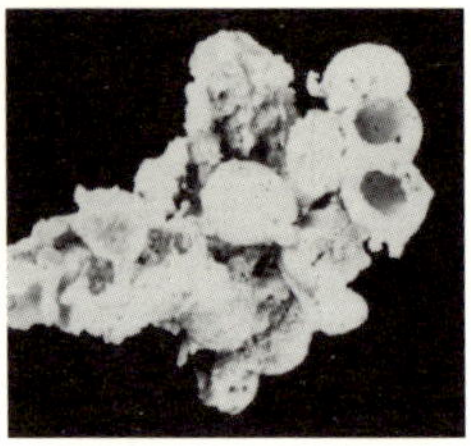

FIG. 90. *Girtyocoelia beedei* (GIRTY, 1908) showing absence of cloaca in juvenile stages (sectioned chambers at right); sponge grew on front of *Guadalupia* GIRTY, Permian Cathedral Mountain Formation, Glass Mountains, Texas, AMNH 504, ×2 (new).

variation is determined by it. One can assert on *a priori* grounds that major differences between species are genetically determined.

FRY (1970) questioned the taxonomic value of external form on the grounds that form and functioning of the whole sponge are determined by interactions at the cellular level and that the sponge should be treated as a population of cells and cell products analogous to a mixed population of whole organisms. This seems to be an extreme view. To reject one whole class of information on the grounds of presumed incompatibility with another class of information is to abandon the principle of multiple working hypotheses. *A priori* considerations aside, most paleontological classifications will have to rely heavily on external form and intermediate-level structures, such as pores, canals, and skeletal organization, because statistically useful populations of spicules are not always available, and of cells not at all. Furthermore, inasmuch as a natural classification is a statement about phylogeny, the more lines of evidence that converge to establish it, the more securely founded it is.

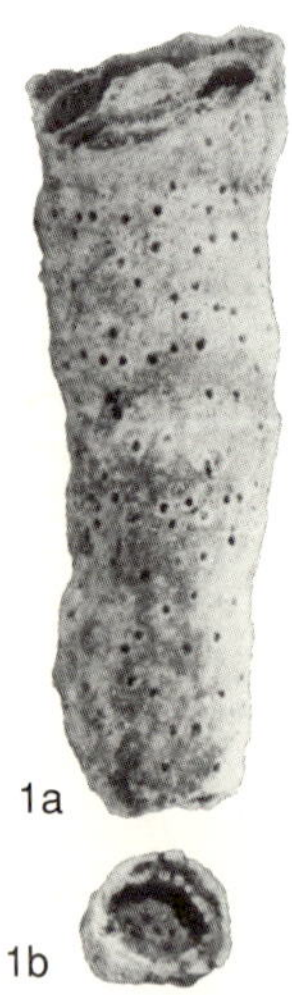

FIG. 91. *Stylopegma* sp. showing absence of cloaca in early stage in separated lower end of specimen (*1a*), as viewed in cross section (*1b*), Permian, Getaway Limestone, Guadalupe Mountains, Texas, AMNH 512, ×2 (new).

TEMPORAL VARIATION

Many individual sponges undergo considerable changes in shape during their lifetimes. Some of these changes may be ontogenetic, that is, a regular sequence characteristic of the life history of the species. Other sponges, however, particularly encrusting forms, appear to undergo constant and often drastic changes in shape of an irregular and unpredictable sort.

Ontogenetic variability may involve changes in proportion. The Permian hexactinellid *Microstaura doliolum* FINKS, 1960, occurs in a range of sizes at a single locality. The very small ones are nearly spherical, while the larger and presumably older are barrel shaped and subprismatic, and the larger ones are more elongate (FINKS, 1960, pl. 34; Fig. 89 herein). Some Permian cateniform Sphinctozoa, such as *Girtyocoelia beedei* (GIRTY, 1908) and a species of *Stylopegma* KING, 1943, lack in the earliest stages the central cloaca characteristic of the genus (Fig. 90–91).

A more irregular type of temporal change has been described by BURTON (1949, fig. 12–13) and SARÀ (1970, fig. 3–4; Fig. 92–93 herein), based on observations of the same sponges over periods of a year or more. Outlines of these encrusting sponges changed, partly by growth, partly by coalescence of neighboring individuals, partly by

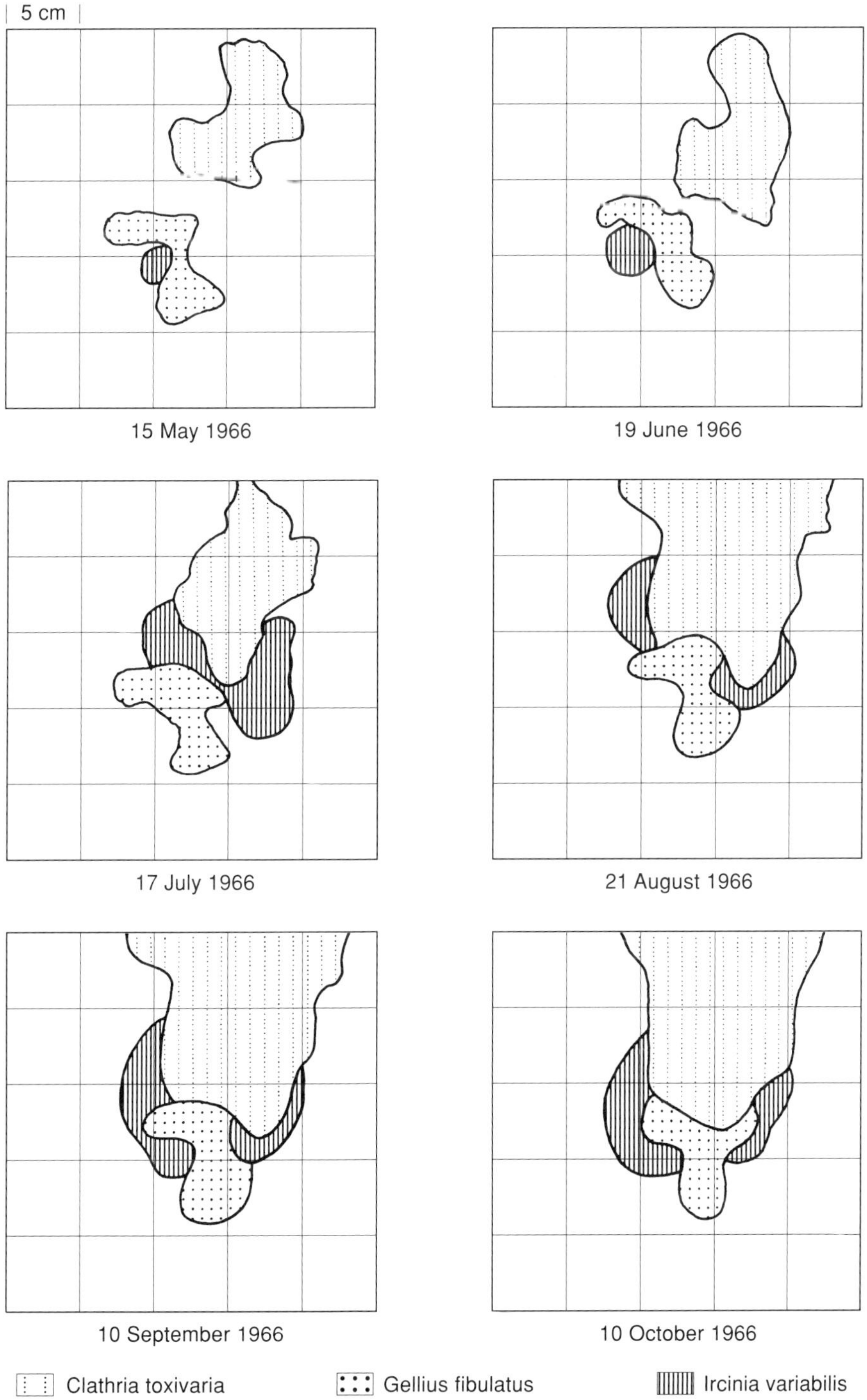

FIG. 92. Variations with time in part of sponge population in Grotta della Regina, near Monopoli, in southern Italy (May–October, 1966) (Sarà, 1970; courtesy of Zoological Society of London).

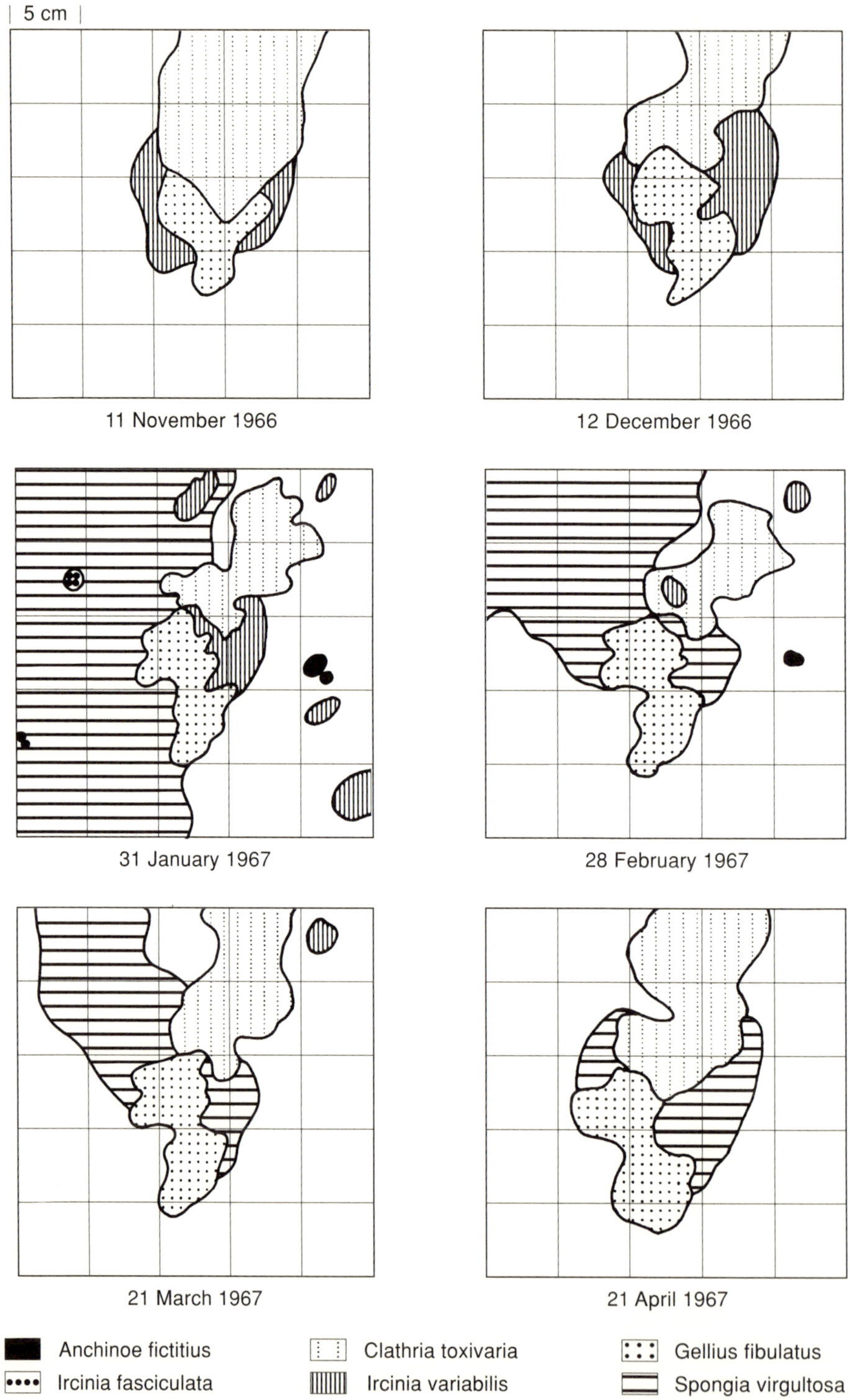

FIG. 93. Variations with time in part of sponge population in Grotta della Regina, near Monopoli in southern Italy (November, 1966–April, 1967) (Sarà, 1970; courtesy of Zoological Society of London).

the presumed dying or disintegration of tissues, or, as BURTON suggested (1949, p. 909), by slow movement of the tissues. BOROJEVIC (oral commununication, 1968) has observed in the laboratory that small, starved sponges will abandon their skeletons and migrate slowly over the substrate; consequently, such motion seems to be possible.

BURTON (1949) made an effort to avoid sites where sponges were altered in form by predation. Predation and mechanical injury will affect obviously the form of a sponge, and in species with a characteristic shape, such teratological changes can be recognized, as in a specimen of the Permian *Stylopegma* KING, 1943 (Fig. 94), which has been injured and healed.

FIG. 94. Healed injuries in *Stylopegma* sp.; *1a,* hole at base of left specimen is an injury that apparently led to constriction of part of specimen above it; *1b,* viewed from above, showing flattening of normally circular outline, at left, where part of side was removed, perhaps the bite of a predator, and then healed over; Permian Getaway Limestone, Guadalupe Mountains, Texas, AMNH 512, ×1 (new).

Seasonal changes have been reported by SIRIBELLI (1961) in species of the demosponge *Axinella.* In *A. verrucosa* (ESPER, 1794) specimens collected in the fall and winter are thinly branched with a slightly hispid surface, while those collected in the summer have progressively thicker branches with a rugose surface and have **anastomoses** between neighboring branches. *A. damicornis* (ESPER, 1794) is flabellate and anastomosing all year round, but the branches become very thin in the fall and winter and thicken in the spring and summer. Internal arrangements of spicules differs between the two species and is apparently constant.

SPICULES

The form and dimensions of spicules and the relative frequency of various types have long been used in sponge taxonomy, apparently not always with proper appreciation of their variability. In a detailed study of the demosponge *Ophlitaspongia seriata* (GRANT, 1826) from four localities, two in Wales and two in northern France (Brittany), FRY (1970) demonstrated differences between the populations and also between oscular and interoscular parts of the sponge, both in relative frequencies of spicules types (tylostyles, subtylostyles, and toxas) and in the size-frequency distribution within each type (FRY, 1970, p. 156, fig. 12 and table IX; Fig. 95, Table 5). Differences between the two Welsh localities, on the one hand, and the two French localities, on the other hand, are readily apparent in both figures. With the relative frequencies of tylostyles, however, this distinction is more marked in interoscular than in oscular samples and among subtylostyles more marked in oscular than in interoscular ones. When oscular and interoscular frequencies are combined (Table 5), discrepancies are compensated largely. Size-frequencies (Fig. 95) (oscular) of toxas are different between the two Welsh populations as well as between the French ones. It is also apparent from the histograms (Fig. 95) that the size-frequency distributions are highly skewed and in some instances bimodal or polymodal.

Polymodality in size-frequency distribution was demonstrated for amphidisc microscleres in several species of the hexactinellid *Hyalonema* by LENDENFELD (1915). In most of his species there were two sizes of amphidiscs, each often separable into two subgroups (LENDENFELD, 1915, fig. 9, 13; Fig. 96–97 herein). Where spicules from more than one individual were plotted separately, position of the modes is more or less the same (LENDENFELD, 1915, fig. 19;

TABLE 5. Mean percentage frequencies of spicules; *O,* oscular sample; *IO,* interoscular sample; *E,* edge sample (Fry, 1970; courtesy of Zoological Society of London).

	O	IO	O+IO	E
subtylostyles				
Church Island	65.88%	69.63%	67.75%	77.33%
Bodorgan	72.80	62.01	67.40	65.76
Menenett	55.47	63.36	59.41	63.48
Le Loup	58.85	61.00	59.92	57.04
tylostyles				
Church Island	6.73%	7.22%	6.98%	4.40%
Bodorgan	3.95	8.39	6.18	9.62
Menenett	4.22	3.20	3.72	3.73
Le Loup	2.37	2.91	2.65	2.26
toxa				
Church Island	27.39%	23.15%	2.27%	18.27%
Bodorgan	23.25	29.60	26.42	24.61
Menenett	40.31	33.44	36.87	32.79
Le Loup	38.78	36.09	37.43	40.70

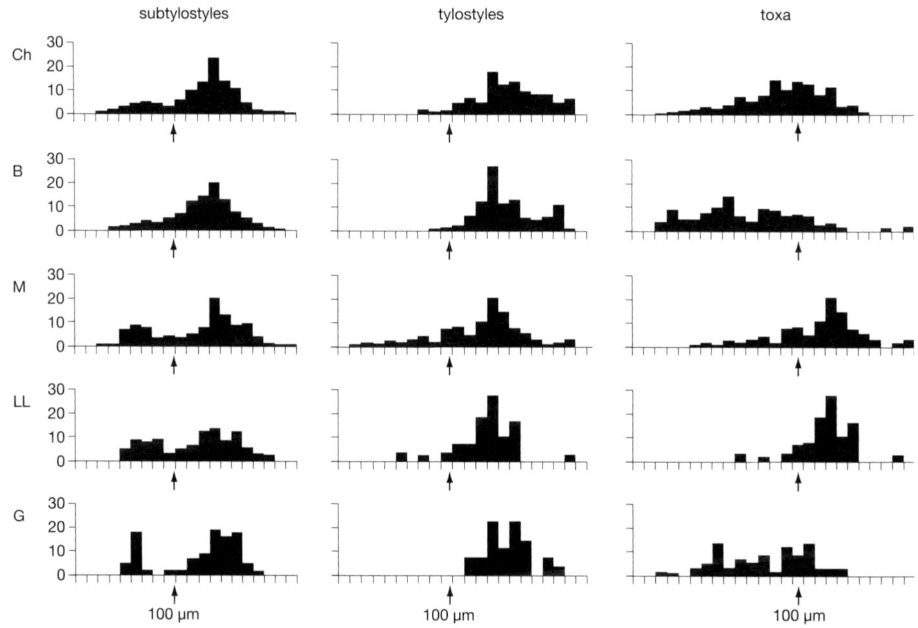

FIG. 95. Percentage size frequency distributions of three spicule types in oscular samples from four populations of *Ophlitaspongia seriata; Ch,* Church Island; *B,* Bodorgan; *M,* Menenett; *LL,* Le Loup; *G,* data for slide preparation that is probably from holotype of *Ophlitaspongia papilla* BOWERBANK, 1866, collected from Guernsey; with exception of G data, histograms based on data from two oscular samples from five specimens from each population; size represents greatest chord length and is shown in class intervals of 6.75 μm (Fry, 1970).

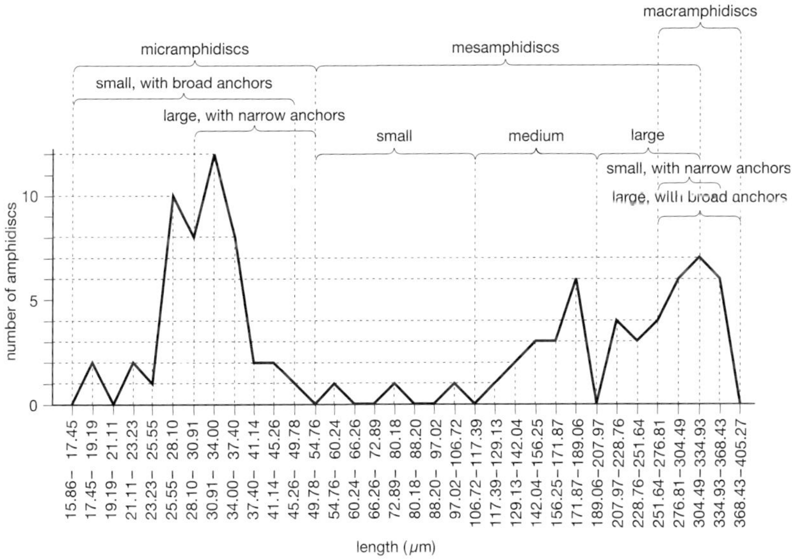

FIG. 96. Length-frequency curve of amphidiscs in *Hyalonema (Hyalonema) placuna* Form B (Lendenfeld, 1915).

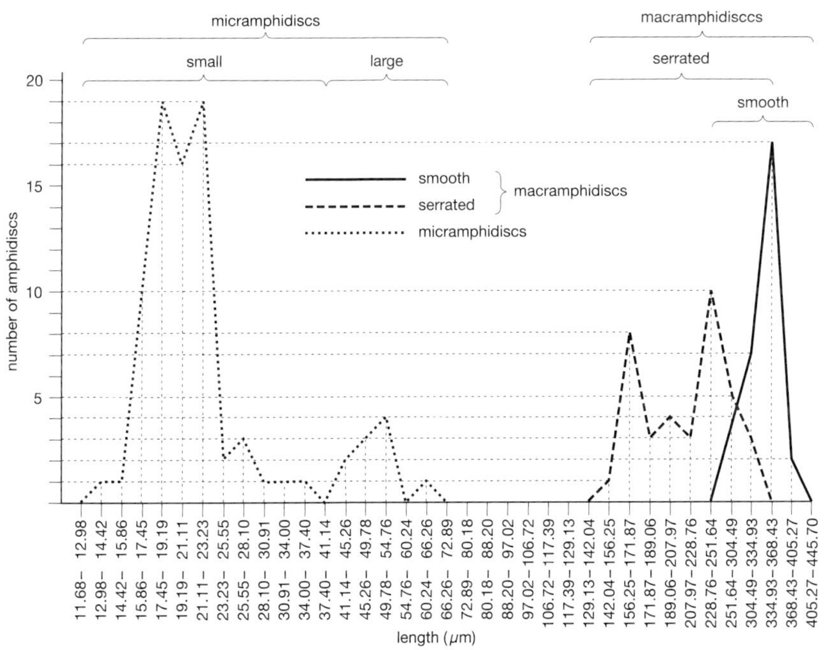

FIG. 97. Length-frequency curve of amphidiscs in *Hyalonema (Prionema) crassum* (Lendenfeld, 1915).

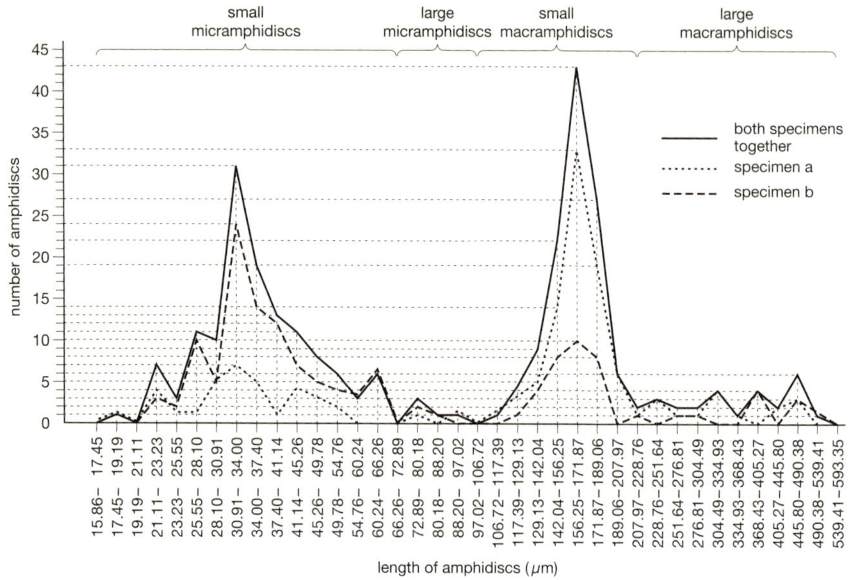

FIG. 98. Length-frequency curve of amphidiscs in *Hyalonema (Oonema) bianchoratum pinulina* (Lendenfeld, 1915).

Fig. 98 herein). Measurements given by SIMON (1953, fig. 12–15) for spicules of the freshwater demosponge *Spongilla lacustris* (LINNÉ, 1759) have a more nearly symmetrical distribution, but they vary between individuals at a given locality (SIMON, 1953, fig. 13; Fig. 99 herein) as well as between distributions of average values from different lakes (SIMON, 1953, fig. 15; Fig. 100 herein).

Causes of size variation of spicules are not known. To the extent that size reflects stages in growth of individual spicules, polymodality may represent cyclicity in spicules production. In many instances size distinction is apparently functional, for spicules of the same sort but of different size occupy special areas of the sponge, such as the dermal membrane. FRY (1970, p. 157) suggested that postlarval stages of different genotypes may fuse to form a single sponge; the spicules produced by descendant cells of each larva differ. This is not known with certainty to occur, but fusion of separate conspecific sponges has been observed. In any case the observed skewness and polymodality in the size-frequency distribution of spicules supports FRY's (1970, p. 145 ff.) assertion that information is lost if only the mean and extreme sizes are given. It is also important to note from what part of the sponge the spicules were obtained.

Variability in frequency of different spicule types between individuals of the same species may occur to the extent that one of the spicule types, sometimes a diagnostic one, may be absent or so reduced in numbers that it is difficult to find on the specimen. Several apparent instances of this sort have been reported by DE LAUBENFELS (1936) and BURTON (1932), among others.

Spicules of a given type may also vary in shape or ornamentation. The example given by BURTON (1932, fig. 23–24; Fig. 101–102 herein) from the demosponge *Iophon proximum* (RIDLEY) may be representative of a number of similar instances. The variability affects both the principal acanthostyles and the chelalike microscleres. Some variability of this sort may be clearly teratological. Such instances have been reported by SIMON (1953) and by TUZET and CONNES (1962) for the freshwater demosponges *Spongilla lacustris* (LINNÉ, 1759) and

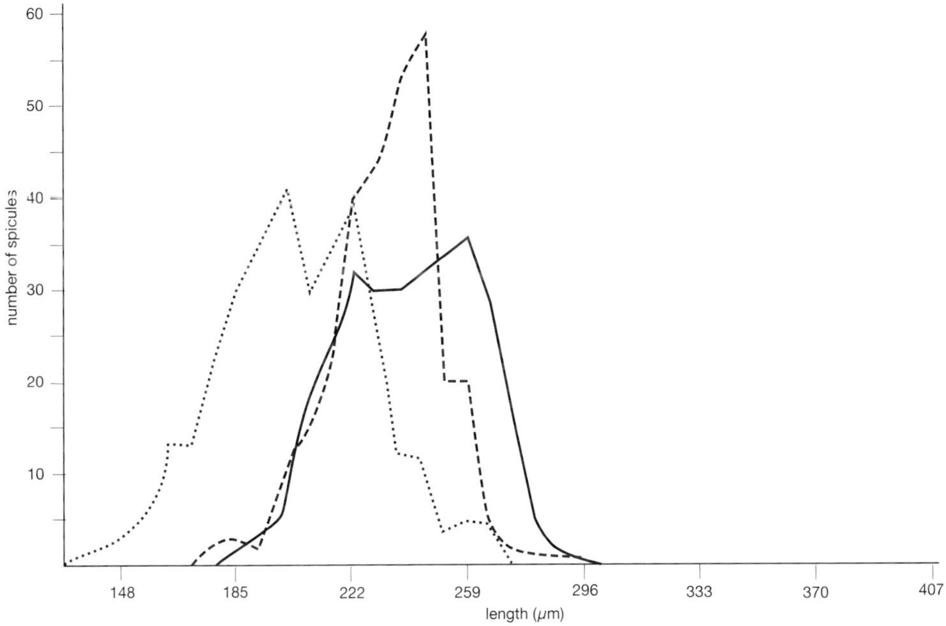

FIG. 99. Spicule length-frequency curves of three different colonies of *Spongillia lacustris* (300 spicules from each colony) from Schleinsee (Simon, 1953).

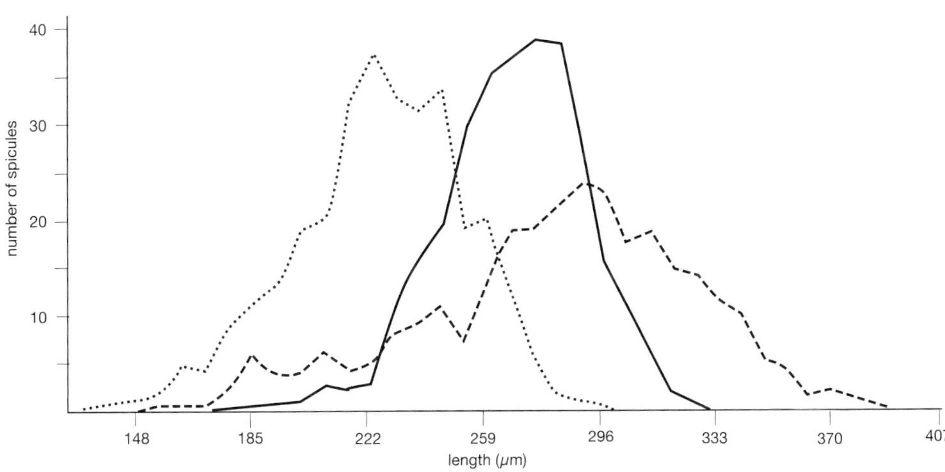

FIG. 100. Average frequencies of spicules of given length in *Spongilla lacustris* from each of three German lakes: Schleinsee (3 colonies: 900 spicules) represented by *dotted line;* Meisinger See (4 colonies: 1200 spicules), *solid line;* Klosterweiher (3 colonies: 900 spicules), *dashed line* (Simon, 1953).

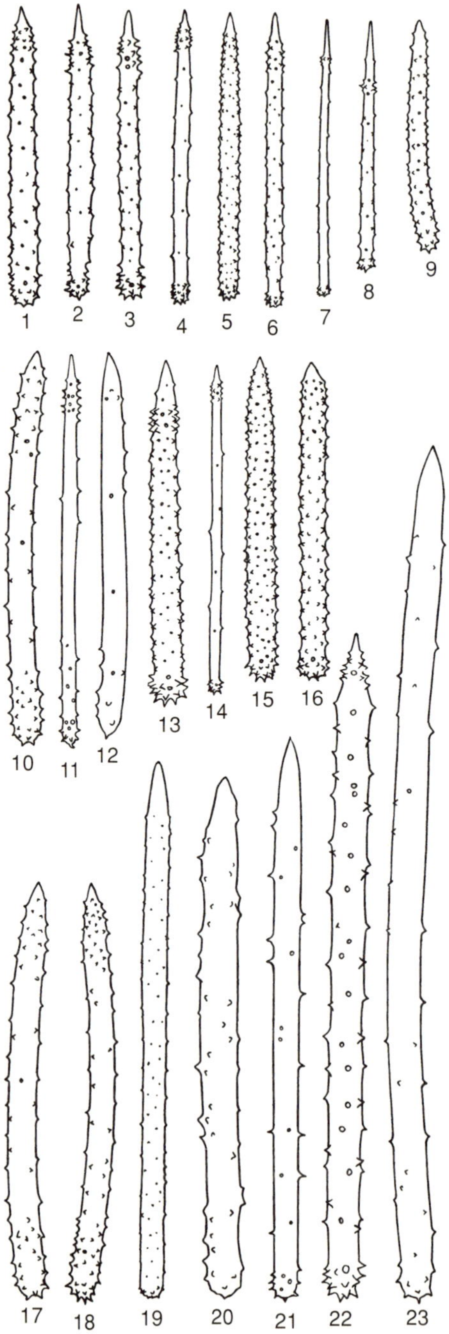

FIG. 101. Variation in size and shape of acanthostyles in *Iophon proximum* (RIDLEY), ×200 (Burton, 1932; courtesy of Cambridge University Press).

Ephydatia fluviatilis (LINNÉ, 1759), respectively. SIMON (1953, p. 220) noted that particular malformations characterized each lake from which the sponges were obtained (*ibid.,* fig. 16; Fig. 103 herein), thus pointing to ecological causes. TUZET and CONNES ascribed the malformations of their sponges to conditions of strong currents (the sponges occur in water passages of a pumping station). Strong currents were previously noted by SIMON (*fide* TUZET & CONNES, 1962) as inducing malformations.

Variability may occur regularly within an individual sponge. Differences in size within a given spicule type, such as oxeas and hexactines, may be related to their position within the sponge. The dermal membrane in particular may contain smaller sizes of such spicules than occur in the principal skeleton (*vis.,* oxeas in the Permian lithistid *Scheiia tuberosa* TSCHERNYCHEV & STEPANOV, 1916 (FINKS, 1971b) or triactines in the Permian hexactinellid *Carphites plectus* FINKS, 1960 (FINKS, 1960, pl. 43,*5–6;* Fig. 104 herein). Other variability may be ontogenetic. In the earliest formed layers of spicules in the Permian lithistids *Anthracosycon* GIRTY, 1908, and *Haplistion* YOUNG & YOUNG, 1877, the monaxonic desmas occur singly in an isodictyal net and bear only terminal zygoses (dendroclones) (Fig. 105). Very soon the spicules were grouped in parallel bundles and bore lateral zygoses for mutual articulation; in *Haplistion* the terminal zygoses are absent in these later spicules, and they have the form of typical rhizoclones (FINKS, 1960, p. 78, 89, pl. 20,*4–5,* pl. 26,*10,12*). Here the variant forms seem to be homologous and their differences related to changing functional needs within the organism. This indicates that spicule form is determined not only by the genotype but also by the internal milieu.

Seeming variability in spicules could result from incorporation of foreign spicules from the sediment by the sponge. This does not seem to be a common occurrence,

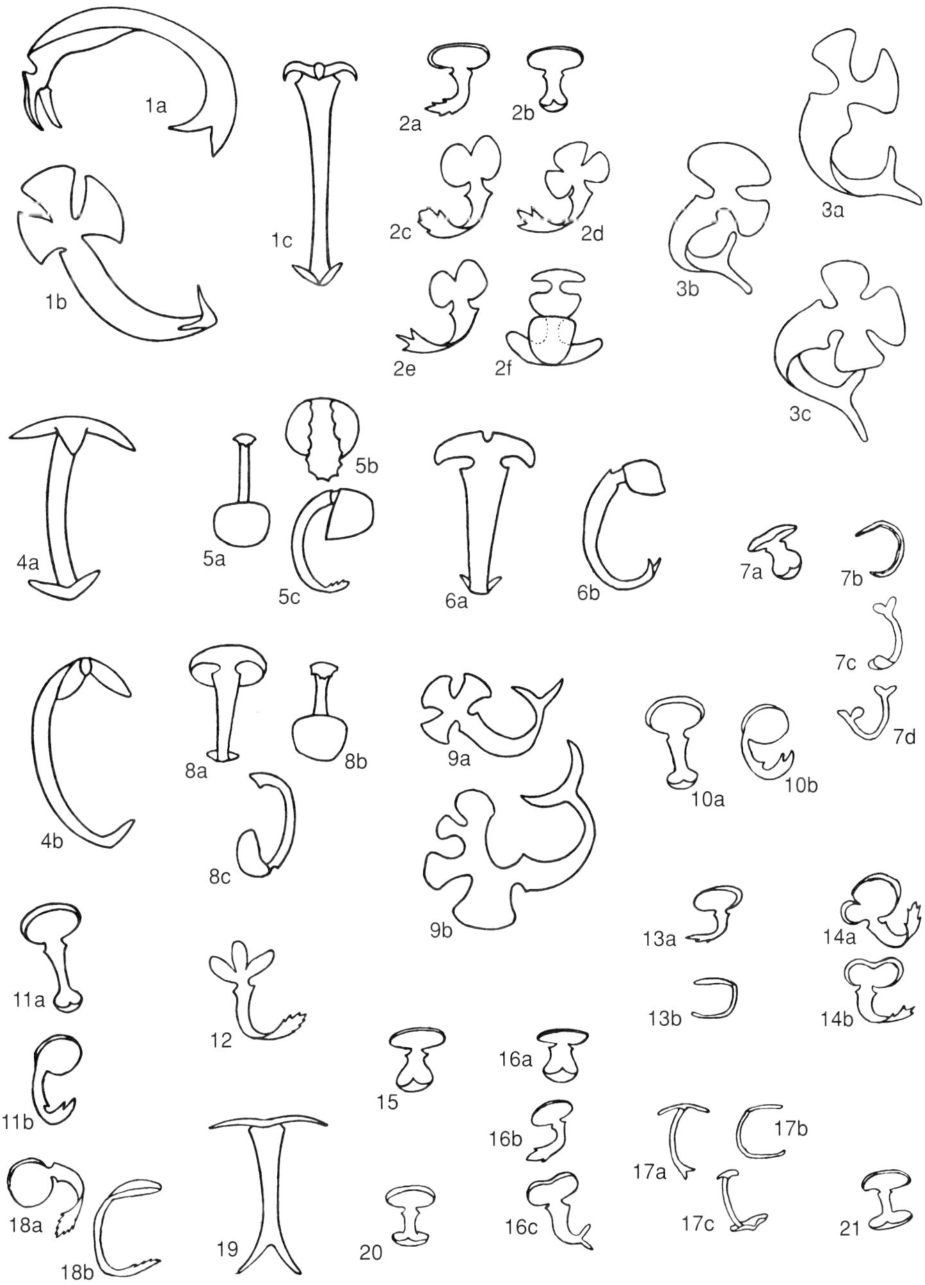

FIG. 102. Variation in size and shape of chelate microscleres in *?Iophon proximum* (RIDLEY), ×1 (Burton, 1932; courtesy of Cambridge University Press).

FIG. 103. Spicule malformations in *Spongilla lacustris; 1–5,* Meisinger See; *6–10,* Schleinsee; *11–13,* Klosterweiher, approximately ×133 (Simon, 1953).

however. Keratose sponges, which do not secrete any spicules of their own, frequently incorporate sand grains and sometimes spicules of other sponges in their spongin fibers (DE LAUBENFELS, 1936). Sponges that secrete their own spicules, however, seem to be discriminatory toward foreign spicules. SIMON (1953) studied this experimentally with the freshwater sponges *Spongilla, Ephydatia,* and *Trochospongilla.* He found that *Spongilla lacustris* (LINNÉ, 1759) accepted spicules of *Ephydatia fluviatilis* (LINNÉ, 1759) only af-

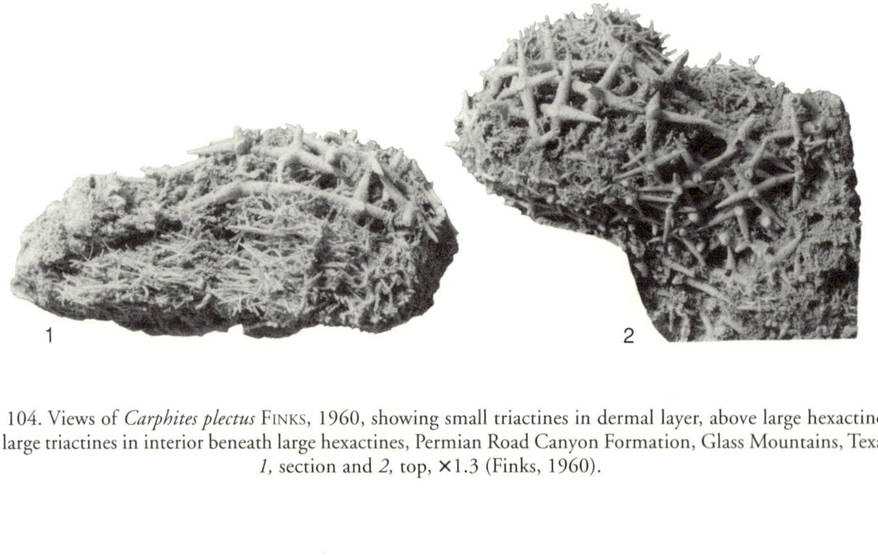

FIG. 104. Views of *Carphites plectus* FINKS, 1960, showing small triactines in dermal layer, above large hexactines, and large triactines in interior beneath large hexactines, Permian Road Canyon Formation, Glass Mountains, Texas; *1,* section and *2,* top, ×1.3 (Finks, 1960).

ter their organic coatings had been removed through treatment with H_2SO_4 and H_2O_2. *Trochospongilla horrida* WELTNER accepted both *Spongilla* and *Ephydatia* spicules to only a limited extent even after such treatment and to a very slight extent when treated with HCl only. *Ephydatia* did not accept foreign spicules at all, even when treated. Such evidence of positive rejection of foreign spicules explains the rarity of such incorporation. Among fossils, of course, there is the possibility that loose foreign spicules were swept into the skeleton after death of the sponge.

SKELETAL NET

Spatial organization of spicules and other skeletal elements is an important familial, generic, and specific character. Although it is relatively constant within a species, usually more so than the external form, individual variation does occur. This is restricted usually to variation in thickness of skeletal fibers or in numbers of spicules lying side-by-side in them (see, for example, BURTON, 1932, p. 268) rather than involving major differences in the geometry of the net. Nevertheless, even this much can be quite constant in some groups. In the Permian lithistid *Anthracosycon* the skeletal fibers are composed of several spicules side-by-side in some populations and of single spicules in others (FINKS, 1960, p. 77 ff.; see also Fig. 105 herein), but each population from a single locality (named as species) is either of one composition or the other. SIRIBELLI (1961, fig. 5–6; Fig. 105 herein) illustrated two types of skeletal net in *Axinella,* characterizing each of two species, which remain constant despite considerable individual variation in external form.

AQUIFEROUS SYSTEM

Form of the canal system and size, spacing, and grouping of pores are also useful taxonomic characters, especially at generic and specific levels. Size and grouping of homologous pores, however, may indicate

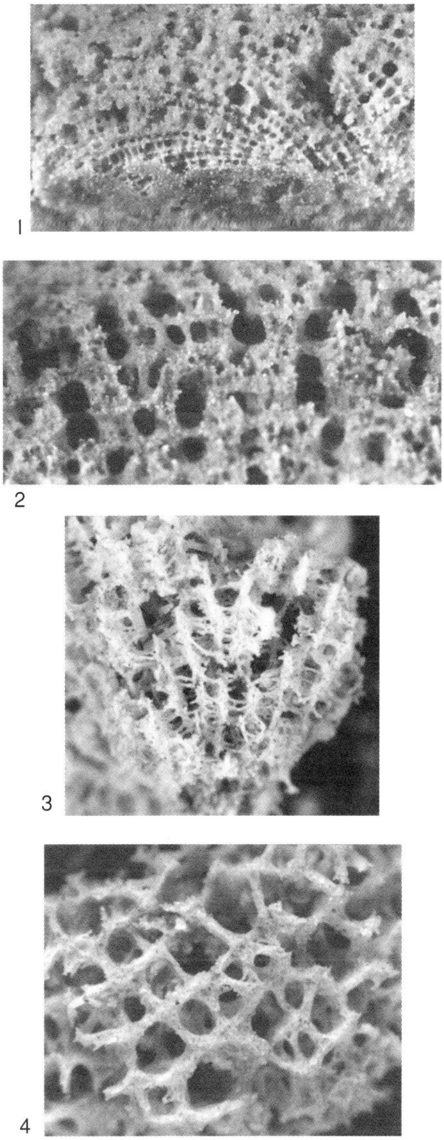

FIG. 105. Ontogenetic change in spicule form and skeletal net. In initial part of skeleton, spicules are dendroclone-like and occur singly, whereas in upper, later-formed part, spicules are rhizoclone-like and grouped in bundles (Finks, 1960); *1–2, Anthracosycon ficus* GIRTY, 1909, holotype, Permian Bone Spring Formation, Guadalupe Mountains, Texas; *1,* earlier and *2,* later parts, ×15; *3–4, Haplistion aeluroglossa* FINKS, 1960, holotype, Permian Road Canyon Formation, Glass Mountains, Texas; *3,* earlier and *4,* later parts, ×15 (Finks, 1960).

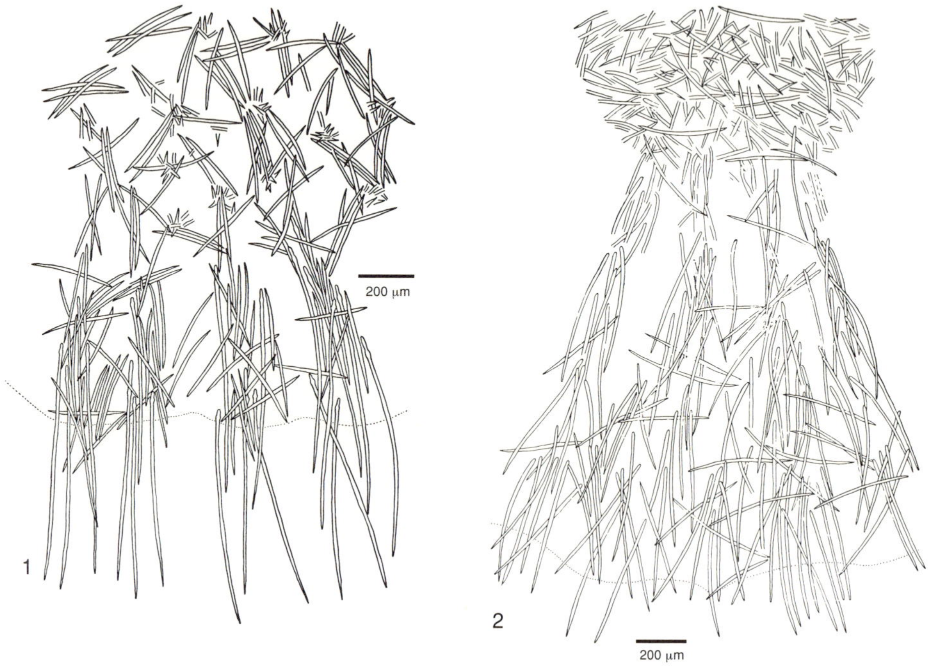

FIG. 106. *1,* Skeleton of *Axinella damicornis,* section perpendicular to axis of frond; *2,* skeleton of *Axinella verrucosa* section perpendicular to axis of branch (Siribelli, 1961).

individual variation within species and also on parts of the same specimen. For example, in the Permian lithistid *Multistella porosa* FINKS, 1960, the number of pores in the exhalant clusters and the spacing of these clusters is variable in the same specimen (FINKS, 1960, pl. 9,*3;* Fig. 107 herein). The number of pores in a cluster varies within narrow limits and is determined probably by functional factors, namely the volume served by each exhalant cluster. Their spacing on the sponge surface, however, is more irregular, and presumably fortuitous or unique events during development were the cause of the irregularities in their dispersion.

FIG. 107. Variability in dispersion of exhalant pore clusters on *Multistella porosa* FINKS, 1960, Permian Getaway Limestone, Guadalupe Mountains, Texas, ×2 (Finks, 1960).

In the Permian lithistid *Collatipora? pyriformis* FINKS, exhalant pores low on the side of the sponge are more widely spaced and have collarlike rims about them (FINKS, 1960, p. 84, pl. 23,*1,3;* Fig. 108 herein). This may be a compensatory adaptation for increasing the velocity of outflow from them under conditions of less agitated ambient water than prevails on the upper part of the sponge, thus carrying the waste water away from the sponge. The oscular collar as an adaptation for quiet water is predicted by

FIG. 108. Views of *Collatipora? pyriformia* FINKS, 1960, showing how oscules in sheltered location on side of sponge (*1,* those at left, and *2,* at right) developed raised lips to direct exhalant currents upwardly, while oscules on exposed top of sponge are flush with surface; smaller pores to right in view *1* presumed to be inhalant; Permian, Road Canyon Formation, Glass Mountains, Texas, *1,* side and *2,* top, ×1.6 (Finks, 1960).

BIDDER's theoretical analysis (BIDDER, 1923). This instance of intra-individual variation may thus be related to functional needs; consequently another specimen of the same species that bore collared exhalant pores over the entire surface might be interpreted as having lived in quieter water.

CONCLUSION

Almost every character used to describe and characterize sponges is subject to considerable intraspecific variation. Consequently it is necessary to indicate the extent of variation when describing any character, whether quantitatively or qualitatively, so that the species definition will conform to or parallel the reality of a natural population. Likewise, it is important to record, when possible, the correlation of variation with environmental and sedimentologic conditions, so that an ecologic interpretation of variation may become possible in conjuction with theoretical models of functional morphology.

ECOLOGY AND PALEOECOLOGY OF SPONGES

ROBERT M. FINKS

[Department of Geology, Queens College (CUNY)]

INTRODUCTION

Sponges are ecologically different from other sessile, **epibenthic suspension feeders** in respect to the following: (1) their major food intake is from smaller particle sizes; (2) they shelter a much larger volume of **inquilines;** (3) predation pressure upon them is relatively less strong; (4) symbiotic relationships with **prokaryotes** and unicellular algae are possibly more frequent and lead, in the most highly developed instances, to bacteria-sponge entities reminiscent of lichens.

Sponges resemble corals and bryozoans, for most of them use space for a considerable distance above the sea bottom, and their mode of occupying this space is an important element of their adaptation. Competition with corals and bryozoans, therefore, is strong and is confirmed by their geologic history and present distribution. Morphologic convergence with corals and bryozoans is also strong, despite differences in food sources and feeding methods.

ECOLOGY OF SPONGES

The functional relationships of sponges with their environment center upon (1) feeding, (2) occupancy of substrate space and of the adjacent water, (3) provision of cover and camouflage to vagile benthos, and (4) provision of food to browsing carnivores. Respiration and reproduction seem less critical interactions, in that they are accomplished by modes common to most marine organisms.

FEEDING

Because digestion is intracellular (and they lack an organ for mechanical breakdown of food) sponges are limited in their use of food to the smaller end of the spectrum of particle sizes available to suspension feeders. These range from molecules through organic detritus and bacteria to the smaller protozoa and one-celled algae. Length of the **choanocyte collar** through which most food is captured is about 5 um (REISWIG, 1975). Amoebocytes of 10 um in diameter (BRIEN, 1973a) represent the upper limit on food particle size. Many one-celled organisms exceed these dimensions. Sponges do not seem to be limited in their distribution by availability of food. Particles of the size required occur in all oceanic waters, as do the sponges. Limited distributions of sponge species may be related partly to food requirements, but nothing is known at present of food-specificity in sponges. In balance, environmental factors other than food must be looked to for an explanation of the distribution of individual species.

Sponge abundance, on the other hand, seems strongly determined by food supply. For example, the area off the North Cape of Norway, where sponges constitute more than 90 percent of a locally very high absolute biomass, is beneath an area where Gulf Stream plankton are killed by Arctic waters. Other examples are the general abundance of sponges in the vicinity of coral reefs, which have a high rate of primary production, and the concentration of sponges near organic-detritus-bearing river mouths in Bermuda (DE LAUBENFELS, 1950). The abundance of sponges near the Antarctic Ross shelf ice, where algal blooms are common, can also be related to the local abundance of food.

In most of these instances successful competition for substrate space is also involved, for the sponges nearly exclude other sessile organisms. Rapid growth consequent upon a rich food supply, leading to preemption of bottom space seems to be the cause. Marginal conditions for competing species may also be involved for the two polar-water examples.

SPACE OCCUPANCY BY SPONGES

Sponges are sessile, benthic, epifaunal suspension feeders that, as a rule, project a considerable distance above the substrate. They present a large surface area of intake of food and metabolites, often by means of an elaborately branched body shape. In this they resemble colonies of bryozoans and corals and differ from brachiopods and bivalves that tap resources of the environment through relatively small areas and do not greatly increase the cross section of capture through individual growth. Such relationships have several ecological consequences. (1) Sponges take food from a considerable distance above the bottom and, thus, compete more directly with bryozoans and corals and less directly with brachiopods and bivalves. (2) An individual sponge captures quantities of food corresponding to that taken by a dense concentration of individual bivalves or brachiopods. (3) Adaptations for obtaining food by sponges is directly related to gross body shape and especially to external surface area, which in bivalves and brachiopods is related more to changes in the process or efficiency of food gathering. (4) A sponge can compete with other organisms for food through sheer growth. (5) Competition for food involves competition for space.

It is not surprising, therefore, that in the geologic past, times of sponge abundance have tended to have an inverse relationship with times of bryozoan and coral abundance. This is especially true of reef-building sponges in relation to reef-building corals. Sponges dominated reef communities in the Ordovician before corals became diverse and abundant and again in the Permian and Triassic, when rugose and tabulate corals were dying out and scleractinians were just beginning. This matter is discussed further below.

It is not surprising that sponges often have a patchy distribution. Where they are abundant they tend to occupy the substrate space to the near exclusion of other sessile epibenthos. When part of a mixed community, on the other hand, they are individually less abundant than the associated brachiopods and bivalves.

The greatest surface area for a given volume of water space occupied is achieved by a branching sponge with numerous, narrow branches. Such a shape is mechanically weak and is, therefore, most efficient in quiet water. Broadly flattened branches improve mechanical strength while maintaining much of the surface area. *Microcionia* and *Axinella* are common sponges with such shapes, as are some species of the Permian *Guadalupia* (see Fig. 85). Stouter cylindrical branches are also mechanically stronger but reduce the amount of surface area relative to internal sponge tissue unless the sponge develops a hollow tubular shape. The Pennsylvanian *Heliospongia excavata* has solid, flattened branches, while *H. ramosa* has cylindrical branches with a central cloaca, as does the later, stouter *H. vokesi.* A thin-walled tube has greater surface area relative to sponge volume but is mechanically weaker. Many such sponges as *Callyspongia* and *Mycale,* living in quieter waters, have thin-walled tubular branches.

Sponges that do not maximize surface area of food capture in the waters above the sea bottom but rather confine themselves to the space close to the sediment surface may assume mechanically stronger shapes. Simple encrusting forms can survive high wave energy. More massive shapes, including spheroidal ones, are also strong but require a more elaborate internal canal system to compensate for reduced surface area relative to internal volume. Spheroidal shapes, such as those of *Tethya, Tetilla,* and *Geodia* have complex aquiferous systems with specialized structures such as chones to regulate internal water flow. Sponges with open-cup or goblet shapes represent a compromise between mechanical strength and increased surface area of food capture.

Sponges in which rapid growth is an important element of adaptation are expected to maximize surface area of food capture. It is not surprising, therefore, to find that

Mycale has a tubular branching form, for REISWIG (1973, 1974) noted that the species of *Mycale* he studied in Jamaica is an opportunistic generalist with a high metabolic rate, high growth rate, and high reproductive rate. On the other hand, the slower growing specialist species, *Verongia gigantea* and *Tethya crypta* (REISWIG, 1973, 1974), have thick-walled goblet and spheroidal shapes, respectively. These species are adapted to protect the life of the individual sponge at the expense of slower growth (REISWIG, 1973).

Unfortunately for the paleoecologist, sponge shape alone cannot be used as a simple environmental indicator. As the above examples show, shape may relate both to mechanical efficiency and to food-getting efficiency; some species may sacrifice one for the other. Nevertheless, shape gives us a working hypothesis about the environment to be tested by independent lines of evidence. Where closely related forms have a systematic variation in shape in adjacent areas, we may accept as a possibility some form of environmental control. As an example, closely similar forms of *Guadalupia* from the same locality in the Road Canyon Formation of the west Texas Permian sometimes vary from minutely branching shapes, to flattened, subparallel branches to open cups, to cups so deep as to be nearly tubular. If mechanical efficiency alone is considered, minutely branched individuals are most effective in quiet water and open cups in rough water. Alternatively, branched forms could represent an attempt to increase food-getting efficiency under conditions of reduced supply or increased competition. In the history of the group, a minutely branched form is unusual and confined to a limited time and place, while the open-cup is most common and occurs earliest. Tubes, on the other hand, dominate in times subsequent to the Permian. Inasmuch as tubes and cups occur in reefs (presumably rough water), while the minutely branched forms occur in less clearly reefal localities (along with some cups), the quiet-water interpretation is strengthened. Likewise, evolutionary development of tubular forms may be related to increased food-getting efficiency in a relatively constant environment.

Indications of paleocurrent direction may be less equivocal. Sponges with subparallel, subhorizontal branches are usually elongate into the current, with the fastest-growing end likely to be pointed upcurrent. Flabellate vertical sheets, on the other hand, are likely to be oriented across the current.

From the point of view of efficiency of waste disposal (BIDDER, 1923), sponges with oscular chimneys or narrow-mouthed, stalked goblet-shaped forms are most efficient in quiet water, while open cups are less effective. Thus, increased food-getting surfaces in quiet waters will play against decreased efficiency of waste disposal and projection. Minutely branched forms may be best in quiet water with a constant current.

ECOLOGICAL DISTRIBUTION OF SPONGES

HOLOCENE SPONGES

Environments in which Holocene sponges occur in particular abundance include the foreslopes of tropical coral reefs below the zone of maximum coral growth, i.e., from about 23 mm to about 150 m (REISWIG, 1974; VACELET, 1981; LEWIS, 1965b); rock bottoms in sheltered quiet-water lagoons behind coral reefs (REISWIG, 1973; HAY, WIEDENMAYER, & MARSZALEK, 1970); dark caves within coral reefs (JACKSON, GOREAU, & HARTMAN, 1971; VACELET & VASSEUR, 1971b); and dark places in very shallow water such as under stones and in coastal caves (VACELET, 1967a, 1967b, 1994; VACELET, BOURY-ESNAULT, & HARMELIN, 1994). Sponge reefs dominated by hexactinellids have been recently reported from depths up to approximately 200 meters on the British Columbia continental shelf in Hecate Strait (KRAUTTER & others, 2001; CONWAY & others, 1991, 2001). Also included in these dark environments are the abyssal plains, in general (ZENKOVITCH, 1963); areas adjacent to the

Antarctic shelf ice (BULLIVANT, 1960; DAYTON, ROBILLIARD, & PAIUNE, 1970); and localized patches in moderately deep water, such as off North Cape of Norway in the Barents Sea at depths of around 300 m (ZENKOVITCH, 1963, p. 145–146). A number of these environments are relatively dark, and various authors have suggested that sponges have generally lived in dark environments, even in the geologic past. Frequent occurrences of abundant sponges with algae in the fossil record (FINKS, 1971a), however, suggest that they have also flourished in lighted environments. Bryozoans and brachiopods co-occur with sponges in present-day dark marine caves (JACKSON, GOREAU, & HARTMAN, 1971).

Another common aspect of most of these environments is a hard rock substrate. This points to two sponge requirements: a hard surface for attachment and sediment-free water. That the latter is the more important of the two is suggested by the abundance of sponges having special adaptations for attachment in soft sediments (root tufts), in environments where a soft bottom is accompanied by a low sedimentation rate, such as on the **abyssal plain.**

The following discussion summarizes those instances of clear dominance of sponges in terms of biomass, in particular ecological situations, or of restriction to particular environments. Undoubtedly many examples have been omitted through ignorance of the literature or because no clear statement was made concerning relative abundance in primary sources. The relatively recent review by SARÀ and VACELET (1973) has been particularly helpful.

Among living Monaxonida, the Poecilosclerida (sigma-bearing sponges with a fibrous skeleton of more than one type of megasclere, stylote, or oxeote) have a wide ecological distribution but are particularly characteristic of deeper waters. Two families, the Cladorhizidae and Chondrocladiidae, are confined to **bathyal,** abyssal, and **hadal** mud bottoms (SARÀ & VACELET, 1973, p. 486, 552). The deepest known occurrences of sponges of any kind are of these families. The Myxillidae and Hymedesmiidae are characteristic of bathyal hard bottoms (SARÀ & VACELET, 1973). The Mycalidae also have abyssal and hadal muddy-bottom representatives (*Abyssocladia*) and are likewise characteristic of bathyal hard bottoms (SARÀ & VACELET, 1973). The Mycalidae also have warm, shallow-water representatives that are often the dominant sponges in their respective environments, such as species of *Mycale* in coral reefs (both on the outer slope and on inner reef flats) (REISWIG, 1973; SARÀ & VACELET, 1973, p. 547) and various species of *Biemna* and *Desmacella* in mangrove swamps. Other shallow-water poecilosclerids that are locally abundant include the Microcionidae (such as *Microciona*) in temperate and tropical waters and especially the Tedaniidae (*Tedania*) in tropical waters (HAY, WIEDENMAYER, & MARSZALEK, 1970). All the shallow-water forms are most abundant on hard bottoms, but *Biemna* may live partially buried in the mud (SARÀ & VACELET, 1973, p. 550). *Crambe* is a strong dominant in the more lighted parts of Mediterranean coastal caves (SARÀ & VACELET, 1973, p. 543).

Hadromerida (=Clavulida) (sponges with spinispires [=spirasters] and tylostyles) have some forms adapted to deep-water mud bottoms, such as the bathyal *Radiella* and abyssal to hadal members of the Polymastiidae (SARÀ & VACELET, 1973, p. 546, 552). The main center of abundance of the Hadromerida, however, is on hard bottoms or shells in shallow water, with a particular tolerance shown to brackish-water **estuarine** conditions, especially by members of the families Clionidae and Suberitidae (SARÀ & VACELET, 1973, p. 558). The Spirastrellidae are especially common in shallow-water lagoons associated with coral reefs (*Speciospongia, Anthosigmella, Spirastrella*) with specialized species found in rough-water environments (*Placospongia*) (HAY, WIEDENMAYER, & MARSZALEK, 1970; SARÀ &

VACELET, 1973, p. 547, 561). The Hadromerida also include all the known boring sponges (Clionidae).

Axinellida (sponges with plumosely arranged styles and no microscleres) are particularly abundant in temperate waters on hard bottoms at moderate to shallow depths (SARÀ & VACELET, 1973, p. 545–546).

Halichondrida (sponges with various monaxonid megascleres, simple raphid microscleres, and a dermal specialization) include some forms particularly abundant on the inner reef flat of coral reefs (*Acanthella*) (SARÀ & VACELET, 1973, p. 547).

The Haplosclerida (sponges with oxeas, usually reticulate skeletons, and no dermal specialization) include the only freshwater sponge families (Spongillidae, Potamolepidae, and Lubomirskiidae) (SARÀ & VACELET, 1973, p. 553, 555). Occurrence of Plio-Pleistocene freshwater forms in central Japan was well documented by MATSUOKA (1987), for example. In addition they include forms abundant in shallow-water lagoons associated with coral reefs (*Haliclona, Gellius)* or in strong-current areas on the reefs themselves (*Strongylophora*) (HAY, WIEDENMAYER, & MARSZALEK, 1970; SARÀ & VACELET, 1973, p. 547, 549). *Petrosia* may be very abundant in darker parts of deeper Mediterranean caves (SARÀ & VACELET, 1973, p. 539–540).

The Lithistida as a whole are adapted by their skeletons to rough-water conditions in which they are found on coral reefs, but their main abundance is either in the totally dark parts of subreef caves and tunnels or in deeper parts of forereef slopes below the major zone of abundance of other sponges (LEWIS, 1965b; SARÀ & VACELET, 1973, p. 478, 486, 544, 548).

Tetractinellida include a number of groups with characteristic environmental tolerances, on the whole tending toward deep, cold, and unlighted waters. The Theneidae (*Thenea*) are adapted to bathyal, abyssal, and hadal soft bottoms (SARÀ & VACELET, 1973, p. 486, 546, 552). The Tetillidae have a wide distribution but are extremely abundant (*Tetilla, Cinachyra*) in the Antarctic (SARÀ & VACELET, 1973, p. 551) and locally (*Craniella*) in the Arctic (off the North Cape of Norway) (ZENKOVITCH, 1963, p. 145–146). The Geodiidae (*Geodia*) are also common in Arctic waters, both shallow and deep (ZENKOVITCH, 1963, p. 145), but are found also in warmer waters and by their special, dense, dermal spiculation of sterrasters are adapted to resist shallow, rough water in general (SARÀ & VACELET, 1973, p. 561). *Geodia* is very abundant in the darkest parts of shallow Mediterranean caves (SARÀ & VACELET, 1973, p. 476). The Chondrillidae (*Chondrilla*) are very common on hard bottoms in shallow, quiet waters associated with coral-reef lagoons (HAY, WIEDENMAYER, & MARSZALEK, 1970). In shallow-water marine caves the tetractinellids tend to occupy the semilighted parts (SARÀ & VACELET, 1973, p. 544).

The Epipolasida are especially common in warm water: in shallow protected lagoons (*Tethya*) or mangrove swamps (*Prostylissa*) (SARÀ & VACELET, 1973, p. 473, 550; REISWIG, 1973, p. 202–203).

Keratosa are another largely warm-water group (SARÀ & VACELET, 1973, p. 473). Their tough resilient skeleton adapts them to rough water (SARÀ & VACELET, 1973, p. 478). They also prefer lighted to unlighted situations, occurring at the light end of marine caves (SARÀ & VACELET, 1973, p. 544).

In shallow lagoons of coral reefs, the genera *Ircinia, Verongia,* and *Spongia* are common (HAY, WIEDENMAYER, & MARSZALEK, 1970), and *Verongia* is also locally abundant on the upper part of the outer reef slope (REISWIG, 1973).

The Sclerospongia are also adapted by their massive skeletons to rough water but are largely confined to totally dark parts of subreef caves and tunnels or to deep but warm waters, as in the Mediterranean (SARÀ

& VACELET, 1973, p. 549–550; JACKSON, GOREAU, & HARTMAN, 1971).

The Pharetronida, among the Calcarea, have a distribution similar to the Sclerospongia (SARÀ & VACELET, 1973, p. 549–550). The non-pharetronid Calcarea are almost entirely confined to very shallow water (see chapter on Geographic and Stratigraphic Distribution, p. 276).

The Hexactinellida are dominantly cold- or deep-water forms, some groups being adapted to bathyal, abyssal, and hadal mud bottoms, others to bathyal hard bottoms. They also occur abundantly on shallower hard bottoms in the Antarctic (especially *Rosella*) (DEARBORN, 1965) and with greater diversity, if not abundance, in moderately shallow tropical waters of Indonesia (see chapter on Geographic and Stratigraphic Distribution, p. 276). Hexactinellid sponges form reefs in the cold water of Hecate Strait on the continental shelf of British Columbia (KRAUTTER & others, 2001; CONWAY & others, 2001).

FOSSIL SPONGES

The record of Precambrian sponges is mainly of isolated spicules from China, Iran, and elsewhere (GLAESSNER, 1962; BRASIER, 1992; STEINER & others, 1993) and impressions of what are interpreted to be hexactinellid sponges from the Ediacaran beds of Australia (GEHLING & RIGBY, 1996). The *Spinther*-like *Dickinsonia* of the Ediacara fauna, if it had the specialized sponge-feeding habits of the living *Spinther,* implies the presence of its prey. By Cambrian times, however, Heteractinida, monaxonid Demospongea, and lyssacine Hexactinellida are well developed, particularly in such black shale facies as the Burgess Shale (WALCOTT, 1920; RIGBY, 1986a), which represent quiet-water environments with much organic detritus. The Archaeocyatha appear in shallow-water limy facies at the very base of the Cambrian, often before the first trilobites. Their systematics and development have been covered in a chapter of the first revision of Part E of the *Treatise on Invertebrate Paleontology* (HILL, 1972, p. 49).

Early and Middle Cambrian true sponges are generally thin-walled, saclike forms, which is apparent both in the completely preserved whole sponges and in isolated spicules that are almost entirely two-dimensional forms such as stauractines, sexiradiates, and oxeas arranged in single layers. Early sponges appear to have been largely low, benthic forms. Tall, thin-walled Lower Cambrian sponges from Anhui Province of China, however, have been interpreted to have been high-tiering organisms (YUAN & others, 2002). By Late Cambrian time thicker-walled sponges appeared, such as *Multivasculatus* with its hexactines and the early lithistids, *Wilbernicyathus* and *Gallatinospongia.* Most of these latter sponges occur in deposits of limy shallow-water facies like that in which undoubted lithistids flourished in Ordovician times. The lithistid type of skeleton seems to have arisen as an adaptation to rough water. Not only do the interlocking zygoses provide rigidity but so does the triangular configuration of dendroclones characteristic of the first lithistids (see Anthaspidellidae), an arrangement that provides maximum resistance to deformation for a skeleton built of rodlike elements. These early lithistids are associated with shelly fossils, reefs, and algae (TOOMEY, 1970; FINKS & TOOMEY, 1969). The algae provide indisputable evidence of the waters' being shallow and well lighted.

In the Early Ordovician (Canadian) the anthaspidellids are largely alone among the sponges in this environment and by Chazyan time the anthaspidellids were more diversified and were joined by sclerosponges (stromatoporoids) (BASSLER, 1941; RAYMOND & OKULITCH, 1940; FINKS & TOOMEY, 1969; PITCHER, 1964; CHURCH, 1974; KAPP, 1975; WYATT, 1979). In the succeeding Black River two new lithistid groups appeared, the Astylospongiidae and the Hindiidae, as well as the sublithistid Dystactospongiidae, of axinellid or poecilosclerid affinity (FINKS, 1967b, 1971a). Some thick-walled hexactinellids (Brachiospongiidae) are also associated with this environment, beginning with the

Trentonian (BEECHER, 1889), as are apparently the rarer heteractinids at least as early as the Chazyan (RIGBY, 1967a).

Thick-walled heteractinids (Astraeospongiidae) became common in this environment during the Silurian, along with anthaspidellids and astylospongiids in the well-known middle Silurian Gotland and Tennessee faunas (RAUFF, 1893, 1894). After the Silurian the Astylospongiidae are known only from Devonian faunas in Australia (PICKETT, 1969; PICKETT & RIGBY, 1983; RIGBY, 1986b), but the rest of the elements persisted in the shallow-water sponge fauna until the end of the Paleozoic.

In the later Paleozoic, anthaspidellid lithistids were slowly replaced in abundance and diversity by their relatives and descendants, the Chiastoclonellidae and Anthracosyconidae. Likewise astraeospongiids were replaced by their descendants the wewokellids, and the brachiospongiids by their possible descendants the stiodermatids and docodermatids. The hindiids persisted with less change.

There was a drastic change in shallow-water sponge faunas beginning with the Pennsylvanian. Calcareous sponges, which had appeared in early Paleozoic assemblages (PICKETT & JELL, 1983; WEBBY & RIGBY, 1985; RIGBY & POTTER, 1986), became major elements in mid-Pennsylvanian times, starting in low paleolatitudes in the Tethyan realm (Des Moinesian of Kansas: FINKS, 1960, 1970) and dominating shallow-water Permian and Triassic sponge faunas. These are the sphinctozoans and pharetrones. In the Mesozoic these calcareous sponges spread more widely and to higher paleolatitudes. The sphinctozoan-pharetronid association is recognizable at least as late as the Cretaceous Farringdon Sponge Gravel of England.

Rhizomorine lithistids are another important element of Mesozoic shoal-water sponge faunas. These first appear in the fossil record in the Ordovician of Australia (RIGBY & WEBBY, 1988) but became common beginning in Carboniferous times (Visean of Scotland: HINDE, 1887b, 1888).

In addition to the shallow-water sponge fauna associated with algae and shelly facies, the Paleozoic record preserves another sponge facies associated with the quiet, if not necessarily deep, black-shale deposits and similar black limestones. They range in estimated depth from a 1-meter deep in Pennsylvanian coastal lagoon in Indiana (ZANGERL & RICHARDSON, 1963) to the 1,800-meter deep in the Permian Delaware basin (NEWELL & others, 1953; NEWELL, 1957). Earlier examples include the Cambrian Burgess Shale (WALCOTT, 1920; RIGBY, 1986a) of British Columbia, the Wheeler Shale (RIGBY, 1978) and Marjum Formation (RIGBY, 1983a) of Utah, and the Ordovician Utica Shale (WALCOTT, 1879; RUEDEMANN, 1925) of New York. Thin-walled hexactinellids of the families Protospongiidae and Teganiidae are especially characteristic of the black shales. Thicker-walled relatives belonging to the family Dictyospongiidae are found in more clastic offshore deposits, such as the delta-front sandstones of the Late Devonian phases of the Catskill Delta (the Chemung glass-sponge fauna), and dark limy shales of the Mississippian of Indiana (the Crawfordsville sponge fauna) that formed the chief sources of HALL and CLARKE's (1899) monograph.

In the Pennsylvanian, *Stioderma* occurs in shallow-water facies; in the Permian it occurs in both deep and shallow water, as do many lithistids (FINKS, 1960). It is possible that the brachiospongiids were facultative occupiers of deeper water at times of greater competition for shallow-water space, such as in the Permian with its proliferation of shallow-water Calcarea. In this they parallel the Permian lithistids; although the lithistids seem to have been sturdier competitors with their principal abundance in shallow water, both within the Texas basin and also outside it, throughout the Cordilleran shelf, and in Arctic Canada and Spitsbergen. In Timor an exclusively lithistid fauna (so far as now known, see GERTH, 1927) of archaic aspect (almost entirely anthaspidellids) is of uncertain facies; the associated fauna of echinoderms and corals (also partly archaic) suggests shallow water. The peculiar nature of this fauna may be a reflection of a high paleolatitude as part of the Australian plate, then much nearer the south pole. (The Permian outcrops are tectonically isolated and need not be part of the same plate or plates as the rest of Timor.) In general the lithistids dominate Permian sponge faunas of high paleolatitudes.

The reef facies is a special development of the shallow-water fauna. Sponges have been an element of reef faunas from Ordovician times up to the present day. Their relative abundance has varied. In general, they were important elements in construction of reefs at times when corals were less prevalent in the world faunas. In Early Ordovician (Canadian) time, reefs were formed of anthaspidellid lithistids and calcareous algae (TOOMEY, 1970; RIGBY, 1971; CHURCH, 1974; WYATT, 1979). By Chazyan time they were joined by stromatoporoids (sclerosponges), bryozoans, and the first tabulate corals. In the Chazy Group of the Lake Champlain area, the relative abundance of these reef builders varies from one reef mound to its neighbor, at the same horizon, suggesting a competitive relationship among these elements (PITCHER, 1964; FINKS & TOOMEY, 1969; TOOMEY & FINKS, 1969; KAPP, 1975).

In reefs of the Silurian, lithistid sponges were already a distinctly subordinate element, present only in early stages of reef succession (LOWENSTAM, 1957). Tabulate corals and bryozoans are distinctly dominant in Silurian reefs. By Devonian time, rugose and tabulate corals formed reefs by themselves (Coeymans and Onondaga formations, see OLIVER, 1951, 1956) or largely by themselves (PLAYFORD & LOWRY, 1966; PLAYFORD, 1967), although sponges do occur as moderately abundant elements in the reefs of Western Australia (RIGBY, 1986b). Stromatoporoids are among the major reef constituents in the Upper Devonian of Alberta (FISCHBUCH, 1970). After the Devonian, however, stromatoporoids become rare, but bryozoans reappear, forming reefs with algae in the Permian of Germany (MÄGDEFRAU, 1933). In the great Permian reefs of Texas, they are joined by sphinctozoans and pharetronid calcareous sponges (NEWELL & others, 1953; NEWELL, 1957; FINKS, 1960; PRAY & ESTEBAN, 1977; YUREWICZ, 1977a, 1977b, RIGBY & SENOWBARI-DARYAN, 1996b; RIGBY, SENOWBARI-DARYAN, & LIU, 1998). This reappearance of sponges coincides with the noticeable decline of tabulate and rugose corals, which became, respectively, nearly and totally extinct by the end of the Permian. The scleractinian corals that replaced them as dominant reef builders in Mesozoic and later times did not appear until the mid-Triassic and did not enter significantly into reefs until the upper successional stages of Upper Triassic reefs in the Alps (SIEBER, 1937; FLÜGEL, 1981; SCHÄFER & SENOWBARI-DARYAN, 1981; SENOWBARI-DARYAN, SCHÄFER, & ABATE, 1982; FLÜGEL & STANLEY, 1984; REID & GINSBURG, 1986). Most earlier Triassic reefs, as well as Late Triassic Tethyan-type reefs on the southern Yukon in North America (SENOWBARI-DARYAN & REID, 1987), are built by a sphinctozoan-pharetronid sponge fauna almost identical to that of the Permian reefs. The cryptostome bryozoan and specialized strophomenid brachiopod elements present in Permian reefs, however, have become extinct; they were replaced by hydrozoan and possible sclerosponge (spongiomorph) elements. Calcareous algae were still important.

An ecologic succession in the Upper Triassic reefs was worked out by SIEBER (1937) and included in a paper that deserves to be better known, for it seems to be the first paper in which ecologic succession was documented from the fossil record. This succession predicts strikingly the future development of Mesozoic reef faunas. Examination of succession in Silurian reefs (LOWENSTAM, 1957; NICOL, 1962) and Ordovician reefs (TOOMEY & FINKS, 1969; CHURCH, 1974; WYATT, 1979) shows a similar predictive power. This can be understood if the assumption is made that the relative competitive relationships between organisms is the principal determinant of ecologic succession rather than a change in water depth as is often assumed. A documented reef succession at a moment in time can then be used as an experimental text of relative competitive relationships between these organisms at that juncture in the history of the earth.

Alpine Upper Triassic reefs began as patches of crinoids on a seabed bearing an epifaunal pelecypod interreef fauna. These crinoid-rich areas provided a harder substrate suitable for the attachment of the sphinctozoan and pharetronid calcisponges that built much of the rest of the reef mass. Toward the top of the reef mound, scleractinian corals appear, increasing in abundance relative to the sponges and ultimately almost crowding them out completely. According to Sieber (1937) this process was repeated in each reef mound at more than one level. Such a succession has also been documented for Permian reefs in China (Fan, Rigby, & Qi, 1990).

In Ordovician reefs of the Chazy Group (Pitcher, 1964; Finks & Toomey, 1969; Toomey & Finks, 1969; Church, 1974) lithistid sponges are replaced upward in most reef mounds by either stromatoporoids or tabulate corals (*Eofletcheria, Billingsaria*), which are the two groups that dominate the subsequent Silurian reefs. Likewise, in the Silurian reefs of the Niagaran beds (Lowenstam, 1957; Nicol, 1962) lithistids are minor elements that appear very early in the succession and die out upward, while stromatoporoids, tabulates, and rugose corals attain their greatest abundance at the top of the reef. It is precisely these three groups that dominate reef structures in the succeeding Devonian (Oliver, 1956; Fischbuch, 1970; Playford, 1967).

Ecologic succession in Jurassic reefs has been best documented from the Oxfordian of Germany (Gwinner, 1958, 1968, 1976; Gaillard, 1983). In these reefs, pharetronid sponges are still an important element; sphinctozoans, which became extinct in the Cretaceous, are less so. Nevertheless, they are subordinate to scleractinians from the reef beginning and become less abundant in higher parts of the reef. Sponges are particularly characteristic of the reef-flank fauna.

In the Upper Jurassic of Germany there are also reeflike structures built of siliceous sponges, principally hexactinellids (Roll, 1934). This was, thus, the second time since the Ordovician that siliceous sponges formed moundlike masses, for Triassic sponge mounds have been described recently from China in the Sichuan Province (Wu & Zhang, 1982). These mounds have been interpreted as deep-water structures formed by sponges that trapped mud moved along the sea floor; thus they are a different sort of community from the shallow-water coral reef and are analogous to the deep-water coral banks described by Teichert (1958).

The coral reef community, as it is known today, with scleractinians and hydrozoans as the dominant frame builders with calcareous algae as binders, was already present by Cretaceous time. As far as sponges go, sphinctozoans died out gradually during the Cretaceous (Finks, 1967b) and likewise diminished in the reefs. The pharetronids were still more abundant in Cretaceous coral reefs than in present ones. Although the Cretaceous was the period of acme for pharetronids, lithistids, and hexactinellids in the world history of sponges, nevertheless, the pharetronids were not as abundant in coral reefs of the Cretaceous as in the non-reef environments. By Holocene time, pharetronids and the sclerosponges became minor constituents of coral reefs and are largely confined to caves and passages under reefs (Jackson, Goreau, & Hartman, 1971; Sarà & Vacelet, 1973), where they may be locally abundant.

Although many non-lithistid demosponges are abundant in the vicinity of present-day reefs, they are most numerous away from the reef, either in back-reef lagoons or on fore-reef slopes (e.g., Reiswig,

1973). Lithistids tend to attain maximum abundance in deeper waters of the fore-reef slope. LEWIS (1965b) studied the ecology of the deeper water adjacent to coral reefs of Barbados and recognized a sponge-coral community between 50 and 150 meters depth, just below the active coral reef. He also recognized a coelenterate-mollusc-echinoderm community between 100 and 300 meters depth, where massive, upright lithistids replace the branching and encrusting non-lithistids of the higher sponge-coral zone. A corresponding concentration of lithistids was noted in deeper waters adjacent to the calcareous-sponge reefs of the Texas Permian (FINKS, 1960). For a discussion of possible origins of deep-sea and hadal sponge faunas, which may postdate the Cretaceous, see the chapter on Geographic and Stratigraphic Distribution (p. 275).

Where sponges are very abundant and conditions are favorable for their preservation, sedimentary deposits consisting almost entirely of sponge spicules may form. Such **spiculites** seem to be formed from spicules of siliceous sponges. They may be cemented with silica probably remobilized from the spicules themselves to form bedded chert, but they are also known with calcareous or argillaceous matrix. Spiculites are known in the fossil record from the Lower Cambrian. They have almost certainly formed in more than one environment and must be interpreted in terms of abundance of sponges rather than a particular environment. They are forming today on the coast of Antarctica at depths between 50 and 400 meters (DEARBORN, 1965). Spiculites in the basin facies of the Texas Permian may have accumulated in water as deep as 1,800 meters (NEWELL & others, 1953; NEWELL, 1957). On the other hand, some appear to have formed in shallow, nearshore environments, such as the chert of the Permian Phosphoria Formation (YOCHELSON, 1968) that intertongues with nearshore sandstone and shoal-water carbonates. A particularly instructive example was reported by CAVAROC and FERM (1968) from the Pennsylvanian of West Virginia and adjacent states. Spiculitic cherts, such as the Kanawha Flint, pass laterally within 5 to 10 miles into seat-earth and coal beds across an intermediate belt of brachiopod-bearing siltstone and shale. CAVAROC and FERM noted an absence of such spiculites where a wide belt of brackish-water beds separates the coal from normal marine faunas and concluded that the sponges were abundant where normal-marine waters closely approached the shoreline. They suggested that abundant silica from the weathering of the seat-earth provided a favorable environment for siliceous sponges. Abundant organic detritus from the coal-swamp, however, is also a likely cause of sponge proliferation.

SILICA AND SPONGE ECOLOGY

For such planktonic organisms with siliceous skeletons as radiolaria and diatoms, silica is a limiting element because surface waters of the oceans are almost totally depleted in dissolved silica through the activities of these organisms (BROECKER, 1974, p. 7). Siliceous sponges, which are the only benthic silica-secreting organisms of any importance, are not so limited because the rain of planktonic skeletal silica, as well as terrigenous silica in sediments, keeps bottom waters well supplied. Nevertheless, all ocean water is strongly undersaturated in silica (BROECKER, 1974, p. 33), and lateral variation in dissolved silica may have an effect on distribution of sponges. In the present deep sea, silica content of bottom water increases from the Atlantic to the Antarctic to the Indian to the Pacific Oceans, so that the Pacific bottom water has five times the silica concentration of the North Atlantic bottom water, as a consequence of bottom water circulation proceeding in this direction (BROECKER, 1974, p. 23). It is of interest to note that Monaxonida dominate the Atlantic deep-sea sponge fauna, the Lyssacinosa that of the Antarctic, and Hexactinosa-Lychniscosa that of the Pacific (see chapter on Geographic and Stratigraphic Distribution, p. 275). If their distribution is controlled by dissolved silica concentration, the fact that the Monaxonida reached an early peak in the Ordovician (Dystactospongiidae, Anthaspidellidae), the Lyssacinosa-Reticulosa somewhat later in the Devonian to Permian, and the Hexactinosa-Lychniscosa not appearing until the Triassic and peaking in the Cretaceous suggests a possible progressive increase in dissolved silica of the oceans. Inasmuch as the Cretaceous diversity peak of the Hexactinosa-Lychniscosa was greater than their present diversity, a post-Cretaceous decrease in dissolved silica is also suggested. Obviously other explanations are possible.

INFAUNAL SPONGES

Sponges are predominantly epifaunal organisms. Nevertheless, a substantial number of species live almost completely buried in loose sediment or in hard substrates. Those living in hard substrates include not only the well-known boring sponges but also those that occupy borings made by other organisms, including other sponges. Those living in soft sediment are less well known but are of particular interest to the paleontologist because it may be difficult to differentiate such a sponge from a spicule-filled burrow or from a concretion when encountered in the rock. Similar problems may be caused by the many epifaunal sponges that incorporate substantial quantities of sediment in their bodies.

Living sponges that characteristically grow buried in sand include species of the poecilosclerid genera *Biemna, Pellina, Siphonodictyon,* and *Ichnodonax* and the hadromerid genus *Anthosigmella* (DE LAUBENFELS, 1936, p. 66; 1954, p. 112, 167, 201; RÜTZLER, 1974, p. 13; WIEDENMEYER, 1974). The buried part of the sponge is either a ramifying, rhizomelike mass (*Anthosigmella, Ichonodonax*) or a massive, cakelike structure (*Biemna, Pellina, Siphonodictyon*), with oscular chimneys that rise from buried parts to the sediment surface. All the above are known from coral sand in shallow water; sometimes the sand is mixed with organic-rich mud, often near mangroves. Sand may be incorporated in the base of some of these sponges (*Anthosigmella, Biemna*) (RÜTZLER, 1974, p. 13).

Fossil sponges with similar habits are less easy to recognize. A species of a pharetronid calcisponge *Virgola* from the Permian of Texas was buried at least partially in shelly sand during life. The lower parts of the ramifying mass incorporated shell hash continuous with that of the surrounding matrix, as is clearly shown when the silicified matrix and sponge were etched from limestone.

Boring sponges, which live infaunally in excavations of their own making in solid calcium carbonate substrates such as shells, coral skeletons, and lithified limestone, are known in both fossil and recent examples. They are an important cause of erosion of living coral reefs. (In this discussion, the term boring is used for sponges that excavate solid substrates, and the term burrowing for those that displace unconsolidated substrates. The term **interstitial** may be used for those that occupy only **interstices** between grains of unconsolidated sediments; these are also known as cementing sponges (RÜTZLER, 1965a). In addition to the living genera *Cliona* and *Cliothosa,* RÜTZLER (1973) has shown that the living poecilosclerid (adociid) *Siphonodictdyon* and the hadromerids (spirastrellids) *Anthosigmella* and *Spheciospongia* also bore by the same process. Besides those known definitely to bore, the tetraxonid *Samus* occupies or shares borings presumed to have been produced by another sponge (DE LAUBENFELS, 1954, p. 132).

Borings of *Cliona* have a ramifying, anastomosing pattern of tubular galleries of small diameter and more or less uniform spacing; often small, globular spaces are connected by finer tubules. Frequent openings to the surface are present. Some fossil examples have been referred to the genus *Entobia* PORTLOCK, 1843, understood as applying to borings rather than to the organism that produced them (BROMLEY, 1970). The record of such borings goes back to the Lower Cambrian (KOBLUK, 1981a) and Ordovician (KOBLUK, 1981b) where carbonate chips associated with borings have the distinctive scalloped surfaces of clionid-produced debris. The Devonian *Clionoides* and similar Paleozoic branching borings in shells are less likely to have been produced by sponges. The characteristically concavely chipped inner surface of borings produced by sponges is visible with scanning electron microscope and offers a possible means of identifying fossil sponge borings. The chips themselves form an appreciable component of the sediment in the vicinity of boring sponges. They may be identified by their characteristic shape when isolated but recognizing them may not be possible in lithified sediment. Their sizes range from 15 to 94 µm (RÜTZLER & RIEGER, 1973, p. 159). Presence of tylostyles or spinispires in a fossil boring is strong although not conclusive evidence of their production by sponges but would not identify them as those of *Cliona* because other hadromerids also bore. It is of interest to note that most of the sponges that live buried in sand belong to families that include the non-clionid boring sponges (Adociidae, Spirastrellidae). It is apparent that the boring habit is related to a general infaunal adaptation and that either burrowing led to boring or vice versa. Many of the infaunal sponges live in the intertidal zone and are thereby enabled to support exposure at low tide (SARÀ & VACELET, 1973, p. 487).

An interstitial habit is related to the burrowing habit in soft substrates and occurs where the sponge occupies and completely fills interspaces between clastic grains. This habit, in turn, grades into the condition in which the sponge incorporates sediment into its body while remaining above the sediment surface. The two modes of life are difficult to separate in a fossil. Interstitial sponges cement and at least temporarily stabilize loose clastic sediments. Living interstitial sponges include species of the keratose genus *Ircinia,* the poecilosclerids *Hiattrochota* and *Tedania,* and the tetraxonid *Geodia* (DE LAUBENFELS, 1954, p. 124, 127–129; RÜTZLER, 1965a, p. 291).

SPONGES INCORPORATING SAND IN THEIR BODIES

Foreign material in a sponge may be of accidental origin, but in a large number of genera the presence of foreign material and its localization in the sponge seems characteristic of the taxon. Most species of Keratosa incorporate sand in their spongin fibers, frequently only in the larger ascending fibers (in the Dictyoceratida). In addition, many species have the entire ectosome or cortex more or less heavily charged with sand; this is particularly characteristic of the genus *Phyllospongia,* but it occurs in many other genera of the Spongiidae. Other keratose genera have the whole endosome more or less filled with grains and clumps of sand, sometimes to the point where protoplasm of the sponge seems like a cement for the sand. The genus *Dysidea* is particularly characterized by large amounts of sand, a circumstance that does not prevent the sponge from assuming a characteristic external shape. In all these instances, spicules of other sponges, often broken, may be included with the inorganic and other bioclastic debris. Spicules and debris may be both siliceous and calcareous.

The Keratosa may be thought of as using sand in place of spicules, but in other demosponges with proper spicules, sand may also occur in large quantities. In the Haplosclerida sand may be incorporated in spicule tracts of some species (e.g., species of *Callyspongia, Iotrochota*) or in the ectosome (*Desmopsamma*). The Poecilosclerida may include many species that incorporate sand, usually in the endosome. One group, sometimes united in the family Psammascidae DE LAUBENFELS, comprises species in which the great bulk of the body is made of sand, along with the sponge's own spicules. Some species of Hadromerida incorporate notable quantities of sand; they belong to the families Suberitidae and Spirastrellidae. Table 6 lists notable recent genera that include sand-bearing species.

Occasional foreign bodies are as common in fossil sponges as in Holocene ones. More thoroughgoing studies of incorporation of sand into fossil sponges have not been published (Table 6).

PREDATION ON SPONGES

Although sponges have comparatively few predators, a considerable number of animals are specialized predators on them, and many others of more generalized feeding habits include sponges in their diets. Among the specialized predators of sponges, the aberrant polychaete annelid *Spinther* is of interest because of its resemblance to the late Precambrian organism *Dickinsonia.* Many species of *Spinther* have the color of the sponge species on which they feed and live (MacGinitie & MacGinitie, 1968). The related and equally aberrant *Euphrosyne* is also specialized for living and feeding on sponges (Ushakov, 1955). Species of *Haplosyllis* and *Typosyllis* are parasitic inquilines of sponges, especially dictyoceratid Keratosa (Reiswig, 1973; Ushakov, 1955) as are some eunicids (Ushakov, 1955). In addition, suspension-feeding spionids have been reported to be opportunistic feeders on the larvae of *Microciona* and *Ophilitaspongia* (Bergquist & Sinclair, 1968).

Among the gastropods, many species of dorid nudibranchs are specialized feeders on particular species of demosponges, which they mimic in color, while others, somewhat less specialized, bestow their attentions on more than one type of sponge (see Table 7 for details).

Among the echinoderms, a number of asteroids are specialized or occasional predators of sponges (see Table 7): *Henricia sanquinolenta* and *Echinaster sepositus* are specialized predators of sponges and prefer species of *Mycale, Ficulina,* and *Hymeniacidon* (Vasserot, 1961). Reiswig (1973) reported that the echinoids *Eucidaris tribuloidea, Lytechinus variegatus,* and *Tripneustes ventricosus* are major predators on *Tethya crypta* in Jamaica.

Among the arthropods, the larvae of the neuropteran fly *Sisyra* and one stage in the life cycle of the mite *Unionicola* are specialized feeders on freshwater sponges (Sarà & Vacelet, 1973). The decapod crustacean *Typton spongicola* appears to be a parasitic inquiline of sponges (Hunt, 1925; Balss, 1927).

Several species of the bony fish *Pomacanthus, Holacanthus, Cantherhines, Acanthostracion,* and *Chaetodipterus* seem to subsist largely on various demosponges, while other genera and species feed more occasionally on sponges (Bakus, 1964; Randall & Hartman, 1968; Sarà & Vacelet, 1973; Wulff, 1994). The hawksbill turtle, *Eretmochelys imbricata,* is also known to be an active sponge browser (Meylan, 1990).

In general, specialized predators of sponges seem to belong to small, restricted groups in various phyla, and it must be admitted that many of them, such as the nudibranchs and pomacanthid fish, are of late origin, despite the possible Precambrian existence of spintherids.

OTHER ORGANISMS IN SPONGES

Living sponges normally play host to a great number of epibiotic and endobiotic commensals. In some instances the relationship is a regular association. When the sponge and a sessile commensal modify one another's growth, the association provides a powerful tool for determining the constituent elements of the local organism community among fossils. Cross associations among species from a single locality can yield a substantial list of coexisting species (16 genera of 5 phyla from one Permian locality, Finks, 1960, p. 30).

So many organisms are associated with sponges that a list would include most of the phyla of animals. Only a few will be mentioned here. Annelids and crustacea are

TABLE 6. Sand-bearing sponges and sponges submerged in sand (new).

TAXON	LOCATION OF SAND	REFERENCE
KERATOSA		
Aplysilla	whole sponge	de Laubenfels, 1954
Aulena	ectosome	de Laubenfels, 1954
Cacospongia	fibers	de Laubenfels 1954; Vacelet & Vasseur, 1971b
Druinella	fibers	de Laubenfels, 1954
Dysidea	fibers, endosome, ectosome	de Laubenfels, 1936, 1954; Hechtel, 1965
Euryspongia	fibers, ectosome	de Laubenfels, 1954
Heteronema	fibers	de Laubenfels, 1954
Hippiospongia	fibers, ectosome	de Laubenfels, 1954
Hircinia	fibers, ectosome	de Laubenfels, 1954
Ianthella	endosome	de Laubenfels, 1954
Ircinia	fibers, ectosome	de Laubenfels, 1954
Oligieras	fibers, ectosome, endosome	de Laubenfels, 1954
Phyllospongia	fibers, ectosome	de Laubenfels, 1954; Vacelet & Vasseur, 1971b
Polyfibrospongia	fibers	de Laubenfels, 1954
Spongia	fibers	de Laubenfels, 1954
Thorectopsamma	fibers, ectosome	de Laubenfels, 1954
Trypespongia	ectosome	de Laubenfels, 1954
HAPLOSCLERIDA		
Callyspongia	ectosome of cloaca	de Laubenfels, 1936, 1954
Desmapsamma	ectosome	de Laubenfels, 1954; Hechtel, 1965
Fibula	buried in sand	de Laubenfels, 1936
Iotrochota	fibers	de Laubenfels, 1954
Protphlitaspongia	ectosome	de Laubenfels, 1954
Rhizochalina	ectosome	de Laubenfels, 1954; Vacelet & Vasseur, 1971b
POECILOSCLERIDA		
Biemna	sand in base, sponge buried	Rützler, 1974
Clathriopsamma	endosome	de Laubenfels, 1954; Vacelet & Vasseur, 1971b
Didiscus	choanosome	de Laubenfels, 1954; Vacelet & Vasseur, 1971b
Hiattrochota	interstitial	de Laubenfels, 1954
Holopsamma	whole sponge	de Laubenfels, 1936
Ichnodonax	no sand in sponge, although buried	de Laubenfels, 1954
Iotrochopsamma	whole sponge	de Laubenfels, 1954
Lissodendroyx	endosome	de Laubenfels, 1954
Meriamium	endosome, tracts	de Laubenfels, 1936
Mycale	endosome	de Laubenfels, 1954
Myrmekioderma	ectosome	de Laubenfels, 1954
Pellina	endosome	de Laubenfels, 1936
Psammascus	whole sponge, especially ectosome	de Laubenfels, 1936, 1954
Psammolchela		de Laubenfels, 1936
Psammodoryx		de Laubenfels, 1936
Psammopemma		de Laubenfels, 1936
Psammotoxa		de Laubenfels, 1936
Stylotrichophora	tracts	de Laubenfels, 1954
Tedania	interstitial	de Laubenfels, 1954
Siphonodictyon	bores in coral skeleton or buries in sand	Rützler, 1974; Vacelet & Vasseur, 1971b
Tedaniopsamma	choanosome	de Laubenfels, 1954
Thalysias	ectosome	de Laubenfels, 1954
HALICHONDRIDA		
Raphisia	endosome?, whole sponge?	de Laubenfels, 1954
HADROMERIDA		
Aaplos	endosome	de Laubenfels, 1954
Anthosigmella	sand in base, sponge buried	Rützler, 1974
Cliona	bores in shells, corals, and limestone	de Laubenfels, 1936
Laxosuberites	ectosome	de Laubenfels, 1954; Vacelet & Vasseur, 1971b
Spheciospongia	sand in pits on surface	de Laubenfels, 1936
Spirastella	whole sponge, interstitial	de Laubenfels, 1936, 1954
Terpios	endosome	de Laubenfels, 1954
CARNOSA		
Samus	lives in *Cliona* borings	de Laubenfels, 1936

frequent endobionts. Many crabs of the superfamily Oxrhyncha invest themselves with living sponges, along with other benthonic organisms, presumably as camouflage (MACGINITIE & MACGINITIE, 1968). More specific is the association between hermit crabs (Paguridae) and species of *Suberites*. The sponge overgrows the gastropod shell in which the crab lives, extending beyond it and providing the growing crab with an ever-expanding home. The original shell apparently dissolves, leaving an external mold within the sponge.

Another specific association is that of the thin-shelled oyster *Ostrea permollis* that occurs only within the sponge *Steletta grubii* (FORBES, 1964). The oyster has a direct inhalant opening to the surface of the sponge but makes use of the sponge's exhalant system for its own exhalant stream. Several species of the scallop *Chlamys* are regularly coated with specific sponges. *Chlamys hericius* and *C. hindsii* of the Puget Sound area are coated on their upper valves with *Myxilla parasitica* or *Mycale adhaerens* (MACGINITIE & MACGINITIE, 1968). The sponges apparently occur preferentially or exclusively on this substrate. In New Zealand, *Chlamys diefenbachi* develops spines only when encrusted by a sponge (unidentified), which is its most common condition (BEU, 1965). The *Chlamys* is completely enclosed by the sponge. Terebratuloid brachiopods have also been found coated with sponges during life (G. A. COOPER, personal communication, 1955).

Among the coelenterates, the scyphozoan *Stephanoscyphus* goes through one state of its life cycle embedded in various sponges (SARÀ & VACELET, 1973). If the conulariids are scyphozoans, this association may have considerable antiquity, for Permian conulariids have been found embedded in several species of lithistid and calcareous sponges (FINKS, 1955, 1960). The hydroid *Dupurena halterata* lives imbedded in suberitids (SARÀ & VACELET, 1973).

Other probable commensal associations known from both living and fossil sponges include ophiuroids from as far back as the Devonian (CLARKE, 1912, 1921) and pleurotomarian archaeogastropods from as far back as the Pennsylvanian (BATTEN, 1958; FINKS, 1960), both occurring in the cloacae of various large sponges.

Many microorganisms inhabit living sponges, but it is not always possible to tell whether they are commensal, symbiotic, or parasitic. There are four groups, however, where a symbiotic relationship seems likely. These are zoochlorellae (unicellular Chlorophyceae) in freshwater spongillids; zooxanthelae (unicellular Chrysophyta) in some marine demosponges; cyanophyta in other marine demosponges; and bacteria in both demosponges and calcisponges (SARÀ & VACELET, 1973). The symbioses with prokaryotes are of particular interest.

Sponges are the only metazoans that have a symbiosis with blue-green algae. Likewise, the extent to which symbiotic bacteria participate in the bulk of some sponges has no parallel among other organisms. The established great antiquity of prokaryotes suggests the possibility of great antiquity of the sponges and of the symbioses between them.

The symbiotic bacteria (*Pseudomonas* sp. and *Aeromonas* sp.) of the keratose sponge *Verongia cavernicola* regularly constitute 38 percent of the tissue volume of the individual sponge. By comparison, cells of the sponge constitute only 21 percent, the remainder (41 percent) being intercellular substance (mesogloea) (SARÀ & VACELET, 1973). Although the same species of bacteria are capable of living outside the sponge, they have special morphological features when they are within the sponge (thicker cell walls); and when the sponge dies, the bacteria also die. The sponge appears to crop constantly its population of bacteria by **phagocytosis**, using them as a food source and maintaining constant population density. This confirms the symbiotic nature of the association (SARÀ & VACELET, 1973). The same authors noted that similar large symbiotic bacteria populations occur in many other demosponges, not only among Keratosa but also in tetractinellids. It also

TABLE 7. Predators, their degree of specialization, and sponge taxa with which they interact (new).

PREDATOR	SPONGE	DEGREE OF SPECIALIZATION	REFERENCE
Mollusca, Gastropoda, Opisthobranchia			
Doris sp.	*Halichondrida panacea*	specialist (mimics color of sponge)	Sarà & Vacelet, 1973
Rosanna rubicund	*Microciona coccinea* *Ophilitaspongia soriata* *Holoplocamium neozelanicum*	specialist	Ayling, 1968
Rosanna pulchra	orange-red sponge	specialist (mimics color of sponge)	MacGinitie & MacGinitie, 1968
Peltodoris atromaculata	*Petrosia ficiformis*	specialist	MacGinitie & MacGinitie, 1968
Tylodina fungina	*Verongia thiona*	major predator	MacGinitie & MacGinitie, 1968
Tylodina perversa	*Verongia aerophoba*	specialist (mimics color of sponge)	Sarà & Vacelet, 1973
Glossodoris tricolor	*Cacospongia scalaris*	specialist	Sarà & Vacelet, 1973
Archidoris stellifera	*Stylotella*	specialist	Morton, 1967
Archidoris pseudoargus	*Halichondrida*	specialist	Morton, 1967
Archidoris britannica	various demosponges	specialist on demosponges but not a particular species	Sarà & Vacelet, 1973
Prosobranchia			
Lamellaria orbiculata	*Lissodendoryx noxiosa*	major predator	MacGinitie & MacGinitie, 1968
Fissurellidae, including *Emarginula conica, E. crassa, E. reticulata*	sponges, especially monaxonida	specialist	Graham, 1971; Fretter & Graham, 1976a
Puncturella noachina	sponges	specialist	Graham, 1971
Diodora apertura	sponges, especially *Halichondrida* and *Hymeniacidon*	specialist	Fretter & Graham, 1976a
Cerithiopsis tubercularis	*Halichondrida, Hymeniacideon, Grantia*	specialist	Fretter & Graham, 1976a
Triphora perversa	*Halicondria, Hymeniacidon*	specialist	Fretter & Graham, 1976a
Drupe ricina	sponges and holothurians	nonspecialist	Wu S. K., 1965
Patella	various sponges	occasional nonspecialist	Sarà & Vacelet, 1973
Theodoxus	green protophytes; sponges	nonspecialist	Fretter & Graham, 1976a
Littorina	various sponges	occasional nonspecialist	Sarà & Vacelet, 1973
Cypraea friendi, C. venusta, C. marginata Citeres, C. cervus	sponges	specialist	Wilson & Gillette, 1971; Taylor & Wells, 1975
Polyplacophora			
Placiphorella velata	various sponges	occasional nonspecialist	MacGinitie & MacGinitie, 1968
chitons	various sponges	occasional nonspecialist	MacGinitie & MacGinitie, 1968; Sarà & Vacelet, 1973

Annelida, Polychaeta			
Haplosyllis spongicola	*Verongia gigantea* and other dictyoceratids	parasitic inquiline	Reiswig, 1973; Ushakov, 1955
Typosyllis sclerolaema	sponges	parasitic inquiline	Ushakov, 1955
a polychaete	*Ianthella ardia*	?	Sarà & Vacelet, 1973
spionid polychaetes	larvae of *Microciona* and *Ophilitaspongia*	occasional but temporary major predator (suspension feeder)	Bergquist & Sinclair, 1968
Spinther citrinus	yellow sponges	specialist (mimics color of sponge)	MacGinitie & MacGinitie, 1968
Spinther spp. (all spp.)	sponges	specialist	Ushakov, 1955
Euphrosyne spp.	sponges	feeds mainly on sponges (parasitic inquiline)	Ushakov, 1955
some Eunisids	sponges	parasitic inquiline	Ushakov, 1955
Arthropoda, Crustacea, Decapoda			
Typton spongicola	?	specialist? (parasitic inquiline)	Hunt, 1925; Balss, 1927 (*fide* Sarà & Vacelet, 1973)
Cammarus	spongillids	occasional?	Sarà & Vacelet, 1973
deep water Mediterranean prawns	sponges	occasional	Sarà & Vacelet, 1973
Copepoda			
copepods	sponges	?	Sarà & Vacelet, 1973
Insecta, Neuroptera			
Sisyra larvae	spongillids	specialist (major predator)	Killian, 1964 (*fide* Reiswig, 1973); Sara & Vacelet, 1973
Arachnida, Acarina			
Unionicola	spongillids	specialist (part of life cycle)	Sarà & Vacelet, 1973
Echinodermata, Asteroidea			
4 spp. of asteroids	Antarctic sponges *Rosella* and other lyssacines	major predator	Dayton, Robilliard, & Paiune, 1970; *fide* Reiswig, 1973; Sarà & Vacelet, 1973
Asterina gibbosa	sponges	major predator	MacGinitie & MacGinitie, 1968
Patiria miniata	*Myxilla noxiosa*	occasional	MacGinitie & MacGinitie, 1968
Henricia sanguinolenta	*Mycale, Ficulina, Hymeniacidon*	specialist	MacGinitie & MacGinitie, 1968; Vasserot, 1961 (*fide* Sarà & Vacelet, 1973)
Echinaster sepositus	*Mycale, Ficulina, Hymeniacidon*	specialist	MacGinitie & MacGinitie, 1968; Vasserot, 1961 (*fide* Sarà & Vacelet, 1973)
Pteraster tessellatus	sponges	occasional	MacGinitie & MacGinitie, 1968
Echinoidea			
Eucidaris tribuloides	*Tethya crypta*	major predator	Reiswig, 1973
Lytechinus variegatus	*Tethya crypta*	major predator	Reiswig, 1973
Tripneustes ventricosus	*Tethya crypta*	major predator	Reiswig, 1973

occurs among Calcarea (*Leuconia* and *Clathrina*), although the bacteria are of a different type (spiral form).

Symbiotic cyanophytes are also widespread among sponges, sometimes lending their blue-green color to the sponge. As with the bacteria, the algae are phagocytosed by the sponge (SARÀ & VACELET, 1973). Multicellular algae also occur in symbiotic association with sponges. Some Keratosa use the algal filaments as partial or total replacement for their own skeletal fibers (SARÀ & VACELET, 1973), for example the alga *Jania* in the sponge *Dysidea fragilis. Gellius cymiformis* is so intertwined with its algal symbiont *Ceratodictyon spongiosum* that the form of the ensemble is controlled by the alga (SARÀ & VACELET, 1973). It would seem likely that such associations with macroscopic algae would be recognizable in the fossil record, although I do not know of any such that have been reported.

SUBSTRATE

It is widely stated in various ecologic discussions that many sponge species are confined to particular substrates. This is obviously of great paleoecological interest. Unfortunately, the species are rarely identified, and it is not possible to make useful generalizations.

LIGHT

Many sponge species require light for symbiotic algae or for other clearly understood reasons. SARÀ and VACELET (1973) summarized much information on the distribution of sponges in marine caves, where light intensity seems to be a principal control. The living sclerosponges, as well as many pharetronids and lithistids, are more abundant in the totally dark parts of such caves than elsewhere. It has been speculated that lack of competition from the light-requiring and faster-growing reef corals and comparable sessile benthos are the chief reasons for the limitations of these sponges in dark environments.

SALINITY

The vast majority of sponges require full marine salinity for survival. Nevertheless, some monaxonid demosponges are adapted to brackish or fresh water. The Spongillidae and Potamolepidae are specialized freshwater dwellers. Some species of clionids (Clavulina: Hadromerida) and some other marine monaxonids are tolerant of estuarine conditions (HARTMAN, 1958b).

TEMPERATURE

Some taxonomic groups of varying levels in the hierarchy are confined to warmer waters, such as the Keratosa. Information on higher categories is summarized in the chapter on Geographic and Stratigraphic Distribution (p. 275), but for genera and species few useful generalizations can be made. Many genera and certainly higher taxa have an extremely wide range of temperature tolerance, although individual species are probably restricted more narrowly.

EVOLUTION AND ECOLOGIC HISTORY OF SPONGES DURING PALEOZOIC TIMES

Robert M. Finks

[Department of Geology, Queens College (CUNY)]

The best sample of earliest sponge faunas is that of the Middle Cambrian Burgess Shale of western Canada, although it is not the earliest record of sponges. Impressions of what are interpreted as hexactinellid sponges have been reported from the Neoproterozoic Ediacaran beds of Australia (Gehling & Rigby, 1996). Isolated spicules have been reported from probably equivalent beds from the Yangtze Gorge area of China (Steiner & others, 1993); hexactines and monaxons have been reported from the Neoproterozoic of India by Tiwari, Pant, and Tewari (2000); and monaxons have been described from late Precambrian rocks from northern Iran (Brasier, 1992). The earliest known body fossils of Early Cambrian sponges are those described by Steiner and others (1993) from the lowermost Cambrian Niutitang Formation in northern Hunan Province, China. No major types have been recovered from Precambrian or Lower Cambrian strata that are not also represented in the Burgess Shale. Three major classes of the phylum Porifera were already present in the Early Cambrian: the Hexactinellida, Heteractinida, and Demospongea (Rigby, 1986a, 1987b).

The hexactinellids (*Protospongia*) and at least some of the heteractinids (*Eiffelia*) and demosponges (*Leptomitus*) of the Burgess Shale (Walcott, 1920) are extremely thinly walled, saclike sponges in which the spicules form essentially a single layer (Fig. 109). Rays of principal spicules are paratangential to the body wall of the sponge. Such a skeleton could not have supported a thick-walled sponge, and among the hexactinellids, there would have been no mesogloea for support either. Thus, these early hexactinellids of the *Protospongia* type may have had an asconoid structure, a possibility strengthened by the fact that embryos of living hexactinellids are asconoid and have a monolayered spiculation of stauractines like that of *Protospongia*. Although the earliest hexactinellids may have been asconoid, they are nearly all thick walled and presumably leuconoid from the Ordovician onward.

All Early Cambrian hexactinellids are of the *Protospongia* type, or consist of isolated stauractines. The first hexactines, implying a thicker body wall, appear in the Middle Cambrian, although it was not until the discovery of the Late Cambrian *Multivasculatus* that a coherent skeleton of hexactines was known as a fossil (Fig. 110). If the earliest hexactinellids were asconoid, it took about a third of Cambrian time, say 30 million years, to develop a leuconoid hydraulic system. Among the hexactinellids, possible asconoid *Protospongia* type forms persisted into the Ordovician, where they are found in black-shale facies.

The earliest demosponges from Lower Cambrian rocks in Vermont and Pennsylvania and from the Middle Cambrian Burgess Shale are thin-walled, tubular forms but obviously thick enough to have been of leuconoid architecture. In view of the fact that living demosponge embryos develop a leuconoid canal system directly, members of this class may never have had asconoid or syconoid forms.

Spiculation of the Burgess Shale demosponges generally consists almost entirely of oxeas. They are arranged either in an isodictyal net, with mesh spaces as wide as one spicule length or in ramifying, subparallel, anastomosing tracts in which fascicles of oxeas may have a slightly plumose arrangement (Walcott, 1920; Rigby, 1986a). Both types of organization occur in the genus *Hazelia* (Fig. 111). It is likely that spongin held the spicules together. Tangential or perpendicular oxeas supporting a dermal

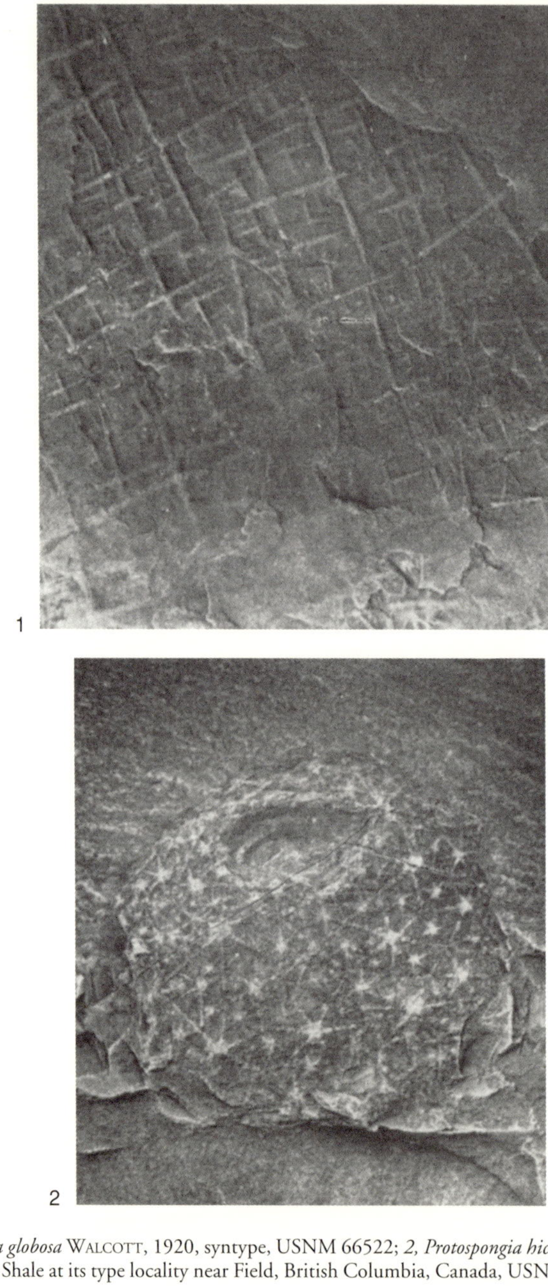

FIG. 109. *1, Eiffelia globosa* WALCOTT, 1920, syntype, USNM 66522; *2, Protospongia hicksi* HINDE (WALCOTT, 1920), hypotype, Burgess Shale at its type locality near Field, British Columbia, Canada, USNM 66502, ×3 (Finks, 1970).

membrane may have been present in some forms, but no specialized spicules are known.

The demosponges were the first group to undergo a major expansion or adaptive radiation in the Ordovician. Perhaps arising from such Middle Cambrian genera as *Rankenella* or *Capsospongia* (KRUSE, 1983; RIGBY, 1986a) or the Late Cambrian *Wilburnicyathus* and *Gallatinospongia* but certainly appearing by the Early Ordovician *Archaeoscyphia,* the lithistid family Anthaspidellidae expanded to some seventeen de-

FIG. 110. *Multivasculatus ovatus* HOWELL & VAN HOUTEN, hypotype, USNM 163626, Windfall Formation, Eureka, Nevada, USA, ×30 (Finks, 1970).

scribed genera in lower Middle Ordovician rocks (Llanvirn–Llandeilo). The skeleton is essentially an isodictyal net of the sort seen in the Cambrian *Hazelia* in which constituent monaxons have developed terminal zygoses. Sponges of the family participated in the building of Lower (Arenig) and Middle (Llanvirn) Ordovician reefs and occur elsewhere in shallow-water, shelly facies. The lithistid modification is undoubtedly adaptive for such an environment, conferring strength and resistance against mechanical disruption in rough water. It may indeed be that the availability of such niches gave direction to this branch of demosponge evolution. Adaptive radiation was cut short in the later Ordovician, as was the participation of lithistid sponges in reefs, although the family persisted into the Permian. The rhizomorine type of lithistid skeleton, in which zygoses develop along the sides as well as at the ends of spicules, also appeared in the Ordovician (*Warrigalia, Taplowia, Boonderooia, Nipterella*).

By the later Middle Ordovician (Caradoc, Trentonian) at least three other major types of demosponges evolved and constituted three separate lineages that continued into later times. One of these lineages is represented by *Saccospongia* and related dystactospongiids, with skeletons built of subparallel plumose bundles of styles coated with monocrepid desmoids. Very similar structures are present in living sublithistid Sigmatosclerophora and a direct line of descent is quite possible (FINKS, 1967a). One could even see the Cambrian roots of this lineage in a form like *Hazelia palmata* or *Hazelia dignata* (WALCOTT, 1920; RIGBY, 1986a).

Two types, rather different from the foregoing, also first appeared at this time, namely *Hindia* and *Astylospongia*. The first is built of concentric shells of tripodal

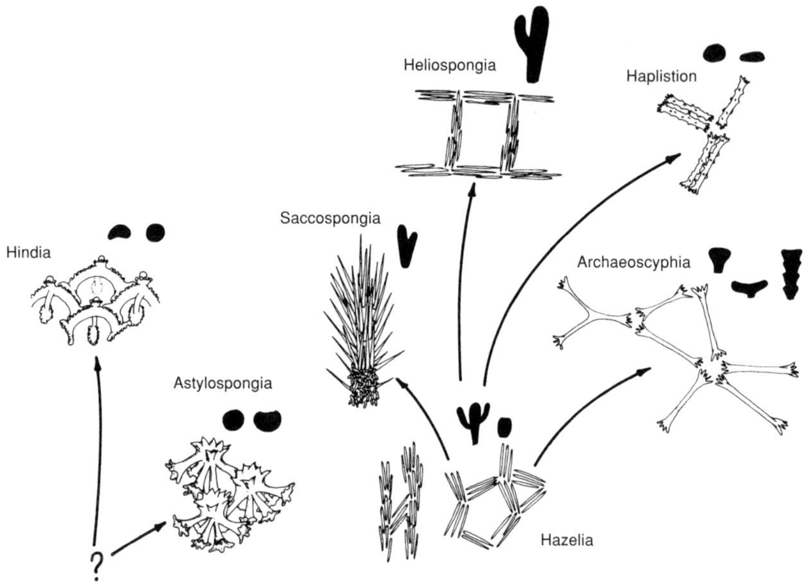

FIG. 111. Principal types of skeletal nets among Paleozoic demosponges showing postulated relationships; forms of whole sponges shown in silhouette (Finks, 1970).

spicules, the second of a three-dimensional net of hexapodal spicules. Both have, in addition, radial and tangential oxeas that probably supported a dermal membrane (Fig. 112). Both evolved into Mesozoic and later forms (Dicranocladina and Sphaerocladina, respectively) that possess tetraxonid dermalia and astrose microscleres.

At this juncture in demosponge history it seems not unreasonable to recognize the already differentiated roots of the Calvaxinellida and Ceractinomorpha, represented by *Saccospongia* and its nonlithistid relatives, and at least some of the roots of the Tetractinomorpha, e.g., the hindiids and astylospongiids (Fig. 113).

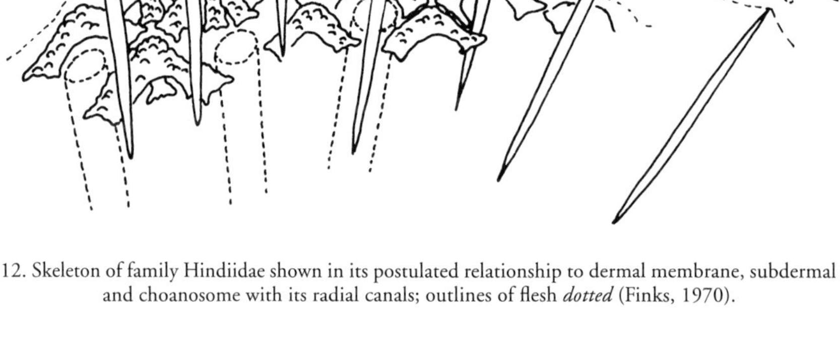

FIG. 112. Skeleton of family Hindiidae shown in its postulated relationship to dermal membrane, subdermal space, and choanosome with its radial canals; outlines of flesh *dotted* (Finks, 1970).

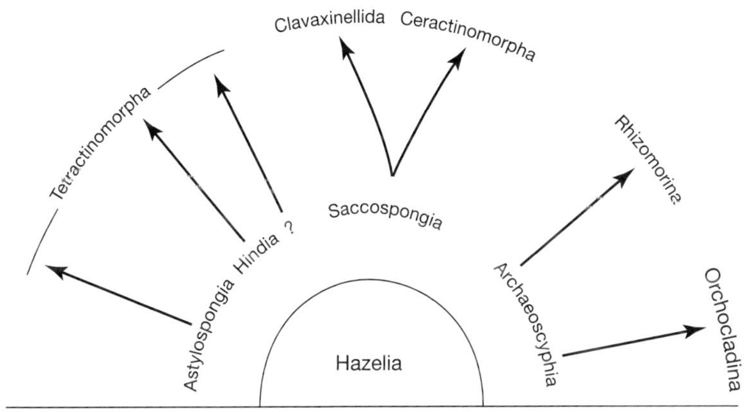

FIG. 113. Postulated lines of descent of living demosponge taxa from Cambrian and Ordovician genera and their relatives (Finks, 1970).

Work by HARTMAN and GOREAU (1966) has strongly suggested that at least some of the Paleozoic Stromatoporoidea were ancestors of the living aberrant, lime-secreting demosponges *Ceratoporella, Astrosclera,* and *Merlia.* If so, this line of sponge evolution diverged early, for the first stromatoporoids appeared in the Cambrian. Furthermore, they predate, by far, any forms that have been referred, even remotely, to the Calcarea (except for the Heteractinida and the Archaeocyatha). The Stromatoporoidea, it

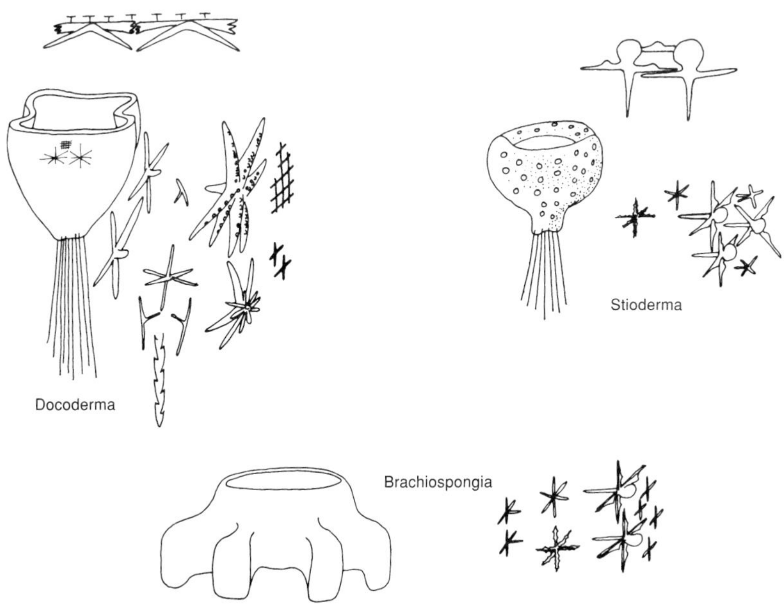

FIG. 114. Paleozoic hexactinellids related to *Brachiospongia; Brachiospongia* is of Ordovician age, *Stioderma* is of Carboniferous and Permian ages, and *Dococerma* is of Permian age; spicular skeleton is shown in exploded view, outside of sponge to right; above *Dococerma* and *Stioderma* are side views of dermal layer showing manner of joining of hypodermalia and autodermalia; scale variable (Finks, 1970).

1

2

FIG. 115. *Docoderma rigidum* FINKS, 1960, showing fusion of hypodermal and autodermal spicules; *1,* outer surface of fused hypodermal spicules; *2,* another part of same specimen in which fused autodermalia overlie hypodermalia; holotype, USNM 127659, Road Canyon Formation, Permian, Word Ranch, Glass Mountains, Texas, USA, ×5 (Finks, 1970).

may be noted, had their acme of development in the Silurian and Devonian, at which time they and the rugose and tabulate corals were the chief reef-building animals.

The mid-Ordovician also marked the first appearance of an advanced hexactinellid *Brachiospongia* (Fig. 114), which has specialized, large, hypodermal spicules, covered by smaller autodermalia. Its body wall is much thicker than that of any of the known Cambrian sponges, and its spicules lack the regular parallel orientation of the Cambrian forms. Silurian forms such as *Oncosella* may have continued this line, and a late Paleozoic form (*Stioderma*) has a remarkably similar complement of spicules, especially in the form of the hypodermalia and the interior spinose hexactines. Other Permian forms such as *Docoderma* have a similar construction. In apparent late Paleozoic end members of this lineage, hypodermalia and autodermalia tended to fuse in the adult sponges, forming a rigid skeleton (Fig. 115). *Docoderma* developed interlocking processes on hypodermal spicules, a function parallel to the skeleton of the lithistid demosponges

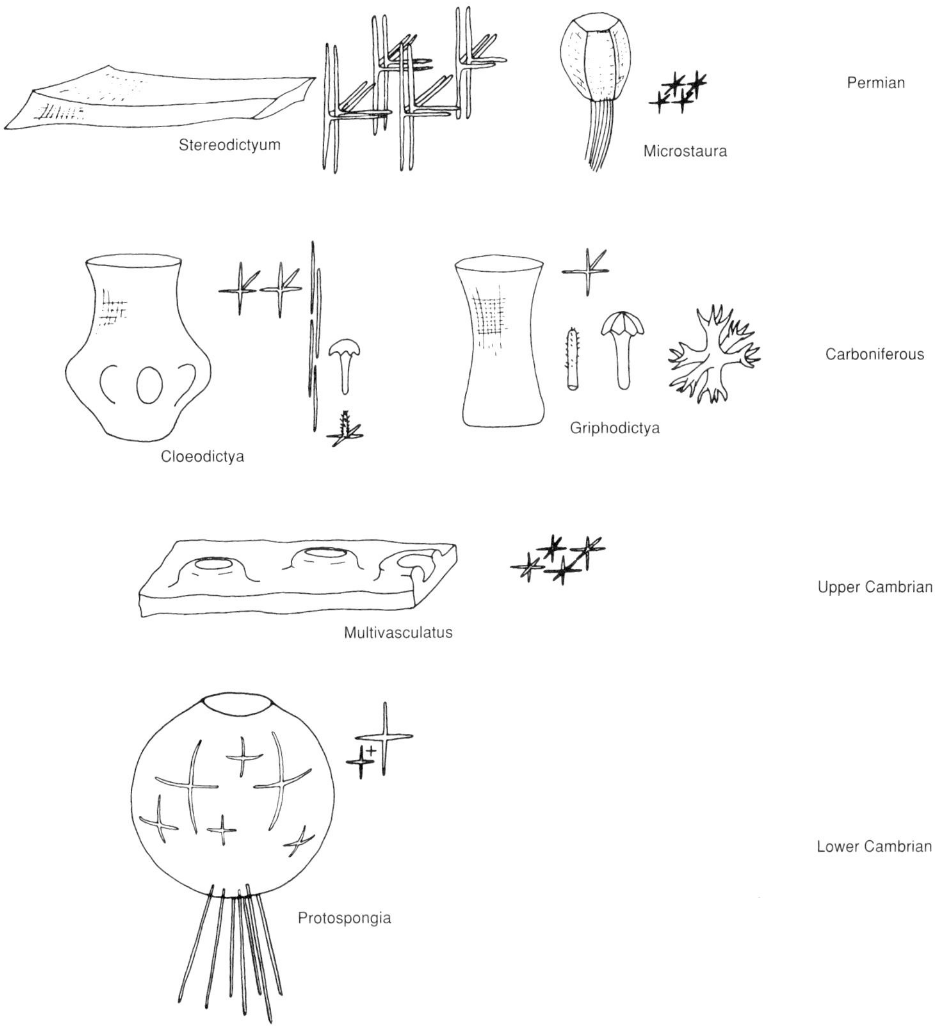

FIG. 116. Paleozoic hexactinellids showing variants of simple, parallel, spicule arrangement characteristic of *Protospongia,* from which they may have descended; geologic periods are shown along side, scale variable; two spicules on far right beside each Carboniferous dictyosponge are microscleres, approximately ×3,000 (Finks, 1970).

and apparently unique among the hexactinellids (see FINKS, 1960).

The Ordovician record also contains the beginnings of a second hexactinellid lineage springing directly from the Cambrian *Protospongia* (Fig. 116). *Cyathophycus* of the Ordovician is like a thicker-walled *Protospongia* and is not too different from the simplest of the Devonian dictyosponges, namely *Dictyospongia* itself (see HALL & CLARKE, 1899). In this line of development the spiculation remains that of simple hexactines and their derivatives, in parallel orientation without enlarged hypodermalia. The structural type and presumed lineage can be traced through to the Permian *Microstaura,* with a curious thick-walled offshoot, *Stereodictyum.* The dictyonine type of rigid net, which first appeared in the Devonian, could have been derived from this lineage, and the presence of hexasters in a Carboniferous dictyosponge (*Griphodictya* HALL

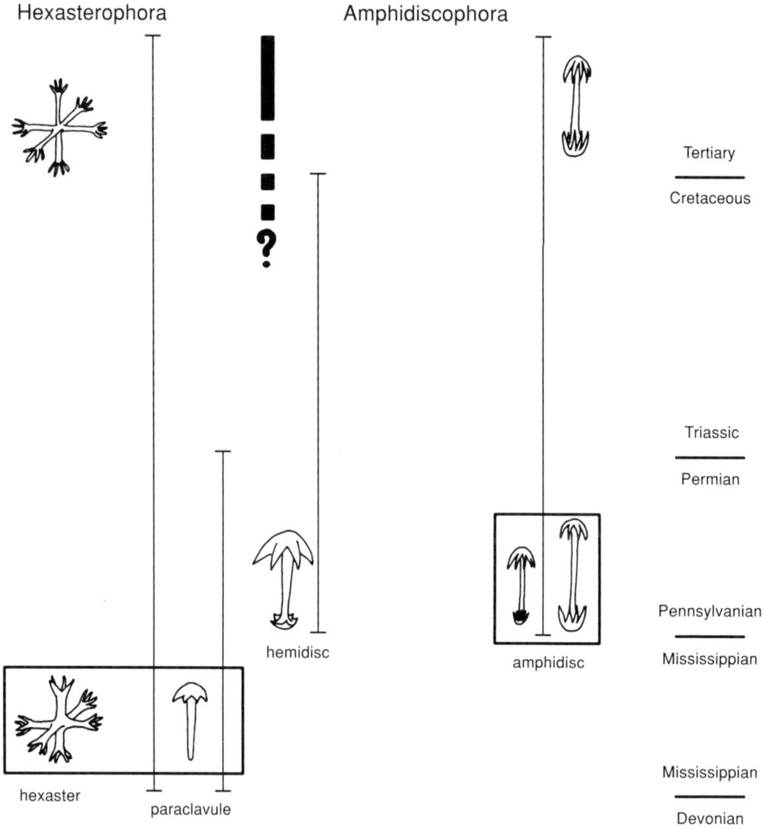

FIG. 117. Ranges of diagnostic microscleres in class Hexactinellida; two spicules enclosed by a rectangle occur in same specimen (Finks, 1970).

& CLARKE, 1899) does not invalidate this hypothesis.

The Paleozoic history of hexactinellid microscleres is worth noting, inasmuch as the classification of living hexactinellids is based on the mutual exclusion of hexasters and amphidiscs. In addition to hexasters, *Griphodictya* possesses paraclavules, as do several other dictyosponge genera. Paraclavules are shaped like one-ended amphidiscs. Unequal-ended amphidiscs, called hemidiscs, are known as isolated spicules from the Cretaceous (ORTMANN, 1912; SCHRAMMEN, 1924a). True amphidiscs are known almost as early as hexasters, having been reported from the Upper Carboniferous (*Uralonema,* LIBROVICH, 1929). KLING and REIF (1969) reported two Late Carboniferous species, one bearing amphidiscs, the other bearing hemidiscs. At least one of the amphidiscs they described is slightly unequal ended (Fig. 117). The graded series paraclavule-hemidisc-amphidisc and the occurrence of paraclavules with hexasters suggest that the Hexasterophora and Amphidiscophora could have diverged from a common stock during the Paleozoic and that they could be the sole survivors of a number of early lineages in some of which both hexasters and amphidiscs could have occurred together.

The Heteractinida, in their relatively modest history, trend in general toward increasing complexity and diversity of the spicules (Fig. 118). The main line of heteractinid evolution seems to start with the

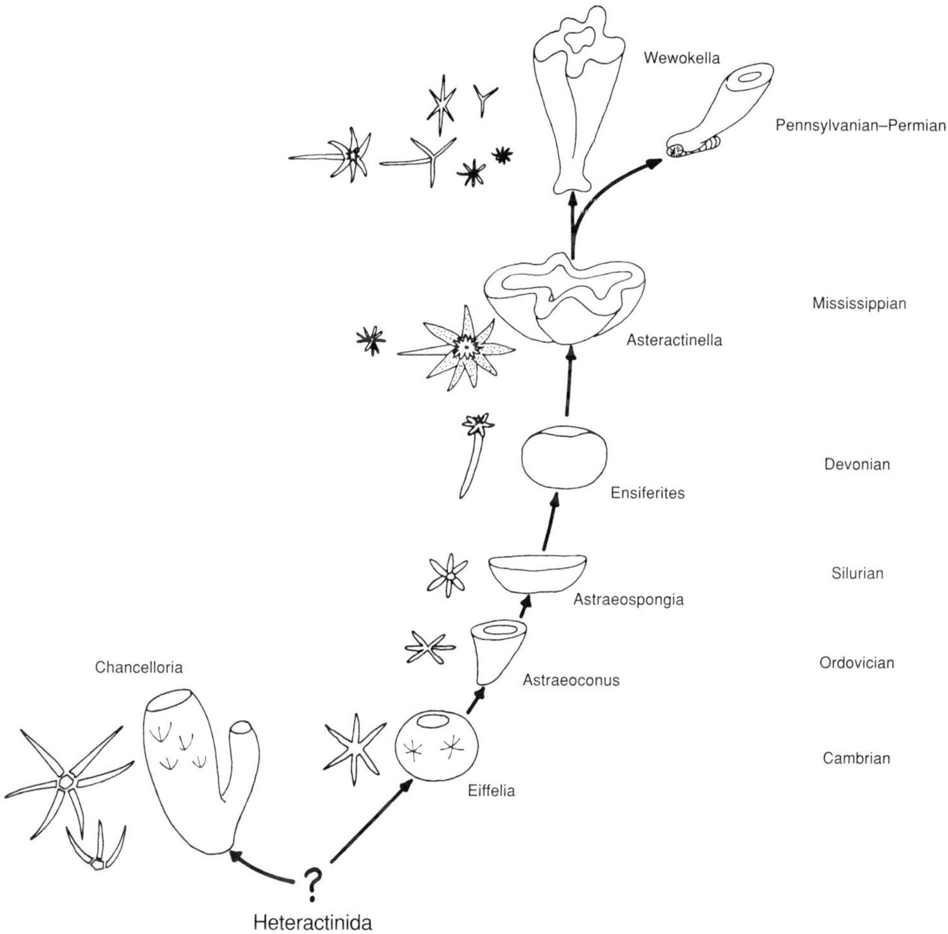

FIG. 118. Evolution of Heteractinida; geologic periods indicated along right side; spicule complements shown to left of each sponge; scale variable (Finks, 1970).

Cambrian *Eiffelia.* The symmetrical, two-dimensional sexiradiates of *Eiffelia* continued in the Ordovician thicker-walled sponge *Astraeoconus* (RIETSCHEL, 1968), but in the Silurian and Devonian *Astraeospongium,* short proximal and distal rays may be developed. The Devonian *Ensiferites* (REIMANN, 1945b) has a greatly enlarged proximal ray, and the distal ray may have dissolved into a rosette of short branches. The Early Carboniferous *Asteractinella,* described by VANDERCAMMEN (1950), continued this trend with spicules having recurved, umbrella-like paratangential rays, often more than six in number and bearing a distal rosette or tubercles, to which was added a new type of spicule, the globular polyactine. The later Carboniferous *Wewokella* and the Carboniferous-Permian *Regispongia* have essentially the same types of spicules as *Asteractinella,* but the paratangential rays of the principal spicules are often reduced to three in *Wewokella.* Simple triradiates and sexiradiates are also present.

The Calcarea in the strict sense (that is, excluding the Archaeocyatha, Aphrosalpingoidea, Heteractinida, and Stromatoporoidea) appeared later in the record than

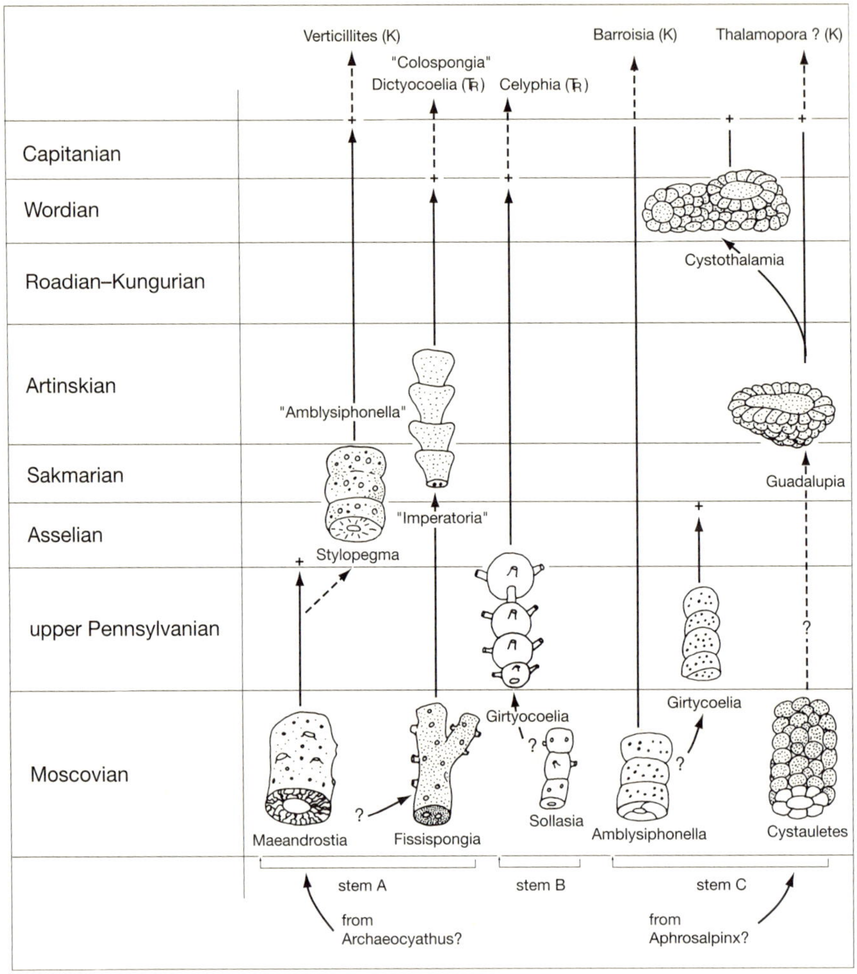

FIG. 119. Postulated lineages among Late Paleozoic Sphinctozoa from Carboniferous until their extinction; relative scale of sponge drawings is approximately uniform (Finks, 1970).

other classes of sponges. The Devonian *Protoleucon* (BOLKHOVITINOVA, 1923) is not clearly a calcareous sponge. The isolated triradiates and tetraradiates from the Lower Carboniferous described as *Peronidella sparse* by HINDE (1888) may belong to the Calcarea and, if so, are the oldest described spicules of the class, although affinity to the Heteractinida cannot be excluded.

The Sphinctozoa have an undoubted appearance in the Ordovician of Australia (WEBBY & RIGBY, 1985) and North America (RIGBY & POTTER, 1986) but have only a limited record until the Middle Carboniferous of Spain (Bashkirian) and North America (Des Moinesian, Moscovian) (STEINMANN, 1882; BARROIS, 1882; FINKS, 1960). Among these early Sphinctozoa (Fig. 118–119) are three major structural types: (1) those built of porous-walled chambers containing internal solid partitions (*Amblysiphonella, Sebargassia, Girtycoelia, Cystauletes*); (2) those built of imperforate-walled, hollow chambers with a small number of spoutlike openings (*Sollasia, Girtyocoelia*); and (3) those containing

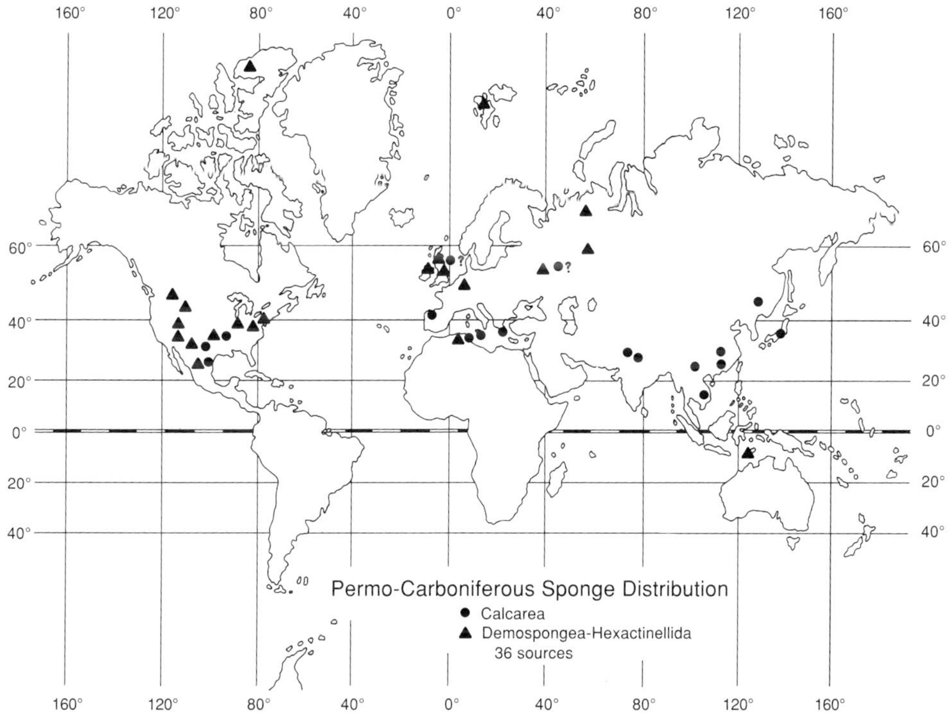

FIG. 120. Geographic distribution of reported occurrences of known Calcarea, Hexactinellida, and Demospongea from Carboniferous and Permian rocks; some localities close to each other have been combined (Finks, 1970).

coarsely perforate dermal and cloacal walls (*Maeandrostia, Fissispongia*). Because in Texas and elsewhere each of the structural types can be traced through continuously developing lineages into the Permian and possibly beyond, they may represent three basic evolutionary stocks among the Sphinctozoa. They appear suddenly in the record, and there is no trace of their prior divergence from a common ancestor. It is worth noting, however, that *Maeandrostia* has an overall structure very much like that of Early Cambrian *Archaeocyathus,* and *Cystauletes* is not too different from that of Silurian-Devonian *Aphrosalpinx.*

The earliest Pharetronida appear to be two genera that enter the record in the Lower Permian (Lenox Hills, Sakmarian) of Texas. Other genera appeared in the Permian in several parts of the world, and the group expanded into the Mesozoic. The Calcarea were active in reef building in Permian time and continued so into the Triassic.

The late Paleozoic Calcarea (Fig. 120) have a more concentrated geographic distribution than the siliceous sponges that could be termed Tethyan. This may have resulted from climatic control and perhaps also from a center of origin.

All the classes of sponges have had at least some representatives that lived in shallow water during the Paleozoic. This can be seen from a list of those groups that have been associated with algae, that is, with the euphotic zone, or with reefs that reveal evidence of having grown in rough water, such as accumulations of talus (Fig. 121). Persistent environmental preferences of some groups of sponges, however, can be demonstrated by noting their association with specific types of sediment. Also, associations of

period	occurring with algae	occurring with reefs
Permian	Hexactinellida Demospongea Heteractinida Calcarea	Hexactinellida Demospongea Calcarea
Carboniferous	Hexactinellida Demospongea Heteractinida Calcarea	Hexactinellida Demospongea
Devonian	Hexactinellida Demospongea? Receptaculitida	Hexactinellida Demospongea? Heteractinida? Receptaculitida
Silurian	Hexactinellida? Demospongea? Heteractinida Receptaculitida	Demospongea Heteractinida Receptaculitida
Ordovician	Hexactinellida Demospongea Receptaculitida	Hexactinellida? Demospongea Receptaculitida
Cambrian	Archaeocyatha	Archaeocyatha

FIG. 121. Shallow-water sponges of each period as shown by occurrences with algae or reefs; *question marks* indicate uncertainty as to reality of association in life (Finks, 1970).

sponges characteristic of specific environments can be traced through time, and changes in the composition of these associations can be noted (Fig. 122).

One such environment is quiet water, identifiable by fine-grained sediment and other geologic criteria. This environment has been the special province of the siliceous sponges (Hexactinellida, Demospongea). Abundant occurrences of Paleozoic hexactinellids are almost always in such sediments, although sporadic occurrences in rougher waters and even in reefs are known. Quiet water is not necessarily deep; where depth can be estimated, estimates range from 500 meters in the basin facies of the Texas Permian (NEWELL & others, 1953; NEWELL, 1957) to 1 m in the Late Carboniferous coastal lagoon represented by a black shale overlying a coal bed (ZANGERL & RICHARDSON, 1963). Black shales form a distinct subgroup within this environment and were the special home of thin-walled, delicate hexactinellids of persistently primitive type.

A different environment of agitated, shallow water with relatively little suspended sediment, identified by deposits built largely of whole or fragmented, coarse skeletal debris, is commonly designated the shelly facies. It has a characteristic sponge fauna from the Ordovician on. Lithistid demosponges and heteractinids are particularly characteristic of this facies, as are the stromatoporoids. In the Ordovician record, anthaspidellid lithistids are particularly dominant among the sponges of this facies. In the Silurian and Devonian record the facies is marked by the presence of hindiid lithistids and the astraeospongiid heteractinids, accompanied in the Silurian by another lithistid group, the astylospongiids. In

period	quiet water	shelly facies	reef builders
Permian	demosponges hexactinellids	Sphinctozoa Pharetronida Chiastoclonellidae Heliospongiidae* Wewokellidae* Rhizomorina Hindiidae (Hexactinellida)	Sphinctozoa Pharetronida
Carboniferous	demosponges hexactinellids	Sphinctozoa Wewokellidae Heliospongiidae (Rhizomorina) (Hexactinellida)	
Devonian	demosponges hexactinellids	Stromatoporoidea Hindiidae Astraeospongiidae Receptaculitida (Hexactinellida)	Stromatoporoidea
Silurian	demosponges hexactinellids	Stromatoporoidea Astylospongiidae Astraeospongiidae Hindiidae (Hexactinellida?)	Stromatoporoidea
Ordovician	demosponges hexactinellids	Stromatoporoidea Anthaspidellidae Hindiidae Dystactospongiidae Astylospongiidae Receptaculitida (Hexactinellida?)	Stromatoporoidea Anthaspidellidae Receptaculitida
Cambrian	demosponges hexactinellids heteractinids	Archaeocyatha Stromatoporoidea	Archaeocyatha

FIG. 122. Sponge taxa associated with three types of environments in different periods; under shelly facies and reef builders, taxa arranged in approximate order of abundance, with most abundant on top; those in *parentheses* are minor elements; *asterisks* indicate extinction before end of period; *question marks* indicate uncertainty as to reality of association (Finks, 1970).

the later Paleozoic, stromatoporoids dropped out. The astraeospongiids evolved into the wewokellids. New demosponge groups became abundant, especially the Heliospongiidae, and Calcarea appeared in force, first the Sphinctozoa, then the pharetronids. The rigid or heavy skeletons of all these groups are undoubtedly adapted to rough water. It should be noted that some of the lithistids of this facies may also be abundant in the quiet-water facies, but not the Calcarea, Heteractinida, and Stromatoporoidea.

Reefs form a special environment that may be regarded as an offshoot of the environment in which the shelly facies accumulated. When sponges participate in reef building, they are represented consistently by groups that are most abundant in the shelly facies. The times of greatest sponge participation in reefs are the Ordovician and the Permian. When principal reef-building organisms are listed by periods (Fig. 123), it is evident that corals (Tabulata and Rugosa) dominated the reefs during the intervening periods, a time that coincided with their

period	reef builders
Permian	algae Sphinctozoa* pharetronids* Bryozoa hydrocorallines?
Carboniferous	corals algae
Devonian	corals stromatoporoids algae
Silurian	corals stromatoporoids Bryozoa algae?
Ordovician	Bryozoa stromatoporoids algae demosponges* corals Receptaculitida
Cambrian	Archaeocyatha algae
Pre-Cambrian	algae

FIG. 123. Principal reef-building organisms in each period of Paleozoic, listed in approximate order of abundance under each period, with most abundant organisms at top; *asterisks* indicate sponges; *question marks* indicate uncertainty of identification of organisms (Finks, 1970).

general peaks of abundance and diversity. In the reef environment, corals were apparently competitively superior to sponges, with the exception of the stromatoporoids that appear to have been the competitive equals of the corals. Because sponges and corals do not use the same food sources (suspension feeding versus carnivorous macrophagy), the food supply was unlikely to have been a factor in either successful or unsuccessful competition. Bryozoans appear to have been the competitive equals of some sponges but also inferior to the corals. That these various groups competed with each other is suggested by the tendency for approximately contemporaneous reefs in the same area to be dominated by one or the other of the reef-building groups, as described for the Ordovician by PITCHER (1964).

GEOGRAPHIC AND STRATIGRAPHIC DISTRIBUTION

ROBERT M. FINKS and J. KEITH RIGBY

[Department of Geology, Queens College (CUNY); and Department of Geology, Brigham Young University]

HOLOCENE DISTRIBUTIONS

Sponges are almost exclusively marine, except two freshwater families, the Spongillidae and the closely related Lubomirskiidae of Lake Baikal, both members of the class Demospongea. At the level of classes sponge distribution is cosmopolitan, extending from the Arctic Ocean on the north to the Antarctic Continent on the south and occupying all oceans. At the ordinal level some groups are restricted. The Lychniscosa of the class Hexactinellida appear to be restricted to low latitudes; 11° N to 26° N in the Atlantic and 11° S to 6° N in the Indo-Pacific. The Lithistida of the class Demospongea are restricted to low and middle latitudes: 10° S to 42° N in the Atlantic and 40° S to 35° N in the Indo-Pacific. The Pharetrones in the class Calcarea occur in the Indo-Pacific between latitudes 33° S and 34° N. In the Atlantic region only one species is known and there is also one from the Mediterranean (39° N to 44° N). The three latter groups are relatively uncommon and their true distribution may be greater than that cited; the paucity of reports on South Atlantic sponges is almost certainly a contributing factor to the lack of records in that area. Nevertheless, the limited distribution must be real and reflects a contraction from their Cretaceous distribution. At the level of families and lower categories, distribution may be even more restricted, although there are many cosmopolitan genera in all three major classes of the phylum.

BATHYMETRY

Depth distributions of various groups of living sponges have been used to interpret water depths of deposits of fossil sponges (e.g., OAKLEY, 1938; WAGNER, 1963). Since the publication of the Challenger Reports (POLÉJAEFF, 1884; RIDLEY & DENDY, 1887; SCHULZE, 1887a; SOLLAS, 1888) the major classes have been recognized as having maximum abundances at different depths, although with strongly overlapping ranges. Figure 124, adapted from NESTLER (1961), shows the relative diversity of the major classes of sponges at various depths. It is based on an earlier compilation (WALTHER, 1893–1894) but is not seriously changed by later data. It seems worthwhile to review, at least in a preliminary way, the bathymetric data now available to provide a frame of reference for further paleoecologic interpretation.

The accompanying charts (Fig. 125–129) show depth distributions of each major order recognized in this volume. The total range of each group is plotted against latitude, because the latter parameter is correlated to some extent with temperature. Depths for Indo-Pacific occurrences have been plotted separately from those for the Atlantic-Arctic-Mediterranean systems to allow some control over effects of provincial faunal differences. Only the major monographs and such smaller reports as yield extremes in depth and latitude were consulted, and of these effort focused only on those that were relatively easily used. Despite these limitations, most of the available data have been gathered, and it is likely that patterns displayed reflect true distributions of most sponges. It is possible that new or missed data may extend greatly the extremes for the less common groups.

Use of total range is more appropriate to the usual sort of paleontological work, although knowledge of maximum abundance would provide a sharper tool. The nature of paleontological samples is generally too limited to provide secure data on abundance, but simple presence or absence can be interpreted in connection with the charts herein.

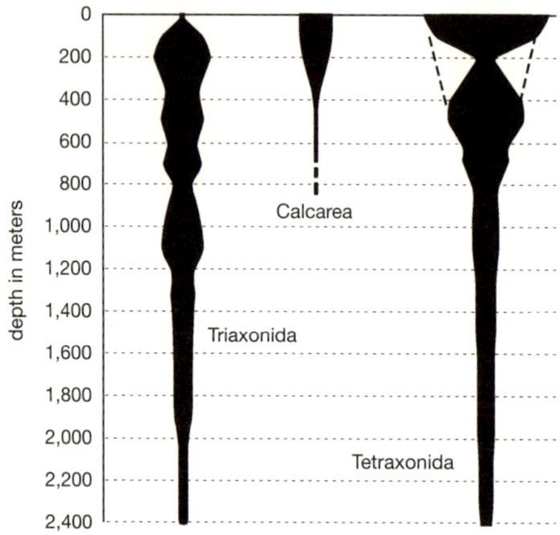

FIG. 124. Relative diversity of major classes of living sponges at various depths in modern oceans (Nestler, 1961).

Assessment of the present distribution of abundance also involves problems of sampling.

The class Calcarea retains its former reputation for a dominantly shallow-water distribution, in general being confined to depths less than 600 m and occurring as shallowly as low tide level (Fig. 125). Recent work (KOLTUN, 1964; BOROJEVIC & GRAAT-KLEETON, 1965), however, has extended the extreme range of deep-sea species from the Arctic to 3,800 m. In the equatorial regions there are few reports below 200 m; whether this is due to deficiencies of sampling or to some other cause is not clear. The Pharetronida range from 5 to 25 m in the Mediterranean (all in poorly lighted caves and hollows) and from 5 to 430 m in the Indo-Pacific.

The class Hexactinellida has a different pattern for each of its subgroups (Fig. 126–127). The deepest record of any sponge is an as yet unidentified hexactinellid from the Kurile-Kamchatka Trench at 8,610 m (VINOGRADOV, 1958, p. 87). The subclass Amphidiscophora is everywhere confined to depths below 200 m. Maximum recorded depth is 7,000 m (*Hyalonema* from Kermadec Trench, LÉVI, 1964, p. 83). Their total latitudinal extent is from the Arctic (80° N) to the vicinity of Antarctica. The upper limit of depth distribution does not rise with increasing latitude (as does that of the Lyssacinosa), suggesting that more is involved than simple temperature control.

Of the subclass Hexasterophora, the Lyssacinosa have upper and lower depth limits that tend to rise toward the poles, suggesting a temperature control of distribution. The pattern is best shown in the Atlantic-Arctic (Fig. 126), with a range of 800 to 5,900 m near the equator, rising to about 70 to 3,000 m at the Arctic end (70° to 80° N) and 50 to 4,000 at the southern end (50° to 60° S). In the Indo-Pacific (Fig. 127) the corresponding figures are 110 to 5,000 m near the equator, 130 to 1,200 m at 55° N, and 50 to 600 m at 72° S. The deepest record is at 32° S where *Caulophacus* has been taken from 7,000 m in the Kermadec trench (LÉVI, 1964, p. 83). The northern end is truncated somewhat arbitrarily by inclusion of the Arctic with the Atlantic (with which it has the most open connection and to which it belongs faunistically (see HENTSCHEL, 1929, p. 1012); the southern end is more significant with Lyssacinosa occurring on the Antarctic shelf, often beneath shelf ice. It may be argued that the lower limit is controlled by bathymetry of the ocean floor, but the upper limit is not controlled and yet shows a significant rise.

The Hexactinosa are almost completely absent from water shallower than 100 m, but a few occurrences in equatorial regions record them as shallow as 10 to 40 m. Their latitudinal distribution extends, again, from Arctic to Antarctic regions. The maximum depth record is a *Farrea* from 6,860 m in the Kurile-Kamchatka trench (VINOGRADOV, 1958, p. 87). The Hexactinosa are common elements in the only known living hexactinosan sponge reefs recently discovered on the continental shelf of British Columbia in Hecate Strait, in water up to 200 m deep (KRAUTTER & others, 2001; CONWAY & others, 2001). A variety of lyssacinosan sponges also occur in the reefs.

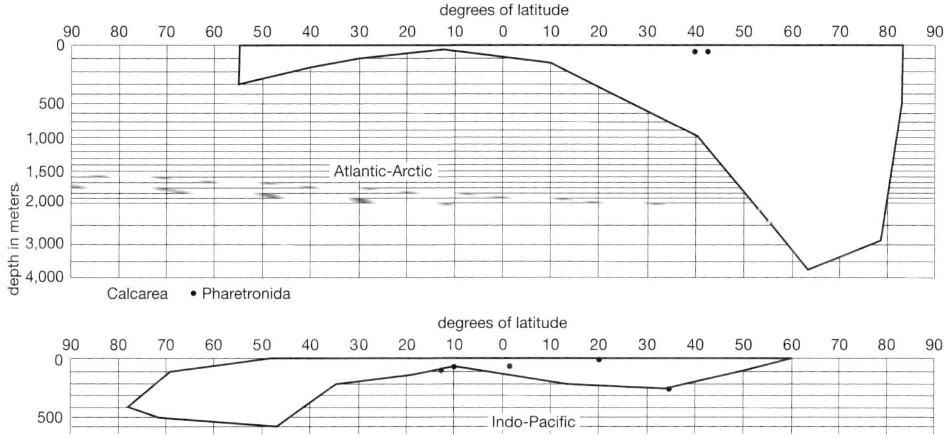

FIG. 125. Distribution of Calcarea, including the Pharetronida, showing depth and latitudinal occurrences in the Atlantic-Arctic Ocean, above, and the Indo-Pacific Ocean, below (new).

The Lychniscosa have the most restricted distribution in the class. They are confined entirely to low latitudes and to moderate depths. In the Atlantic they occur between 11° N and 26° N at depths between 200 and 700 m. In the Indo-Pacific they occur between 11° S and 6° N at depths between 80 and 600 m. The latitudinal restriction may be related to a temperature requirement, the bathymetric restriction perhaps to some form of competitive exclusion that seems to keep all living Hexactinellids out of shallowest water.

The class Demospongea, which includes the great majority of living sponge species, also has noteworthy differences in distribution between the orders recognized here (Fig. 128–129). Representatives of all orders have been reported at the very shallowest depths, low tide, or even slightly above it. The Monaxonida have the deepest range, extending to 7,000 m (*Asbestopluma hadalis* LÉVI, 1964, from the Kermadec trench); they appear to be the only demosponges to populate the deep sea below about 4,200 m. They also include the only freshwater

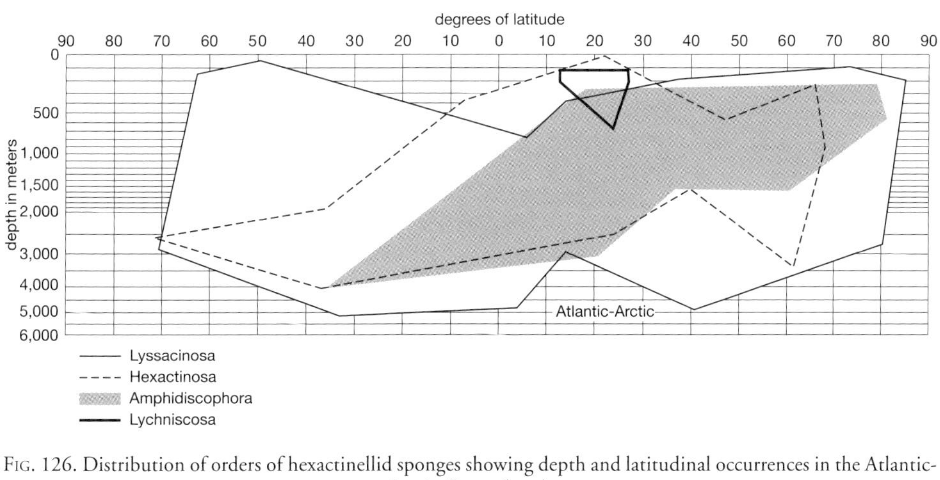

FIG. 126. Distribution of orders of hexactinellid sponges showing depth and latitudinal occurrences in the Atlantic-Arctic Ocean (new).

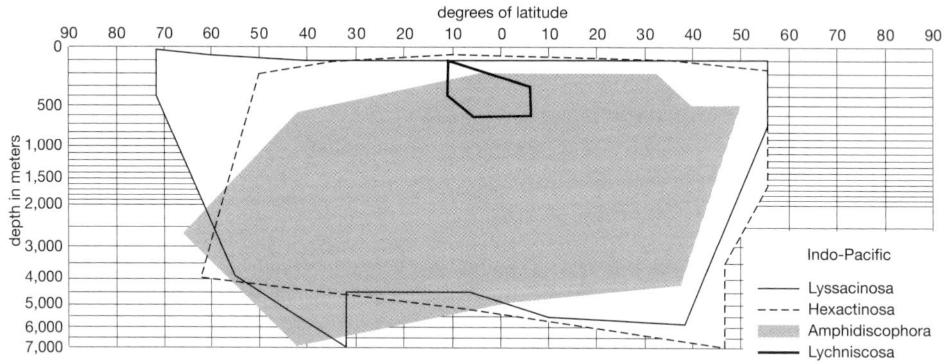

FIG. 127. Distribution of orders of hexactinellid sponges showing depth and latitudinal occurrences in the Indo-Pacific Ocean (new).

sponges. The Keratosa, which appear to be most closely related to the Monaxonida (and especially to the sigma-bearing forms that comprise most of the deep-sea monaxonids), on the other hand, are the most limited in depth. They are not known below 2,000 m, and most of them not below 200 m. The Choristida extend down to 4,160 m (*Sphinctrella horrida* SCHMIDT from the Azores, TOPSENT, 1913). The Lithistida range from 110 to 2,200 m in the Atlantic and 0 to 1,300 m in the Indo-Pacific, being the group most restricted in depth after the Keratosa. The lithistids are also the only demosponge order to have a latitudinal restriction. They are confined in the Indo-Pacific between 40° S and 35° N (Fig. 128). In the Atlantic they range from 10° S to 42° N (Fig. 129), although one report of *Macandrewia azorica* from about 60° N was given by SOLLAS, 1888. Temperature may limit the distribution of the lithistids.

Considering the phylum as a whole, the monaxonid demosponges (or more precisely the order Cornacuspongida of LENDENFELD and other authors) have been the most adaptable. They include most living sponge species; are the only ones to live in fresh water; and range more deeply in the sea than any other sponges except amphidiscophoran

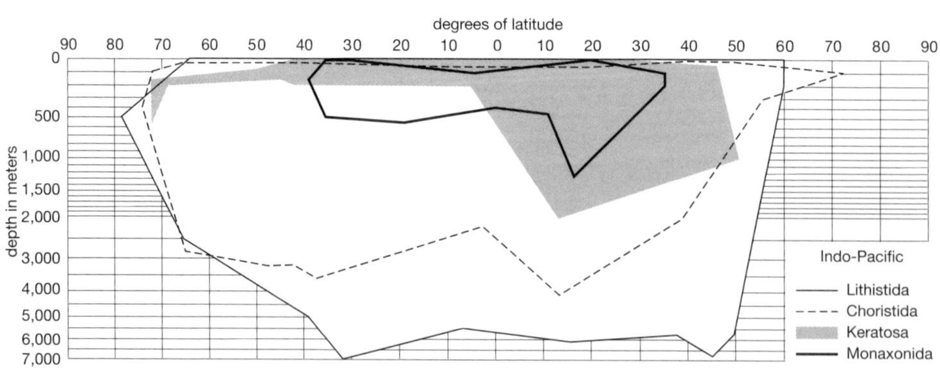

FIG. 128. Distribution of orders of demosponges showing depth and latitudinal occurrences in the Indo-Pacific Ocean (new).

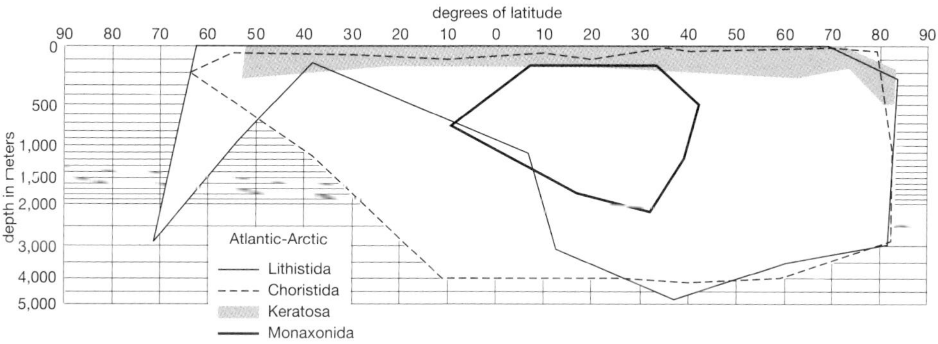

FIG. 129. Distribution of orders of demosponges showing depth and latitudinal occurrences in the Atlantic-Arctic Ocean (new).

Hexactinellida. Their microscleres include the most complex and asymmetrical ones, have the greatest development of spongin, and are characterized by a sort of ovoviviparity or incubated larval stage. The class Hexactinellida, with its low-efficiency hydraulic system, is largely excluded from present shallow waters, possibly because of competition related to feeding. In the Mesozoic and Paleozoic this restriction was much less noticeable. The class Calcarea is largely excluded from deeper waters, possibly because of increased solubility of calcium carbonate at greater depths. Three archaic groups, one from each class, are restricted to low latitudes: the Lithistida, the Lychniscosa, and the Pharetrones. Their present restriction seems to be most readily explained by limited tolerance to low temperatures. Because these groups might be characterized as those with the most massive skeletons, the postulated intolerance to low temperature may be related to a higher metabolic rate. Their greater diversity, abundance, and geographic distribution in Mesozoic times may have resulted from a warmer world climate.

VINOGRADOV (1958) reviewed records of sponge species occurring at depths below 2,000 m (abyssal and hadal zones) and identified two well-defined depths at which there is a major turnover in species (Fig. 130–131). That is, at these depths there is a strong maximum in both the number of species that appear for the first time at this depth and in the number of species that disappear. Overlapping of ranges leads also to a maximum in the total number of species at that depth (Table 8).

For sponges as a whole and for Hexactinellida alone (Fig. 130), the upper maximum is at 2,500 to 3,000 m and the lower at 4,000 to 4,500 m. The upper maximum for the Demospongea alone (Fig. 131) is slightly higher (2,000 to 2,500 m) and the lower maximum for the Choristida ("Tetraxonida") alone is likewise higher (3,500 to 4,000 m). The differences reflect the smaller number of specifically abyssal and hadal species among the demosponges; the lower maximum for choristids is scarcely above the lower limit of their distribution. It is not clear what critical factors determine the two depths of faunal turnover. VINOGRADOV (1958) found corresponding changes in many other groups of animals.

ORIGIN OF THE ABYSSAL SPONGE FAUNA

Sponges form an important element of populations on the sea floor below 2,000 m. Over vast areas of the abyssal plain of the central Pacific, well below 3,000 m, sponges have constituted more than 50 percent of trawl catches (SOKOLOVA, 1964; and Fig. 132 herein). Nevertheless, very few genera

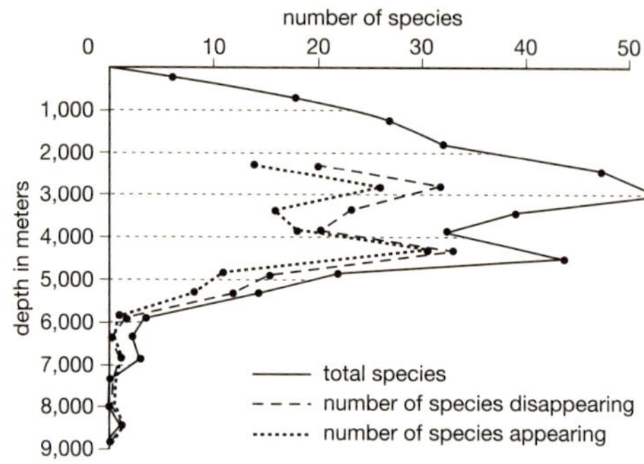

FIG. 130. Numbers of species of Hexactinellida, per depth, in modern oceans, and numbers of species appearing and disappearing at particular depths (Vinogradov, 1958).

are confined to abyssal depths, most abyssal sponges being eurybathic genera. On the assumption that stenobathic abyssal genera originated as adaptations to the abyssal environment, their first appearance in the fossil record should give a minimal age for the origin of that environment, and the numbers of species in each genus living today should provide a minimal rate for evolution of the genera. Following is a list taken from the compilation of LÉVI (1964) of stenobathic abyssal genera with more than one living species, which is in essential agreement with the similar data of VINOGRADOV (1958). Their geologic range is also added.

Class Demospongea
- Order Monaxonida
  - *Cladorhiza* 14 living spp. (*?Eocene–Holocene*)
  - *Asbestopluma* 6 spp. (*Holocene*)
  - *Chondrocladia* 10 living spp. (*?Eocene–Holocene*)
  - *Esperiopsis* 3 living spp. (*?Eocene–Holocene*)
- Order Choristida
  - *Thenea* 12 spp. (*Holocene*)

Class Hexactinellida
- Order Amphidiscophora
  - *Hyalonema* 1 living sp. (*?Eocene–Holocene*)
  - *Oonema* 11 spp. (*Holocene*)
  - *Prionema* 4 spp. (*Holocene*)
  - *Leptonema* 9 spp. (*Holocene*)
- Order Hexasterophora
  - Suborder Lyssacinosa
    - *Holascus* 10 spp. (*Holocene*)
    - *Malacosaccus* 6 spp. (*Holocene*)
    - *Caulophacus* 15 living spp. (*?Eocene–Holocene*)

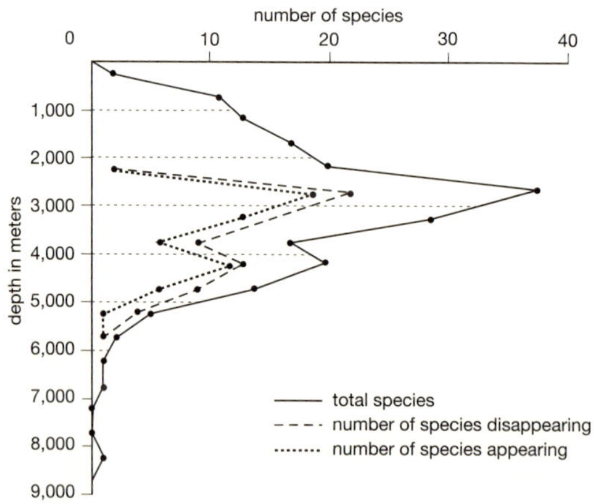

FIG. 131. Numbers of species of Demospongea, per depth, in modern oceans, and numbers of species appearing and disappearing at particular depths (Vinogradov, 1958).

TABLE 8. Depth ranges of various taxa of sponges (new).

GROUP	UPPER MAXIMUM			
		Number of species		
	Depth (in m)	Total	Appearing	Disappearing
Triaxonida (Hexactinellida)	2,500–3,000	38	18	22
Tetraxonida (approx=Choristida)	2,000–2,500	ND	ND	ND
Cornacuspongida (approx=Monaxonida)	2,000–2,500	ND	ND	ND
Total sponges	2,500–3,000	55	26	32
GROUP	LOWER MAXIMUM			
		Number of species		
	Depth (in m)	Total	Appearing	Disappearing
Triaxonida (Hexactinellida)	4,000–4,500	20	12	13
Tetraxonida (approx=Choristida)	3,500–4,000	ND	ND	ND
Cornacuspongida (approx=Monaxonida)	4,000–4,500	ND	ND	ND
Total sponges	4,000–4,500	45	32	33

Unfortunately the first Eocene records of *Cladorhiza, Chondrocladia, Esperiopsis, Hyalonema,* and *Caulophacus* are from from a paper by HINDE and HOLMES (1892) on Eocene sponges from Oamaru, Otago, New Zealand, in which identifications were made on the basis of isolated spicules, mainly microscleres. At that time, shape of individual spicules was considered to be more characteristic of particular genera than has turned out to be so. Consequently we cannot rely on these, the only fossil records of the genera involved. Nevertheless, we can assert for the first four Eocene genera, on the basis of the microscleres described, that reasonably close relatives were present, as they were also in the Cretaceous (ZITTEL, 1876; SCHRAMMEN, 1910, 1912).

Even if these stenobathic abyssal genera were actually present in the Eocene, they were associated in the fauna described by HINDE and HOLMES (1892) with spicules of lithistids that are today entirely excluded from the abyssal fauna. The same is true of the Cretaceous chalk. Therefore, the known Cretaceous and early Tertiary sponge faunas that have the closest affinities to the living abyssal fauna appear to be of a shallower water nature, possibly within the bathyal zone. If the abyssal sponge fauna had not already originated at this time, a matrix for it was at least present in the bathyal fauna; and differentiation of abyssal adaptations may conceivably date from the early Tertiary.

Indeed, in present seas, in addition to living stenobathic abyssal forms, there are several eurybathic genera whose main centers of development are in the bathyal or even sublittoral zone but which have wholly or partly abyssal species. Included are several Hexactinosa of the class Hexactinellida, namely *Eurete, Farrea, Aphrocallistes, Bathyxiphus, Auloplax, Chonelasma,* and *Aulocalyx* (LÉVI, 1964). Of these seven genera, the first five are known from the Upper Cretaceous. The same is true of the lyssacinosan *Regadrella* (REID, 1958a). Many more genera of Demospongea with abyssal representatives were listed by LÉVI (1964), but they are almost all unknown as fossils. Two that are recognizable by characteristic spicules are *Latrunculia* and *Geodia,* and these are both present in the New Zealand

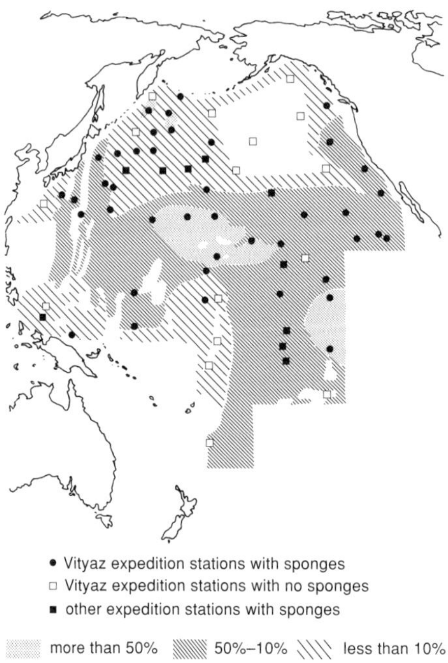

FIG. 132. Distribution of sponges as a percentage of trawl catches from central and northern Pacific Ocean (Sokolova, 1964).

Eocene deposit described by HINDE and HOLMES (1892). *Geodia* is also known from the Cretaceous (HURCEWICZ, 1966) and *Geodia*-like sterrasters as far back as the Upper Jurassic *Rhaxella* (HINDE, 1890). Thus, several of the eurybathic generic elements of the present abyssal sponge fauna can be traced back at least to the early Tertiary or Cretaceous.

An analysis of genera of abyssal sponges given by LÉVI (1964) is presented in Table 9 (Calcarea from KOLTUN, 1964, and BOROJEVIC & GRAAT-KLEETON, 1965).

The small number of exclusively abyssal genera may have resulted from a relatively recent origin of the fauna (and the environment). Or it may reflect the uniformity or the extreme nature of the environment or both. The greater percentage of exclusively abyssal Hexactinellid genera may indicate the deeper water focus of development of this class.

HISTORY OF DEPTH DISTRIBUTIONS

To trace depth distributions of sponges into the geologic past we must start with fixed points of reference that are known with the greatest certainty and from these build an interlocking net of evidence. Identification of the shallow-water environment seems to be the most securely founded and the one that can be determined within the narrowest limits. Among indicators of shallow water, the presence of benthic calcareous algae is least subject to alternative interpretations and determines depth most precisely.

Algae must have light for photosynthesis, and the depth of light penetration sufficient for photosynthesis is known in modern seas to extend from the surface down to a maximum of about 200 m. The Rhodophyceae, which are able to use the more penetrating shorter wavelengths of light, occur down to this depth in clear waters of low latitudes, where light penetration is greatest owing to the high angle of incidence (POLUNIN, 1960, p. 513, 530). At higher latitudes and in more turbid waters, maximum depths may be considerably less. The lower limit of "the more abundant attached plants" is only about 40 to 60 m (POLUNIN, 1960, p. 512). Consequently fossil sponges directly associated with sessile algae must have lived at depths no greater than 200 m and probably no greater than 100 m (assuming no radical changes in insolation). It should be noted that these are limits and that the greatest abundance of algae is commonly within 20 m of the surface (EMERY, TRACY, & LADD, 1954); fossil sponges associated with large numbers of algae may have lived in very shallow water indeed.

A somewhat more equivocal criterion of shallow water environment is occurrence in or upon a fossil reef. Such a structure, when surrounded by an aureole of talus beds that clearly interfinger with the reef mass, must have formed in water shallow enough for wave action to break up the fabric of the reef while it was actively growing. It is difficult to place a numerical value upon a maximum depth for such an occurrence since it depends both on the force of the waves and on the construction of the reef. The mere occurrence of a reef structure, in the sense of a solid mass of autochthonous skeletal material that stood above the sea floor, does not guarantee shallow depth, for deep-water coral reefs are known from present seas and have sponges as members of the associated fauna (TEICHERT, 1958). Present-day tropical scleractinian-algal-foraminiferal reefs are limited to depths shallower than about 30 m because of limits on the algae and symbiotic zooxanthellae of the corals (WELLS, 1957). There is no secure basis for assuming that fossil reefs formed by organisms other than Scleractinia were confined to the same depths. Most fossil reefs, however, contain substantial quantities of calcareous algae and, thus, must have been limited to the photic zone as discussed above.

Lithologic criteria of shallow-water deposition include oolites, which form in water turbulent enough to suspend the ooliths while they received their successive coatings of calcium carbonate. Other criteria are of much more sporadic occurrence, although they may furnish quite precise indications of depths; an example is a limestone-boulder conglomerate of Miocene age deposited in localized depressions on a surface of unconformity and in which the boulders are extensively bored by various lithophages, including *Cliona* (RADWANSKI, 1964).

Finally, we may seek corroborative evidence from the sponges themselves. Three groups that are common as fossils have a present distribution confined to shallow and intermediate water depths. These are the Calcarea (0 to 600 m, rarely to 3,800 m), the Lithistida (0 to 2,200 m), and the Lychniscosa (80 to 700 m). At least some of the Calcarea have been associated with both algae and reefs continuously since the Permian. The Lithistida also have a nearly continuous association with algae and reefs since their first significant appearance in the Ordovician. They are also associated with both Calcarea and Lychniscosa since the first appearance of these groups. The Lychniscosa participated in construction of reefs in the Upper Jurassic, although associated with Calcarea and Lithistids, and may not be associated with algae. This would be in line with their present exclusion from waters shallower than 80 m. It is also in line with the conclusion of ROLL (1934) that talus associated with the German Jurassic sponge reefs was formed from later erosion of the dead reef following a fall in sea level. He based his conclusion on the fact that no binding organisms were present, the sponges merely serving as sediment traps; thus the living reef surface was not solid enough to be broken up by wave action. He discounted reports of algae by SCHRAMMEN (1924b). It should be noted that REID (1958b, p. 265) reported specimens of the Upper Cretaceous lychniscosan *Rhizopoterion tubiforme* to have healed breakage during life, presumably broken by rough water.

TABLE 9. Distribution of abyssal genera of sponges (new).

Group	Abyssal genera	Exclusively abyssal genera	Present %
Demospongea	68	6	8.8
Monaxonida	58	5	8.6
Choristida	10	1	10.0
Hexactinellida	44	10	22.7
Amphidiscophora	14	4	28.6
Hexasterophora	30	6	20.0
Lyssacinosa	23	6	26.1
Hexactinosa	7	0	0.0
Lychniscosa	0	0	0.0
Calcarea	2	0	0.0

The Calcarea, Lithistida, and Heteractinida have had shallow-water representatives essentially throughout their history. The Amphidiscophora, on the contrary, have not been clearly associated with shallow water, although as early as the Pennsylvanian and as late as the Eocene they have been associated with lithistids (not necessarily shallow-water ones but probably not abyssal ones). The Lyssacinosa or Reticulosa have had shallow-water representatives at least from Devonian to Permian times but only doubtfully earlier (Precambrian to Silurian) or later (Triassic to Holocene). The Hexactinosa and Lychniscosa are associated with oolites and reefs in Jurassic deposits and with Calcarea and Lithistida in Cretaceous ones, but never or rarely with algae and possibly, therefore, never or rarely occurring in shallowest water. The Monaxonida have very shallow-water representatives at least as early as the Permian (probably earlier, but isolated monaxons cannot be assigned with certainty) and have freshwater representatives (Spongillidae) from the Jurassic on (NALETOV, 1961; HINDE, 1890). The Choristida have shallow-water representatives as early as their first appearance in the Mississippian.

Here and there we catch glimpses of more detailed patterns of distribution. In the Upper Cretaceous of northern France and Belgium a transgressive, shoreward conglomeratic facies is dominated by the lithistid group Rhizomorina with subordinate Sphaerocladina. A more seaward, quartz-sandy facies is dominated by Tetracladina, a limy-argillaceous facies by Hexactinosa, and the chalk by Lychniscosa (DEFRETIN-LEFRANC, 1961). Dominance is expressed both in variety of species and in number of individuals. It is apparent that the pattern is, in part, a reflection of bathymetry, but other environmental factors may be involved, most obviously the nature of the sediment substrate. For example, lithistids of the conglomeratic facies are encrusting forms that coat the boulders of the conglomerate (DEFRETIN-LEFRANC, 1961).

Contemporaneous beds of Northern Ireland discussed by REID (1958b, p. 261) differ in some respects because the glauconitic sandstone facies does not contain a rich development of tetractinellid Demospongea and because the chalk facies Hexactinosa and Lychniscosa of the chalk facies appear to occur in equal variety. REID (1958b, p. 264) suggested that nonbathymetric factors are involved in the distribution of Hexactinellida, inasmuch as Mesozoic Hexactinellida are predominant in calcareous sediments but are absent or rare in clays, such as in the Lias, that seems to have been deposited in similarly deep water. NESTLER (1961) noted the presence of the Pharetronid calcisponge *Porosphaera* accompanying Lychniscosa and Hexactinosa of the latest Cretaceous chalk of northern Germany (island of Rügen). Because modern Pharetronida range from 5 to 430 m; and, as REID (1958b) pointed out, modern Hexactinellida have as great a variety and abundance at 200 m as at any greater depth, a depth of the sponge-bearing chalk of 200 to 400 m accords with present day distributions.

Although the pharetronid Calcarea occur sporadically in facies where lithistids and hexactinellids dominate, their greatest development in the Cretaceous is in the glauconitic or greensand facies, but only very locally, as at the famous occurrences of Faringdon, England (Aptian), and Essen, Germany (Cenomanian). There they dominate to the virtual exclusion of other sponges. These are usually considered to represent very shallow-water, nearshore deposits, either formed in a rough-water environment or washed into offshore depressions from a rough-water source area (CASEY, 1961).

The Cretaceous bathymetric scheme can be traced back into the Jurassic (GAILLARD, 1983). WAGNER (1963) cited a coral reef from the upper Malm of Germany in which numerous silicisponges are present, namely, Hexactinosa, Rhizomorina, and Tetracladina. One layer in this reef, however, contains a large number of pharetronid calcisponges together with abundant encrusting calcareous algae that are notably absent from most of the reef. The association of Calcarea with algae makes a bathymetric explanation of the sponge distribution likely. In coral reefs of identical age and from the same general region that were described by FRENTZEN (1932), algae, Calcarea, and silicisponges (Rhizomorina and Sphaerocladina) occur together in the reef, and Calcarea dominate the talus fauna. In this seemingly shallower-water reef it is noteworthy that Hexactinellida and Tetracladina are absent from the sponge fauna, while the Rhizomorina and Sphaerocladina remain, the very silicisponges that are characteristic of the shallowest water in the Cretaceous (DEFRETIN-LEFRANC, 1961). In reefs of the lower Malm of the same region, corals and algae are absent; and the reef fabric is composed of Hexactinellid (Hexactinosa and Lychniscosa) with very subordinate numbers of Calcarea (DORN, 1932). Both the absence of algae and the presence of the Cretaceous deeper-water sponge fauna is consistent with a deeper-water site for these earlier reefs. The same conclusion about bathymetry was reached by ROLL (1934) from an independent line of reasoning, namely that the lack of binding organisms in the reef implies that the living reef could not have been wave resistant and therefore must have grown below the zone of rough water.

In the Triassic, records of siliceous sponges are too few for any bathymetric patterns to emerge, but abundant sphinctozoans and pharetronid Calcarea associated with algae are locally present as reef builders, and almost certainly lived in a shallow-water environment (SIEBER, 1937; FLÜGEL, 1981; SCHÄFER & SENOWBARI-DARYAN, 1981). Similar reefs of shallow and somewhat deeper environments also occur in Upper Triassic rocks of the southern Yukon (REID & GINSBURG, 1986; SENOWBARI-DARYAN & REID, 1987). The absence of siliceous sponges from these reefs is consistent with the same type of depth zonation seen in younger rocks.

The calcareous sponges of the Triassic provide a link with the Permian, for the same general fauna of demosponge sphinctozoans and Pharetronida is found in reefs and associated with algae (NEWELL & others, 1953; FINKS, 1971a). The Permian reefs of Texas, however, contain also subordinate numbers of Lithistida (Rhizomorina, Orchocladina, Eutaxicladina), Monaxonida (Heliospongiidae), Heteractinida, and two genera of lyssacine Hexactinellida. The lyssacines are a Paleozoic shallow-water element seen here for the first time, as are the exclusively Paleozoic groups, Heteractinida, Orchocladina, and Eutaxicladina. The same tendency for a bathymetric separation of Calcarea and siliceous sponges is nevertheless present. A basin facies of dark limestones in front of or marginal to the reefs contains a fauna dominated by siliceous sponges, in which there is a great proliferation of species of lyssacine and reticulosid hexactinellids at some localities (FINKS, 1960).

There is independent stratigraphic evidence that the basin lithofacies, in general, was deposited in water significantly deeper than that of the reefs. In the upper Guadalupian Stage the depth of water in the basin may have attained a maximum of about 545 m below the reef's surface, although it was apparently less during most of the Permian (NEWELL & others, 1953; NEWELL, 1957). In the vicinity of some patch reefs an autochthonous fauna of reef-type Calcarea has been collected in proximity to a rich basin-type fauna (FINKS, 1960, p. 33). This suggests that the bathymetric difference was not always large and that such factors as water agitation or nature of the sea bottom may have controlled distribution. The lithistids are common to both facies.

Beyond this point it becomes difficult to follow the thread. The presence of hexactinellids in lagoonal black shale above a Pennsylvanian coal seam (ZANGERL & RICHARDSON, 1963) and in Devonian reefs (LECOMPTE, 1936) reveals a persistent shallow-water strain in this group.

The most comprehensive data from the middle Paleozoic come from the middle Silurian reefs of Illinois (LOWENSTAM, 1948, 1957). Here we find the germ of later patterns. Sphaerocladines (*Astylospongia, Palaeomanon*), eutaxicladines (*Hindia*), and heteractinids (*Astylospongia*) occur in the interreef beds and in the lower levels of reefs. *Hindia* and *Astraeospongia* persist longer to midlevels of the growing reefs and, therefore, are considered to have been more tolerant of shallow water. In the rough-water stage of the mature reef their place is taken by the algal Receptaculitida (*Ischadites*). They are the bathymetric equivalents of the later Calcarea.

Still farther back, in mid-Ordovician (Chazy) times, orchocladine lithistids (*Zittelella*) are associated with algae in reefs, while lophophytous hexactinellids occupy the somewhat deeper water around them (PITCHER, 1964, and field observations of RMF, 1968). Here indeed is the earliest example of the persistent bathymetric separation of lithistids and hexactinellids. Earlier than this, secure data on relative bathymetry are not available, although one of the earliest lithistids, the Lower Ordovician *Archaeoscyphia,* is associated with algae in shallow-water deposits in the El Paso region of Texas and New Mexico (TOOMEY & NITECKI, 1979).

HOLOCENE SPONGE FAUNAL PROVINCES

Most living sponge genera are cosmopolitan or at least widely distributed in the world's oceans. Most species, on the other hand, are provincial. Arctic (HENTSCHEL, 1929) and Antarctic (BURTON, 1932) sponge faunas each have characteristic species, and their faunas appear to have differentiated out of faunas of the middle latitudes of neighboring seas. BURTON (1932) pointed out that Antarctic species of the monaxonid sponges *Tedania* and *Iophon* appear to be more specialized relatives of lower-latitude species. HENTSCHEL (1929) noted that the Arctic fauna has 60 percent of its species in common with the fauna of the North Atlantic, with which the Arctic Ocean is in open communication, while it has only 8.3 percent in common with the North Pacific, where the narrow Bering Strait (or at times in the past, a land bridge) forms a barrier. Opposite sides of the Atlantic appear to have different faunal assemblages (BURTON, 1934), with those of the European-African side (including the Mediterranean) more closely related to that of the Arctic. The cooler marine climate of the east side of ocean basins, as a result of the upwelling of deeper waters brought about by the circulation pattern of surface currents, may account for this distribution. The Arctic monaxonid *Stylocordyla borealis,* on the other hand, has a deep-water distribution down the east coast of North America to the West Indies. BURTON (1934) suggested that the deeper currents may distribute this species.

The Caribbean area forms a distinct province with a proliferation of species. It is related to a Brazilian province on the south and both of them to the vast Indo-Pacific province, whose center of species proliferation is in the Indonesian-Malayan region (SOLLAS, 1888). These three provinces form a nearly continuous circumtropical belt and undoubtedly owe their resemblances to common climatic conditions. BURTON (1934) called attention to the similarity of Keratosa in the Caribbean and in coastal regions of northern Australia. This is perhaps not surprising inasmuch as both areas are on the western sides of ocean basins, where warm currents begin to turn poleward. The horseshoe crab *Limulus* is a well-known example of a genus with a similarly disjunct distribution in climatically analogous areas.

SOLLAS (1888) recognized, in addition, three temperate zone provinces in the southern hemisphere, associated with the three main continental masses, namely, a South African, a South Australian, and a Patagonian province. A separate Magellanic province occupies the cool-water antiboreal region of the southern end of South America.

Lithistids are most diversified in the Caribbean and Indo-Pacific provinces (SOLLAS, 1888). Data from the HMS *Challenger* expedition (SOLLAS, 1888) show that the Atlantic has fewer species of siliceous sponges than the Indo-Antarctic and the Pacific; the

TABLE 10. Numbers of species of deep-sea sponges (data from Lévi, 1964; new).

	ANTARCTIC	ARCTIC	ATLANTIC	INDIAN	PACIFIC	TOTAL*
Choristida	0	0	14	5	5	24
Monaxonida	4	12	75	15	23	114
Lyssacinosa	15	1	18	17	23	66
Hexactinosa	2	0	10	2	3	16
Amphidiscophora	1	0	4	14	41	61
Total	22	13	121	53	95	281

* Each species is counted only once.

Indo-Antarctic has a somewhat greater variety of Hexactinellida and Monaxonida than the Pacific; and the Pacific has a greater variety of Choristida and Lithistida than the Indo-Antarctic.

Deep-sea sponges have a rather different pattern. LÉVI (1964) compiled records of all species reported to that time from depths below 2,000 m. A summary of his data are presented in Table 10.

The Arctic deep-sea sponge fauna is dominated by Monaxonida, while that of Antarctica by Lyssacinosa. The Monaxonida also dominate the Atlantic deep-sea fauna, having four times as many deep-sea species as the next most diversified group, the Lyssacinosa. The Monaxonida are also much more diversified in the Atlantic deep sea than in either the Indian or Pacific deep sea or than in the Indian and Pacific together. The Choristida, Hexactinosa, and Amphidiscophora are virtually absent from the polar deep sea. The Hexactinosa are much more abundant in the Atlantic deep sea, having very few deep-sea species in the Indian or Pacific Oceans. The reverse is true of the Amphidiscophora, which have few species in the Atlantic but many more in the Indian and Pacific. The Amphidiscophora are by far the most diversified element of the Pacific deep-sea fauna. The Lyssacinosa are the most evenly distributed, except for their virtual absence from the Arctic.

The Lyssacinosa appear to have a southern center of distribution, spreading northward into all three seas, perhaps with Antarctic bottom water. The deep-sea Monaxonida, by contrast, appear to have a more northerly focus or perhaps an Atlantic one and dominate the interconnected bottom waters of the Atlantic and Arctic. The Amphidiscophora appear to have their center of distribution in the lower latitudes of the Pacific, perhaps the eastern part, to judge from most of the records (LÉVI, 1964). The Choristida and Hexactinosa are also mainly lower-latitude forms and not especially diversified in deep water. In general the bottom sponge fauna of the Pacific is mainly hexactinellid, especially amphidiscophoran; the Atlantic fauna is mainly monaxonid. As expected, faunas near the poles have less diversity than ones from lower latitudes; since we are dealing with deep-sea forms, these differences must be due to factors other than temperature or light. It is perhaps unexpected that the Atlantic has greater diversity of deep-water sponges than the other oceans, especially in view of the smaller diversity of its shallow-water sponges (SOLLAS, 1888).

GEOGRAPHIC DISTRIBUTION IN THE PAST

Provincial faunas probably existed in the past, but the limited distribution of rich sponge faunas in the fossil record makes it difficult to delineate such provinces. Sporadic examples show that contemporaneous local faunas often differ at the species level. It is obvious, for example, when studying collections from the rich middle Silurian sponge faunas of Tennessee and Gotland that although both faunas are dominated by the lithistid family Astylospongiidae, the specimens from Gotland are more robust and differ in details of surface ornamentations from those of Tennessee (see also RAUFF, 1894). On a more restricted geographic scale, a few genera of sponges out of a large fauna appear to have specific differences on opposite sides of the Delaware basin (approximately 60 km apart) in Leonardian beds of the west Texas Permian (FINKS, 1960). Local differences in environment or slight differences in time may account for these examples, rather than the separate evolution of isolated populations.

The tying of some sponge faunas to specific lithofacies and therefore to specific environments is frequently demonstrable in the fossil record. Their geographic distribution as preserved, however, may indicate only the chance preservation of the appropriate lithofacies and not the original pattern of their distribution. In the Upper Cretaceous, Hexactinellida are almost entirely confined to calcareous facies, while Calcarea dominate in sandy or glauconitic facies (REID, 1958b). In Europe, their distributions are related to the form of the advancing Upper Cretaceous sea, and a rich fauna is developed in both facies. Yet there is a possible wider pattern involved here, which is difficult to identify as to cause. The Atlantic Coastal Plain of North America contains an extensive series of sandy and glauconitic beds and the Gulf Coastal Plain a well-developed chalk facies, both being contemporaneous with the Santonian to Maastrichtian sponge-rich beds of Europe. Yet sponges are seemingly extremely rare throughout this vast region except for the calcisponge *Porosphaera* in the Gulf Coast chalk. Those sponges that do occur can be referred to European genera. There may be here the shadowy outline of some large geographic pattern, as yet impossible to delineate.

In the Permian, however, a truly global pattern occurs. The Calcarea and calcareous sphinctozoans are confined largely to the Tethyan realm (RIGBY & SENOWBARI-DARYAN, 1995). Abundant sponges occur at many localities at higher paleolatitudes outside the Tethyan realm, in the northern Laurentian-European realm or in the southern Indo-Australian realm, but there the sponges are entirely siliceous. The same pattern seems to be present in the preceding Carboniferous Period, although the number of localities available is fewer. A climatic zonation is clearly indicated by this pattern. Inasmuch as the Carboniferous was also a time of great diversification of the calcareous sponges, it is tempting to see the distribution as a reflection of their focus of origin. In view of the rapidity with which living sponge faunas are known to migrate, however, it is likely that these calcareous sponges would have spread to the limits of the available congenial environments within less than a geological period.

In the Triassic the Calcarea were more widely distributed than in the preceding two periods. One may argue that the distribution results merely from the distribution of shallow-water deposits because the Calcarea have always been concentrated in shallow waters. It is not possible to refute entirely this point of view because depth control on most of the exclusively siliceous sponge deposits is lacking, but at least one of them, the Park City Formation of the Wyoming Permian, appears to be of shallow-water origin (FINKS, YOCHELSON, & SHELDON, 1961). It is worth noting that the Permian and Triassic mark the acme of the Sphinctozoa and that both they and the Pharetronida are sufficiently dominant in shallow-water faunas during these periods that they are important reef builders. It is also worth noting that the geographic expansion of the calcareous sponges between the Permian and the Triassic may be related to the reduction of competition.

Siliceous sponges are almost completely absent from Triassic deposits, although they occur abundantly with calcisponges before the Late Permian extinctions. This cannot be weighed too heavily, however, because in the subsequent Jurassic and Cretaceous the

Calcarea not only lived together again with siliceous sponges but the Pharetronida also attained their greatest diversity. Perhaps world environmental conditions in the Triassic were more marginal than before or after and favored Calcarea over siliceous sponges. It is obvious that the calcareous sphinctozoan sponges survived and indeed thrived in environmental conditions that caused extinction of archaic Paleozoic groups during the later Permian. If the Permian geographic distribution of the calcareous sponges indicates a tropical adaptation, the adverse conditions could have been related to high temperatures.

PRECAMBRIAN

Sponges are known to occur in upper Proterozoic sequences from Mongolia, China, India, and Australia, but their discoveries have been reported only recently, and their record is so scattered and incomplete that suggestion of faunal provinces is impossible at this time.

Precambrian demosponges were reported from Vendian beds by LI, CHEN, and HUA (1998) from central Guizhou (South China), where small fossils have skeletons of siliceous monaxial spicules that are randomly oriented. These small fossils are mostly globular, with a few that are tubular, and reportedly have preserved evidence of former soft parts in the phosphatic replacements. Other possibly Vendian spicules have been reported by TIWARI, PANT, and TEWARI (2000) from the Lesser Himalayas, northern India, where isolated hexactines have been observed in the Gangolihat Dolomite.

Somewhat younger unequivocal hexactines were reported by BRASIER, GREEN, and SHIELDS (1997) from beds of late Ediacaran age from Mongolia. Other hexactinellid spicules have been reported from late Proterozoic Ediacaran beds from China by STEINER and others (1993). Complete skeletons of possible hexactinellid sponges were described by GEHLING and RIGBY (1996) from Ediacaran beds of South Australia, but identification of hexactines in the moldic preservation of these latter sponges was not possible.

CAMBRIAN

Cambrian sponge faunas are almost entirely Hexactinellida, Heteractinida, and Monaxonida and are widely distributed in broad outlines, cosmopolitan. Few entire sponges have been preserved, and regional differences cannot be identified. Isolated stauractines, pentactines, and hexactines are widespread in dark limestones of Middle and Late Cambrian age in North America (PALMER, personal communication, 1966). Several genera have been described from the Burgess Shale (WALCOTT, 1920; RIGBY, 1986a) that are not known elsewhere, but it is hardly possible to state that such a distribution represents an endemic development (WALCOTT, 1920).

ORDOVICIAN

Some Ordovician genera have a worldwide distribution, such as the Lower Ordovician orchocladine lithistid *Archaeoscyphia* known from all over North America (TOOMEY, 1964; TOOMEY & NITECKI, 1979; JOHNS, 1994; CARRERA & RIGBY, 1999), Argentina (BERESI & RIGBY, 1993; CARRERA & RIGBY, 1999), northern Europe (HACHT & RHEBERGEN, 1997; VAN KEMPEN, 1978; RHEBERGEN & others, 2001), Australia (RIGBY & WEBBY, 1988), and China (GRABAU, 1932; LIU & others, 1997). In North America the lithistid Orchocladina are the dominant element in the sponge faunas at most localities with abundant sponges (Antelope Valley Limestone, Nevada; Pogonip Group, Utah; Platteville Limestone, Illinois; Crown Point Limestone, New York). Possible endemism may be shown by the Nevada and Utah faunas (JOHNS, 1994), which contain several genera not known elsewhere, along with other genera that are more widespread, and by the abundance of Dystactospongiidae (Demospongea) in north-central United States

(Tennessee to Minnesota). Since the hexactinellid *Brachiospongia* occurs in the central United States and in Scotland, however, much more will have to be known about Ordovician sponges before conclusions can be drawn concerning provincial faunas. Hexactinellida are most abundant in black shales (Utica, Little Métis), the orchocladines may have lived in reefs (New York, Texas, Utah, and Nevada).

SILURIAN

The lithistid Astylospongiidae and Hindiidae and the heteractinid Astraeospongiidae seem to dominate Silurian centers of sponge abundance. Possible provincial differences between North America and Sweden have been discussed above. The North American astylospongiids and astraeospongiids appear to have their maximum abundance in a somewhat deeper-water facies between and outside of reefs (Lowenstam, 1957). Diverse assemblages of orchocladine, megamorine, and rhizomorine lithistid sponges occur with astylospongiids and hindiids in middle Silurian slope to basin assemblages in Arctic Canada (Rigby & Chatterton, 1989) and in the Northwest Territories of northwestern Canada (Rigby & Chatterton, 1999).

DEVONIAN

Although sporadic occurrences of heteractinids (Astraeospongiidae) and lithistids (Hindiidae) are widespread in limy facies, as for example in the reefs of the Canning basin in western Australia (Rigby, 1986b), the most diversified and widespread sponges are Hexactinellida of the family Dictyospongiidae. These are known from many parts of the world (Europe, Africa, North America) and occur mainly but not exclusively in sandstone. The well-known fauna from the Upper Devonian of New York (Hall & Clarke, 1899) contains many genera not known elsewhere and may represent an endemic development, as do the largely undescribed Devonian sponge faunas from Poland (Rigby, Racki, & Wrzolek, 1982; Rigby, Wrzolek, & Racki, 2001).

CARBONIFEROUS

The Carboniferous was marked by the first emergence of geographic patterns of distributions on a global scale, as discussed above. The calcareous sphinctozoan demosponges expanded in the late Paleozoic part of their record, as seen in the Upper Carboniferous in south-central United States, Spain, and Manchuria. Bashkirian representatives are known from Spain, and a Tethyan origin of later lineages is possible. They are not known from any of the many sponge-bearing localities to the north of the Tethyan belt, and since this pattern persists into the Permian, it indicates a genuine geographic restriction. Carboniferous siliceous sponges are mainly reticulosid Hexactinellida, rhizomorine and orchocladine Lithistida, and isolated spicules of the first Choristida. There is also a variety of Heteractinida that follow the distribution of the siliceous sponges and locally may be quite abundant. All these sponges occur with the sphinctozoans in the south-central United States, so the restricted distribution is mainly on the part of the later sponges.

PERMIAN

The Permian had a tremendous development of sphinctozoan demosponges along with the first Pharetronida. They are locally abundant in a belt that runs from the Mediterranean area including Tunisia, through the Himalayan region, southeast Asia, southern China, and Japan. They are also very well developed in south-central United States and northern Mexico. Outside these areas they are completely absent, although numerous Lithistida occur at many localities to the north (northern Rocky Mountains, Arctic Canada, Spitsbergen, Ural Mountains). There is less control in the southern hemisphere (Rigby & Senowbari-Daryan, 1995), but a rich lithistid fauna occurs in Timor. Siliceous sponges are associated with the calcareous sphinctozoans in Mexico, USA (Texas), and Tunisia. There is some indication of endemism among the calcisponges, for there is a great proliferation of the Guadalupiidae in Texas that are not common elsewhere, as well as several individual genera of sphinctozoans or Pharetronida that are confined to Texas, Sicily, Tunisia, or China. *Amblysiphonella,* on the other hand, is found everywhere that calcisponges occur. Endemism also seems to occur among the lithistids, with a variety of quite different genera of Orchocladina in Timor and Texas. A diversified lyssacine hexactinellid fauna is also known from the dark limestone facies of Texas that has not been recognized elsewhere. On the other hand cosmopolitan genera of lithistids, such as the eutaxicladine *Scheiia* and the rhizomorine *Haplistion,* occur almost wherever there are siliceous sponges. It is noteworthy that the sphinctozoans, along with other Demospongea and a few Pharetronida, participated in construction of reefs in Texas, China, and Tunisia, a habit that was continued into the Triassic.

TRIASSIC

Very sporadic occurrences of Hexactinellida are known from the Triassic, including the earliest representatives of the Hexactinosa (Anisian) and Lychniscosa (Carnian). Lithistida appear to be completely absent from published records. This paucity of silicisponges is striking when compared with the abundance of the calcareous sphinctozoans and Pharetronida, which continued their Permian abundance and diversity, in some instances even being assigned to Permian genera. The continuity of calcisponges contrasts with the sharp break in silicisponges, where nearly all the Paleozoic families were extinguished at the end of the Permian, and where the forms that straggle back into the record, in the later Triassic and Jurassic, belong mostly to new orders. The Calcarea extend their geographic distribution beyond that of the Permian. In addition to the rich Alpine European faunas, diverse Upper Triassic sphinctozoan faunas have been described by BOIKO, BELYAEVA, and ZHURAVLEVA (1991) from the Caucasus and Pamir regions of Russia, Tajikistan, and the far-eastern part of Russia. New Triassic faunas of calcareous sponges from central Iran have also been reported by SENOWBARI-DARYAN, SEYED-EMAMI, and AGHANABATI (1997). Calcareous sponges have also been found in Upper Triassic patch reefs in the Yukon Territory of western Canada (SENOWBARI-DARYAN & REID, 1987) and in at least two places in Indonesia. One of these latter places is Timor, which is of particular interest because calcareous sponges have not been recognized in the Permian section, although a rich fauna of archaic lithistids was developed. Another place to which the calcareous sponges extended their range is Peru. Marine Lower Permian faunas are known from Peru (NEWELL, CHRONIC, & ROBERTS, 1948), and they have considerable affinity to the contemporary faunas of Texas. Lower Permian sponges are not known from Peru, however, although Triassic ones are.

The Triassic sphinctozoans appear to have a decided development of provincial faunas, Peru, USA (Nevada), and the Canada (Yukon) forming one province, the Alps another, and Indonesia a third. The American province is more distinctive than the other two, although all three differ at the generic level as originally described. The Alpine and Indonesian faunas also have more continuity with the preceding Permian fauna and contain several Permian genera.

JURASSIC

Jurassic sponges are known almost entirely from Europe, where a rich fauna is developed in the limy beds, especially in the oolites of England, the Jurassic of France, and the Malm of Germany (GAILLARD, 1983). All the modern orders are present, some for the first time. No geographic patterns of distribution can be recognized in this small area save those imposed by bathymetry, discussed elsewhere above (p. 283). In the Jurassic the first record of the invasion of fresh water by sponges occurs also. Spongillidae occur in the Purbeck beds of England (Hinde, 1893b), in the region of Lake Baikal in Siberia (Naletov, 1961), and in the Morrison Formation of Colorado in the western United States (Dunagan, 1999). A record of Hexactinosa from Australia suggests the probable worldwide distribution of many sponges in the Jurassic despite the paucity of records.

CRETACEOUS

The Cretaceous appears to have been the peak of diversity of the phylum as a whole and of each of the three living classes individually. Yet this great richness is known to us almost entirely from the chalk facies and greensand facies of Europe. Contemporaneous facies in North America and elsewhere have not been nearly so productive. Part of this may be due to the longer history of intensive paleontological work in Europe, but it must reflect surely a real pattern of distribution as well. If the cause of the pattern is environmental, its nature is not apparent in the lithology of the sediments. No provincial patterns are apparent from the data; on the contrary there seems to be an essential uniformity.

The most significant feature of Cretaceous distribution is the presence at high latitudes in northern Europe of such groups as the Lychniscosa, Lithistida, and Pharetronida, which are today confined to lower latitudes (tropical in the instance of the Lychniscosa). Because the relative position of poles and continents was close to the present arrangement by the Late Cretaceous according to most interpretations, the distribution suggests either a warmer climate or greater tolerances on the part of the sponges. Because many of the sponges are of living genera and because the same expanded distribution of present warm climate types are shown by many other marine animals and land plants, the first alternative seems more likely.

TERTIARY

The Tertiary record of the Porifera is less abundant but more widely dispersed. The presence of Lychniscosa in the Eocene of northern Europe suggests the persistence of warmer conditions there than those of today. Paleocene pharetronids in Denmark and Eocene lithistids in Europe and New Zealand point to the same conclusion. Beyond this, little pattern emerges from the scanty data. In the Lake Baikal region, the endemic, still-living, freshwater sponges, *Lubomirskia* and *Baicalospongia* have been identified in sediments possibly as early as the Paleogene and certainly by the Miocene (Naletov, 1961).

DETERMINANTS OF SPONGE DISTRIBUTION

The distribution of any organism is determined by both positive and negative factors. Negative factors are (1) physical and chemical conditions that inhibit metabolism of the organism, and (2) competition from other organisms that preempt metabolites and space or that create unfavorable physical and chemical conditions. Positive factors are (1) presence of appropriate food species or other sources of energy and (2) presence of other organisms that facilitate metabolism, either through direct symbiotic relationships or through favorably modifying the physicochemical environment.

Limits of geographical distribution are set mainly by limiting physical and chemical conditions and by the availability of food or energy. In the instance of suspension-feeders such as sponges, the availability of food is a minor factor in limiting distribution, for usable suspended particles are available almost everywhere in the aquatic environment. Therefore, mainly physicochemical conditions limit the latitudinal and bathymetric distribution of sponges. Abundance, on the other hand, is determined by both optimum physicochemical and food conditions and by interactions with other organisms. Symbioses do not play a vital role in sponges, and local areas of abundance are probably determined by absence of competition as well as by optimum conditions.

It is well known that sponges tend to have a spotty distribution with local areas of great abundance. This is evident in the fossil record where some formations and localities are crowded with sponges (e.g., the Ordovician Antelope Valley Limestone of Nevada, USA; the Devonian Chemung sandstones of New York, USA: the Permian limestones of west Texas, USA; the Cretaceous Faringdon sponge gravel of England; and the Upper Cretaceous chalk facies of northwestern Europe). Such a spotty distribution occurs also in present seas.

A particularly instructive example is a large patch of Demospongea on the bottom of the Barents Sea, north of the North Cape of Norway, where the sponges constitute as much as 95 to 98 percent of the total biomass of the benthos (ZENKOVITCH, 1963, p. 145–146 and fig. 56). The sponges *Geodia barretti, Craniella cranium,* and *Thenea maricata* lie free on a mainly sandy bottom at 150 to 350 m depth and occur in such numbers that they seriously interfere with operations of fishing vessels. The biomass of this community, which averages 351 g/m^2 but locally reaches 5,000 to 6,000 g/m^2, is considerably higher than that of any other bottom community in the Barents Sea (the next highest averages 221 g/m^2, see ZENKOVITCH, 1963, fig. 56, reproduced herein as Fig. 133). The high biomass of the sponge-dominated community relative to those of the surrounding communities implies an optimum food supply. MANTEUFEL (1938) has suggested that debris from the warm-water plankton brought in by the Gulf Stream and killed by the cold waters at the entrance to the Barents Sea may be the source of this food. Nevertheless, the fact that it is sponges rather than some other suspension-feeding organisms that take advantage of this food supply implies that some other factor is also involved that favors the sponges.

The absence of competition might be a reasonable explanation and is consistent with the following data. The sponge *Geodia barretti* was considered by BLACKER (1965) to be an indicator of Atlantic water and presently attains the northernmost limit of its distribution, so far as the vicinity of Spitsbergen goes, along the west coast of West Spitsbergen. A recent compilation of water temperatures (SCHRÖDER, 1963) at a depth of 200 m in the North Atlantic, which is the approximate depth of occurrence of *Geodia barretti* off both Spitsbergen and North Cape, shows that the average temperature is approximately the same in the two localities, namely around 3° C. Furthermore, distribution of *Geodia barretti* appears to be controlled by temperature, because it has extended its range in the last few decades from south of Spitsbergen northward along its entire west coast, consequent to worldwide warming of the climate (BLACKER, 1965). Since *Geodia barretti* is at the northern limit of its range at this temperature near Spitsbergen, it seems likely that it is also near the limit of its range near North Cape. Thus the abundance of *Geodia barretti* off North Cape is not likely to be due to optimum physical conditions, but rather to be in spite of the physical conditions. As pointed out above, an optimum food supply will not account for development of sponges to the virtual exclusion of other suspension feeders it must be the reduced viability of possible competitors in that environment that has permitted sponge numbers to expand so profusely, so close to the limits of their own temperature tolerances. Indeed, it may be the possession of a slight edge over competitors in a marginal environment that determines the abundant occurrence of sponges elsewhere, such as in the Antarctic. Such an explanation may account for many of the sponge concentrations in the fossil record as well.

One aspect of the food supply should be pointed out, however, for the food of sponges may differ from that of most suspension feeders. It has been shown (MADRI & others, 1967) that sponges can consume great quantities of bacteria, and that prosopyles of flagellated chambers of most demosponges are scarcely large enough to

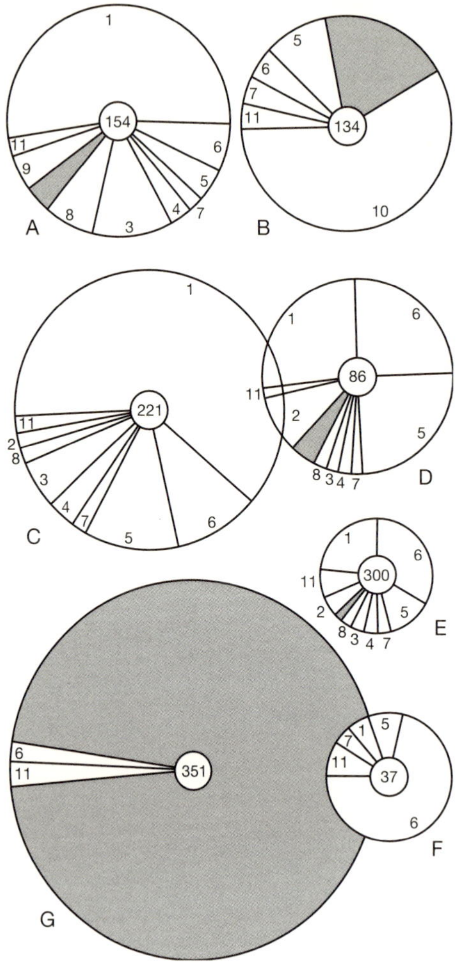

FIG. 133. Graphs showing composition of bottom faunas from seven localities in Barents Sea; *A,* Eastern (littoral); *B,* Northern (deep water); *C,* Eastern (medium depths); *D,* Central; *E,* Northern (littoral); *F,* Waldheimia-Brisaster; *G,* Porifera; *1,* Lamellibranchiata; *2,* Gephyrea; *3,* Crustacea; *4,* Coelenterata; *5,* Polychaeta; *6,* Echinodermata; *7,* Sipunculoidea; *8,* Gastropoda; *9,* Tunicata; *10,* Brachiopoda; *11,* Varia; *shaded,* Porifera; *numbers* in center circles refer to biomass in terms of g/m^2 (Zenkovitch, 1963).

admit a larger organism (see chapter on Physiology, p. 201 herein). Sponges are also capable of absorbing directly dissolved amino acids (EFREMOVA, 1965).

Data are apparently unavailable on bacterial content of waters near North Cape, but an abundance of bacteria and dissolved organic matter, if it occurs, would explain the concentrations of sponges near the outlets of rivers. DE LAUBENFELS (1958) has suggested this explanation for the concentrations of sponges near river mouths in Bermuda as well as for the concentration near the west coast of Florida.

Sponges in general are concentrated in two main environments of present seas, one in the sublittoral environment immediately adjacent to the land and the other in centers of ocean basins. Such a distribution is shown in Figure 134, a map of the northwestern Pacific by FILATOVA and BARSANOVA (1964). A survey of the Pacific as a whole (SOKOLOVA, 1964) shows that in the center of the Pacific basin, sponges form more than 50 percent by weight of trawl catches. This focus of their relative abundance lies within a larger zone in which more than 50 percent by weight of trawl catches is composed of suspension feeders (Fig. 135). Detritus feeders and carnivores dominate a zone outside of this, closer to the land masses but still in deep water. It is apparent that in the center of the Pacific the organic matter available to benthic animals is chiefly in the form of suspended particles and that under such conditions sponges tend to become abundant. It appears, furthermore, that their concentration in the center of the suspension-feeder zone implies a greater ability to survive on lower concentrations of suspended matter than other suspension feeders and that they may thrive here because competition is reduced. This interpretation assumes that the concentric pattern indicates a gradient of decreasing particle density toward the center of the oceanic area, an assumption supported by figures for plankton productivity, which are 0.2 g dry organic matter/m^2/day in the open Pacific, versus 3.2 g for shallow, inshore waters (ODUM, 1959). The sponges in question consist of the taxonomically rather limited abyssal fauna, mainly Amphidiscophora. It should be noted that total biomass of benthos in the central ocean area is very much less (by a factor of 10^5) than in the coastal zone (0.010 g/m^2 versus 1,000–5,000 g/m^2, ZENKOVITCH, 1963, p.

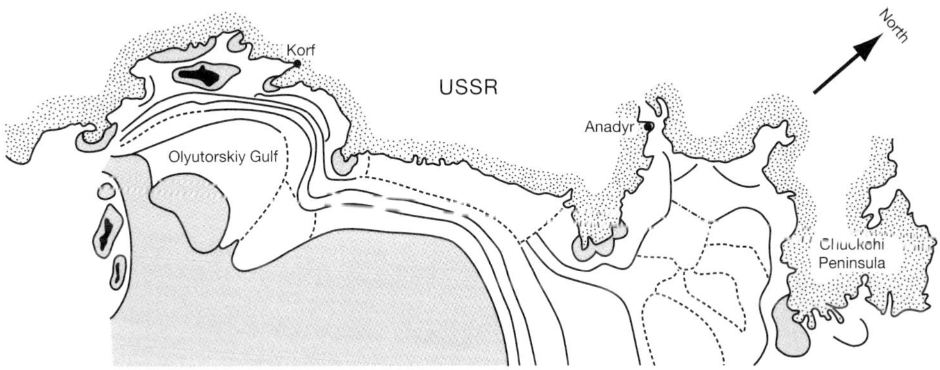

FIG. 134. Distribution of modern modern sponges concentrated in sublittoral environments adjacent to the Russian shore and in separated northwestern Pacific Ocean basin (Filatova & Barsanova, 1964).

723). Abundance of sponges in the coastal zone is probably related to the general abundance of food.

The Hexactinellida are most abundant today in deeper or colder waters. Their very open canal system and flagellated chambers provide lower hydraulic efficiency than that of most other sponges and seemingly require the passage of large quantities of water through the body at a slow rate (BIDDER, 1923; JØRGENSEN, 1955, 1966; see also chapter on Physiology, p. 201, herein). This implies a lower rate of metabolism than other sponges, which is already lower than that of most other animals (see chapter on Physiology, p. 203, herein). As mentioned above, their present distribution may reflect their ability to survive under conditions that are adverse for suspension feeders with higher metabolic rates. Their past distribution seems to have been in warmer and shallower waters, but it is tempting to speculate whether their times and places of abundance even then may have been relatively poor in suspended particles, sufficiently so to give them an advantage over suspension feeders with higher food requirements.

The Calcarea seem always to have been most abundant in the photic zone or a little below it. The greater solubility of calcium carbonate in colder water may be a limiting factor in relation to their calcium metabolism, although the existence of abyssal Arctic species shows that this can be circumvented if it is indeed operative. It is possible that the key to their distribution lies in their larval ecology, possibly involving a phototropic element, but knowledge on this point is insufficient to decide the question.

Distribution of Demospongea has been so wide and their apparent adaptability so great that any generalization about controls of distribution would have to be made at the level of species and genus. It is apparent that their tolerance of reduced salinity is greater than that of other sponges. Besides including the only freshwater sponges, several species are tolerant of brackish water. For example, *Cliona truitti* occurs in brackish waters with a salinity range as low as 10 to 16‰, whereas the marine species *C. celata* rarely lives in waters with a salinity below 25‰ (HARTMAN, 1958b). The Demospongea also include abyssal species that range as deeply as any of the Hexactinellida.

As for determining factors in the formation of faunal provinces, historical factors of isolation and connection are sometimes invoked, but in view of the radical changes in climate during the Pleistocene and the rapidity with which faunal boundaries can shift, this must be done with caution. BLACKER (1965) pointed out that the Atlantic fauna, including the sponge *Geodia barretti,* has spread northward for a distance of nearly 200 km along the west coast of

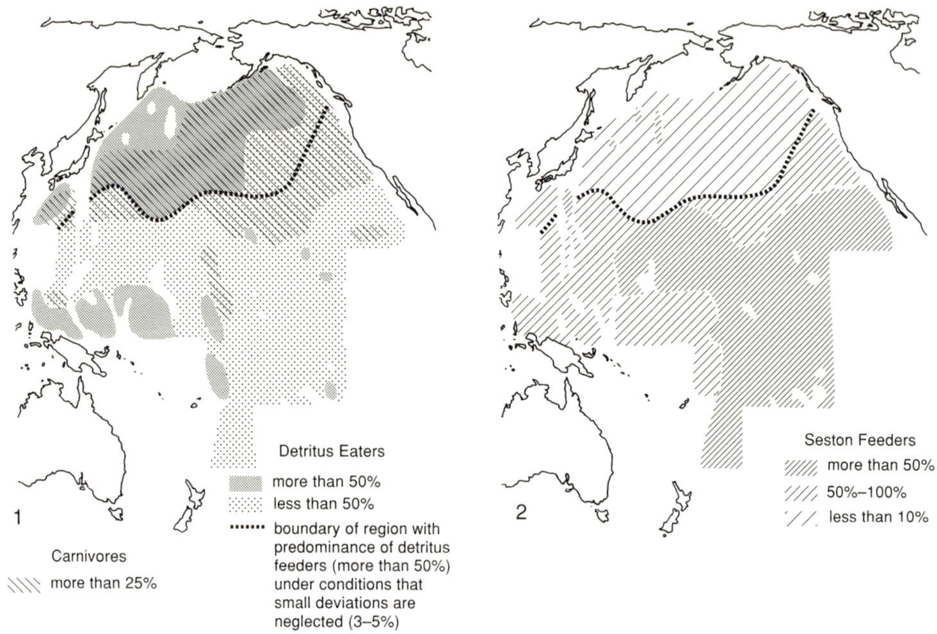

FIG. 135. Composition of trawl catches, by percentage weight, in northern and central Pacific basin; sponges make up more than fifty percent of suspension feeders in the central Pacific basin; *1,* detritus feeders; *2,* seston, or suspension, feeders (Sokolova, 1964).

Spitsbergen during the past few decades, presumably as a result of climatic warming, and replaced the Arctic fauna to as far north at 79°. *Geodia barretti* and other Atlantic-water indicators were not present at all along the west coast of Spitsbergen, as recorded in reports of oceanographic expeditions from 1878 to 1931, but were present in reports of expeditions made between 1949 and 1959. Inasmuch as the temperature changes in recent decades are very much less than those that accompanied glacial periods, fluctuations of faunal boundaries in the past must have covered much wider areas. Consequently there must have been considerable opportunity for mixing of latitudinally separate faunas, at least in the two major ocean basins. The present provinces may be largely climatically determined. Nevertheless, there is a degree of endemism in the Arctic and Antarctic (HENTSCHEL, 1929) and on the two sides of the North Atlantic (BURTON, 1934), which suggests that propagation of many faunal elements is coastal and that the open ocean may serve as a barrier.

TECHNIQUES OF STUDY

Robert M. Finks

[Department of Geology, Queens College (CUNY)]

INTRODUCTION

Observations of fossil sponges include such matters as spacing, density, diversity, life position, and orientation to be determined in the field; and growth form, size and arrangement of pores and canals, form of skeletal net, and form, size, and arrangement of spicules to be determined in the laboratory. The soft parts best studied in living sponges by histological and cytological techniques, can be studied as well in a limited way through skeletal correlates (e.g., the presence of spicules that normally support a dermal membrane, or skeletal canals that correspond to known structures of the aquiferous system); but much is lost, especially in the study of sponges whose skeletons fall part after death. The analysis of organic molecules concentrated in the vicinity of sponge fossils has not so far been published.

FIELD OBSERVATIONS

Where large bedding surfaces or a laterally extensive vertical section of a single bed are exposed, and sponges are sufficiently numerous, one can estimate the proportion of each species to the total fauna and their volume per unit area of sea bottom (biovolume of Walker & Bambach, 1974). Both estimates can be obtained simultaneously by the -linear-transect method; that is by stretching a line or a series of lines in a grid pattern over the bedding surface (or along it in a section) and measuring the linear intercept of each species. When converted to a percentage, this is proportional to their basal area or to their volume. This is also true of point counts of their occurrences at intersections of a regular grid. Illustrations of these techniques were given by Ager (1963, fig. 14.3, 14.7, 14.8).

A test of spatial dispersion (that is, clumped, random, or even distributions) can be made by counting numbers of specimens in each square of a grid and testing its fit to a Poisson distribution. Many even more sophisticated distribution functions have also been developed in recent years. Detailed mapping of extensive bedding surfaces to show occurrence of larger patches of sponges can be used in favorable exposures. Gridded data can be subjected to many other statistical techniques, for example, trend-surface analysis of sponge size (for specimens clearly in place) to determine patterns of environmental conditions).

Orientations of oscules or of branches of sponges in life position are always worth recording, as a significant common orientation may indicate current directions (oscules tend to face down current and flabellate branches across current) (Bidder, 1923; Warburton, 1960). Methods of determining statistical parameters of orientation data were given by Reyment (1971). When feasible, it is desirable to record stratigraphic top and a compass orientation of each specimen collected.

Where well preserved, sponges should be collected as completely as possible with enough of the surrounding matrix to insure that such delicate, outlying structures as root tufts, dermal spicules, and prostalia have been included. Because sponges are irregular and complex in outline, care is necessary to see that the entire specimen has been removed.

LABORATORY STUDY

PREPARATION

Special problems in preparation of sponges arise from the fact that their surfaces are often irregular, and as a result mechanical separation from the matrix is not easy. Dermal spicules are often unattached to the rest of the skeleton and are delicate. If the sponge is siliceous or silicified and the matrix is a carbonate, the sponge may be freed by etching with acid. Where large numbers of specimens are to be processed or large blocks contain a mixed fauna, a less careful technique may be tolerated; but there is the risk of losing loose spicules or delicate structures. If specimens are unusually complete, rare, or delicate, as with lyssacine hexactinellids or nonlithistid demosponges, great care in using this method can be very rewarding.

The less careful method of processing is as follows. Completely cover the base of the limestone block containing the siliceous fossils with moderately viscous cellulose lacquer, continuing the coating about halfway up the sides of the block. This will prevent the acid from undermining the block, which would cause crushing of the fossils in the lower part. Then place the block in an acid-proof vat on a tray of acid-proof screening (fiberglass or plastic), just cover the block with water, and add an equal amount or half this amount of concentrated hydrochloric acid (technical grade will do). Adding a small amount of surfactant to the water will prevent the formation of foam and reduce the violence of the bubbling. Use of a weaker organic acid, such as acetic acid or formic acid will cause less violent bubbling, which is desirable if spicules are delicate, but it will lengthen the time required. (In such instances, however, it is better to use the more careful technique described below.) If the matrix is partly argillaceous, it will be necessary to wash gently the etched surface of the block with water from time to time to lay bare fresh matrix. If bubbling stops before the fossils are freed, more acid must be added, but it may be desirable to leave a delicate specimen only partly freed from the matrix. After etching is complete, neutralize any remaining acid with sodium carbonate and remove the screening with the contained fossils to a container through which a gentle current of water is made to flow for a period of several hours. After a thorough rinse, the fossils may be air dried or dried in an oven at low temperature.

For delicate or rare sponge specimens it is desirable to etch them separately. The acid may be applied with a brush or dripped, washing immediately with a dropperful of water and repeating until as much of the specimen as necessary is freed. If the spicular net shows signs of disintegrating, polystyrene dissolved in xylene, acetone, or some other thin lacquer may be applied to it with a small brush before proceeding further with the etching. The polystyrene solution should not be so thick that it obscures the spicules. Areas of matrix may be blocked off in the same way to prevent undermining by the acid. As unattached spicules appear they may be removed after their position has been noted.

A delicate specimen, completely freed from matrix by whatever method, may be strengthened by dipping in a very thin solution of polystyrene in xylene or acetone, or if extremely delicate, by spraying with clear acrylic resin obtainable in aerosol cans.

Calcareous specimens in a soft argillaceous or limy matrix or siliceous specimens in a soft, noncalcareous matrix are best prepared mechanically. An air-abrasive machine (e.g., that produced by S.S. White & Company) using powdered dolomite abrasive may work in some instances, but delicate spicules are frequently abraded or blown away. Using sharpened needles (dissecting needles or sewing needles in a pin vise) or dental drills permits the greatest control, but even the finest needles or burrs are often too coarse for a fine spicular net.

In the study of sponges without a rigid skeletal net or of sponges that do not differ chemically or mechanically from the enclosing matrix, polished sections, acetate peels (if calcareous), and thin sections are the only means of study. One section should be in an axial or sagittal plane, if such exists, and another at right angles to the axis; but the choice of orientations may need to be varied according to the often complex form of the sponge. Tangential sections at various points on the outer and inner surfaces may be needed to show the distribution of pores and of skeletal material in a surficial layer. Sections tangent to internal structures, such as canals and spiculo-fibers, may also be needed.

EXAMINATION AND DOCUMENTATION

In addition to the qualitative description of external form, patterns of skeletal canals on the interior, patterns of skeletal pores on exterior and cloacal surfaces, forms of spicules, localization of spicules, patterns of spicular arrangements, and microstructure of aspicular parts of the skeleton (sclerosome), there are also quantitative data that may be determined. These include size distributions of pores, canal diameters, spicule lengths, spicule-ray thickness, skeletal mesh spacing, thickness of trabeculae or spiculo-fibers, thickness of cortex, size of micro-ornament on spicules, and size of spherulites or other microelements of sclerosome. The extent to which statistical treatment of these measures is done depends on the purpose of the study, but for usual taxonomic purposes range, mean or mode, and possibly standard deviation are sufficient. Down to the generic level, shapes and patterns of distribution of morphologic features are sufficient for taxonomic discrimination.

For both study and illustration of spicules and of the microstructure of sclerosome, the scanning-electron microscope is the instrument of choice. The production of photographic stereopairs is relatively easy by rotating the specimen. Unfortunately the use of the instrument and the preparation of specimens is expensive and time consuming. For the study and photography of thin-sections and acetate peels, a biological microscope with high magnification (400X) and good optics is indispensable. For ordinary study of whole specimens, especially to examine the spicular network in three dimensions, a binocular microscope of considerable range of magnification (from 5X to 100X) is essential, preferably mounted on a beam with swiveling capabilities so as to permit the positioning of large, irregular sponges. Photography through such a microscope, with the camera replacing the eyepiece, usually gives better results than the use of close-up lenses on the camera. Careful adjustment of lighting and coating of the specimen with sublimated ammonium chloride often bring out fine spicular detail. Photography of specimens in black shale while they are submerged in xylene or water will often bring out good detail, but it should be done in a ventilated hood because of the poisonous nature of the xylene vapors.

REFERENCES CITED

Ager, D. V. 1963. Principles of Paleoecology–An introduction to the study of how and where animals and plants lived in the past. McGraw-Hill. New York. 371 p.

Aleotti, G., G. Dieci, & F. Russo. 1986. Éponge Permiennes de la Vallé de Sosio (Sicile). Révision systématique des Sphinctozoaires. Annales de Paléontologie 72(3):211–246, 1 fig., 8 pl.

Altman, P. L., & D. S. Dittmer. 1962. Growth including reproduction and morphological development. Committee on Biological Handbook, Federation of American Societies for Experimental Biology. Washington, D.C. 608 p.

———. 1964. Biology Data Book. Committee on Biological Handbook, Federation of American Societies for Experimental Biology. Washington, D.C. xix + 633 p.

Ankel, W. E. 1948. Über Fragen der Formbildung und der Zelldetermination bei Süsswasserschwämmen. Deutsche Zoologische Gesellschaft Verhandlungen, Leipzig 1948:58–66, 8 fig.

Ankel, W. E., G. Wintermann-Kilian, & E. F. Kilian. 1955. Fehlbildungen bei *Ephydatia fluviatilis* und ihre Bedeutung für das Verständnis der normalen Korrelationen. Deutsche Zoologische Gesellschaft Verhandlungen 1954:362–371, 11 fig.

Annandale, Nelson. 1911. Freshwater sponges, hydroids and Polyzoa. *In* A. E. Shipley, ed., The Fauna of British India, including Ceylon and Burma. Taylor and Francis. London. p. 27–126, 241–245, pl. 1–2.

d'Archiac, E. J. A. 1843 [1842]. Description géologique du Départemente de l'Aisne. Mémoire de la Société Geologique de France, Paris (series 2, part 2) 5:129–419 (1–290), pl. 21–31.

———. 1846. Description des fossiles recueillis par M. Thorent dans les couches à Nummulites des environs de Bayonne. Mémoire de la Société Géologique de France, Paris (series 2) 2(1), 4:197.

Arendt, Y. A. 1959. [A new calcareous sponge from the Lower Carboniferous of the Moscow Basin]. Paleontologicheskii Zhurnal 2:46–52, 1 pl. In Russian.

Armstrong, J., J. Young, & J. Robertson. 1876. Catalogue of Western Scottish Fossils. British Association for the Advancement of Science. Blackie & Son. Glasgow.

Arndt, W. 1941. Bibliographia Spongiologica. I. Spongiae utiles (1940). W. Junk. The Hague. 114 p.

Ayling, A. M. 1968.The feeding behavior of *Rostanga rubicunda* (Mollusca, Nudibranchia). Tane 14:25–42, 7 fig.

Baer, Leopold. 1906 [1905]. Silicispongien von Sansibar, Kapstadt and Papeete. Archiv für Naturgeschichte 72(1):1–32, 5 pl.

Bagby, R. M. 1965. The contractile system of marine sponges. Dissertation Abstracts 25:6721.

Bakus, G. J. H. 1964. The effects of fish-grazing on invertebrate evolution in shallow tropical waters. Allan Hancock Foundation Occasional Papers 27:1–29, 1 fig., 1 table.

Balss, H. 1927. Decapoda. *In* Kukenthal und Krumbach, ed., Handbuch der Zoologie, de Gruyter, Berlin 3(1):840–1038, 217 fig.

Barboza du Bocage, J. V. 1869. Éponges siliceuses nouvelles de Portugal et de l'ile Saint-Iago (archipel de Cap-vert). Jornal de Sciencias mathematicas, physicas et naturaes, publicado sob os auspicios da Academia real das sciencias de Lisboa (May 1869) 4:159–162, pl. 10–11.

Barrois, Charles. 1882. Recherches sur les terrains anciens des Asturies et de la Galice. Association pour l'Avancement des Sciences, Congrès de Rouen. Lille. 630 p., 20 pl.

Bartolomaus, Werner, & Martin Lange. 1998. A new Ordovician sponge from the Kaolinsand Formation of the Isle of Sylt. Archiv fuer Geschieberkunde 2(6):398–402.

Bassler, R. S. 1927. A new Ordovician sponge fauna. Journal of the Washington Academy of Science 17(15):390–394.

———. 1941.The Nevada Early Ordovician (Pogonip) sponge fauna. Proceedings of the United States National Museum 91(3126):91–102, pl. 19–24.

Batten, R. L. 1958. Permian gastropods of the southwestern United States. 2. Pleurotomariacea; Portlockiellidae, Phymatopleuridae, and Eotomariidae. American Museum of Natural History Bulletin 114:153–246, 17 fig., pl. 32–42, 29 tables.

Beauvais, L. 1980. Les Calcarea (spongiaires) du Lias du Maros. Annales de Paléontologie (Invertebrata) 66(1):21–41.

Bechstädt, T., & R. Brandner. 1970. Das Anis zwichen St. Vigil und dem Hohlensteintal (Prager- und Olanger Dolomiten, Südtirol). Festband Geologische Institut. 300 Jahr-Freier Universitet Innsbruch. p. 9–103, 4 fig., 18 pl.

Bedford, R., & W. R. Bedford. 1934. New species of Archaeocyathinae and other organisms from the Lower Cambrian of Beltana, South Australia. Kyancutta Museum Memoir 1:1–7, fig. 1–36, pl. 1–6.

———. 1936. Further notes on Archaeocyathi (Cyathospongia) and other organisms from the Lower Cambrian of Beltana, South Australia. Kyancutta Museum Memoir 2:9–20, pl. 14–20.

Bedford, R., & J. Bedford. 1937. Further notes on Archaeos (Pleospongia) from the Lower Cambrian of South Australia. Kyancutta Museum Memoir 4:27–38, fig. 106–159, pl. 27–41.

Beecher, C. E. 1889. Brachiospongidae: A memoir on a group of Silurian sponges; with six plates. Peabody Museum of Natural History. Yale University, Memoir 2, part 1. 28 p., 4 fig., 6 pl.

Beede, J. W. 1899. New fossils from the Kansas Coal Measures. Kansas University Quarterly 8:123–130, pl. 32–33.

Belyaeva, G. V. 2000. Novye taksony sfinktozoa iz Permskikh rifov iugo-vostochnogo Kitaia [New taxa of Sphinctozoa from the Permian reefs of southeastern China]. Paleontologicheskii Zhurnal 2000(2):41–46, 3 fig.
In Russian. English translation: Paleontological Journal 34:155–160, 4 fig.

Benett, E. 1831. A catalogue of the organic remains of the county of Wilts. Warminster, England. p. i–iv, 1–9, 15 pl.

Bengtson, Stefan. 1986. Siliceous microfossils from the Upper Cambrian of Queensland. Alcheringa 10:195–216, 11 fig.

———. 1990a. Spicules. *In* Stefan Bengtson, Simon Conway Morris, B. J. Cooper, P. A. Jell, & B. N. Runnegar, eds., Early Cambrian fossils from South Australia. Association of Australasian Palaeontologists, Memoir 9. Brisbane. p. 24–37, fig. 11–21.

———. 1990b. Chancelloriids. *In* Stefan Bengtson, Simon Conway Morris, B. J. Cooper, P. A. Jell, & B. N. Runnegar, eds., Early Cambrian fossils from South Australia. Association of Australasian Palaeontologists, Memoir 9. Brisbane. p. 45–68, fig. 23–40.

Beresi, M., & J. K. Rigby. 1993. The Lower Ordovician sponges of the San Juan Formation, Argentina. Brigham Young University Geology Studies 39:1–64, 8 fig., 13 pl.

Berg, C. 1899. Substitucion de nombres genericos; III Comm. Museo Nacional. Buenos Aires. p. 1:77–80.

Bergmann, W., & R. J. Feeney. 1949. Contributions to the study of marine products, XXIII. Sterols from sponges of the family Haliclonidae. Journal of Organic Chemistry 14(6):1078–1084.

———. 1950. The isolation of a new thymine pentoside from sponges. Journal of the American Chemical Society 72:2805.

Bergmann, W., & W. J. McAleer. 1951. The isolation of metanethole from the sponge *Spheciospongia vesparia.* Journal of the American Chemical Society 73(10):4,969–4,970.

Bergmann, W., & F. H. McTigue. 1949. Contributions to the study of marine products, XXI. Chondrillasterol. Journal of Organic Chemistry 13(5):738–741.

Bergmann, W., F. H. McTigue, E. M. Low, W. M. Stokes, & R. J. Feeney. 1950. Contributions to the study of marine products, XXVI. Sterols from the sponges of the family Suberitidae. Journal of Organic Chemistry 15(1):95–105.

Bergquist, P. R. 1961. A collection of Porifera from northern New Zealand, with descriptions of seventeen species. Pacific Science 15:33–48.

———. 1965. The sponges of Micronesia, Part 1: the Palau Archipelago. Pacific Science 19:123–204, 34 fig.

———. 1967. Additions to the sponge fauna of the Hawaiian Islands. Micronesia 3:159–174.

———. 1968. The marine fauna of New Zealand, Porifera, Part 1 (Tetractinomorpha and Lithistida). New Zealand Department of Scientific and Industrial Research Bulletin 188:1–106.

———. 1978. Sponges. University of California Press. Berkeley & Los Angeles. 268 p., 81 fig., 12 pl.

Bergquist, P. R., & W. D. Hartman. 1969. Free amino acid patterns and the classification of the Demospongiae. Marine Biology 3:247–268.

Bergquist, P. R., & J. J. Hogg. 1969. Free amino acids in Demospongiae: a biochemical approach to sponge classification. Cahiers de Biologie Marine 10:205–220.

Bergquist, P. R., & M. E. Sinclair. 1968. The morphology and behaviour of larvae of some intertidal sponges. New Zealand Journal of Marine and Freshwater Research 2:426–437.

Beu, A. G. 1965. Ecological variations of *Chlamys diefenbachi* (Reeve) (Mollusca, Lamellibranchiata). Transactions of the Royal Society of New Zealand (Zoology) 7:93–96, 1 pl.

Bidder, G. P. 1893. On the flask-shaped ectoderm and spongoblasts in one of the Keratosa. Royal Society Proceedings, London 52:134–139.

———. 1898. The skeleton and classification of calcareous sponges. Proceedings of the Royal Society, London 64:61–76.

———. 1923. The relation of the form of a sponge to its currents. Quarterly Journal of Microscopical Science 67:293–323.

———. 1928. Some sponges of the South-West Coast. Proceedings of the Southwestern Naturalist Union 1928:12–20, 2 pl.

———. 1929. Sponges. Encyclopaedia Britannica, 14th ed., vol. 21. Encylopaedia Britannica Company. New York. p. 254–261.

Billings, Elkanah. 1859. Fossils of the Calciferous sandrock, including some of the deposits of White limestone at Mingan, supposed to belong to the formation. Canadian Naturalist and Geologist and Proceedings of the Natural History Society of Montreal 4(27):345–346.

———. 1861. New species of lower Silurian Fossils. Geological Survey of Canada Pamphlet 21:24 p.

———. 1865. On some new or little-known species of lower Silurian fossils from Potsdam Group (Primordial Zone). *In* Palaeozoic Fossils, vol. 1. Containing Descriptions and Figures of New or Little Known Species of Organic Remains from the Silurian Rocks. Geological Survey of Canada. Baillière. London, New York, & Paris. 426 p., 399 fig.

———. 1875. On some new or little known fossils from the Silurian and Devonian rocks of Ontario. Canadian Naturalist (new series) 7:230–240.

von Bistram, A. F. 1903. Beiträge zur Kenntnis der Fauna des unteren Lias in der Val Solda, Geologisch-paläontologische Studien in den Comasker Alpen, I. Berichte der Naturforschenden Gesellschaft zu Freiburg I. Br. 13:1–99[116–214], pl. 1–8.

Bizzarini, Fabrizio, & Franco Russo. 1986. A new genus of Inozoa from S. Cassiano Formation (Dolomiti di Braies, Italy). Memorie di Scienze Geologiche 38:129–135.

Blacher, L. J. 1965. Esquisse de l'histoire des stations biologiques maritimes russes. (Résumé). Vie Milieu, supplement 19:261–263.

Blacker, R. W. 1965. Recent changes in the benthos of the West Spitsbergen fishing grounds. International Commission of the Northwest Atlantic Fisheries, Special Paper 6(H-1):791–794.

de Blainville, H. M. D. 1830. Zoophytes. *In* F. G. Levrault, ed., Dictionnaire des Sciences Naturelles 60:546.

———. 1834. Manuel d'Actinologie ou de Zoophytologie. F. G. Levrault. Paris and Strasbourg. vol. 1, p. i–viii, 1–644; vol. 2, atlas, 100 pl.

Blumenbach, J. F. 1815. Specimen archaeologicae telluris terrarumque inprinis Hannoveranarum alterum. Commentationes Societas Scientiarum, Göttingen 3:3–25.
Sponge on p. 24.

Bocage—see Barboza du Bocage

Bodzioch, Adam. 1993. Sponges from the epicontinental Triassic of Europe. *In* Hans Hagdorn & Adolf Seilacher, eds., Muschelkalk: Schöntaler Symposium 1991. Goldschneck-Verlag Werner K. Weidert. Stuttgart. p. 235–244, 18 fig.

Böhm, J. 1927. Beitrag zur kenntnis der Senonfauna der Bithynischen Halbinsel. Palaeontographica, Stuttgart 69:187–222, 3 fig., pl. 11–18.

Boiko, E. V. 1979. O semeistve Verticillitidae Steinmann, 1882, ego sestare i sistematischeskom polozhenii [On the family Verticillitidae Steinmann, 1882, its composition and systematic position]. Trudy Instituta Geologii i Geofiziki Nauk SSSR, Sibirskoe Otdelenie 481:74–82.
In Russian.

———. 1984. Nekotorye pozdnetriasovye izvestkovye gubki iugo-vostochnogo Pamira [Certain Late Triassic calcareous sponges from southeastern Pamira]. *In* M. R. Dzhalilov, ed., Namudkhoi navi 'okimondakhoi khaivonot va na'ototi tochikiston [New species of fossil fauna and flora of Tadjhikistan]. Donish. Dushambe. p. 28–41, pl. V–IX.
In Russian, unpaginated plates.

———. 1990. [On the diversity of skeletal structures of Porifera Camerata]. Akademiya Nauk SSSR. Siberskoe Otdelenie Institut Geologii i Geofiziki Trudy 783:119–129, pl. XXXVII–XLVII.
In Russian.

Boiko, E. V., G. V. Belyaeva, & I. T. Zhuravleva. 1991. Sfinktozoa fanerozoya territorii SSSR [Sphinctozoa of the Phanerozoic of the USSR]. Nauka. Moskova. 224 p., 35 fig., 64 pl.

Bolkhovitinova [Bolkovitinoff], M. A. 1923. O kamennougol'nykh gubkakh Moskovskoi gubernii [Carboniferous sponges of Moscow Province]. Vestnik Moskovskoi Gornoi Akademii 2(1):61–72, 1 pl.

Borojevic, R. 1967. Éponges calcaires recueillies en Nouvelle-Calédonie par la Mission Singer-Polignac. Expédition Française sur les réçifs coralliens de la Nouvelle-Calédonie, Paris 2:1–10.

———. 1968. Éponges calcaires des côtes de France, IV: Le genre *Ascaltis* Haeckel emend. Archives de Zoologie Experimentale et Géneral 109:193–210.

———. 1979. Evolution des spongiaires Calcarea. Colloques Internationaux, Centre National de la Recherche Scientifique 291:527–530.

Borojevic, R., & N. Boury-Esnault. 1986. Une nouvelle voie d'évolution chez les éponges Calcinea: description des genera *Burtonella* n. g. et *Levinella* n. g. Bulletin du Muséum national d'Histoire naturelle (series 4, section A) 8(3):443–455.

Borojevic, R., N. Boury-Esnault, & J. Vacelet. 1990. A revision of the supraspecific classification of the subclass Calcinea (Porifera, class Calcarea). Bulletin du Muséum National d'Histoire Naturelle de Paris 12:243–276.

Borojevic, R., L. Cabioch, & C. Lévi. 1958. Inventaire de la faune marine de Roscoff, spongiaires. Cahiers de Biologie Marine 9(1):1–44.

Borojevic, R., & G. Graat-Kleeton. 1965. Sur une nouvelle aspic de *Sycon* et calces demosponge récoltées par le 'Cirrus' dans l'Atlantique Nord. Beaufortia 13:81–85, 1 fig.

Borojevic, R., & C. Lévi. 1964. Métamorphose artificielle de larves d'éponges, après dissociation et réaggrégation des cellules larvaires. Compte Rendus, Académie des Sciences, Paris 259:4,364–4,366, 1 fig.

Bouchet, P., & K. Rützler. 2002. Case 3211. Clionidae Rafinesque. 1815 (Mollusca, Pteropoda) and Clionidae d'Orbigny 1851 (Porifera, Hadromerida): proposal to remove homonymy by emending the sponge family-group name to Clionaidae; and Clionidae Rafinesque, 1815 and Clioidae Jeffreys, 1869 (Mollusca, Pteropoda): proposed confirmation that they are based on *Clione* Pallas, 1774 and *Clio* Linnaeus, 1767, respectively. Bulletin of Zoological Nomenclature, in press.

Bowerbank, J. S. 1842. *Halichondria johnstoni; Pachymatisma johnstoni. In* G. Johnston, History of British sponges and lithophytes. W. H. Kizars. Edinburgh. p. 198, 244.

———. 1845. Observations on the Spongiadae, with descriptions of some new genera. Annals and Magazine of Natural History (series 1) 6:400–410.

———. 1858. On the anatomy and physiology of the Spongiadae, Part I: On the spicula. Philosophical Transactions of the Royal Society of London 148:279–332.

———. 1861. List of British sponges. *In* R. McAndrew, ed., List of the British marine invertebrate fauna. Reports of the British Association for the Advancement of Science 30:235–236.

———. 1862. On the anatomy and physiology of the Spongiadae, Parts II, III. Philosophical Transactions of the Royal Society of London 152:747–836, 1,087–1,135, pl. 27–36, 72–74.

———. 1863. A monograph of the Spongillidae. Proceedings of the Zoological Society of London 1863:440–472, pl. 38.

———. 1864. A monograph of the British Spongiadae, vol. I. Ray Society. London. 290 p., 37 pl.

———. 1866. A monograph of the British Spongiadae, vol. II. Ray Society. London. 388 p.

———. 1869. A monograph on the siliceo-fibrous sponges. Proceedings of the Scientific Meetings of the Zoological Society, London 1869:66–108, pl. 3–6, part 1; p. 323–351, pl. 20–25, part 3.

———. 1874a. Contributions to a general history of the Spongiadae, part 6. Proceedings of the Zoological Society of London 1874:298–305, pl. 41–42.

———. 1874b. A monograph of the British Spongiadae, vol. III. Ray Society. London. xvii + 167 p., 92 pl.

———. 1882. A monograph of the British Spongiadae, vol. IV. Ray Society. London. xvii + 250 p., 17 pl.

Brasier, M. D. 1992. Nutrient-enriched waters and the early skeletal fossil record. Journal of the Geological Society, London 149:621–629.

Brasier, M. D., O. Green, & G. Shields. 1997. Ediacaran sponge spicule clusters from southwestern Mongolia and the origin of the Cambrian fauna. Geology 25:303–306.

Breistroffer, M. 1949. Note de nomenclature paléontologique: Spongiaires crétacés. Bulletin de la Société Scientifique du Dauphine, Grenoble 62(2):103.

Brien, Paul. 1967. Formation des statoblastes dans le genre *Potamolepis*: *P. symoensi* (Marshall), *P. pechuelli* (Marshall), *P. schoutedeni* (Burton). Bulletin de la Classe des Sciences, Académia royale de Belgique 53:573–590.

———. 1969. À propos de deux éponges du Cameroun appartenant au genre *Corvospongilla* (Annandale). Embryogénèse—la Parenchymula—la Gemmule. Revue de Zoologie et de Botanique Africaines 80:121–156, 15 fig., 3 pl.

———. 1970. Let potamolépides africaines nouvelles du Luapula et du la Moero. Symposium of the Zoological Society of London 25:163–187, 2 fig., 2 pl.

———. 1973a. Les Démosponges: Morphologie et réproduction. *In* P. P. Grassé, ed., Traité de Zoologie, Anatomie, Systématique, Biologie, III, Spongiaires. Masson et Cie. Paris. p. 133–461, fig. 90–351.

———. 1973b. *Malawispongia echinoides* Brien: Etudes complémentaires—histologie—sexualité—embryologie—affinitiés systématiques. Revue de Zoologie et de Botanique Africaines 87(1):50–76, 12 fig.

Brimaud, C. 1984. Étude biosedimentologique des gisements à Spongiaires du Tortonien des Cordillères betiques orientals. Thèse, Université de Marseille. 482 p., 32 pl.
Unedited.

Brimaud, Claudine, & Daniel Vachard. 1986. Les Spongiaires siliceux du Tortonian des Bétiques (Miocène de l'Espagne du Sud): espèces nouvelles ou peu connues, I. Choristides et Lithistides. Bulletin du Muséum National d'Histoire Naturelle, Paris (series 4, section C) 8:293–341, 9 pl.

———. 1987 [1986]. Les Spongiaires siliceux du Tortonian des Bétiques (Miocène de l'Espagne du Sud): espèces nouvelles ou peu connues, II. Hexactinellides. Bulletin du Muséum National d'Histoire Naturelle, Paris (series 4, section C) 8:415–445, 7 pl.

Broecker, W. S. 1974. Chemical Oceanography. Harcourt Brace Jovanovich, Inc. New York. 214 p.

Bromley, R. G. 1970. Borings as trace fossils and *Entobia cretacea* Portlock, as an example. Geological Journal, Special Issue 3:49–900.

Bromley, R. G., & Ulla Asgaard. 1993. Endolithic community replacement on a Pliocene rocky coast. Ichnos 2:93–116, 17 fig.

Brongniart, Adolphe. 1828. Histoire des végétaux fossiles, ou recherches botaniques et géologiques sur les végétaux renfermés dans les diverses couches du globe. G. Dufour et Ed. D'Ocagne, Printers. Paris. vol. 1, 488 p.; vol. 2, plates.

Brönn, H. G. 1825. System der urweltlichen Pflanzenthiere. J. C. B. Mohr. Heidelberg. i–iv, 1–47 p., 7 pl.

———. 1837–1838. Lethaia geognostica oder Abbildungen und Beschreibungen der für die Gebirgs-Formationen bezeichnendsten Versteingerungen. E. Schweizerbart. Stuttgart. 1,350 p.

———. 1848. Index Palaeontologicus, Ubersicht der bis jetzt bekannten fossilen Organismen. Erste Abtheilung, Nomenclator Palaeontologicus, vol. 1, no. 2. E. Schweizerbart. Stuttgart. 1,260 p.

Brydone, R. M. 1912. The Stratigraphy of the Chalk of Hants, with map and palaeontological notes. Dulau and Co, Ltd. London. 116 p., 3 pl.

Buckland, W. 1817. Description of the Paramoudra, a singular fossil body, that is found in the Chalk of the north of Ireland, with some general observations upon flint in chalk, tending to illustrate the history of their formation. Transactions of the Geological Society of London 4:413–423.

van Budden-brock. 1939. Über die Abhängigkeit der Atmung vom Sauerstoffdruk: Zugleich ein Beitrag zur Diffusionstheorie der Atmung. Nova Acta Leopoldina Carol, Halle 6(1939):557–565.

Bullivant, J. S. 1960. Photographs of the bottom fauna in the Ross Sea, New Zealand Oceanographic Institute, Department of Scientific and Industrial Research, Wellington. New Zealand Journal of Science 2:485–497, 10 fig.

Bullock, T. H., & G. A. Horridge. 1965. Structure and function in the nervous systems of invertebrates. W. H. Freeman & Co. San Francisco & London. xx + 798 p., vol. 1; vii + p. 799–1,719, vol. 2.

Burton, M. 1928. Report on some deep-sea sponges from the Indian Museum collected by R.I.M.S. "Investigator," Part II. Tetraxonida (concluded) and Euceratosa. Records of the Indian Museum, Calcutta 30(1):109–138, pl. 1–2.

———. 1929. Description of South African sponges collected in the South African Marine Survey, Part II. The Lithistidae. Union of South Africa Fisheries and Marine Biology Survey South Africa Report 7, Special Report 2:1–12.

———. 1932. Sponges. Discovery Reports 6:237–392, pl. 48–57.

———. 1934. Sponges. Further Zoological Research, Swedish Antarctic Expedition 1901–1903, Stockholm 3(2):1–58, 16 fig., 8 pl.

———. 1948. Ecology of sponges. Nature 162(4,106):73–74.
———. 1949. Observations on littoral sponges, including supposed swarming of larvae, movement and coalescence in mature individuals, longevity and death. Proceedings of the Zoological Society of London 118:893–915, 27 fig.
———. 1956. The sponges of West Africa. Atlantide Reports, Danish Science Press IV:111–147, 4 fig.
———. 1959. Sponges. Scientific Reports of the John Murray Expedition, 10(5):151–281, 41 fig.
———. 1963. A Revision of the Classification of the Calcareous Sponges. British Museum of Natural History. London. 693 p., 375 fig.
Carrera, M. G. 1994. An Ordovician sponge from the San Juan Formation, Precordillera Basin, western Argentina. Neues Jahrbuch für Geologie und Paläontologie, Abhandlungen 191:201–220, 6 fig.
———. 1996. Ordovician megamorinid demosponges from San Juan Formation, Precordillera, western Argentina. Geobios 29:643–650, 4 fig., 1 pl.
———. 1998. First Ordovician sponge from the Puna region, northwestern Argentina. Ameghiniana 35(2):205–210, fig. 1–3, pl. 1.
Carrera, M. G., & J. K. Rigby. 1999. Biogeography of Ordovician sponges. Journal of Paleontology 73:26–37, 4 fig.
Carter, H. J. 1849. A descriptive account of the freshwater sponges (genus *Spongilla*) in the island of Bombay, with observations on their structure and development. Annals and Magazine of Natural History (series 2) 4:81–100.
———. 1871. On fossil sponge-spicules of the Greensand compared with those of existing species. Annals and Magazine of Natural History (series 4) 7:112–141, pl. 7–10.
———. 1872. On two new sponges from the Antarctic Sea, and on a new species of *Tethya* from Shetland; together with observations on the reproduction of sponges commencing with zygosis of the sponge-animal. Annals and Magazine of Natural History (series 4) 9(54):409–435, pl. 20–22.
———. 1873. On the Hexactinellidae and Lithistidae generally, and particularly on the Aphrocallistidae, Aulodictyon, and Farreae, together with facts elicited from their deciduous structures, and descriptions respectively of three new species. Annals and Magazine of Natural History (series 4) 12:349–373, 437–472, pl. 13–17.
———. 1874. Descriptions and figures of deep-sea sponges and their spicules from the Atlantic Ocean, dredged up on board H.M.S. "Porcupine," chiefly in 1869; with figures and descriptions of some remarkable spicules from the Agulhas shoal and Colon, Panama. Annals and Magazine of Natural History (series 4) 14:207–221, 245–257, pl. 13–15.
———. 1875. Notes introductory to the study and classification of the Spongida. Annals and Magazine of Natural History (series 4) 16:1–40 (Part 1, Anatomy and Physiology), 126–145, 177–200 (Part 2, Proposed classification of the Spongida).
———. 1876. Descriptions and figures of deep-sea sponges and their spicules, from the Atlantic Ocean; dredged on board H.M.S. "Porcupine," chiefly in 1862. Annals and Magazine of Natural History (series 4) 18:226–240, 307–324, 388–410, 458–473, pl. 12–16.
———. 1878. Mr. James Thomson's fossil sponges from the Carboniferous System of the south-west of Scotland. Annals and Magazine of Natural History (series 5) 1:128–143.
———. 1879. On a new species of excavating sponge (*Alectona Millari*); and on a new species of *Rhaphidotheca* (*R. affinis*). Journal of Royal Microscopical Society 2:493–499, pl. 17.
———. 1880a. On fossil sponge-spicules from the Carboniferous of Ben Bulbel near Sligo. Annals and Magazine of Natural History (series 5) 6:209–214, pl. 14B.
———. 1880b. Report on specimens dredged up from the Gulf of Manaar and presented to the Liverpool Free Museum by Capt. W. H. Cawne Warren. Annals and Magazine of Natural History (series 5) 6(31):35–61, pl. 4–6; p. 129–156, pl. 7–8.
———. 1881a. History and classification of the known species of *Spongilla*. Annals and Magazine of Natural History (series 5) 7:77–107, pl. 5–6.
———. 1881b. On *Spongilla cinerea*. Annals and Magazine of Natural History (series 5) 7:263–264.
———. 1882. Some sponges from the West Indies and Acapulco in the Liverpool Free Museum described, with general and classificatory remarks. Annals and Magazine of Natural History (series 5) 9(52):266–301, 346–368, pl. 11–12.
———. 1885. Report on a collection of marine sponges from Japan made by Dr. J. Anderson, F.R.S. (Hexactinellida). Annals and Magazine of Natural History (series 5) 15:387–406.
———. 1886. Descriptions of sponges from the neighborhood of Port Phillip Heads, South Australia, continued. Annals and Magazine of Natural History (series 5) 17:40–53, 112–127, 431–441, 502–516; vol. 18:34–55, 126–149.
Casey, R. 1961. The stratigraphical palaeontology of the Lower Greensand. Palaeontology 3:487–621, 8 pl.
Caster, K. E. 1939. Siliceous sponges from Mississippian and Devonian strata of the Penn-York Embayment. Journal of Paleontology 13:1–20, 8 fig.
———. 1941. The Titusvillidae: Paleozoic and Recent branching Hexactinellida. Palaeontographica Americana 2:470–523, 5 pl.
Cavaroc, V. V., & J. C. Ferm. 1968. Siliceous spiculites as shoreline indicators in deltaic sequences. Geological Society of America Bulletin 79:263–271.
Chapman, Frederick. 1940[1939]. On a new genus of sponges from the Cambrian of the Flinders Range, South Australia. Transactions of the Royal Society of South Australia 64(1):101–108, pl. 9–12.
Chapman, F., & Irene Crespin. 1934. The Palaeontology of the Plantagenet Beds of Western Australia. Journal of the Royal Society of Western Australia 20:103–136, pl. 6–11.
Charlesworth, E. 1848. On the mineral condition and general affinities of the Chalk at Flamborough and Bridlington. Proceedings of the Yorkshire Philosophical Society, York 1848:73–77.

Chen J.-Y., Hou X.-G., & Li G.-X. 1990. New Lower Cambrian demosponges—*Quadrolaminiella* gen. nov. from Chengjiang, Yunnan. Acta Palaeontologica Sinica 29(4):402–414, 4 fig., 6 pl.
In Chinese with English summary.

Chen J-Y., Hou X.-G., & Lu H.-Z. 1989. Lower Cambrian leptomitids (Demospongea), Chengjiang, Yunnan. Acta Palaeontologica Sinica 28:17–31, 6 fig., 6 pl.
In Chinese with English summary.

Chen Meng-e, & Xiao Zong-Zheng. 1992. Macrofossil biota from Upper Sinian Doushantuo Formation in eastern Yangtze Gorges, China. Acta Palaeontologica Sinica 31(5):513–529, 6 pl.
In Chinese with English summary.

Christ, J. 1925. Ein neues fossilen Spongiengattung *Asteriscosella,* im Unterdevon des Nassauischen Hunsrückschiefers, *Asteriscosella nassovica.* Jahrbuch des Vereins Naturkunde, Wiesbaden 77:1–12.

Church, S. B. 1974. Lower Ordovician patch reefs in western Utah. Brigham Young University Geology Studies 21(3):41–62.

Clarke, J. M. 1900. Dictyonine hexactinellid sponges from the Upper Devonic of New York. New York State Museum Bulletin 39:187–194, pl. 10–11.

———. 1908. Early Devonic history of New York and eastern North America. New York State Museum Memoirs 9:1–252.

———. 1912. Early adaptation in the feeding habits of starfishes. Academy of Natural Sciences of Philadelphia Journal 15(art. III):113–118.

———. 1918a. Contributions to the paleontology of New York: Devonian glass sponges. I. The ontogeny of *Hydnoceras.* New York State Museum Bulletin 196:177–198, pl. 1–6.

———. 1918b [1917]. Devonian glass sponges. New York State Museum Bulletin 196:177–198.

———. 1920. *Armstrongia,* a new genus of Devonian glass sponges. New York State Museum Bulletin 219–220:143–146, pl. 1.

———. 1921. Organic dependence and disease, their origin and significance. New York State Museum Bulletin 221–222:1–133, 105 fig.

———. 1924. Eighteenth report of the director of the State Museum and Science Departments. New York State Museum Bulletin 251:192 p., 20 fig., 31 pl.

Claus, C. F. W. 1872. Grundzuge der Zoologie, 2nd ed. N. G. Elwert. Marburg, Leipzig, & Jena. 1,170 p.

Clausen, C. K. 1982. *Wienbergia,* new genus for *Barroisia faxensis* (Porifera: Demospongia) from the Middle Danian of Denmark. Bulletin of the Geological Society of Denmark 30:111–115, 11 fig.

Conrad, D. A. 1842. Observation on the Silurian and Devonian Systems of the United States, with descriptions of new organic remains. Journal of the Academy of Natural Sciences, Philadelphia 8:267–268, l fig., pl. 16.

Conway, K. W., J. V. Barrie, W. C. Austin, & J. L. Lauternauer. 1991. Holocene sponge bioherms on the western Canadian continental shelf. Continental Shelf Research 11:771–790.

Conway, K. W., M. Krautter, J. V. Barrie, & M. Neuweiler. 2001. Hexactinellid sponge reefs on the Canadian Continental Shelf: a unique "living fossil." Geoscience Canada 28(2):71–78, 8 fig.

Cossman, M. 1909. Rectifications de nomenclature. Revue Critique Paléozoologie 13:67.

Courtiller, A. 1861. Éponges fossiles des sables du terrain crétacé supérieur des environs de Saumur (étage Sénonien de d'Orbigny). Annales de la Société Linnéenne de Maine-et-Loire 4:117–142, pl. 1–40.

Croneis, Carey, & D. F. Toomey. 1965. Gunsight (Virgilian) wewokellid sponges and their depositional environment. Journal of Paleontology 39:1–16, 2 fig., 7 pl.

Cuif, J.-P. 1973. Histologie de quelques sphinctozoaires (Porifères) Triasiques. Geobios 6:115–125, 4 fig., pl. 8–10.

———. 1974. Role des sclérosponges dans la faune récifale du Trias des Dolomites (Italie du Nord). Geobios 7:139–153, 5 fig., pl. 29–31.

———. 1979. Caractères morphologiques et microstructuraux de trois sclérosponges triasiques association avec des Chaetetida. *In* Claude Lévi & Nicole Boury-Esnault, eds., Biologie des Spongiaires, Colloques Internationaux du Centre National de la Recherche Scientifique 291:475–481, 1 pl.

Cuif, J. P., F. Debrenne, J. G. Lafuste, & J. Vacelet. 1979. Comparaison de la microstructure du squelette carbonaté non-spiculare d'éponges actuelles et fossiles. *In* Claude Lévi & Nicole Boury-Esnault, eds., Biologie des Spongiaires, Colloques Internationaux du Centre National de la Recherche Scientifique 291:459–465.

Cuif, J. P., & P. Gautret. 1991. Taxonomic value of microstructural features in calcified tissue from recent and fossil Demospongiae and Calcarea. *In* J. Reitner & H. Keupp, eds., Fossil and Recent Sponges. Springer-Verlag. Berlin. p. 159–169.

Cullison, J. S. 1944. The stratigraphy of some Lower Ordovician formations of the Ozark Uplift. University of Missouri School of Mines and Metallurgy Bulletin (technical series) 15(2):1–112, 35 pl.

Dawson, J. W. 1883. Preliminary notice of new fossils from the Lower Carboniferous of Nova Scotia and Newfoundland. McGill University Peter Redpath Museum, Report 2:10–15.

———. 1888. Preliminary note on new species of sponges from the Quebec Group at Little Métis. The Canadian Record of Science 3(2):49–59.

———. 1896. Additional notes on fossil sponges and other organic remains from the Quebec Group at Little Métis, on the Lower St. Lawrence, with notes on some of the specimens by Dr. G. J. Hinde. Transactions of the Royal Society of Canada, Section 4:91–121, 4 pl.

Dawson, J. W., & G. J. Hinde. 1889. On new species of fossil sponges from the Siluro-Cambrian at Little Métis on the Lower St. Lawrence; including notes on the specimens by Dr. G. J. Hinde, F.G.S. Transactions of the Royal Society of Canada (section IV, Geological and Biological Sciences) 7(4):31–55, 27 fig., pl. III.

Dayton, P. K., G. A. Robilliard, & R. T. Paiune. 1970. Benthic faunal zonation as a result of anchor ice at McMurdo Sound, Antarctica. *In* M. W. Holdgate, ed., Antarctic Ecology, vol. 1, part 5. Academic Press. London. p. 244–258, 5 fig., 2 tables.

Dearborn, J. H. 1965. Ecologic and faunistic investigations of the marine benthos at McMurdo Sound. Ph.D Dissertation. Stanford University. 180 p.

Debrenne, Françoise, & J. Lafuste. 1972. Nouvelle données sur la microstructure du squelette de quelque sphinctozoaires. Bulletin de la Société Géologique de France 14:325–330.

Debrenne, Françoise, Henri Termier, & Geneviève Termier. 1971. Sur de noveaux représentants de la classe des *Radiocyatha.* Essai sur l'évolution des Métazoaires primitifs. Bulletin de la Société Géologique de France (series 7) 13(3–4):439–444, pl. 29–30.

Debrenne, Françoise, & Rachel Wood. 1990. A new Cambrian sphinctozoan sponge from North America, its relationship to archaeocyaths and the nature of early sphinctozoans. Geological Magazine 127:435–443, 5 fig.

Debrenne, Françoise, & A. Zhuravleva. 1994. Archaeocyathan affinities: how deep can we go into the systematic affiliation of an extinct group? *In* R. W. M. van Soest, T. M. G. van Kempen, & J. C. Braekman, eds., Sponges in Time and Space. A. A. Balkema. Rotterdam. p. 3–12.

De France, M. J. L. 1816. Alcyonium. *In* F. G. Levrault, ed., Dictionnaire des Sciences Naturelles, 2nd ed., I. F. G. Levrault. Paris. Supplement, p. 107.

———. 1829. Verticillites. *In* F. G. Levrault, ed., Dictionnaire des Sciences Naturelles, vol. 58. F. G. Levrault. Paris. p. 5–6.

Defretin-Lefranc, Simone. 1961. Contribution à l'étude des spongiaires siliceux du Crétacé supérieur du Nord de la France. Thèse Présentée à la Faculté des Sciences de Lille pour obtenir le Grade de Docteur es Sciences Naturelles. Université de Lille. Lille. 178 p., 47 fig., 27 pl.

Delage, Y. 1892. Embryogénie des éponges; developpement postlarvaire des éponges siliceuses et fibreuses marine et d'eau douce. Archive de Zoologie Expérimentale et Générale (series 2) 10(3):345–498, 8 pl.

Dendy, Arthur. 1893. Synopsis of the Australian Calcarea Heterocoela, with a proposed classification of the group, and descriptions of some new genera and species. Proceedings of the Royal Society of Victoria (new series) 5:69–116.

———. 1905. Report on the sponges collected by Professor Herdman, at Ceylon, in 1902. Report on the Pearl Oyster Fisheries of the Gulf of Manaar, Royal Society, London, supplement 18:57–246, 16 pl.

———. 1910. Porifera, Part I–Non-Antarctic sponges. British Antarctic (Terra Nova) Expedition, Zoology 6(3):269–392, 15 pl.

———. 1916. Report on the Homosclerophora and Astrotetraxonida collected by H. M. S. "Sealark" in the Indian Ocean. Transactions of the Linnean Society, Zoology 17:225–271, 4 pl.

———. 1917. On the occurrence of gelatinous spicules and their mode of origin in a new genus of siliceous sponges. Proceedings of the Royal Society of London (section B) 89:315–321.

———. 1921. The tetraxonid sponge spicule: a study in evolution. Acta Zoologica, Stockholm 2:95–152.

———. 1922. Report on the Sigmatotetraxonida collected by H. M. S. "Sealark" in the Indian Ocean. Transactions of the Linnean Society, London 18:1–164, 18 pl.

———. 1924a. Porifera, Part 1. Non-Antarctic Sponges. British Antarctic ("Terra Nova") Expedition, 1910, Zoology 6(3):269–392, 15 pl.

———. 1924b. On an orthogenetic series of growth-forms in certain tetraxonid sponge spicules. Proceedings of the Royal Society of London (section B) 97:243–250, pl. 9–10.

Dendy, Arthur, & W. H. Row. 1913. The classification and phylogeny of the calcareous sponges, with a reference list of all the described species, systematically arranged. Proceedings of the Zoological Society, London 1913:704–813, 1 fig.

Deng Zhan-Qiu. 1981. Upper Permian sponges from Laibin of Guangxi. Acta Palaeontologica Sinica 20(5):418–427, 4 pl.
In Chinese with English summary.

———. 1982. Paleozoic and Mesozoic sponges from Southwest China. *In* Stratigraphy and Palaeontology in western Sichuan and eastern Xizang, China, Part 2. Sichuan Renmin Chuban She. Chengdu. p. 245–258, pl. 1–6.
In Chinese with English summary.

———. 1990. New materials of Permian sponges. Acta Paleontologica Sinica 29(3):315–320, pl. 1.
In Chinese with English summary.

De Saporta, L. C. J. G. 1887. Nouveaux documents relatifs aux organismes problématiques des anciennes mers. Bulletin de la Société Géologique de France (series 3) 15:286–302, pl. 3–7.

Dieci, G., A. Antonacci, & R. Zardini. 1968. Le Spugne cassiane (Trias medio-superiore) della regione dolomitica attorno a Cortino d'Ampezzo. Bollettino della Società Paleontologica Italiana 7(2):94–155, pl. 18–33.

Dieci, G., A. Russo, & F. Russo. 1974a. Nota preliminare sulla microstruttura di spugne aragonitiche del Trias medio-superiore. Bollettino della Società Paleontologica Italiana 13(1–2):99–107, pl. 32–37.

———. 1974b. Revisione del genere *Leiospongia* d'Orbigny (Sclerospongia triassica). Bollettino della Società Paleontologica Italiana 13(1–2):135–146, pl. 51–53.

Ding L., Li Y., Hu X., Xiao Y., Su C., & Huang J. 1996. Sinian Miaohe Biota. Geological Publishing House. Beijing. 221 p.

Ding Wei-ming, & Qian Yi. 1988. Late Sinian to Early Cambrian small shelly fossils from Yangjiaping, Shimen, Hunan. Acta Micropalaeontologica Sinica 5(1):39–55, 4 pl. In Chinese with English summary.

Doederlein, L. 1892. Über *Petrostoma schulzei* n. g., n. sp., der Kalkschwamme. [Description of *Petrostoma schulzei* of Calcarea, representing a new order of Lithones]. Verhandlungen Deutsche Zoologische Gesellschaft 2:143–145.

Dong Xiping, & A. H. Knoll. 1996. Middle and Late Cambrian sponge spicules from Hunan, China. Journal of Paleontology 70:173–184, 7 fig.

Doré, Francis, & R. E. H. Reid. 1965. *Allonia tripodophora* nov. gen., nov. sp., nouvelle Éponge du Cambrien inferior de Carteret (Manche). Comptes Rendus Somme Seances, Société Géologique de France 1:20–21, 1 fig.

Dorn, P. 1932. Untersuchungen über fränkische Schwammriffe. Abhandlungen Geologisches Landesamt Bayern Oberbergamt, Munchen, p. 13–44, 6 pl.

du Dresnay, Renaud, Geneviève Termier, & Henri Termier. 1978. Les hexactinellides (lyssakides et dictyonines) du lias marocain. Geobios 11(3):269–295, 4 fig., 6 pl.

Duchassaing de Fonbressin, P., & G. Michelotti. 1864. Spongiaires de la mer Caraïbe. Natuurkundige verhandelingen van de Hollandsche maatschappij der wetenschappen te Haarlem 21(2):1–124, pl. I–XXV.

Dumortier, E. 1871. Sur calces gisements de l'Oxfordian inferieur de l'Ardeche. Paris-Lyon, 4, 85 p.

Dunagan, S. P. 1999. A North American freshwater sponge (*Eospongilla morrisonensis* new genus and species) from the Morrison Formation (Upper Jurassic), Colorado. Journal of Paleontology 73:389–393, 3 fig.

Duncan, P. M. 1879. On some spheroidal lithistid Spongida from the Upper Silurian Formation of New Brunswick. Annals and Magazine of Natural History (series 5) 4:84–91.

Duner, H., & B. Pernow. 1963. Chapter 22, Histamine. *In* U. S. von Euler and H. Heller, eds., Comparative Endocrinology, vol. 2. Academic Press, Inc. New York. p. 239–257, 1 fig.

von Dunikowski, Emil. 1883. Die Pharetronen aus dem Cenoman von Essen, und die Systematische Stellung der Pharetronen. Paleontographica 29:283–323, pl. 37–40 (1–4).

———. 1884. Über Permo-Carbon-Schwämme von Spitzbergen. Koniglia Svenska Vetenskaps Akademiens, Handligar, Stockholm (series 4) 21:1–18.

———. 1892. Die Pharetronen aus dem Cenoman von Essen und die Systematische Stellung der Pharetronen. Palaeontographica 29:281–348.

Dunn, M. T., R. H. Mapes, & J. K. Rigby. 2003. A land plant not a sponge: a re-evaluation of the Mississippian demosponge *Vintonia* and the family Vintoniidae. Journal of Paleontology 77:397–399, 1 fig.

Edwards, M. A., & A. T. Hopwood, eds. 1966. Nomenclator Zoologicus, vol. 6, 1946–1955. The Zoological Society of London. London. 329 p.

Edwards, M. A., & H. G. Vevers, eds. 1975. Nomenclator Zoologicus, vol. 7, 1956–1965. The Zoological Society of London. London. 374 p.

Efremova, S. M. 1965. Experiments on the use of glycin C^{14} for the study of the nutrition of sponges *Sycon lingua* Haeck. Vestnik Leningradskogo Gosudarstvennogo Universiteta 3:17–23, 6 fig. In Russian.

Eichwald, E. 1860. Lethaea rossica ou paléontologie de la Russie, décrite et figurée par Eduard d'Eichwald, vol. 1. E. Schweiserbart. Stuttgart. 359 p.

Elliott, G. F. 1963. Problematical microfossils from the Cretaceous and Paleocene of the Middle East. Palaeontology 6:293–300, 3 pl.

Emery, K. O., J. L. Tracy Jr., & H. S. Ladd. 1954. Geology of Bikini and nearby atolls. U. S. Geological Survey Professional Paper 260-A:1–265.

Engeser, T. S. 1986. Nomenklatorische notiz zur gattung *Dictyocoelia* Ott, 1967 ("Sphinctozoa", Porifera). Neues Jahrbuch für Geologie und Paläontologie, Monatshefte 10:587–590.

Engeser, T. S., & Dorte Mehl. 1993. Corrections and additions to the nomenclature of the Porifera in *The Treatise on Invertebrate Paleontology, (Part E)*. Berliner Geowissenschaftliche Abhandlungen (Reihe E) 9:183–198.

Engeser, T. W., & H.-H. Neumann. 1986. Ein neuer verticillitider "sphinctozoe" (Demospongiae, Porifera) aus dem Campan der Krappfeld-Gosau (Kärnten, Österreich). Mitteilungen, Geologisch-Paläontologische Institut, Universität Hamburg 61:149–159, 1 pl.

Engeser, T. S., & P. D. Taylor. 1989. Supposed Triassic bryozoans in the Klipstein Collection from the Italian Dolomites redescribed as calcified demosponges. Bulletin of the British Museum (Natural History, Geology series) 45:39–55, 9 fig.

Esper, E. J. C. 1791. Oryctographiae Erlangensis specimina quaedam imprimis spongiarium petrefactarum. Nova Actas Physikal-Medecinische Abhandlungen der Akademie Caesarea Leopold-Carol. Naturae Curios, vol. 8. p. 194–204, pl. 8.

———. 1794. Die Pflanzethiere in Abbildungen nach der Natur mit Farben erleuchtet nebst Beschreibungen. Zweiter Theil. Raspe. Nürnberg. 303 p.
This second part appeared 1791–1794 and contains lieferung 7–12, about sponges, p. 102, 165–282, 289–294.

Étallon, M. A. 1859a [1858]. Études paléontologiques sur le Haut-Jura. Additions et Rectifications S. 24 im Separat-Abdruck der Mémoires de la Société Jurassienne d'Émulation du Départment du Doubs 3:1–153.

———. 1859b. Études paléontologiques sur le Haut-Jura, Rayonnes du Corallien. Part III. Mémoires de la Société Jurassienne d'Émulation du Départment du Doubs 3:401–553.

———. 1860. Sur la classification des Spongiaires du Haut-Jura, et leur distribution dans les étages. Actes de la Société Jurassienne d'Émulation, pendant l'année 1858. Porrentruy. p. 129–160.

———. 1863. Lethea bruntrutana. Denkschriften der Schweizerischen naturforschenden Gesellschaft (Mémoires de la Société helvétique des sciences naturelles) 20:357–454.

———. 1864. Lethea bruntrutana. Siehe No. 222. Études paleontologiques sur le Jura graylois. Mémoires de la Société d'Émulation du Départment du Doubs 8:221–506.

Fabricius, O. 1780. Fauna Groenlandica, systematice sistens, Animalia Groenlandiae occidentalis hactenus indagata, quoad nomen specificum, triviale, vernaculumque; synonym auctorum plurium, descriptionem, locum, victum, generationem. Mores, usum, capturamque singuli; prout detegendi occasio fuit, maximaque parti secundum proprias observationes. Hafniae & Lipsiae. xvi + 1–452, 1 pl.

Fan Jiasong, J. K. Rigby, & Qi Jingwen. 1990. The Permian reefs of South China and comparisons with the Permian reef complex of the Guadalupe Mountains, West Texas and New Mexico. Brigham Young University Geology Studies 36:15–55, 16 fig., 11 pl.

Fan Jiasong, J. K. Rigby, & Zhang Wei. 1991. "Hydrozoa" from Middle and Upper Permian reefs of South China. Journal of Paleontology 65:44–68.

Fan Jiasong, Wang Yu-Mao, & Wu Ya-Sheng. 2002. Calcisponges and hydrozoans from Permian reefs in western Guangxi (China). Acta Palaeontologica Sinica 41(3):334–348, 2 fig., pl. 1–4.
In Chinese with English summary.

Fan Jiasong, & Zhang Wei. 1985. Sphinctozoans from Late Permian reefs of Lichuan, West Hubei, China. Facies 13:1–44, pl. 1–8.

Fedorov, A. B. 1987. Tip Gubki [Phylum Sponges]. *In* Yu. Ya. Shabanov & others, Nizhniy paleozoy yugozpapdnogo sklona Anabarskoy anteklizy (po materialam bureniya) [Lower Paleozoic of the southwestern slope of the Anabar Anteclise (according to boring data)]. Nauka. Novosibirsk. 208 p.

Felix, J. 1913. Über ein cretaceïsches Geshiebe mit *Rhizocorallium* Gläseli n. sp. aus dem Diluvium bei Leipzig. Sitzungsberichte der Naturforschenden Gesellschaft zu Leipzig 39:19–26.

Fenton, C. L., & M. A. Fenton. 1932a. Boring sponges in the Devonian of Iowa. American Midland Naturalist 13:42–54, pl. 6–9.

———. 1932b. A new species of *Cliona* from the Cretaceous of New Jersey. American Midland Naturalist 13:54–55, pl. 7.

Filatova, Z. A., & N. G. Barsanova. 1964. [The communities of bottom fauna of the western part of the Bering Sea.] Trudy Instituta Okeanologiya, Moscova 69:6–97, 4 fig.
In Russian with English summary.

Finks, R. M. 1955. *Conularia* in a sponge from the West Texas Permian. Journal of Paleontology 29:831–836, pl. 82.

———. 1960. Late Paleozoic sponge faunas of the Texas region: the siliceous sponges. American Museum of Natural History Bulletin 120(1):1–160, 77 fig., pl. 1–50.

———. 1967a. The structure of *Saccospongia laxata* Bassler (Ordovician) and the phylogeny of the Demospongea. Journal of Paleontology 41:1137–1149, 5 fig., pl. 145–146.

———. 1967b. Phylum Porifera Grant 1836. *In* W. B. Harland, ed., The Fossil Record. Geological Society of London. London. p. 333–341, fig. 9.

———. 1970. The evolution and ecologic history of sponges during Palaeozoic times. *In* W. G. Fry, ed., The Biology of the Porifera. Symposia of the Zoological Society of London 25:3–22, fig. 1–15.

———. 1971a. Sponge zonation in the west Texas Permian. Smithsonian Miscellaneous Contributions, Paleontology 3:285–300, 3 fig.

———. 1971b. A new Permian eutaxicladine demosponge, mosaic evolution, and the origin of the Dicranocladina. Journal of Paleontology 45:977–997, 5 fig., pl. 117–122.

———. 1983a. Pharetronida: Inozoa and Sphinctozoa. *In* T. W. Broadhead, ed., Sponges and Spongiomorphs. Notes for a short course organized by J. K. Rigby and C. W. Stearn, Studies in Geology 7. University of Tennessee. Knoxville. p. 55–69.

———. 1983b. Fossil Hexactinellids. *In* T. W. Broadhead, ed., Sponges and Spongiomorphs. Notes for a short course organized by J. K. Rigby and C. W. Stearn, Studies in Geology 7. University of Tennessee. Knoxville. p. 101–115, 4 fig.

———. 1990. Late Paleozoic pharetronid radiation in the Texas region. *In* Klaus Rutzler, ed., Perspectives in Sponge Biology. 3rd International Sponge Conference 1985. Smithsonian Institution Press. Washington, D.C. p. 17–24.

———. 1995. Some new genera of Paleozoic calcareous sponges. The University of Kansas Paleontological Contributions (new series) 6:9 p., 11 fig.

———. 1997. New name for a Permian calcareous sponge and some related corrections. Journal of Paleontology 71:352.

Finks, R. M., & D. F. Toomey. 1969. The paleoecology of Chazyan (lower Middle Ordovician) "reefs" or "mounds." New York State Geological Association Guidebook, 41st Annual Meeting. Plattsburgh. p. 93–120, 4 fig., 6 pl.

Finks, R. M., E. L. Yochelson, & R. P. Sheldon. 1961. Stratigraphic implications of a Permian sponge occurrence in the Park City Formation of western Wyoming. Journal of Paleontology 35:564–568.

Fischbuch, N. R. 1970. Devonian reef-building stromatoporoids from western Canada. Journal of Paleontology 44:1,071–1,084, pl. 145–149.

Fischer, A. G. 1962. Fossilien aus Riffkomplexen der alpinen Trias: *Cheilosporites* Wähner, ein Foraminifere? Paläontologische Zeitschrift 36:118–124, pl. 13–14.

Fischer, J. C. 1970. Révisions et essai de classification des Chaetetida (Cnidaria) post-Paléozoiques. Annales de Paléontologie des Invertébrés 56:151–220.

Fischer, P. 1867. Note sur calces Spongiaires fossiles de la Craie, appartenent au groupe des Géodies. Actes Société Linnéenne de Bordeaux (series 3) 6:233–238.

Fischer von Waldheim [de Waldheim], G. F., MS in C. E. d'Eichwald [Eduard von]. 1829. Zoologia specialis quam expositis animalibus tum vivis, tum fossilibus potissimum Rossiae in universum, et Poloniae in species, in usum lectionum, vol. 1. J. Zawalski. Vilna. vi + 314 p., 5 pl.

Fischer de Waldheim, G. F. 1830–1837. Oryctographie du Gouvernement de Moscou. Ouvrage publié aux frais de la Société Impériales des Naturalistes de Moscou, 2nd ed. A. Semen. Moscow. v + 202 p., 62 pl.

Fjerdingstad, E. J. 1961. The ultrastructure of choanocyte collars in *Spongilla lacustris* (L.). Zeitschrift für Zellforschung 53:645–657.

Fleming, J. 1828. A history of British Animals, exhibiting the descriptive characters and systematical arrangement of the genera and species of quadrupeds, birds, reptiles, fishes, mollusca and radiata of the United Kingdom. Bell and Bradfuste. Edinburgh & London. xxiii + 565 p.

Flügel, E. 1966. Algen aus dem Perm der Karnischen Alpen. Carintha II, Sonderheft, Klagenfurt 25:3–76, 15 fig., 11 pl.

———. 1981. Paleoecology and facies of Upper Triassic reefs in the Northern Calcareous Alps. *In* D. F. Toomey, ed., European Fossil Reef Models. SEPM Special Publication 30:291–359, 26 fig.

Flügel, E., R. Lein, & B. Senowbari-Daryan. 1978. Kalkschwämme, Hydrozoen, Algen und Mikroproblematika aus des Cidarisschichten (Karn, Ober-Trias) der Mürztaler Alpen (Stiermark) und des Gosaukammes (Oberösterreich). Mitteilungen Gesellschaft des Geologie- und Bergbaustudenten in Österreich 25:153–195.

Flügel, E., & G. D. Stanley. 1984. Reorganization, development, and evolution of post-Permian reefs and reef organisms. Palaeontographica Americana 54:177–186, 5 fig.

Foerste, A. F. 1916. Notes on Cincinnatian fossil types. Bulletin of the Scientific Laboratories of Denison University 18:285–355, pl. 1–7.

Fontaine, H. 1962. Nouveau nom pour le genre *Steinmannia* Waagen et Wentzel. Comptes Rendus, Société Géologique de France 7:205.

Forbes, M. L. 1964. Distribution of the commensal oyster, *Ostrea permollis,* and its host sponge. Bulletin of Marine Science of the Gulf and Caribbean 14:453–464, 2 fig.

Fox, H. M., & Hugh Ramage. 1930. Spectrographic analyses of animal tissues. Nature 126(3,183):682.

———. 1931. A spectrographic analysis of animal tissues. Proceedings of the Royal Society, London (section B) 108(755):157–173.

Fraipont, Charles. 1911. Use hexactinellide nouvelle du Dévonien belge (Calcaire Frasnien) *Pseudopemmatites Fourmarieri* nov. g. et. n. sp. Annales de la Société Géologique de Belgique 38:197–206, pl. 13–15.

de Freitas, Tim A. 1987. A Silurian sphinctozoan sponge from east-central Cornwallis Island, Canadian Arctic. Canadian Journal of Earth Sciences 24:840–844, 3 fig.

———. 1989. Silurian *Archaeoscyphia* from the Canadian Arctic: a case for simplified generic taxonomy in the anthaspidellid lithistids. Canadian Journal of Earth Sciences 26:1,861–1,879.

———. 1991. Ludlow (Silurian) lithistid and hexactinellid sponges, Cape Phillips Formation, Canadian Arctic. Canadian Journal of Earth Sciences 28:2,042–2,061, 9 fig.

Frentzen, Karl. 1932. Paleobiologisches ueber die korallevorkommen im oberen weissen Jura bei Nattheim, O.-A. Heidenheim. Bodische Geologische Abhandlungen 4:43–57.

Fretter, V., & A. Graham. 1976a. The prosobranch molluscs of Britain and Denmark, Part 1, Pleurotomariacea, Fissurellacea, and Patellacea. Journal of Molluscan Studies (supplement) 1:1–37, 25 fig.

———. 1976b. Sponges. *In* V. Fretter & A. Graham, eds., A Functional Anatomy of Invertebrates. Academic Press. London. p. 44–52, 2 fig. Page information for complete book: vii + 589 p.

de Fromentel, M. E. 1860a [1859]. Introduction à l'étude des éponges fossiles. Mémoires de la Société Linnéenne de Normandie 11:1–50, pl. 1–4.

———. 1860b. Catalogue raisonné des Spongitaires de l'étage Néocomien. Bulletin de la Société des Sciences historiques et naturelles de l'Yonne (series 4) 14:1–19, pl. 1–4.

———. 1861 [1860]. Catalogue raisonné des Spongitaires de l'étage Néocomien. Bulletin de la Société des Sciences historiques et naturelles de l'Yonne (series 4) 14(2):356–372.

———. 1864. Monographie des Polypiers jurassiques supérieurs (étages portlandien et kimmeridgien). Mémoires da la Société Linnéenne de Normandie, Paris 13:1–53, 8 pl.

———. 1865. Polypiers coralliens des environs de Gray, considérés dans leurs rapports avec ceux des bassins coralliens de la France, et dans leur developpement pendant la durée de cet étage. Mémoires de la Société Linnéenne de Normandie, Paris 14:1–43.

Fry, W. G. L. 1970. The sponge as a population: a biometric approach. *In* W. G. Fry, ed., Biology of the Porifera. Zoological Society of London Symposium 25:135–162, 12 fig.

Fursich, R. T., T. J. Palmer, & K. L. Goodyear. 1994. Growth and the disintegration of bivalve-dominated patch reefs in the Upper Jurassic of southern England. Palaeontology 37:131–171.

Gaillard, C. 1983. Les bioherms à spongiaires et leur environnement dans l'Oxfordien du Jura méridional. Documents des Laboratoires de Géologie de la Faculté des Sciences de Lyon 90:1–515, 187 fig., 42 pl.

García-Bellido Capdevila, D., & J. K. Rigby. 2004 (in press). Middle and Late Paleozoic sponges from the Iberian Peninsula. Journal of Paleontology 77:77 p., 19 fig.

Gatehouse, C. G. 1968 [1967]. The first record of lithistid sponges in the Cambrian of Australia. Bulletin of the Bureau of Mineral Resources, Geology and Geophysics, Australia 92:57–67, pl. 7–8.

Gatenby, H. B., & T. M. Tahmisian. 1959. The contractile vacuoles and Golgi apparatus of *Ephydatia fluviatilis:* an electron microscope study. Journal of the Royal Microscopical Society (series 3) 77(3–4):107–115, 5 pl.

Gautret, P. 1985. Organisation de la phase minerale chez *Vaceletia crypta* (Vacelet) démosponge, sphinctozoaire actuelle. Comparaison avec de formes aragonitiques du Trias de Turquie. Geobios 18(5):553–562, 2 fig., 4 pl.

Gautret, P., & J. P. Cuif. 1989a. Les démosponges calcifiées des biohermes du Jurassique supérieur du Sud Tunisien (Oxfordian de la region of Tataouine). Geobios 22(1):49–63.

———. 1989b. Microstructure granulaire calcitique de trois sphinctozoaires du Trias supérieur des Dolomites et le Turquie. Annales de Paléontologie 75(4):171–186, 1 pl.

Gautret, P., J. Vacelet, & J.-P. Cuif. 1991. Caractéristiques des spicules et du squelette carbonaté des espèces actuelles du genre *Merlia* (démosponge, Merliida), et comparaison avec des chaetétides fossiles. Bulletin du Muséum nationale de l'histoire naturelles de Paris (section A) 13:289–307.

Gehling, J. G., & J. K. Rigby. 1996. Long expected sponges from the Neoproterozoic Ediacara fauna of South Australia. Journal of Paleontology 70:185–195, 7 fig.

Geinitz, H. B. 1864. Zwei Arten von *Spongillopsis, S. dyadica, S. carbonica.* Neues Jahrbuch für Mineralogie und Geologie 1864:517–519.

Gerassimov, P. R. 1960 [1957]. Gubki podmoskovnoi yury i nizhnego mela [Sponges of the Jurassic and Lower Cretaceous of the Moscow area]. Materialy po Geologii i Poleznym Iskopaemym Tsentral'nykh Rainov Evropeiskoi Chasti SSSR, v. 3.

Gerth, H. 1909 [1907]. *Timorella permica* n. gen. n. sp., eine Neue Lithistide aus dem Perm von Timor. Centralblatt für Mineralogie, Geologie und Paläontologie 22:695–700, 5 fig.

———. 1927. Die Spongien aus dem Perm von Timor. Jaarboek van het Mijnwesen Nederland-Oest Indien, Verhandelungen für 1926:99–132.

———. 1929. XXVII. Die Spongien aus dem Perm von Timor. *In* Joh. Wanner, ed., Paläontologie von Timor, nebst kleineren beiträgen zur paläontologie einiger anderen inseln des ostindischen archipels. E. Schweizerbart'sche verlagsbuchhandlung. Stuttgart. p. 1–35, 6 pl.
Essentially a reprint of the systematics presented in Gerth, 1927.

Giattini, G. B. 1909 [1908]. "*Manzonia aprutina*" Nuona esattinellidae del Miocene Medio di S. Valentino (Chieti). Revista Italiana di Paleontologia 14:57–63.

Giebel, C. G. 1850 [1849]. Über *Scyphia uvaeformis,* n. sp. Jahresbericht ueber Naturwissenschaftlichen Verein in Halle, 2 jahrg. Berlin. p. 57–60.

———. 1852. Deutschlands Petrefacten. Ein systematisches Verzeichniss aller in Deutschland und den angrenzenden Landern vorkommenden Petrefacten nebst Angabe der Synonymen und Fundorte. A. Abel. Leipzig. 706 p.
Sponges on p. 169–183.

Gilbert, J. J., & T. L. Simpson. 1976. Sex reversal in a freshwater sponge. Journal of Experimental Zoology 195:145–151.

Girty, G. H. 1895. A revision of the sponges and coelenterates of the Lower Helderberg Group of New York. New York State Geologist Annual Report 14(64):259–322, pl. 1–7.

———. 1908. On some new and old species of Carboniferous fossils. Proceedings of the U.S. National Museum 34:281–303, pl. 14–21.

———. 1909 [1908]. The Guadalupian Fauna. U.S. Geological Survey Professional Paper 58:1–641, 31 pl.

———. 1912 [1911]. On some new genera and old species of Pennsylvanian fossils from the Wewoka Formation of Oklahoma. Annual Report of the New York Academy of Sciences 21:119–156.

Glaessner, M. F. 1962. Pre-Cambrian fossils. Biological Reviews of the Cambridge Philosophical Society 37(4):467–494.

Goldfuss, A. 1826. Petrefacta Germaniae oder Abbildungen und Beschreibungen der Petrefacten Deutschlands und der angrenzenden Lander, Band 1, Heft 1. Unter Mitwirkung des Grafen George zu Münster. Düsseldorf. p. 1–76, pl. i–xxv.

———. 1829. Petrefacta Germaniae oder Abbildungen und Beschreibungen der Petrefacten Deutschlands und der angrenzenden Lander, Band 1, Heft 2. Unter Mitwirkung des Grafen George zu Münster. Düsseldorf. p. 77–164, pl. xxvi–l.

———. 1831. Petrefacta Germaniae oder Abbildungen und Beschreibungen der Petrefacten Deutschlands und der angrenzenden Lander, Band 1, Heft 3. Unter Mitwirkung des Grafen George zu Münster. Düsseldorf. p. 165–240, pl. li–lxxi.

———. 1833. Petrefacta Germaniae oder Abbildungen und Beschreibungen der Petrefacten Deutschlands und der angrenzenden Lander, Band 1, Heft 4. Unter Mitwirkung des Grafen George zu Münster. Düsseldorf. p. 241–252.

Goryansky, V. Y. 1977. Lervaia Nakhodka ostatkov gubki v nizhnem kembrii vostochnoi sibiri [First discoveries of sponge remains from the Lower Cambrian in eastern Siberia]. Ezhegodnik Vsesoyuznyi Paleontologiya Obshchestvo 20:274–278, pl. 1.
In Russian.

Goreau, T. F., & W. D. Hartman. 1963. Boring sponges as controlling factors in the formation and maintenance of coral reefs. American Association for the Advancement of Science 75:25–54.

Grabau, A. W. 1932. Paleozoic centers of faunal evolution and dispersion. Pan-American Geologist 58(4):273–284.

Graham, A. 1971. British prosobranch and other operculate gastropod molluscs. Key and notes for the identification of the species. Synopses of the British Fauna, new series, no. 2. The Linnean Society of London. London. p. 1–112, 118 fig.

Grant, R. E. 1826. Observations on the structure of some siliceous sponges. Edinburgh New Philosophical Journal 1:341–351.

———. 1833. On the classification of the organs of animals and on the organs of support in animacules and porpherous animals. Lecture IV, University of London lectures on comparative anatomy and animal physiology. The Lancet (1833–1834, Nov. 2, 1833) 1:194–200.

———. 1835–1841. Porifera. *In* H. Bailliere, ed., Outlines of Comparative Anatomy, vol. 1. London. p. 5–9, 310–313, pl. II–IV.

———. 1836. Animal Kingdom. *In* R. B .Todd, ed., The Cyclopaedia of Anatomy and Physiology, vol. 1. Sherwood, Gilbert, & Piper. London. p. 107–118.

———. 1841. Outlines of Comparative Anatomy. Hippolyte Bailliere. London. 656 p., 147 fig.

Gray, D. I. 1980. Spicule pseudomorphs in a new Palaeozoic chaetetid, and its sclerosponge affinities. Palaeontology 23(4):803–820, pl. 102–103.

Gray, J. E. 1832. Synopsis of the contents of the British Museum (London). British Museum Publication, edition 27. London. 212 p.

———. 1835. On the coral known as the glass plant. Proceedings of the Zoological Society of London 1835:492–558, pl. 27–28.

———. 1837. A synoptical catalogue of the species of certain tribes or genera of shells contained in the collections of the British Museum and the author's cabinet, with descriptions of new species. Magazine of Natural History 1:370–376.

———. 1848. List of specimens of British sponges in the collection of the British Museum (London). British Museum Publication viii:1–24.

———. 1857. Synopsis of the axiferous Zoophytes or barked corals. Proceedings of the Zoological Society of London 25:278–294, pl. 9.

———. 1858. On *Aphrocallistes,* a new genus of Spongiadae from Malacca. Annals and Magazine of Natural History (series 3) 2:224.

———. 1859. Description of *MacAndrewia* and *Myliusia,* two new forms of sponges. Proceedings of the Zoological Society of London 1859:437–440, pl. 15–16.
Also published in 1860 in the Annals and Magazine of Natural History (series 3) 5:495–498.

———. 1867. Notes on the arrangement of sponges, with the description of some new genera. Proceedings of the Scientific Meetings of the Zoological Society of London 1867:492–558.

———. 1868. Note on *Hyalonema schultzei* Semper. Annals and Magazine of Natural History (series 4) 2:373–377.

———. 1872a. Notes on the classification of the sponges. Annals and Magazine of Natural History (series 4) 9:442–462.

———. 1872b. On a new genus of hexiradiate and other sponges discovered in the Philippine Islands by Dr. A. B. Meyer. Annals and Magazine of Natural History (series 4) 10:134–139.

———. 1872c. *Crateromorpha. In* H. J. Carter, Description of two new sponges from the Phillippine Islands. Annals and Magazine of Natural History (series 4) 10(56):110–113.

Gregorio, A. 1883. Coralli Giurresi di Sicilia. Il Naturalista Siciliano, Palermo 2(6):121–126.

———. 1908. Sul generi *Zittelspongia.* Il Naturalista Siciliano, Palermo 20:83.

———. 1930. Sul Permiano di Sicilia (Fossili del calcare con Fusulina di Palazzo Adriano). Annals of Geology and Palaeontology 52:1–70.

Gutschick, R. G., & T. G. Perry. 1959. Sappington (Kinderhookian) sponges and their environment. Journal of Paleontology 33:977–985, 3 fig., 2 pl.

Gwinner, M. P. 1958. Schwämmbänke, Riffe und submarines relief im oberen Weissen Jura der Schwäbischen Alb (Württembrug). Geologische Rundschau 47(1):408–418.

———. 1968. Palaeogeographic und Landeschaftsentwicklung im Weissen (ober) Jura der Schwäbischen Alb (Baden-Württemberg). Geologische Rundschau 58(1):32–41.

———. 1976. Origin of the Upper Jurassic limestones of the Swabian Alb (southwest Germany). Contributions to Sedimentology 5:1–75.

Haas, Otto. 1909. Bericht über neue Aufsammlungen in den Zlambach-mergeln der Fischerwiese bei Alt-Aussee. Beiträge zur Paläontologie und Geologie von Österreich-Ungarns und des Orients. Mitteilungen des geologischen und paläontologischen Institutes der Universität Wien 22:143–167, 2 pl.

von Hacht, Ulrich. 1981. *Syltrochos pyramidoidalis*—eine neu oberordovizische Spongie aus der Braderuper Serie der Kaolinsande von Sylt. Grondboor & Hamer 35(6):154–155.

———. 1990. Wenig bekannte Spongien von Sylt. *In* U. von Hacht, ed., Fossilien von Sylt III. Verlag und Verlagbuchhandlung Inge-Maria von Hacht. Hamburg. p. 43–57, 4 pl.

———. 1994. Sponzentelling op Sylt. Grondboor & Hamer 48(4/5):76–80, 8 fig.

von Hacht, Ulrich, & F. Rhebergen. 1997. Ordoviziche Geschiebespongien Europas. *In* M. Zwanzig & H. Löser, Berliner Beiträge zur Geschiebeforschung. Dresden. p. 51–63.

Haeckel, E. 1870. Prodromus eines Systems der Kalkschwämme. Jenaer Zeitschrift für Medicin und Naturwissenschaft 5:236–254.
Translated in Annals and Magazine of Natural History (series 4) 5:176–191.

———. 1872a. Prodromus eines Systems der Kalkschwämme. Zeitschrift für die gesammten Naturwissenschaften, Berlin 6:507–515.

———. 1872b. Die Kalkschwämme. Ein Monographie. Verlag von Georg Reimer. Berlin. Band I, Biologie der Kalkschwämme (Calcispongien oder Grantien), 484 p.; Band II, System der

Kalkschwämme (Calcispongien oder Grantien), 418 p.; Band III, atlas, 60 pl.

Hajdu, E., R. W. M. Van Soest, & J. N. A. Hooper. 1994. Proposal for a phylogenetic subordinal classification of poecilosclerid sponges. *In* R. W. M. van Soest, T. M. G. van Kempen, & J. C. Braekman, eds., Sponges in Time and Space. Balkema. Rotterdam. p. 123–139.

Hall, James. 1863. Observations upon the genera *Uphantaenia* and *Dictyophyton*, with notice of some species from the Chemung Group of New York and the Waverly sandstone of Ohio. New York State Cabinet 16:84–91, pl. 3–5.

———. 1884. Descriptions of the species of fossil reticulate sponges, constituting the family Dictyospongidae. New York State Museum of Natural History, 35th Annual Report. p. 465–481, pl. 17–21.

———. 1890a. On new genera and species of the family Dictyospongidae. New forms of Dictyospongidae from the rocks of the Chemung Group. Geological Society of America Bulletin 1:22–23.

———. 1890b. Fossil Dictyospongidae of the Devonian and Carboniferous formations; new forms of Dictyospongidae from rocks of the Chemung Group. Ninth Annual Report, New York State Geologist. p. 56–60.

Hall, James, & J. M. Clarke. 1899 [1898]. A memoir of the Paleozoic reticulate sponges constituting the family Dictyospongidae. New York State Museum Memoir 2:350 p., 45 fig., 70 pl.
Variously dated as 1898, 1899, and 1900.

Hancock, Albany. 1849. On the excavating powers of certain sponges belonging to the genus *Cliona;* with descriptions of several new species, and an allied generic form. Annals and Magazine of Natural History (series 2) 3:321–348, pl. 12–15.

Häntzschel, Walter. 1962. Trace fossils and Problematica. *In* W. W. Hass, W. Häntzschel, D. W. Fischer, B. F. Howell, F. H. T. Rhodes, K. J. Müller, & R. C. Moore, eds., Treatise on Invertebrate Paleontology, Part W, Miscellanea. University of Kansas & Geological Society of America. Lawrence & Boulder. p. 177–245.

———. 1972. Trace fossils and problematica. *In* Curt Teichert, ed., Treatise on Invertebrate Paleontology, Part W, Miscellanea, Supplement 1. University of Kansas & Geological Society of America. Lawrence & Boulder. 269 p.

Hara, J. 1894. On a new species of calcareous sponge, *Lelapia nipponica.* Zoological Magazine 6:369–370. In Japanese.

Hartman, W. D. 1958a. A re-examination of Bidder's classification of the Calcarea. Systematic Zoology 7:97–110.

———. 1958b. Natural history of the marine sponges of southern New England. Bulletin of the Peabody Museum of Natural History 12:1–155, 46 fig., 12 pl.

———. 1969. New genera and species of coralline sponges (Porifera) from Jamaica. Postilla, Peabody Museum of Natural History 137:1–39, fig. 1–32.

———. 1979. A new sclerosponge from the Bahamas and its relationship to Mesozoic stromatoporoids. *In* C. Lévi & N. Boury-Esnault, eds., Biologie des Spongiaires, Colloques Internationaux du Centre Nationale de la Recherche Scientifique 291:467–474.

———. 1982. Porifera. *In* S. P. Parker, ed., Synopsis and classification of living organisms, vol. 1. McGraw-Hill Book Co. New York, London, Toronto. p. 641–666.

Hartman, W. D., & T. F. Goreau. 1966. *Ceratoporella,* a living sponge with stromatoporoid affinities. American Zoologist 6:262.

———. 1970. Jamaican coralline sponges; their morphology, ecology and fossil relatives. Zoological Society of London Symposium 25:205–243.

———. 1975. A Pacific tabulate sponge, living representative of a new order of Sclerosponges. Postilla, Peabody Museum of Natural History 167:1–14, fig. 1–15.

Haswell, W. A. 1882. On Australian freshwater sponges. Proceedings of the Linnean Society of New South Wales 7:208–210.

Hay, W. W., F. Wiedenmayer, & D. S. Marszalek. 1970. Modern organism communities of Bimini Lagoon and their relation to the sediments. *In* Supko, P., D. Marszalek & W. D. Bock, eds., Sedimentary environments and carbonate rocks, Bimini, Bahamas. Miami Geological Society 4th Annual Field Trip. p. 19–30.

Hechtel, G. J. 1965. A systematic study of the Demospongiae of Port Royal, Jamaica. Peabody Museum of Natural History Bulletin 20:1–94, 15 fig., 8 pl.

Hentschel, Ernst. 1909. Tetraxonida, tl. 1. *In* W. Michaelsen & R. Hartmeyer, Die Fauna Sudwest-Australiens, bd. 2, lfg. 21. G. Fischer. Jena. p. 347–402, 2 pl.

———. 1923–1924. Erste Unterabteilung der Metazoa: Parazoa, Porifera-Schwämme. *In* W. Kükenthal & T. Krumbach, eds., Handbuch der Zoologie, Eine Naturgeschichte de Stämme des Tierreiches, vol. 1, Protozoa, Porifera, Coelenterata, Mesozoa. Walter de Gruyter und Co. Berlin & Leipzig. p. 307–408, fig. 288–377.

———. 1929. Die Kiesel- und Hornschwämmen des Nordlichen meeres. Fauna Arctica, Jena (series 4) 5(4):859–1,042, 4 pl.

Herak, M. 1943. Zur Kenntnis triadischer Kalkschwämme (Sycones). Neues Jahrbuch für Geologie und Paläontologie 88:107–135, 5 fig., 2 pl.

———. 1944. Zur kenntniss triadischer Kalkschwämme (Sycones). Neues Jahrbucher für Mineralogie, Geologie und Paläontologie 88:107–135, pl. 13–14.

Hérenger, Lucette. 1942. Contribution à l'étude des spongiaires du Jurassique et du Crétacé de Catalogne. Travaux du Laboratoire de Géologie de la Faculté des Sciences de l'Université de Grenoble 23:143–192, 13 fig., 4 pl.

———. 1944. Spongiaires siliceux du Crétacé du Sud-Est de la France. Travaux du Laboratoire de Géologie de la Faculté des Sciences de l'Université de Grenoble 24:79–110, 9 fig.

———. 1945 (1944). Nouvelle genre d'éponge siliceuse et remarque sur la classification des Hexactinellides. Comptes Rendus de la Société Géologique de France, Paris 218:688–690.

———. 1946a. Sur calces nouveaux gisements de Spongiaires du Crétacé de Provence et sur une Mégamorine bien conservée provenant de l'un d'entre eux. Travaux du Laboratoire de Géologie de la Faculté des Sciences de l'Université de Grenoble 25:3–7.

———. 1946b. Description de nouvelles formes de spongiaires du Crétacé supérieur des Petites Pyrénees. Comptes Rendus sommaire de la Société Géologique de France 3/4:46–48.

Herrmann-Degen, Wolfgang. 1980. Eine hexactinelliden-faunula aus dem "Chalk" (Maastricht?, Paläozän) Sudwest-Ägyptens. Berliner geowissenschaftliche Abhandlungen (Reihe A, Geologie und Paläontologie) 24:1–29, 3 fig., 10 pl.

Herzer, H. 1901. A new fossil sponge from the Coal Measures. Annual Report of the Ohio State Academy of Science 9:30–31.

Hicks, H. 1869. Notes on a species of *Eophyton* (?) from the Lower Arenig rocks of St. Davids. The Geological Magazine 6:534–535.

Hickson, S. J. 1911. On *Ceratopora,* the type of a new family of Alcyonaria. Proceedings of the Royal Society of London (series B) 84:195–200.

Hill, Dorothy. 1972. Part E, Archaeocyatha. *In* Curt Teichert, ed., Treatise on Invertebrate Paleontology, Part E, revised, vol. 1. Geological Society of America & The University of Kansas Press. Boulder & Lawrence. xxx + 158 p., 107 fig.

Hilmer, Gero, & B. Senowbari-Daryan. 1986. Sphinctozoa aus dem Cenoman von Mühlheim-Broich, SW-Westfalen. Mitteilungen aus dem Geologisch-Paläontologischen Institut der Universität Hamburg 61:161–187, 8 pl.

Hinde, G. J. 1875. [untitled abstract]. Proceedings of the Geological Society of London, Quarterly Journal 31:88.

———. 1882. Notes on fossil Calcispongiae, with descriptions of new species. Annals and Magazine of Natural History 10:185–205, pl. 10–12.

———. 1884a [1883]. Catalogue of the fossil sponges in the Geological Department of the British Museum (Natural History). British Museum (Natural History). London. p. viii + 248 p., 38 pl.

———. 1884b. On fossil calcisponges from the well-boring at Richmond. Quarterly Journal of the Geological Society of London 40:778–783, pl. 35.

———. 1885. On beds of sponge-remains in the Lower and Upper Greensand of the South of England. Philosophical Transactions of the Royal Society, London 1885:403–453, pl. 40–45.

———. 1887a. On the genus *Hindia* Duncan, and the name of a typical species (*fibrosa*) Roemer. Annals and Magazine of Natural History (series 5) 19:67–79.

———. 1887b. A monograph of the British fossil sponges, part 1. Palaeontographical Society Monograph. London. p. 1–92, pl. 1–8.

———. 1888. A monograph of the British fossil sponges, part 2, Sponges of the Palaeozoic Group. Palaeontographical Society Monograph. London. p. 93–188, pl. 9.

———. 1889a. On a new genus of siliceous sponge from the Trenton Formation at Ottawa. Canadian Record of Science 3:395–398.

———. 1889b. On *Archaeocyathus* Billings, and on other genera, allied to or associated with it, from the Cambrian strata of North America, Spain, Sardinia, and Scotland. Quarterly Journal of the Geological Association of London 45:125–148, pl. 5.

———. 1890. On a new genus of siliceous sponge from the lower Calcareous Grit of Yorkshire. Quarterly Journal of the Geological Society of London 46:54–61, pl. 6.

———. 1891. Notes on a new fossil sponge from the Utica Shale formation (Ordovician) at Ottawa, Canada. Geological Magazine (new series, decade III) 8:22–24, 1 fig.

———. 1893a. On *Palaeosaccus dawsoni* Hinde, a new genus and species of hexactinellid sponge from the Quebec Group (Ordovician) at Little Métis, Quebec, Canada. Geological Magazine (new series, decade III) 10(1):56–59, pl. 4.

———. 1893b. A monograph of the British fossil sponges, part 3, Sponges of Jurassic strata. Palaeontographical Society Monograph. London. p. 189–254, pl. 10–19.

———. 1900. On some remarkable calcisponges from the Eocene Strata of Victoria, Australia. Quarterly Journal of the Geological Society (London) 56:50–66, pl. 3–5.

———. 1904. On the structure and affinities of the genus *Porosphaera* Steinmann. Journal of the Royal Microscopical Society 1904:1–25, 2 pl.

———. 1912. A monograph of the British fossil sponges, Index. Palaeontographical Society Monograph. London. p. 255–265.

Hinde, G. J., & W. M. Holmes. 1892 [1891]. On the sponge-remains in the Lower Tertiary strata near Oamaru, Otago, New Zealand. The Journal of the Linnean Society, Zoology 24:177–262, pl. 7–15.

Howell, B. F. 1952. Four new Devonian sponges from Western Australia. Bulletin of the Wagner Free Institute of Science 27(1):1–8, 3 pl.

———. 1956. New Permian sponges from Western Australia. Bulletin of the Wagner Free Institute of Science 31(4):29–38, 3 pl.

———. 1957a. A new Devonian sponge, *Striataspongia cylindrica,* from Western Australia. Bulletin of the Wagner Free Institute of Science 32(1):1–3, 1 pl.

———. 1957b. Four new Cretaceous sponges from Texas. Bulletin of the Wagner Free Institute of Science 32(1):4–10, 2 pl.

———. 1957c. The Australian Devonian sponge, *Devonospongia clarkei* (de Koninck). Bulletin of the Wagner Free Institute of Science 32(2):13–16, 1 pl.

Howell, B. F., & F. B. van Houten. 1940. A new sponge from the Cambrian of Wyoming. Bulletin of the Wagner Free Institute of Science 15(1):1–8, 3 pl.

Howell, B. F., & R. W. Landes. 1936. New monactinellid sponges from the Ordovician of Wisconsin. Journal of Paleontology 10:53–59.

Hozawa, S. 1923. On a new genus of calcareous sponge. Annotationes Zoological Japonense, Tokyo (Article 18) 10:183–190, 1 pl.

Hudson, R. G. S. 1929. A Carboniferous lagoon deposit with sponges. Proceedings of the Yorkshire Geological Society 21:181–195, pl. 6.

Hughes, G. W. 1985. *Silicosphaera asteroderma* (Porifera), a new siliceous microfossil from the South China Sea. Neues Jahrbuch für Geologie und Paläontologie, Monatshefte 10:599–604, 3 fig.

Hunt, O. D. 1925. The food of the bottom fauna of the Plymouth fishing grounds. Marine Biological Association Journal, Plymouth 1925:568.

Hurcewicz, Helena. 1966. Siliceous sponges from the Upper Cretaceous of Poland. Part 1. Tetraxonia. Acta Palaeontologica Polonica 11:15–129, 38 fig., 24 pl.

———. 1975. Calcispongea from the Jurassic of Poland. Acta Palaeontologica Polonica 20:223–291, 32 fig., pl. 29–41.

———. 1983 [1982]. Permian sponges from brachiopod cherts at Hornsund, Spitzbergen. Acta Palaeontologica Polonica 27:85–114, pl. 30–40.

———. 1985. [Frasnian sponges from Wietrznia and Kowala, Góry Swietokrzyskie Mts.]. Kwartalnik Geologiczny 29:271–300, 6 pl.
English summary, p. 288.

———. 1986. *Astylospongia* Roemer (Porifera) from the Givetian at Gorno (Poland). Prezglad Geologiczny 34(2):89–91.

———. 1993. Sponge spicules from the Middle and Upper Devonian of the Holy Cross Mts., and Silesian uplands. Acta Palaeontologica Polonica 37(1,992):291–296.

Hurcewicz, H., & S. Czarniecki. 1986. Lyssakidae sponges from the Carboniferous limestone and the Culm of southern Poland and their environmental differentiation. Annales Societatis Geologorum Poloniae 55:333–354, fig. 1–18, pl. I–IX.

Hyman, L. H. 1925. Respiratory differences along the axis of the sponge *Grantia.* Biological Bulletin 48:379–389.

———. 1940. The Invertebrates: Protozoa through Ctenophora. McGraw-Hill. New York. 726 p.

ICZN (International Commission on Zoological Nomenclature). 1999. International Code of Zoological Nomenclature, 4th ed. The International Trust for Zoological Nomenclature. London. xxix + 306 p.

Ijima, Isao. 1903. Studies on the Hexactinellida, III (*Placosoma,* a new euplectellid; *Leucopsacidae* and *Caulophacidae*). Imperial University of Tokyo, Journal of the Science College Tokyo 18(1):124 p., 8 pl.

———. 1904. Studies on the Hexactinellida, IV, Rossellidae. Imperial University of Tokyo, Journal of Science College 18(7):307 p., 23 pl.

———. 1927 [1926]. The Hexactinellida of the Siboga Expedition. Siboga-Expeditie 6:1–383, 36 fig., 26 pl.

Inai, Yutaka. 1936. *Discosiphonella,* a new ally of *Amblysiphonella.* Proceedings of the Imperial Academy of Japan 12(6):169–171, 4 fig.

Jablonsky, E. 1975. *Colospongia andrusovi* n. sp., eine neue Art von segmentierten Kalkschwämmen (Sphinctozoa). Geolicky Zbornik-Geologica Carpathica Slovenska Akademie Ved 26(2):269–273, pl. 1–3.

Jackson, J. B. C., T. F. Goreau, & W. D. Hartman. 1971. Recent brachiopod-coralline sponge communities and their paleoecological significance. Science 173:623–625, 2 fig.

Jakowska, S., & R. F. Nigrelli. 1960. Antimicrobial substances from sponges. Annals of the New York Academy of Sciences 90:913–916.

Jansa, L. F., G. Termier, & H. Termier. 1983. Les biohermes à algues, spongiaires et coraux des séries carbonatées de la flexure bordière du "paleoshelf" au large du Canada oriental. Revue de Micropaleontologie 25:181–219, 13 pl.

Jepps, M. W. 1947. Contribution to the study of the sponges. Proceedings of the Royal Society of London 134B(896):408–417, 5 fig.

Jeuniaux, C. 1963. Distribution in chitin in animals (Protozoa, Porifera, Cnidaria, Rhynchocoela, Acanthocephala, Aschelminthes, Sipunculoidea, Mollusca, Annelida, Crustacea, Insecta, Echinodermata). Proceedings of the 16th International Congress of Zoology16(2):78.

Johns, R. A. 1994. Ordovician lithistid sponges of the Great Basin. Nevada Bureau of Mines and Geology, NBMG Open-file Report 94-1:vi + 140 p., 16 fig., 16 pl.

Johnson, J. E. 1899. Notes on some sponges belonging to the Clionidae obtained at Madeira. Journal of the Royal Microscopical Society 1899:461–463.

Johnston, G. 1842. History of British sponges and lithophytes. Edinburgh. vii + 1–264 p., 25 pl.

Jones, W. C. 1957. The contractility and healing behaviour of pieces of *Leucosolenia complicata.* Quarterly Journal of Microscopical Science 98:302–217.

———. 1962. Is there a nervous system in sponges? Biological Reviews 37:1–50, 4 fig.

———. 1964. Photographic records of living oscular tubes of *Leuconosolenia variabilis.* I. The choanoderm boundary, the choanocytes and the pore arrangement. Journal of the Marine Biological Association, United Kingdom 44:67–85, 3 pl., 2 fig.

Jørgensen, C. B. 1955. Quantitative aspects of filter feeding in invertebrates. Biological Revue 30:391–454.

———. 1960. Efficiency of particle retention and rate of water transport in undisturbed lamellibranchs. Journal du Conseil International pour l'Exploration de la Mer 26:94–116, 12 fig.

———. 1966. Biology of suspension feeding. Pergamon Press. Oxford. i–xv + 1–845, 660 fig.

Jux, Ulrich. 1992. Schwämme aus dem obersten Mitteldevon der Bergisch Gladbach-Paffrather Mulde (Bergisches Land). Decheniana 145:302–311, 3 pl.

———. 1994. *Schizorhabdus libycus* Zittel, 1877—a lithistid sponge from the late Maastrictian of Egypt. Courier Forschungs-institut Senckenberg 172:299–306, 3 fig.

Kapp, U. S. 1975. Paleoecology of Middle Ordovician stromatoporoid mounds in Vermont. Lethaia 8:195–207, 3 fig., 6 pl.

Kaye, H. R. 1990. Reproduction in West Indian commercial sponges: oogenesis, larval development, and behavior. *In* Klaus Rützler, ed., New Perspectives in Sponge Biology. Smithsonian Institution Press. Washington, D.C. p. 161–169.

Kayser, E. 1885. *Lodanella mire,* ein unterdevonische Spongie. Zeitschrift der Deutschen Geologischen Gesellschaft 37:207–213.

Kazmierczak, J. 1984. Favositid tabulates: evidences for poriferan affinity. Science 225:835–837.

———. 1991. Further evidence for poriferan affinities of favositids. *In* J. Reitner & H. Keupp, eds., Fossil and Recent Sponges. Springer-Verlag. Berlin. p. 212–223.

van Kempen, T. M. G. 1977. *Roepella solanensis,* new sponge from the mid-Tertiary of S.E. Spain; First fossil of a gymeniacidonid, with remarks on other fossilized sponges with monaxonid megascleres. Proceedings Koninklijke Nederlandse Akademie van Wetenschappen (series B, Palaeontology, Geology, Physics, and Chemistry) 80(2):106–131.

———. 1978. Anthaspidellid sponges from the Early Paleozoic of Europe and Australia. Neues Jahrbuch für Geologie und Paläontologie, Abhandlungen 156(3):305–337, 7 fig.

———. 1989. On a new anthaspidellid sponge from the Baltic Early Paleozoic. Mitteilungen aus dem Geologisch-Paläontologischen Institut der Universität Hamburg 68:131–157, fig. 1–2, pl. 1–5.

———. 1990. Two Baltic Ordovician chiastoclonellids (Porifera) from the island of Sylt (NW Germany). *In* Ulrich von Hacht, ed., Fossilien von Sylt III. Inge-Maria von Hacht, Verlag und Verlagsbuchhandlung. Hamburg. p. 151–178, 9 fig.

Kent, W. S. 1870. On the "Hexactinellidae" or hexaradiate spiculed siliceous sponges taken in the "Norna" expedition off the coast of Spain and Portugal. Monthly Microscopical Journal 4:241–252.

Keupp, Helmut, & Dorte Mehl. 1994. *Ammonella quadrata* Walther 1904 (Porifera, Hexactinellida) aus dem solnhofener Plattenkalk von Pfalzpaint: Relikt aus dem Altpaläozoikum? Archaeopteryx 12:45–54, 3 fig.

———. 1995. *Ammonella quadrata* Walther, 1904 (Porifera, Hexactinellida) of the Solnhofen Plattenkalks: Systematics and environmental deductions. Extended Abstracts, II International Symposium on Lithographic Limestones. Ediciones de la Universidad Autonoma de Madrid. Madrid. p. 93–94, 1 fig.

Khorshunov [also Korshunov], V. I. 1968. *Gonamispongia,* novily rod gubok seneystva Chancelloriidae [*Gonamispongia,* a new genus of the family Chancelloriidae]. Paleontologicheskii Zhurnal 3:127–129, 1 fig.
English translation in Palaeontological Journal 1968(3):398–400, 1 fig.

Kilian, E. F. 1952. Wasserströmung und Nahrungsanfrahme beim Süsswasserschwamm *Ephydatia fluviatilis.* Zeitschrift für Vergleichende Physiologie 34(5):407–447, 24 fig.

———. 1964. Zur Biologie der einheimischen Spongilliden Engebniss und Probleme. Unter besonderer Berücksichtigung eigener Untersuchungen. Zoologische Beiträg 10:85–159, 16 fig.

King, R. H. 1933. A Pennsylvanian sponge fauna from Wise County, Texas. The University of Texas Bulletin 3,201:75–85, pl. 7–8.

———. 1938. Pennsylvanian sponges of north-central Texas. Journal of Paleontology 12:498–504, 14 fig.

———. 1943. New Carboniferous and Permian sponges. State Geological Survey of Kansas Bulletin 47:1–36, 2 fig., pl. 1–3.

King, W. 1850. A monograph of the Permian fossils of England. Palaeontograpical Society. London. p. 11–14, fig.1–7, pl. 2.

Kirkpatrick, R. 1900. Description of sponges from Funafuti. Annals of Natural History 7:345–362, pl. 13–15.

———. 1908. On two new genera of recent pharetronid sponges. Annals and Magazine of Natural History (series 8) 12:503–514, pl. 13–15.

———. 1910a. On a remarkable pharetronid sponge from Christmas Island. Proceedings of the Royal Society of London (series B) 83:124–133, pl. 10–11.

———. 1910b. On the affinities of *Astrosclera willeyana* Lister. Annals and Magazine of Natural History (series 8) 5:380–383.

———. 1912. *Merlia normani* and its relation to certain Palaeozoic fossils. Nature 89:502–503.

Kling, S. A., & W.-E. Reif. 1969. The Paleozoic history of amphidisc and hemidisc sponges: new evidence from the Carboniferous of Uruguay. Journal of Paleontology 43:1,429–1,434, 1 fig., pl. 176.

Klipstein, A. von. 1843–1845. Beitrage zur geologischen Kenntnis der östlischen Alpen. Georg Friedrich Heyer's Verlag. Giessen. 311 p., 20 pl.

Kobluk, D. R. 1981a. Lower Cambrian cavity-dwelling endolithic (boring) sponges. Canadian Journal of Earth Sciences 18:972–980.

———. 1981b. Middle Ordovician (Chazy Group) cavity-dwelling boring sponges. Canadian Journal of Earth Sciences 18:1,101–1,108.

Koenig, C. D. E. 1820. Icones fossilium sectilis, centuria prima. C. Koenig. London. 4 p., 19 pl.

Kolb, Rudolf. 1910–1911. Die Kieselspongien des schwäbsichen Weissen Jura. Palaeontographica 57:141–256, fig. 1–27, pl. 11–21.

Koltun, V. M. 1964. Sponges of the Antarctic. 1. Tetraxonida and Cornucospongida. *In* Biological results of the Soviet Antarctic Expedition (1955–1958). Issledovaniya Fauny Morei 2:6–131, 25 fig., 15 pl.

de Koninck, L. G. 1842–1844. Description des animaux fossiles, qui se trouvent dans le terrain carbonifère de Belgique, vol. I–II. H. Dessain. Liège. iv + 650 p., pl. A–H + 1–53.
Volume I contains plates, volume II contains text.

———. 1877. Recherches sur les fossiles paléozoïques de la Nouvelle-Galles du Sud (Australie). Mémoires de la Société Royale des Sciences de Liège (series 2) 6:1–135.

Korshunov, V. I. 1968. *Gonamispongia,* noviy rod gubok seneystva Chancelloriidae [*Gonamispongia,* a new genus of the family Chancelloriidae]. Paleontologicheskii Zhurnal 3:127–129, 1 fig.
English translation in Palaeontological Journal 1968(3):398–400, 1 fig.

Kovács, Sándor. 1978. New sphinctozoan sponges from the North Hungarian Triassic. Neues Jahrbuch für Geologie und Paläontologie, Monatshefte 1978:685–697, 6 fig.

Kozur, H. W. 1991. Erster Nachweis von Polyactinellidae Mostler (Calcispongea, Porifera) im Perm und Revision der Gattung *Phobetractinia* Reif. Zeitschrift für Geologische Wissenschaften 19(5):585–591, 2 fig.

Kozur, H. W., H. Mostler, & J. E. Repetski. 1996. 'Modern' siliceous sponges from the lowermost Ordovician (early Ibexian–early Tremadocian) Windfall Formation of the Antelope Range, Eureka County, Nevada, USA. Geologische-Paläontologische Mitteilungen, Innsbruck 21:201–221, 1 fig., 5 pl.

Krainer, Karl, & Helfried Mostler. 1992. Neue Hexactinellide Poriferen aus der Südalpinen Mitteltrias der Karawanken (Kärnten, Österreich). Geologische-Paläontologische Mitteilungen, Innsbruck 18:131–150, 7 pl.

Krasnopeeva, P. S. 1937. Vodorosli i arkheotsiaty drevneishikh tolshch potekhinskogo plansheta Khakassi [Algae and Archaeocyatha of the most ancient strata on Potekhin's topographic map for the Khakass Autonomous Region]. Materialy po Geologii Krasnoyarskogo Kraya 3:1–51.

———. 1940 [1939]. Al'gonskaya flora i fauna Saralinskogo raiona Kuznetskogo Alatau [Algonkian (pre-Cambrian) flora and fauna of the Saralynsky region of the Kuznetsky Ala Tau]. Material po Geologii Krasnoyarskogo Karya 8:1–32.

———. 1962. *Occultus* (new name for *Archaeospongia* Krasnopeeva, 1937). *In* P. D. Rezvoi, I. T. Zhuravleva, & V. M. Koltun, Phylum Porifera, p. 17–74, fig. 1–107. *In* B. S. Sokolov, ed., Osnovy Paleontologii [Fundamentals of Paleontology], vol. 1, number 2, Porifera, Archaeocyatha, Coelenterata, Vermes. Izdatel'stvo Akademii Nauk SSSR. Moscow. p. 58.

Krautter, Manfred. 1994. Observations on *Eudea clavata* Lamouroux (Calcarea) from the Upper Jurassic of Portugal. *In* R. W. M. van Soest, T. M. G. van Kempen, & J.-C. Braekman, eds., Sponges in Time and Space: Biology, Chemistry, Paleontology. A. A. Balkema. Rotterdam. p. 29–34, 12 fig.

———. 1996. Kieselschwämme aus dem unterjurassischen Misonekalk der Trento-Plattform (Südalpen): Taxonomie und phylogenetische Relevanz. Paläontologische Zeitschrift 70(3–4):301–313, 6 fig.

Krautter, M., K. W. Conway, J. V. Barrie, and M. Neuweller. 2001. Discovery of a "living dinosaur:" globally unique modern hexactinellid sponge reefs off British Columbia, Canada. FACIES 44:265–282, 10 fig., pl. 42–45.

Kravtsov, A. G. 1968. Maastrichtian spiral sponges of the Crimean Highlands. Paleontologischeskii Zhurnal 1968(3):124–127, 2 fig.
Translated in Paleontological Journal 3:401–404.

Kruse, P. D. 1983. Middle Cambrian '*Archaeocyathus*' from the Georgina Basin is an anthaspidellid sponge. Alcheringa 7:49–58, fig. 1–6.

———. 1987. Further Australian Cambrian sphinctozoans. Geological Magazine 124:543–553, 3 fig.

———. 1990. Cambrian palaeontology of the Daly Basin. Northern Territory Geological Survey Report 7:1–58, 23 fig., 27 pl.

Kruse, P. D., & F. Debrenne. 1989. Review of archaeocyath microstructure. Memoirs of the Association of Australasian Palaeontologists 8:133–141.

Kucera, M. 1993. Sponge spicules from the Lower Devonian (Pragian) of the Barrandian. Journal of the Czech Geological Society 38(3–4):193–200.

Kügel, H-W. 1987. Sphinctozoen aus dem Auernigschichten des Nassfeldes (Oberkarbon, Karnische Alpen, Österreich. Facies 16:143–156, pl. 33–35.

Lachasse, Jacques. 1943. Contribution à l'étude des spongiaires fossiles du Campanien des Charentes. Bulletin de la Société géologique de France 1943:43–66, 2 pl.

Lagneau-Hérenger, Lucette. 1955. Remarques sur la classification des spongiaires fossiles et essai do nouvelle classification. Comptes Rendus de l'Académie des Sciences, Paris 240:1,563–1,564.

———. 1961. Contribution à l'étude des Spongiaires siliceux du Crétacé inférieur. Comptes Rendus de la Société Géologique de France 6:168–169.

———. 1962. Contribution à l'étude des Spongiares siliceux du Crétacé inférieur. Mémoires de la Société Géologique de France (new series) 95:252 p., 30 fig., 16 pl., 7 tables.

———. 1967. Les Spongiaires Turoniens du Synclinal d'Eygaliers (Baronnies, Drôme). Travaux du Laboratoire de Géologie de la Faculté des Sciences de l'Université de Grenoble 43:79–93, 3 pl.

de Lamarck, J. P. B. A. 1815 [1814]. Suite des Polypiers empâtés: Téthies, Alcyon, Géodie, Botrylle, et Polycycle (dont l'exposition commence au 20ième vol. des Annales, p. 294). Mémoires du Muséum d'Histoire naturelle, Paris 1:69–80, 162–168, 331–340.

———. 1816a. Histoire des polypiers coralligènes flexibiles, vulgairement nommés Zoophytes. F. Poisson. Caen. 560 p., 19 pl.

———. 1816b. Histoire naturelles des animaux sans vertèbres, présentant les caractères genéraux et particuliers de ces animaux, vol. 2. Verdière. Paris. p. 1–586.

Lamont, Archie. 1935. The Drummuck Group, Girvan; a stratigraphical revision, with descriptions of new fossils from the lower part of the group. Transactions of the Geological Society of Glasgow 19(2):288–332, 4 fig., pl. 7–8.

Lamouroux, J. V. F. 1813. Dictionnaire Classique. Annales du Muséum National d'Histoire Naturelle, Paris 20(1):285.

———. 1816. Histoire des polypiers coralligènes flexibiles, vulgairement nommés Zoophytes. F. Poisson. Caen. 560 p., 19 pl.

———. 1821. Exposition méthodique des genres de l'ordre des Polypiers, des Zoophytes d'ellis et Solander. Chez Mme. Veuve Agasse. Paris. 115 p., 84 pl.

———. 1822. Dictionnaire classique d'histoire naturelle, par Messieures J. V. Audouin, Barron et Bory de Saint-Vincent, vol. I–IV. Lamouroux et Cie. Bourda. 450 p.

Lang, B. 1985. Die ersten Sphinctozoen (segmentierte Kalkschwamme) aus dem Ober-Jura der Frankenalb (Suddeutschland). Mitteilungen Bayerische Staatssammlung für Paläontologie und Historische Geologie 25:3–15.

Laube, G. C. 1864. Bemerkungen über die Münsterschen Arten von St. Casian. Jahrbuch für Koenigliche Geologischen Reichs-Anstalt, vol. 14. Wien. p. 402–412.

———. 1865. Die Fauna der Schichten von St. Cassian. Ein beitrag zur Paläontologie der alpinen Trias, I Abtheilung. Spongitarien, Corallen, Echiniden und Crinoiden. Denkschriften der Kaiserlichen Akademie der Wissenschaften, Mathematisch-naturwissenschaftliche Klasse 24:223–296, 10 pl.

de Laubenfels, M. W. 1932a. The marine and freshwater sponges of California. Proceedings of the U.S. National Museum, Washington, D.C. 81(4):1–140, 79 fig.

———. 1932b. Physiology and morphology of Porifera exemplified by *Iotrochia birotulata* Higgin. Carnegie Institution of Washington Publication 435:37–66, 6 fig., 2 pl.

———. 1934. New sponges from the Puerto Rican deep. Smithsonian Museum Miscellaneous Collections 91(17):1–28.

———. 1936. A discussion of the sponge fauna of the Dry Tortugas in particular, and the West Indies in general, with material for revision of the families and orders of the Porifera. Carnegie Institution of Washington Publication 467:iii + 1–225, 22 pl., 1 map.
Also published as Papers from Tortugas Laboratory, vol. 30, 225 p., 22 pl.

———. 1947. *Monarchopemmatites,* a new genus name for a fossil sponge. Journal of Paleontology 21:187.

———. 1950. An ecological discussion of the sponges of Bermuda. Transactions of the Zoological Society of London 27:155–201, 4 fig.

———. 1953a. Sponges of the Gulf of Mexico. Bulletin of Marine Science of the Gulf and Caribbean 2(3):511–557, 17 fig.

———. 1953b. Fossil sponges of West Australia. Journal of the Royal Society of Western Australia 37:105–117, 8 fig.

———. 1954. The sponges of the west-central Pacific. Oregon State Monographs, Studies in Zoology 7:306 p., 12 pl.

———. 1955. Porifera. *In* R. C. Moore, ed., Treatise on Invertebrate Paleontology, Part E, Archaeocyatha and Porifera. Geological Society of America & The University of Kansas Press. New York & Lawrence. p. 21–112, fig. 14–89.

———. 1957. New species and records of Hawaiian sponges. Pacific Science 11:236–251, 15 fig.

———. 1958. Nomenclature of Porifera, especially concerning the so-called "Glass-sponges" or Hyalosponges. Journal of Paleontology 32:611–616.

Lecompte, Marius. 1936. Contribution à la connaissance des "Recifs" du Frasnian de l'Ardenne. Mémoires de l'Institut Geologique, Université de Louvain 10:93–97, sponges.

———. 1956. Stromatoporoidea. *In* R. C. Moore, ed., Treatise on Invertebrate Paleontology, Part F, Coelenterata. Geological Society of America & The University of Kansas Press. New York & Lawrence. p. 107–144, fig. 87–114.

Lees, J. H., & A. O. Thomas. 1919. The Ste. Genevieve marls near Fort Dodge and their fauna. Des Moines Proceedings of the Iowa Academy of Science 25:599–616, pl. 12.

Leidy, J. 1851. *Spongilla fragilis.* Proceedings of the Academy of Natural Sciences of Philadelphia 5:278.

———. 1868. Description of a new sponge: *Phaeronema Annae.* Proceedings of the Academy of Natural Sciences of Philadelphia 1868:9–11.

Lejal-Nicol, A. 1976. *Platyphyllum* sp. *In* C. Babin, D. Goujet, H. Lardeux, A. Lejal-Nicol, F. Lethiers, P. Morzadec, Y. Plusquellec, & M. Weyant, eds., La Formation des Schistes de Porsguen (Dévonien supérieur de la rade de Brest, Massif Armoricain), Lithologie, flore, faune. Annales de la Société géologique du Nord 96(4):341.

von Lendenfeld, Robert. 1887. On the systematic position and classification of sponges. Zoological Society of London Proceedings 1886:558–662.

———. 1888. Descriptive catalogue of the sponges in the Australian Museum, Sydney. The Australian Museum. Taylor and Francis. London. 260 p., 12 pl.

———. 1889a. A monograph of the horny sponges. Royal Society of London. Trübner and Co. London. 936 p., 50 pl.

———. 1889b. Die Gattung Stelletta. Abhandlungen der Königlich Akademie der Wissenschaften zu Berlin (Anh.) 2:75 p.

———. 1889c. Das System der Spongien. Biologisches Centralblatt 9(4):113–127.
———. 1903. Tetraxonia: Das Tierreich. Konigliche Preussische Akademie der Wissenschaften, Berlin 19:xv + 168 p.
———. 1904a. Über die Herstellung von Nadelpraparaten von Kieselschwämmen. Zeitschrift für Wissenschaftliche Mikroskopie und für Mikroskopische Technik 21:23–24.
———. 1904b. Über die deszendenztheoretische Bedeutung der Spongiosa. Biologisches Zentralblatt 24:635–636.
———. 1904c. Porifera. Tetraxonia. *In* F. E. Schulze, ed., Das Tierreich, vol. 19. Friеländer. Berlin. p. vi–xv, 1–168, 44 fig.
———. 1907. Die Tetraxonia. Wissenschaftliche Ergebnisse der Deutschen Tiefsee-Expedition auf dem Dampfer "Valdivia" 1898–1899, vol. 11, lf. 2. Gustav Fischer. Jena. 373 p., 38 pl.
———. 1910. The sponges. 1. The Geodidae. Reports on the Scientific results the expedition to the Eastern Tropical Pacific, in charge of Alexander Agassiz, by the U.S. Fish Commission Steamer "Albatross," from October 1904 to March 1905, Lieut.-Commander L. M. Garrett, U.S.N., Commanding, and of other expeditions of the "Albatross," 1888–1904. No. 21, Museum of Comparative Zoology, Harvard College, Memoir 41(1):1–259, pl. 1–48.
———. 1915. Reports on the scientific results of the expedition to the Eastern Tropical Pacific in charge of Alexander Agassiz, by the U.S. Fish Commission Steamer "Albatross," from October 1904 to March 1905, Lieut.-Commander L. M. Garrett U. S. N., Commanding, and of other expeditions of the "Albatross" 1891–1899, no. 29, The Sponges, 3: Hexactinellida. Museum of Comparative Zoology, Harvard College Memoir 42:1–396 in 2 vol., 109 pl.
Lentz, T. L. 1966. Histochemical localization of neurohumors in a sponge. Journal of Experimental Zoology 162:171–180.
Leonhard, Richard. 1897. Die Fauna der Kreideformation in Oberschlesien. Palaeontographica, Beitraege zur Naturgeschichte der Vorzeit, Stuttgart 44:11–70, pl. 3–6.
Lesquereux, Leo. 1884. Descriptions of the coal flora of the Carboniferous formations in Pennsylvania and throughout the United States. Geological Survey of Pennsylvania, Second Report of Progress. 977 p.
Lévi, Claude. 1953. Sur une nouvelle classification des Démosponges. Académie des Sciences (Paris), Comptes Rendus des séances 236:853–855.
———. 1955. Les Clavaxinellides: demosponges tetractinomorphes. Archives de Zoologie Expérimentale et Générale 92:78–87.
———. 1956. Étude de *Halisarca* de Roscoff. Embryologie et systématique des Démosponges. Archives de Zoologie Expérimentale et Générale 93:1–181, 62 fig.
———. 1957a. Spongiaires des côtes d'Israel. Research Council of Israel Bulletin 6B(3):201–212, 13 fig.
———. 1957b. Ontogeny and systematics in sponges. Systematic Zoology 6:174–183.
———. 1960. Les Spongiaires à desmas astéroides. Bulletin de l'Institut Océanographique, Monaco 1179:1–8.
———. 1964. Spongiaires des zones bathyale, abyssale et hadale. Galathea Report 7:63–112, 62 fig., 10 pl.
———. 1966. Le glycogène chez les Spongiaires. Comptes Rendus des Séances, Société de Biologie 160:651–652, 2 pl.
———. 1968. Spongiaires. Inventaire de la Faune marine de Roschoff. Travaux de la Station biologique de Roschoff. Robin & Mareuge. Paris. p. 1–28.
———. 1973. Systématique de la classe des Demospongiaria (Démosponges). *In* P.-P. Grassé, ed., Traité de Zoologie. II. Spongiaires. Masson et Cie. Paris. p. 37–631.
Lewis, J. R. 1960. The fauna of the rocky shore of Barbados, West Indies. Canadian Journal of Zoology 38:391–435, 20 fig.
———. 1965a. A preliminary description of some marine benthic communities from Barbados, West Indies. Canadian Journal of Zoology 43(6):1049–1060.
———. 1965b. The ecology of rocky shores. English University Press, Ltd. London. p. i–xii, 1–323.
———. 1968. Water movements and their role in rocky shore ecology. Sarsia 34:13–36, 3 fig., 8 pl., 3 maps.
Li Chia-Wei, Chen Jun-Yuan, & Hua Tzu-En. 1998. Precambrian sponges with cellular structures. Science Magazine 279:879–882.
Liaci, L. 1963. Ricerche morfologiche e citochimiche sui tesociti de *Apatos apatos* O.S. (Demospongiae). Atti, Societa Peloritana di Scienze Fisiche, Matematiche, e Naturali 9:189–197, 3 fig., 1 pl.
Librovich, L. S. 1929. *Uralonema karpinski* nov. gen., nov. sp. i drugie kremnevye gubki iz kamennougol'nykh otlozhenii vostochnogo sklona Urala [*Uralonema karpinskii* nov. gen. nov. sp. and other Silicispongia from the Carboniferous of the eastern slope of the Urals]. Trudy Geologischeskogo Komiteta [Mémoire Committee Geology, new series], Leningrad 179:1–57, 1 fig., 3 pl.
In Russian with English translation.
Lieberkühn, N. 1856. Zusätze zur Entwicklungsgeschichte der Spongillen. Müller Archiv 1856:496–514.
———. 1859. Neue Beiträge zur Anatomie der Spongien. Archiv für Anatomie, Physiologie und Wissenschaftliche Medicin 1859:353–382, 515–529, pl. 9–11.
Linck, G. 1883. Zwei neue Spongiengattungen. Neues Jahrbuch für Mineralogie, Geologie und Palaeontologie 2:59–62, pl. 2–3.
von Linné, C. 1759. Systema Naturae, Vegetabilia, tome 2, 10th ed. Holmiae. Salvii. 1,384 p.
———. 1767. Systema Naturae, 12th ed., vol. 1–2. Holmiae. Salvii. 1,327 p.
Lister, J. J. 1900. *Astrosclera willeyana.* The type of a new family of sponges. *In* A. Willey's Zoological Results, part 4. Cambridge University Press. Cambridge. p. 459–482, fig. 1–3, pl. 45–48.

Little, F. J. 1963. The sponge fauna of the St. Georges sound, Apalachee Bay, and Panama City regions of the Florida gulf coast. Tulane Studies in Zoology 11:31–71.

Liu, Bingli, J. K. Rigby, Jiang Yanwen, & Zhu Zhongde. 1997. Lower Ordovician lithistid sponges from the eastern Yangtze Gorge area, Hubei, China. Journal of Paleontology 71:194–207, 7 fig.

Liu Bingli, J. K. Rigby, & Zhu Zhongde. 2003. Middle Ordovician lithistid sponges from the Bachu-Kalpin area, Xinjiang, northwestern China. Journal of Paleontology 77:430–441, 6 fig.

Lonsdale, W. 1849. Notes on fossil zoophytes found in the deposits described by Dr. Fitton in his memoir entitled "A Stratigraphical Account of the Section from Atherfield to Rocken End." The Quarterly Journal of the Geological Society of London 5:55–103, pl. 4–5.

Low, E. M. 1951. Halogenated amino acids of the bath sponge. Journal of Marine Research 10:239–245, 1 fig.

Lowenstam, H. A. 1948. Biostratigraphic studies of the Niagaran inter-reef formation in northeastern Illinois. Illinois State Museum Scientific Papers 4:1–146, pl. 1–7.

———. 1957. Chapter 10. Niagaran reefs in the Great Lakes area. *In* H. S. Ladd, ed., Treatise on Marine Ecology and Paleoecology, volume 2, Paleoecology. Geological Society of America Memoir 67:215–248, 4 fig.

Lutfy, R. G. 1960. Histochemical studies on glycogen in the cells of the freshwater sponge *Ephydatia fluviatilis.* Cellule 61:145–149, 1 pl.

MacGinitie, G. E., & N. MacGinitie. 1968. Natural History of Marine Animals. McGraw-Hill Book Company. New York, Toronto, London. 523 p., 286 fig.

Madri, P. P., G. Claus, S. M. Kunen, & E. E. Moss. 1967. Preliminary studies on the *Escherichia coli* uptake of the redbeard sponge (*Microcionia prolifera* Verrill). Life Sciences 6:889–894, 1 fig.

Mägdefrau, Karl. 1932. Über einig Bohrgänge aus dem unteren Muschelkalk von Jena. Paläontologisches Zeitschrift 14:513–523.

———. 1933. Zur Entstehung der mitteldeutscher Zechstein-Riffe. Centralblat für Mineralogie, Geologie, und Paläontologie (Abt. B) 11:621–624.

Maithy, P. K., & R. Babu. 1987 (1986). *Misraea,* a new body fossil from the Lower Vindhyan supergroup (Late Precambrian) around Chopan, Mirzapur, U.P. (India). Geophytology 16(2):223–226.

Makiyama, Jirô. 1931. Stratigraphy of the Kakegawa Pliocene in Tôtômi. Memoirs of the College of Science, Kyoto Imperial University (series B) 7(1):1–52, 4 fig., 3 pl.

Maldonado, M. 2002. Family Pachastrellidae Carter, 1875. *In* J. N. A. Hooper & R. W. M. van Soest, eds., Systema Porifera: A Guide to the Classification of the Sponges. Kluwer Academic/Plenum Publishers. New York. p. 141–162, 16 fig.

Malecki, Jerzy. 1996. Die Gattungen *Hyalotragos, Aretotragos, Pyrgochonia* und *Leiocarenus* (Demospongia, Rhizomorina) nach dem Material aus dem Weissen Jura von Zalas. Bulletin of the Polish Academy of Sciences, Earth Sciences 44(1):1–16, 9 pl.

Malfatti, Paolo. 1901 [1900]. Contributo alla spongiofauna del cenozoico italiano. Palaeontographia Italica, Memorie di Paleontologia 6:267–302, pl. 20–25.

Mantell, G. A. 1815. Description of a fossil *Alcyonium* from the Chalk strata of Lewes. Transactions of the Linnéan Society 11:401–402.

———. 1822. The fossils of the South Downs or illustrations of the Geology of Sussex. Lupton Relfe. London. 327 p., 42 pl.

———. 1838. The Wonders of Geology, or, a familiar exposition of geological phenomena, 2 vol. Relfe and Fletcher. London. 689 p., 6 leaves of plates.

Manteufel, B. P. 1938. Kratkaya Khasakteristika osnovnkh Zakonomesnostej Izmeneniyakh plantona Barentseva Morya [A brief characterization of the principal irregularities in the changes of plankton in the Barents Sea]. Trudy Polyarnj Nauchno-Issledovatel'skii Proektnyi Institut Morskogo Rybnogo Khozyajstave i Okeanografii imeni, N. M. Knipovicha (PINRO) 1:134–148, fig. 1–8.

Marek, Jaroslav. 1984 [1982]. *Runia runica* ichnogen. et ichnosp. nov., a new sponge boring from the Silurian of Bohemia. Acta Universitatis Carolinae (Geologica) 1982(4):401–408, 2 pl.

Marsh, O. C. 1867. Notice of a new genus of fossil sponge from the lower Silurian. American Journal of Science and Arts (series 2) 44(no. 30, article 11):88.

Marshall, W. 1875. Untersuchungen über Hexactinelliden. Zeitschrift für Wissenschaftliche Zoologie, Leipzig 27:142–243.

———. 1876. Ideen über die Verwandtschaftverhältnisse der Hexactinelliden. Zeitschrift für Wissenschaftliche Zoologie, Leipzig 27:113–136.

Marshall, W., & A. B. Meyer. 1879. Über einige Neues und wenig bekannte Philippinische Hexactinelliden. Mitteilungen aus dem kaiserliche zoologischen Museum zu Dresden 2:261.

Martin-Wismar, K. 1878a. Untersuchungen über die Organization von *Astylospongia* Ferd. Roemer, und Bemerkungen über die Natur der Wallsteine, Meyn. Archiv des Vereins der Freunde der Naturgeschichte in Mecklenburg 31:1–32.

———. 1878b. Niederländische und nordwestdeutsche Sedimentärgeshiebe, ihre Uebereinstimmung, gemeinschaftliche Herkunft und Petrefacten. (*Silurispongia* n. g.). Leiden.

Masse, D., & D. Vachard. 1979. Le Carbonifère de Libye occidentale: biostratigraphie et micropaléontologie. Position dans le domaine téthysien d'Afrique du Nord. Revue de l'Institut Français du Pétrole 34(1):3–65, fig. 1–19, pl. 1–9.

Masse, P. J. L., G. Termier, & H. Termier. 1989. Nouvelles formes de Spongiaires dans l'Albien de la Sainte-Baume (Provence, France). Geobios 22(6):825–839, 6 fig., 4 pl.

Mastandrea, A., & F. Russo. 1995. Microstructure and diagenesis of calcified demosponges from the Upper Triassic of Northeastern Dolomites (Italy). Journal of Paleontology 69:416–431.

Matsuoka, Keiji. 1983. Pleistocene freshwater sponges (Porifera: Spongillidae) from the Katata Formation of the Kobiwako Group, Shiga Prefecture, Central Japan. The Journal of Earth Sciences, Nagoya University 31:1–16, 6 fig., 3 pl.

———. 1987. Malacofaunal succession in Pliocene to Pleistocene non-marine sediments in the Omi and Ueno Basins, Central Japan. The Journal of Earth Sciences (Nagoya University) 35(1):23–115, 27 fig.

Matsuoka, Keiji, & Yoshiki Masuda. 2000. A new potamolepid freshwater sponge (Demospongiae) from the Miocene Nakamura Formation, central Japan. Paleontological Research 4(2):131–137, 6 fig.

Matthew, G. F. 1886. Illustrations of the fauna of the St. John Group continued. No. III.–Descriptions of new genera and species, (including a description of a new species of *Solenopleura* by J. F. Whiteaves). Transactions of the Royal Society of Canada 3(section IV):29–84, pl. 5–7.

———. 1891. On Cambrian organisms in Acadia. Transactions of the Royal Society of Canada 7(section IV):135–160.

M'Coy, F. 1846. A synopsis of the Silurian fossils of Ireland, collected from the several districts by Richard Griffith, F.G.S., the whole being named, and the new species drawn and described by Frederick M'Coy, F.G.S.D. M. H. Gill at University Press. Dublin. 72 p., 5 pl.
Authorship variously listed as M'Coy, as M'Coy in Griffith, as Griffith and M'Coy, and as M'Coy, Griffith, and Salter).

———. 1848. On some new Mesozoic Radiata. Annals and Magazine of Natural History (series 2) 2:397.

———. 1849. On some new genera and species of Palaeozoic corals and foraminifera. Annals and Magazine of Natural History (series 2) 3:1–20, 119–136.

———. 1850. On some new genera and species of Silurian *Radiata* in the collection of the University of Cambridge. Annals and Magazine of Natural History (series 2) 6:270–290.

———. 1855. *Tetragonis danbyi, Vioa prisca. In* A. Sedgwick & F. M'Coy, Synopsis of the classification of British Palaeozoic rocks by the Rev. Adam Sedgwich, M.A., F.R.S., with a systematic description of the British Palaeozoic fossils in the Geological Museum of the University of Cambridge by Frederick M'Coy F.G.S. J. W. Parker & Son. London & Cambridge. p. 62, 260.

Meek, F. B., & A. H. Worthen. 1860. Descriptions of new Carboniferous fossils from Illinois and other western states. Academy of Natural Sciences of Philadelphia, Proceedings for 1860:447–472.

Mehl, Dorte. 1992. Die entwicklung der Hexactinellida seit dem Mesozoikum- Paläobiologie, Phylogenie und Evolutionsökologie. Berliner Geowissenschaftliche Abhandlungen (series E):164 p., 35 fig., 22 pl.

———. 1996. Phylogenie und Evolutionsökologie der Hexactinellida (Porifera) im Paläozoikum. Geologisch-Paläontologische Mitteilungen Innsbruck 4:1–55, 15 fig., 7 pl.

Mehl, Dorte, & B.-D. Erdtmann. 1994. *Sanshapentella dapingi* n. gen., n. sp.—a new hexactinellid sponge from the Early Cambrian (Tommotian) of China. Berliner Geowissenshaftliche Abhandlungen (series E) 13:315–319, 1 pl.

Mehl, Dorte, & F. T. Fürsich. 1997. Middle Jurassic Porifera from Kachchh, western India. Paläontologische Zeitschrift 71(1/2):19–33, 8 fig.

Mehl, D., & Norbert Hauschke. 1995. *Hyalonema cretacea* n. sp., erste körperlich erhaltene Amphidiscophora (Porifera, Hexactinellida) aus dem Mesozoikum. Geologie und Paläontologie des Westfalen 38:89–97.

Mehl, Dorte, & O. Lehnert. 1997. Cambro-Ordovician sponge spicule assemblages in the Ordovician of the Argentine Precordillera and paleoenvironmental ties. Neues Jahrbuch für Geologie und Paläontologie, Abhandlungen 204(2):221–246.

Mehl, Dorte, & Helfried Mostler. 1993. Neue Spicula aus dem Karbon und Perm: Konsequenzen für die Evolutionsökologie der Hexactinellida (Porifera), Strategien ihrer Gerüstbildung im Spätpaläozoikum und frühen Mesozoikum. Geologisch-Paläontologische Mitteilungen Innsbruck 19:1–28, 11 fig., 6 pl.

Mehl, D., J. K. Rigby, & S. R. Holmes. 1993. Hexactinellid sponges from the Silurian-Devonian Roberts Mountains Formation in Nevada and hypotheses of hexactine-stauractine origin. Brigham Young University Geology Studies 39:101–124.

Meylan, A. 1990. Nutritional characteristics of sponges in the diet of the hawksbill turtle, *Eretmochelys imbricata. In* K. Rutzler, ed., New Perspectives in Sponge Biology, Third International Conference on the Biology of Sponges. Smithsonian Institution Press. Washington, D.C. p. 472–477.

Michelin, H. 1840–1847. Iconographie Zoophytologique, description par localités et terrains des Polypiers fossiles de France et pays environnants. P. Bertrand, ed. Paris. 348 p., 79 pl.

Miller, S. A. 1882. Description of two new genera and eight new species of fossils from the Hudson River Group, with remarks upon others. Journal of the Cincinnati Society 5:34–44.

———. 1889. Class Porifera. *In* North American Geology and Palaeontology. Published by the author. Cincinnati. p. 152–167, fig. 89–127.

———. 1892. Paleontology. Indiana Department of Geologic and Natural Resources, Annual Report 17:611–705.

Miller, S. A., & C. B. Dyer. 1878. Contributions to Palaeontology. Cincinnati Society of Natural History Journal 1:24–39.

Milne-Edwards, Henri, & Jules Haime. 1848. Reserches sur les polypes. Annales des Sciences Naturelles, Paris (series 3) 9:37–89, pl. 4–6, 7–10; 10:65–114, 209–320, pl. 1, 5–9.

———. 1850. Monograph of the British fossil corals, part 1. Palaeontographical Society Monograph. London. Introduction, p. i–lxxv; Description of fossil corals, p. 1–322, 11 pl.

Minchin, E. A. 1889. Éponges Calcaires. La Clathrine coriace, *Clathrina coriacea* (Montagu). Zoologie Descriptive des Invertebres, tome I, chapter 5. p. 107–147, fig. 35–72.
———. 1900. Sponges—Phylum Porifera. *In* E. R. Lankester, ed., A Treatise on Zoology, part 2, chapter 3. The Porifera and Coelenterata. Adam and Charles Black. London. p. 1–178, 97 fig.
———. 1905. A speculation on the phylogeny of the hexactinellid sponges. Zoologische Anzeiger 28:439–448.
Moiseev, C. R. 1939. New data on Upper Triassic of North Caucasus and the Crimea. Doklady Akademii NAUK SSSR 23(1):816–817.
———. 1944. Vodoroslii, gubki, gidroidnye polipy i korally verkhnego triasa Kavkazskogo khrebtra [Algae, sponges, aqueous polyps and corals of the Upper Trias of the Caucasus]. Uchenye Zapiski Leningradskogo Gosaudarstvennogo Universiteta, Seriya Geologo-Pochvenno-Geograficheskaya [Scientific Publications of the Leningrad State University] 11(70):15–28, 5 pl.
Molineaux, Ann. 1994. A Late Pennsylvanian encruster: terminal Paleozoic calcified demosponge? Canadian Society of Petroleum Geologists, Memoir 17:967–982, 16 fig.
Montagu, George. 1818. An essay on sponges, with descriptions of all the species that have been discovered on the coast of Great Britain. Memoirs of the Wernerian Natural History Society 2:67–122, pl. 3–16.
Montanaro-Gallitelli, Eugenia. 1956. Trachypsammiacea. *In* R. C. Moore, ed., Treatise on Invertebrate Paleontology, Part F. University of Kansas Press & Geological Society of America. Lawrence, KS & New York. p. 190–192.
Moore, J. A., ed. 1965. Ideas in modern biology. XVI International Congress of Zoology, Proceedings 6:ix + 563 p., fig.
Moret, Léon. 1924. Contribution à l'étude des spongiaires siliceux du Miocène de l'Algérie. Mémoires de la Société Géologique de France (new series) 1:5–27, 4 pl.
———. 1925. Appendix. *In* C. H. Regnard, Notice sur les Spongiaires cénomaniens. Bulletin de la Société Géologique, Paris 25:486–487.
———. 1926a [1925]. Sur quelques Spongiaires de Catalogne (Argovian, Senomanien, Eocène). Bulletin Societat de Ciencias Naturales de Barcelona 4:8–18, 1 pl.
———. 1926b [1925]. Contribution à l'étude des Spongiares siliceux du Crétacé supérieur français. Mémoires de la Société Géologique de France (new series) 5(2):1–120; 3(1):121–338, 87 fig., 24 pl.
———. 1927. Note préliminaire sur les spongiaires Jurassiques de la Voulte (Ardeche) et de Trept (Isère). Comptes Rendus de l'Association Française pour l'Avancement des Sciences, Paris 50:291–292.
———. 1928. Les spongiaires siliceux du Callovien de la Voulte-sur-Rhone (Ardeche). *In* F. Roman, Études sur le Callovien de la Vallée du Rhone. Travaux du Laboratoire de Géologie de la Faculté des Sciences de Lyon, Fascicule 13, Mémoire 2:123–140, fig. 23–29, pl. 6–9.
Morris, S. C., & Chen Menge. 1990. *Blastulospongia polytreta* n. sp., an enigmatic organism from the Lower Cambrian of Hubei, China. Journal of Paleontology 64(1):26–30, 4 fig.
Morton, J. E. 1967. Mollusks: An introduction to their form and function. Harper and Brothers. New York. 232 p., 23 fig.
Mostler, Helfried. 1985. Neue heteractinide Spongien (*Calcispongea*) aus dem Unter- und Mittelkambrium Südwestsardiniens. Berichte des Naturwissenschaftlich-Medezinischen Vereins in Innsbruck 72:7–32, 11 fig., 5 pl.
———. 1986. Neue Kieselschwämme aus den Zlambachschichten (Obertrias, Nördliche Kalkalpen). Geologische-Paläontologische Mitteilungen, Innsbruck 13:331–361, 8 fig., 9 pl.
———. 1994. Der erste Nachweis von agelasiden Schwämmen (Demospongiae) aus dem Jungpaläozoikum. Abhandlungen der Geologischen Bundesanstalt 50:341–352, 1 fig., 3 pl.
———. 1996a. Demospongien mit aussergewöhnlich Gebauten anatriaenen Megaskleren. Geologisch-Paläontologische Mitteilungen Innsbruck 21:153–171, 3 fig., 6 pl.
———. 1996b. Polyactinellide Schwämme, eine auf des Paläozoikum beschränkte Calcispongien-Gruppe. Geologisch-Paläontologische Mitteilungen Innsbruck 21:223–243, 15 fig., 3 pl.
———. 1996c. Demospongien mit aussergewöhnlich gebauten anatriaenen megaskleren. Geologisch-Paläontologische Mitteilungen, Innsbruck 21:153–171, 3 fig., 6 pl.
Mostler, Helfreid, & Z. Balogh. 1993–1994. Zur skelettarchitektur, entwicklung und stratigraphischen bedeutung ausgewählter lithistider Schwämme aus dem oberjura. Geologisch-Paläontologische Mitteilungen, Innsbruck 19:33–153.
Mostler, Helfried, & A. Mosleh-Yazdi. 1976. Neue Poriferen aus oberkambrischen Gesteinen der Milaformation im Elburzgebirge (Iran). Geologisch-Paläontologische Mitteilungen Innsbruck 5(1):1–36, 25 fig., 5 pl.
Müller, O. F. 1776. Zoologiae Danicae Prodromus seu Animalium Daniae et Norvegiae indigenarium characters, nomina et synonima imprimis popularium. Icones, fasc. 1 XXXII. Havniae. Copenhague. Typis Halligeriis. 274 p.
Müller, W. 1984. Die Kalkschämme der Unterordung Inozoa Steinmann aus dem Oberen Jura von Wurttemberg (SW-Deutschland). Stuttgarter Beiträge Naturkunde (Series B, Geologie und Paläontologie) 100:1–85, 24 pl.
Munier-Chalmas, E. 1882. *Barroisia,* nouvelle genre des éponges. Bulletin de la Société Géologique de France (series 3) 10:425.
Münster, G. Graf zu. 1841. Beschreibung und Abbildung der in den Kalkmergelschichten von St. Cassian gefundenene Versteinerungen. Beiträge zur Geognosie und Petrefactenkunde des südöstlichen

Tirols, vorzieglich des Schichten von St. Cassian. Bayreuth. 152 p., 16 pl.

Murchison, R. I. 1839. Silurian System, founded on geological researches in the counties of Salop, Hereford, Radnor, Montgomery, Caermarthen, Brecon, Pembroke, Monmouth, Gloucester, Worcester and Stafford; with descriptions of the coal-fields and overlying formations. John Murray. London. Part 1, p. 1–576; Part 2, p. 577–768, 53 pl.

Myagkova [Miagkova], E. I. 1955a. K kharakteristike klassa Aphrosalpingoida Miagkova, 1955 [On the characteristics of the Class Aphrosalpingoida, Miagkova, 1955]. Akademiya Nauk SSSR Doklady 104:478–481, 2 fig.
In Russian.

———. 1955b. Novye predstaviteli tipa Archaeocyatha [New representatives of the phylum Archaeocyatha]. Akademiya Nauk SSSR, Doklady 104:638–641, 2 fig.

Naletov, P. I. 1961. Katalog mestonakhozhdeny iskopaemykh fauny, flory, pyl'tsy i slor tsentral'noi chasti Buryatskoi ASSR. Gosudarstvennoe Nauchno-Tekhnicheskoe Izdatel'stvo Literatury po Geologii i Okhrane Nedr, Moskova 1961:1–64.
In Russian.

Nardo, G. D. 1833. Auszug aus einem neuen System der Spongiarien, wornach bereits die Aufstellung in der Universitäts-Sammlung zu Poadua gemacht ist. 73, Isis Journal Collection. Oken. Jena. p. 519–523.

———. 1834. De Spongiis. Isis Journal Collection. Oken. Jena. p. 714–716.

———. 1847a. Observazioni anatomiche sopra l'animale marino detto rognone di mare. Atti del Instituto Veneto di Scienze, lettere ed Arti, Venice 6:221.

———. 1847b. Prospetto della fauna marina volgare del Veneto-Estuario con cenni sulle principali specie commestibili dell'Adriatico, sulle venete pesche, sulle valli, etc. *In* Venezia e le sue lagune. G. Antonelli. Venezia. p. 113–156.
p. 1–45 in reprint.

Nazarov, B. B., & L. E. Popov. 1976. Radiolyarii, bezzamkovye brakhiopody i organizmy neyasnogo sistematischeskogo polozheniya iz srednego ordovika vostochnogo Kazakhstana [Radiolarians, inarticulate brachiopods, and organisms of uncertain systematic position from the Middle Ordovician of eastern Kazakhstan]. Paleontologicheskii Zhurnal 1976 (4):33–42.
In Russian; English translation published in 1977 by the American Geological Institute.

Neave, S. A. 1936–1940. Nomenclator Zoologicus, 4 vol. The Zoological Society of London. London. vol. 1, A–C, p. 1–957; vol. 2, D–L, p. 1–1025; vol. 3, M–P, p. 1–1065; vol. 4, Q–Z, p. 1–758.

———. 1950. Nomenclator Zoologicus, vol. 5. The Zoological Society of London. London. 308 p.

Nekvasilová, O. T., & D. Stemproková. 1960. Die Schwämmenadeln der Branik-Schichten (Unterdevon-Böhmen). Casopis pro Mineralogii a Geologii 5:400–405, 2 pl.

Nestler, H. 1961. Spongien aus der weissen Schreibkreide (Unt. Maastrict.) der Insel Rugen (Ostsee). Paläontologische Abhandlungen, Berlin 1:13–70.

Newell, N. D. 1957. Paleoecology of Permian reefs in the Guadalupe Mountains area. *In* H. S. Ladd, ed., Treatise on Marine Ecology and Paleoecology, vol. 2, Paleoecology. Geological Society of America Memoir 67:407–436, 11 fig.

Newell, N. D., J. B. Chronic, & T. G. Roberts. 1948. Pennsylvanian and Permian of Peru. Geological Society of America Bulletin 58(12):1212.

Newell, N. D., J. K. Rigby, A. G. Fischer, A. J. Whiteman, J. E. Hickox, & J. S. Bradley. 1953. The Permian reef complex of the Guadalupe Mountains region, Texas and New Mexico. W. H. Freeman and Co. San Francisco. xix + 239 p., 85 fig., 32 pl.

Nicol, D. 1962. The biotic development of some Niagaran reefs—an example of an ecological succession or sere. Journal of Paleontology 36:172–176.

Nicol, J. A. C. 1967. The biology of marine animals. John Wiley & Sons, Inc. New York. 699 p.

Nitecki, M. H., & F. Debrenne. 1979. The nature of radiocyathids and their relationship to receptaculitids and archaeocyathids. Géobios 12(1):5–27, 5 pl.

Nutsubidze, K. Sh. 1965. [Liassic sponges of the Dzirulbskogho massif]. Trudy Geologicheskogo Instituta, Tbilisi 14:5–36, 15 pl.
In Russian.

Oakley, K. P. 1938. Some facts about Cretaceous sponges. South-Eastern Naturalist, London 43:58–61, 1 pl.

Odum, E. P. 1959. Homeostasis of the ecosystem in relation to animal populations. Proceedings of the International Congress of Zoology 15:783–784.

Okada, Y. 1928. On the development of a hexactinellid sponge, *Farrea sollasii.* Tokyo University, Journal of the Faculty of Sciences (Section 4, Zoology) 2:1–27.

Oken, L. 1815. Lehrbuch der Naturgeschichte. 3 Band., Lehrbuch der Zoologie, 1st Abtheilung, Zoologie, Fleischlose Thiere. C. H. Reclam. Leipzig. 846 p., 40 pl.

Okulitch, V. J. 1935. Cyathospongia–a new class of Porifera to include the Archaeocyathinae. Transactions of the Royal Society of Canada 29:75–106, 2 fig., 2 pl.

———. 1937. Some changes in nomenclature of Archaeocyathi (Cyathospongia). Journal of Paleontology 11:251–253.

———. 1955. Archaeocyatha. *In* R. C. Moore, ed., Treatise on Invertebrate Paleontology, Part E, Archaeocyatha and Porifera. Geological Society of America & The University of Kansas Press. New York & Lawrence. p. 1–20, fig. 1–13.

Okulitch, V. J., & W. G. Bell. 1955. *Gallatinospongia,* a new siliceous sponge from the Upper Cambrian of Wyoming. Journal of Paleontology 29:460–461, pl. 48–49.

Oliver, W. A., Jr. 1951. Middle Devonian coral beds of central New York. American Journal of Science 249(10):705–728.

———. 1956. Biostromes and bioherms of the Onandaga Limestone in eastern New York. New York State Museum and Science Service Circular 45:1–23.

Olivi, G. 1792. Zoologia Adriatica ossia Catalogo ragionato degli Animali del Golfo e delle Lagune di Venezia; preceduto de una Dissertazione sulla Storia fisica e naturale del Golfo; e accompagnato da Memoria, ed Osservazioni di Fisica Storia naturale ed. Economia dell' Abate. Bassano, Italy. p. xxxi + 1–334.

Öpik, A. A. 1961. The geology and palaeontology of the headwaters of the Burke River, Queensland. Bureau of Mineral Resources, Geology and Geophysics, Bulletin 53:249 p., 24 pl.

Oppliger, F. 1907. Spongien aus dem Argovien I (Birmenstorferschichten) des Département du Jura, Frankreich. Abhandlungen der Schweizerischen Paläontologischen Gesellschaft 34:1–19.

———. 1915. Die Spongien der Birmensdorferschichten des schweizerischen Jura. Abhandlungen der Schweizerischen Paläontologischen Gesellschaft (Mémoires de la Société Paléontologique Suisse), Geneva 40:1–84, 12 pl.

———. 1921a. Über Neues Juraspongien. Actes de la Société Helvetique des Sciences Naturelles, Geneve 101:204–205.

———. 1921b. Über Neues Juraspongien. Eclogae Geologicae Helvetiae 16:133–134.

———. 1926. Kieselspongien des schweizerischen weissen Jura. Abhandlungen der schweiserischen Paläontologischen Gesellschaft (Mémoires de la Société Paléontologique Suisse), Geneva 46:1–76, fig. 1–6, pl. 1–5.

d'Orbigny, A. D. 1849 [1848]. Note sur la classe des Amorphozoaires. Revue et Magazine de Zoologie pure et appliquée (series 2) 1:545–550.

———. 1850–1852 [1849]. Prodrome de Paléontologie stratigraphique universelle des animaux mollusques et rayonnés faisant suite au cours élémentaire de Paléontologie et de géologie stratigraphiques. Victor Masson. Paris. vol. 1, 394 p. (Jan., 1850); vol. 2, 427 p. (Nov. 1850); vol. 3, p. 1–196, and index, p. 1–190 (1852).

———. 1851. Cours élémentaire de Paléontologie et du Géologie stratigraphiques, vol. 2, no. 1. Victor Masson. Paris. 392 p., pl. 1–17.

Ortmann, P. 1912. Die Mikroscleren der Kiesellspongien in Schwämmgesteinen der Senonen Kreide. Neues Jahrbuch für Mineralogie, Geologie und Paläontologie 2:127–149.

Oswald, F. 1847. Über die Petrifacten von Sadewitz. Uebersicht der Arbeiten und Veränderungen. Schlesischer Gesellschaft für Väterlandische Cultur im Jahre 1846. Breslau. p. 56–65.

———. 1850. Über *Aulocopium* und andere Spongien der Sadewitzer Geschiebe. Brieflich Mitth. an Herrn Beyrich. Zeitschrift der Deutschen Geologischen Gesellschaft 2:83–86.

Ott, Ernst. 1967a. Segmentierte Kalkschwämme (Sphinctozoa) aus der alpinen Mitteltrias und ihre Bedeutung als Riffbildner im Wettersteinkalk. Bayerische Akademie der Wissenschaften Mathematisch-Naturwissenschaftliche Klasse, Abhandlungen (new series) 131:96 p., fig. 1–9, 10 pl.

———. 1967b. Die Beziehung zwischen *Colospongia* Laube, *Takreamina* Fontaine, *Girtycoelia* King und *Dictyocoelia* n. gen. (segmentierte Kalkschwämme). Neues Jahrbuch für Geologie und Paläontologie, Monatshefte 1967:44–58, 3 fig.

———. 1974. *Phragmocoelia* n. g. (Sphinctozoa), ein segmentierter Kalkschwämme mit neuem Füllgewebetyp aus der Alpinen Trias. Neues Jahrbuch für Geologie und Paläontologie, Monatshefte 12:712–723, 4 fig.

Ott, Ernst, & Wolfgang Volkheimer. 1972. *Palaeospongilla chubutensis* n. g. et n. sp.—ein Süsswasserschwamm aus der Kreide Patagoniens. Neues Jahrbuch für Geologie und Paläontologie, Abhandlungen 140:49–63, 6 fig.

Owen, D. D. 1858 [1857]. Second report on the geological survey in Kentucky, made during the years 1856 and 1857. Frankfurt, Kentucky. 391 p.

Owen, R. 1841. On the new genus and species of sponge (*Euplectella aspergillum*). Proceedings of the Zoological Society 1841:3.

Pallas, P. S. 1766. Elenchus zoophytorum, sistens generum abdumbrations generaliores specierum cognitarum succinctas descriptiones oum selectes auctorum synonymia. Hagae-Comitum. Petrum van Cleef. The Hague. 451 p.

———. 1776. Reise durch verschiedene Provinzen des Russischen Reichs. Band 3. Kaiserliche Academie des Wissenschaften. St. Petersburg. p. 453–504, pt. 1; p. 701–744, pt. 2; p. 691–760, pt. 3.

Pantic, S. 1975. *Ceotinella mirunae* gen. nov. (Spongia, Familia 'incertae sedis') from the Middle Triassic of Montenegro. Geoloski Anali Balkanskogo Poluostrva 39:153–158.

Parkinson, James. 1822. Outlines of Orycytology. An introduction to the study of fossil organic remains, especially those found in the British strata. Published by the author, printed by J. Compton, Printer. London. p. 35–61, pl. 1.

Parona, C. F. 1933. Le spugne della fauna permiana di Palazzo Adriano (Bacino del Sosio) in Sicila. Memorie della Società Geologica Italiana 1:1–58, 7 fig., 12 pl.

Pavans de Ceccatty, M. 1960. Les structures cellulaires de type nerveux et de type musculaire de l'éponge siliceuse *Tethya lyncurium* Lamarck. Comptes Rendus de l'Académie des Sciences, Paris 2,451:1,818–1,819.

———. 1966. Connections cellulaires et jonctions polarisées du reséau intramésenchymateaux, chez l'éponge *Hippospongia communis* Lamarck. Comptes Rendus Hebdomadaires des Séances de l'Académie des Sciences, Paris 263D:145–147, 2 fig.

Penny, J. T., & A. A. Racek. 1968. Comprehensive revision of a worldwide collection of freshwater sponges (Porifera: Spongillidae). United States National Museum Bulletin 272:1–184, 15 pl.

Phillips, John. 1829–1836. Illustrations of the geology of Yorkshire, Part 1. A description of the strata and organic remains of the Yorkshire coast, accompanied by a geological map, sections and plates of the fossil plants and animals, p. 1–193, 15 pl., T. Wilson and Sons, London; Part 2, The Mountain Limestone District, p. i–xx, + 1–253, 25 pl., John Murray, London.

———. 1875. Illustrations of the Geology of Yorkshire; or a Description of the Strata and Organic Remains, Part I. *In* R. Etheridge, ed., The Yorkshire Coast, 3rd ed. John Murray. London. x + 354 p., 28 pl.

Pickett, John. 1969. Middle and Upper Palaeozoic sponges from New South Wales. Memoirs of the Geological Survey of New South Wales, Palaeontology 16:1–24, 11 pl.
Dated 1967 but not published until 1969.

———. 1982. *Vaceletia progenitor,* the first Tertiary sphinctozoan (Porifera). Alcheringa 6:241–247, fig. 1–6.

———. 1983. An annotated bibliography and review of Australian fossil sponges. Association of Australasian Palaeontologists Memoir 1:93–120, 13 fig.

———. 2002. Order Heteractinida Hinde, 1877. *In* J. N. A. Hooper & R. W. M. Van Soest, eds., Systema Porifera: A guide to the classification of the sponges. Kluwer Academic/ Plenum Publishers. New York. p. 1121–1139, 13 fig.

Pickett, John, & P. A. Jell. 1983. Middle Cambrian Sphinctozoa (Porifera) from New South Wales. Memoir Association of Australasian Palaeontologists 1:85–92, 4 fig.

Pickett, John, & Yves Plusquellec. 1998. Éponges siliceuses du Dévonien Supérieur de la Rade de Brest (France). Geobios 31(6):715–723, 2 fig.

Pickett, John, & J. K. Rigby. 1983. Sponges from the Early Devonian Garra Formation, New South Wales. Journal of Paleontology 57:720–741, 9 fig.

Pisera, A. A. 1997. Upper Jurassic siliceous sponges from the Swabian Alb: taxonomy and paleoecology. Palaeontologia Polonica 57:3–216, 35 fig., 54 pl.

———. 2002. Fossil 'Lithistids,' an overview. *In* J. N. W. Hooper & R. W. M. van Soest, eds., Systema Porifera: A Guide to the Classification of Sponges. Kluwer Academic/Plenum Press. New York. p. 388–402, 22 fig.

Pisera, Andrzej, & Adam Bodzioch. 1991. Middle Triassic lyssacinosan sponges from Upper Silesia (southern Poland), and the history of hexactinosan and lychniscosan sponges. Acta Geologica Polonica 41(3–4):193–207, 4 fig.

Pisera, A. A., & Pere Busquets. 2002. Eocene siliceous sponges from the Ebro Basin (Catalonia, Spain). Geobios 35:321–346, 16 fig.

Pitcher, M. G. 1964. Evolution of Chazyan (Ordovician) reefs of eastern United States and Canada. Bulletin of Canadian Petroleum Geology 12:632–691, 49 fig., 3 pl.

Playford, P. E. 1967. Devonian reef complexes in the northern Canning Basin, Western Australia. International Symposium on the Devonian System, Alberta Society of Petroleum Geologists 2:351–364.

Playford, P. E., & D.C. Lowry. 1966. Devonian reef complexes of the Canning Basin, Western Australia. Geological Survey of Western Australia Bulletin 118:1–150.

Pocta, P. 1883. Beiträge zur Kentniss der Spongien der Böhmischen Kreideformation, part 1, Hexactinellidae. Abhandlungen der Königlisch Böhmisch Gesellschaft der Wissenschaften (series 6) 12:1–45, 3 pl.

———. 1884. Beiträge zur Kentniss der Spongien der Böhmischen Kreideformation, part 2, Lithistidae. Abhandlungen der Königlisch Böhmisch Gesellschaft der Wissenschaften (series 6) 12(9):1 42, 2 pl.

———. 1885. Beiträge zur Kentniss der Spongien der Böhmischen Kreideformation, Tetractinellidae, Monactinellidae, Calcispongiae, Certospongiae, Nachtrag. Abhandlungen der Königlisch Böhmisch Gesellschaft der Wissenschaften (series 7) 1(3):1–46, 1 pl.

———. 1903a. O nekterych nvch houbachz kridoeho utraru. Rozpravy Ceske Akademie Cisare Frantiska Josefa pro vedy slovesn ost a umeni 7(14):8–9.

———. 1903b. Beiträge zur Kenntnis der Calcispongien aus der Kreideformation. Académie des Sciences de l'Empereur François Joseph 1, Bulletin international Résumés des Travaux présentés (Sciences, Mathématiques et Naturelles) VII Année:118–124, 2 fig., 2 pl.

———. 1907. Sur quelques éponges du Sénonian de Nice. Bulletin de la Société Géologique de France, Paris (series 4) 7:163–173.

de Poléjaeff, N. 1883. Report on the Calcarea collected by H.M.S. Challenger during the years 1873–1876. Report on the Scientific Results of the Voyage of H.M.S. 'Challenger,' vol. 8, no. 24. London, Edinburgh, & Dublin. p. 1–76, 9 pl.

———. 1884. Report on the Keratosa collected by H.M.S. Challenger during the years 1873–1876. Report on the Scientific Results of the Voyage of the H.M.S. 'Challenger' 1873–1876. Zoologie 11. London, Edinburgh, & Dublin. p. 1–88, 10 pl.

Polunin, N. V. 1960. Introduction to plant geography and some related sciences. McGraw-Hill. New York. 640 p.

Pomel, A. 1872. Paléontologie ou description de animaux fossiles de la Province d'Oran, Zoophytes, fascicule 5, Spongiaires. Perrier. Oran. 256 p., 36 pl.

Portlock, J. E. 1843. Report on the Geology of Londonderry and parts of Tyrone and Fermanagh. Andrew Milliken, Dublin; Hodges and Smith, College-Green; and Longman, Brown, Green, and Longmans. London. 784 p., pl. 1–38 and A–I.

Potts, E. 1880. On freshwater sponges. Proceedings of the Academy of Natural Sciences of Philadelphia 1880:356–357.
Also published in 1881, Annals and Magazine of Natural History (series 5) 8:387–388.

———. 1881. Some new genera of freshwater sponges. Proceedings of the Academy of Natural Sciences of Philadelphia 1881:149–150.
Also published in Annals and Magazine of Natural History (series 5) 8:387–388.

Pourbaix, N. 1931. Contribution à l'étude de la nutrition chez les Spongiaires (éponges siliceuses). Bulletin des Station Océanographique Salammbô, Tunis 23:3–19, 2 pl.

———. 1932. Note sur la nutrition bactérienne des éponges. Annales de la Société Royale Zoologique de Belgique 63:11–15, 5 fig.

———. 1933. Mécanisme de la nutrition chez les Spongillidae. Annales de la Société Royale Zoologique de Belgique 64:11–20, 1 pl.

———. 1939. Activité respiratoire chez les Spongiaires. Annales de la Société Zoologique de Belgique 79:197–199.

Pray, L. C., & M. Esteban. 1977. Upper Guadalupian Facies, Permian Reef Complex, Guadalupe Mountains, New Mexico and West Texas. 1977 Field Conference Guidebook, vol. 2, Permian Basin Section, Society of Economic Paleontologists and Mineralogists, Publication 17-16:194 p.

Prosser, C. L. 1960. Mechanical responses of sponges. Anatomical Record 138(3):37.

Putter, A. 1914. Der stoffwechsel der Kieselschwämme. Zeitschrift für Allgemeine Physiologie 16:65–114.

Qian Jianxin, & Bing Xiao. 1984. An Early Cambrian small shelly fauna from Aksu-Wushi region, Xinjiang. Professional Papers on Stratigraphy and Palaeontology 13:65–90, 4 pl.
In Chinese with English summary.

Qian Yi, & Yin Gongzheng. 1985. Small shelly fossils from the lowerest Cambrian in Guizhou. Professional Papers on Stratigraphy and Palaeontology 14:91–121, 6 pl.

Quenstedt, F. A. 1843. Das Flötzgebirge Würtembergs. H. Laupp'schen, Buchhandlung. Tubingen. p. 407–427, 466–467.

———. 1852. Handbuch der Petrefactenkunde. H. Laupp'schen Buchhandlung. Tubingen. 982 p. Sponges on p. 666–678, 62 pl.

———. 1858. Der Jura. H. Laupp'schen Buchhandlung. Tübingen. 842 p., 100 pl.

———. 1877–1878. Petrefactenkunde Deutschlands. Der ersten Abtheilung, funfter Band. Die Schwamme. Leipzig, (1. Leif., p. 1–96, pl. 115–118, April,1877; 2, Leif., p. 97–224, pl. 119–124, 1877; 3. Leif., 225–320, pl. 125–130; 1877; 4. Leif., p. 321–448, pl. 131–136, Autumn, 1877; 5. Leif., p. I–VIII, 449–612, pl. 137–142, July 1878).

Radwanski, A. 1964. Boring animals in Miocene littoral environments of southern Poland. Bulletin de l'Académie Polonaise des Sciences, Série des Sciences Géologiques et Géographiques 12:57–62, 6 pl.

Rafinesque-Schmaltz, Constantine Sam. 1839. Descriptions des genres fossiles Ditaxopus, Trianisitis, Trioxites, Menepites, et Trianistes. Bulletin de la Société Géologique, Paris 10:378–381.

Ramond de Carbonnière, L. F. É. 1801. Nouveau genre de Polypiers fossiles, *Ocellaria*. Scientifique Société Philomèle, Paris, Bulletin 2:177.

Randall, J. E., & W. D. Hartman. 1968. Sponge-feeding fishes of the West Indies. Marine Biology 1:216–225.

Rasmont, R. 1959. L'ultrastructure des choanocytes d'éponges. Annales des Sciences Naturelles, Zoologie (series 12) 1(2):253–263, 2 pl.

Rauff, Hermann. 1891a. Vorläufige Mitteilung über das Skelet der Anomocladinen, sowie über eine eigenthümliche Gruppe fossiler Kalkschwämme (Polysteganinae), die nach dem Sycones-Types gebaut sind. Neues Jahrbuch für Mineralogie, Geologie und Paläontologie 1:278–284.

———. 1891b. Über *Palaeospongia prisca* Bornem., *Eophyton* z. Th., *Chondrites antiquus, Haliserites* z. Th. und ähnliche Gebilde. Neues Jahrbuch für Mineralogie, Geologie und Paläontologie 2:92–98.

———. 1891c. Über den Bau des Stützskeletes bei den Anomocladinen und Tetracladinen. Sitzungberichte Niederrheinische Gesellschaft 48:33–37.

———. 1891d. Über eine eigenthümliche Gruppe fossiler Kalkschwämme (Polysteganinae). Sitzungberichte Niederrheinische Gesellschaft 48:45–50.

———. 1892. Untersuchungen über die Organisation und systematische Stellung der Receptaculitiden. Abhandlungen der Bayerische Akademie der Wissenschaften, Mathematisch-Physikalische Klasse 17:645–722, 12 fig., 7 pl.

———. 1893. Paleospongiologie, Erster oder allgemeiner Theil, und Zweiter Theil, erste Halfte. Palaeontographica 40:1–232, fig. 1–48.

———. 1894. Paleospongiologie, Erster oder allgemeiner Theil, und Zweiter Theil, erste Halfte. Palaeontographica 41:233–346, fig. 49–75, pl. 1–17.

———. 1895. Paleospongiologie. Zweiter Theil. Fortsetzung. Spongien des Silurs. Palaeontographica 43:223–272, fig. 76–124, pl. 20–26.

———. 1913. *Barroisia* und die Pharetronenfrage. Palaeontologische Zeitschrift 13:74–144, 2 fig., 2 pl.

———. 1933. Spongienreste aus dem (oberteronen) Grünsand vom Kassenberg in Mülheim-Broich an der Ruhr. Preussischen Geologischen Landesanstalt, Abhandlungen (new series) 158:75 p., 5 pl.

———. 1938. Über einige Kalkschwämme aus der Trias der peruanischen Kordillere, nebst einem Anhang Über *Stellispongia* und ihre Arten. Palaeontologische Zeitschrift 20:177–214, 10 fig., pl. 18–21.

Ravn, J. P. J. 1899. Et par danske Kridtspongen. Meddelanden fra Dansk Geologisk Forening 1(5):24–32.

Raymond, P. E. 1931. Notes on invertebrate fossils, with descriptions of new species. Bulletin of the Museum of Comparative Zoology at Harvard College 55:165–213, 6 pl.

Raymond, P. E., & V. J. Okulitch. 1940. Some Chazyan sponges. Bulletin of the Museum of Comparative Zoology at Harvard College 86:197–214, 4 fig., 7 pl.

Regnard, C.-H. 1926 [1925]. Notice sur les spongiaires Cénomanians de Coulonges-les-Sablons (Orne). Bulletin de la Société Géologique de France 25:469–488, pl. 18–21.

Reid, R. E. H. 1957a. On Hexactinellida, "Hyalospongia," and the classification of siliceous sponges. Journal of Paleontology 31:282–286.

———. 1957b. Notes on hexactinellid sponges—II. *Dactyocalyx* Stutchbury and the Family Dactylocalycidae Gray. Annals and Magazine of Natural History (series 12) 10:821–826.

———. 1958a. A monograph of the Upper Cretaceous Hexactinellida of Great Britain and Northern Ireland, part I. Palaeontographical Society Monograph. London. p. i–xlvi.

———. 1958b. Remarks on the Upper Cretaceous Hexactinellida of Country Antrim. The Irish Naturalists Journal 12(9,10):236–268.

———. 1959. A monograph of the Upper Cretaceous Hexactinellida of Great Britain and Northern Ireland, part II. Palaeontographical Society Monograph. London. London. p. xlvii–xlviii, 1–26, pl. 1–4.

———. 1961. A monograph on the Upper Cretaceous Hexactinellida of Great Britain and Northern Ireland, part III. Palaeontographical Society Monograph. London. p. 27–48, 7 pl.

———. 1962. Notes on hexactinellid sponges—IV. Nine Cretaceous Lychniscosa. Annals and Magazine of Natural History (series 13) 5:33–45.

———. 1963a. Notes on a classification of the Hexactinosa. Journal of Paleontology 37:218–231.

———. 1963b. Hexactinellida or Hyalospongea. Journal of Paleontology 37:232–243.

———. 1963c. *Spiractinella* Hinde is a demosponge. Irish Naturalists Journal 14:129–131.

———. 1963d. A classification of the Demospongia. Neues Jahrbuch für Geologie und Paläontologie, Monatshefte 4:196–207, 2 fig.

———. 1963e. Preliminary notice of a classification of the Demospongia. The Irish Naturalists Journal 14:90–94.

———. 1964. Upper Cretaceous Hexactinellida of Great Britain and Northern Ireland, part IV. Palaeontographical Society (London), Monograph. Palaeontographical Society. London. p. xlix–cliv, fig. 25–61.

———. 1968a. Microscleres in demosponge classification. University of Kansas Paleontology Contributions Paper 35:11–37, fig. 1–10.

———. 1968b. *Hyalostelia smithii* (Young & Young) and the sponge genus *Hyalostelia* Zittel (Class Hexactinellida). Journal of Paleontology 42:1,243–1,248, pl. 162.

———. 1968c. The Carboniferous sponge "*Doryderma*" *dalryense* Hinde and the origin of heloclones and megaclones. Journal of Paleontology 42:1,249–1,254.

———. 1968d. *Tremacystia, Barroisia,* and the status of Sphinctozoida (Thalamida) as Porifera. The University of Kansas Paleontological Contributions 34:1–10.

———. 1969. Notes on hexactinellid sponges; 5, *Verrucocoelia* gen. nov., with a discussion of the genera *Verrucolia* Étallon and *Periphragella* Marshall. Journal of Natural History 3:485–492.

———. 1970. Tetraxons and demosponge phylogeny. Zoological Society of London Symposium 25:63–69.

Reid, R. P., & R. N. Ginsburg. 1986. The role of framework in Upper Triassic patch reefs in the Yukon (Canada). Palaios 1:590–600, 10 fig.

Reif, W.-E. 1968. Schwammreste aus dem oberen Ordovizium von Estland und Schweden. Neues Jahrbuch für Geologie und Paläontologie, Monatshefte 12:733–744, 3 fig.

Reimann, I. G. 1935. *Pseudohydnoceras,* a new Hamilton dictyosponge. Bulletin of the Buffalo Society of Natural Sciences 17(1):13–17, 2 pl.

———. 1945a. New Middle Devonian octactinellids. Paleontological Contributions, Bulletin of the Buffalo Society of Natural Sciences 19(2):16–21, pl. 2–4.

———. 1945b. New Hamilton lyssacine sponges. Paleontological Contributions, Bulletin of the Buffalo Society of Natural Sciences 19(2):43–49, pl. 8–9.

Reiswig, H. M. 1971. The axial symmetry of sponge spicules and its phylogenetic significance. Cahiers de Biologie Marine 12:505–514, 1 fig., 1 table.

———. 1973. Population dynamics of three Jamaican Demospongiae. Bulletin of Marine Science 23:191–226.

———. 1974. Water transport, respiration and energetics of three tropical marine sponges. Journal of Experimental Marine Biology and Ecology 14(3):231–249, 6 fig., 4 tables.

———. 1975. Bacteria as food for temperate-water marine sponges. Canadian Journal of Zoology 53(5):582–589, 5 fig.

———. 2002. Family Dactylocalycidae Gray, 1867. *In* J. N. A. Hooper & R. W. M. van Soest, eds., Systema Porifera: A Guide to the Classification of Sponges. Kluwer Academic/Plenum Press. New York. p. 1293–1300, 4 fig.

Reitner, Joachim. 1987a. A new calcitic sphinctozoan sponge belonging to the Demospongiae from the Cassian Formation (Lower Carnian; Dolomites, northern Italy) and its phylogenetic relationship. Géobios 20(5):571–589, 1 fig., 3 pl.

———. 1987b. Phylogenie und konvergenzen bei Rezenten und fossilen Calcarea (Porifera) mit einem kalkigen basalskelett ("Inozoa," Pharetronida). Berliner geowissenschaftliche Abhandlungen (Reihe A, Beiträge zur Paläontologie) 86:87–125, 8 fig., 8 pl.

———. 1987c. *Euzkadiella erenoensis* n. gen. n. sp. ein Stromatopore mit spikulärem skelett aus dem Oberapt von Ereño (Prov. Guipuzcoa, Nordspanien) und die systematische stellung der Stromatoporen. Paläontologische Zeitschrift 61:203–222, 11 fig.

———. 1991. Phylogenetic aspects and new descriptions of spicule-bearing hadromerid sponges with a secondary calcareous skeleton (Tetractinomorpha, Demospongiae). *In* J. Reitner & H. Keupp, eds., Fossil and Recent Sponges. Springer-Verlag. Berlin. p. 179–211.

———. 1992. "Coralline Spongien:" Der Versuch einer phylogenetisch-taxonomischen Analyse ["Corraline Sponges:" An attempt of a phylogenetic-taxonomic analysis]. Berliner Geowissenschaftliche Abhandlungen (Reihe E) 1:1–325, 90 fig., 62 pl.

Reitner, Joachim, & Theo Engeser. 1985. Revision der Demospongier mit einem Thalamiden, aragonitischen Basalskelett und trabekulärer Internstruktur ("Sphinctozoa" pars). Berliner Geowissenschaftliche, Abhandlungen (Reihe A) 60:151–193, 10 fig., 6 pl.

Reitner, J., & H. Keupp. 1991. The fossil record of the haplosclerid excavating sponge *Aka* de Laubenfels. *In* J. Reitner & H. Keupp, eds., Fossil and Recent Sponges. Springer-Verlag. Berlin. p. 102–120, 17 fig.

Reitner, Joachim, & Felix Schlagintweit. 1990. *Calcisuberites stromatoporoides* n. gen, n. sp., ein neue Taxon der Hadromerida (Demospongiae, Porifera) mit einem kalkigen Basalskelett aus der tethyalen Unterkreide. Berliner Geowissenschaftliche Abhandlungen (Reihe A, Geologie und Paläontologie) 124:247–257, 2 fig., 2 pl.

Rendel, J. M. 1965. The effect of genetic change at different levels. *In* J. A. Moore, ed., Ideas in modern biology. XVI International Congress of Zoology, Proceedings 6:285–295.

Reuss, A. Em. 1840. Geognostische Skizzen aus Böhem, Bd. 1, Die umgebungen von Toplitz und Belin, Ein Beitrag zur Physiographie des böhmischen Mittlegebirges. G. W. Medau & Company. Prague, Leitmeritz, & Toplitz. 298 p., 9 pl.

———. 1844. Geognostische Skizzen aus Böhem. Bd. 2, Die Kreidegebilde des westlichen Böhmens, ein monographischer Versuch., Nebst Bemerkungen über die Braunkohlenlager jenseits der Elbe un eine Uebershcht der fossilen Fischreste Böhmens. G. W. Medau & Company. Prague, Leitmeritz, & Toplitz. 304 p., 3 pl.

———. 1845–1846. Die versteinerungen der Böhmischen Kreideformation, 2 Abtheilungen. E. Schweizerbart'sche Verlagsbuchhandlung. Stuttgart. 148 p., 51 pl.

———. 1867. Die Bryozoen, Anthozoen und Spongarien des braunen Jura von Balin bei Krakau. K. K. Hof- und Staatsbruckerei. p. 117–214, 11 pl.

Reyment, R. A. 1971. Multivariate normality in morphometric analyses. Journal of the International Association for Mathematical Geology 3(4):357–368.

Rezvoi, P. D., I. T. Zhuravleva, & V. M. Koltun. 1962. Phylum Porifera. *In* B. S. Sokolov, ed., Osnovy Paleontologii [Fundamentals of Paleontology], vol. 1, number 2, Porifera, Archaeocyatha, Coelenterata, Vermes. Izdatel'stvo Akademii Nauk SSSR. Moscow. p. 17–74, fig. 1–107.
Translation published in 1972, Israel Program for Scientific Translations Ltd., Jerusalem, p. 5–97.

Rhebergen, F., R. Eggink, T. Koops, & B. Rhebergen. 2001. Staringia 9, Ordovicische zwerfsteensponzen. Grondboor & Hamer, Tweemaandelijks tijdscjhrist van de Nederlandse Geologisch Vereniging, Jaargang 55 (2001), nummer 1, 143 p., 68 fig., 43 pl.

Rhebergen, Freek, & Ulrich von Hacht. 2000. *Schismospongia syltensis* gen. n. sp. n. (Porifera) ein neuer Geschiebeschwamm aus plio/pleistozänen Kaolinsanden von Sylt (Nordwest-Deutschland). Archiv für Geschiebekunde 2(1):797–804, 1 fig., 3 pl.

Rhebergen, F., & T. M. G. van Kempen. 2002. An unusual Silurian erratic astylospongiid (Porifera) from Gotland, Sweden. GFF (Geologiska Föreningens i Stockholm Förhandlingar) 124:185–192, 13 fig.

Richardson, L., & A. G. Thacker. 1920. On the stratigraphical and geographical distribution of the sponges of Inferior Oolite of the West of England. Proceedings of the Geologists' Association, London 31:161–186, pl. 12–13.

Richter, Gotthard, & Michael Wuttke. 1999. *Lutetiospongilla heili* n. gen. n. sp. und die eozäne Spongillidenfauna von Messel. Courier Forschungsinstitut Senckenberg 216:183–195, 3 pl.

Ridley, S. O. 1881. XI. Spongida. Horny and siliceous sponges of Magellan Straits, S.W. Chili, and Atlantic off SW Brazil. *In* A. Gunther, ed., Account of the zoological collections made during the survey of H.M.S. 'Alert' in the Straits of Magellan and on the coast of Patagonia. Proceedings of the Zoological Society of London 1881:107–137, 140–141, pl. 10–11.

Ridley, S. O., & Arthur Dendy. 1886. Preliminary report on the Monaxonida collected by H.M.S. 'Challenger.' Annals and Magazine of Natural History (series 5) 18:325–351, 470–493.

———. 1887. Report on the Monaxonida collected by H.M.S. Challenger during the years 1873–1876. Reports of the Scientific Results of the Voyage of H.M.S. Challenger, Report 20(59):lxviii + 275 p., 51 pl.

Riedel, R., & B. Senowbari-Daryan. 1991. Pharetronids in Triassic reefs. *In* J. Reitner & H. Keupp, eds., Fossil and Recent Sponges. Springer-Verlag. Berlin & Heidelberg. p. 465–476, 4 fig.

Rietschel, Siegfried. 1968a. Die Octactinellida und ihnen verwandte paläozoische Kalkschwämme (Porifera, Calcarea). Paläontologische Zeitschrift 42(1/2):13–32, 4 fig., 1 pl.

———. 1968b. *Devonoscyphia* n. g. und "*Scyphia constricta* Sandberger," Kieselschwämme (Eutaxicladina) aus dem Mitteldevon der Lahnmulde (Rhein. Schiefergebirge). Jahrbücher des Nassauischen Vereins für Naturkunde 99:98–106, 2 fig., 2 pl.

———. 1970. Beitrage zur Sedimentation und Fossilführung des Hunsrückschiefers. 28. *Rectifungus rudens* n. g., n. sp., ein dictyospongiider Kieselschwamm aus dem Hunsrückschiefer. Notizblatt des Hessischen Landesamtes für Bodenforschung zu Wiesbaden 98:30–35, 2 fig., pl. 4.

Rigby, J. K. 1966a. *Protospongia hicksi* Hinde from the Middle Cambrian of western Utah. Journal of Paleontology 40:549–554, 7 fig., pl. 66.

———. 1966b. Microstructure and classification of an Ordovician sponge, *Dystactospongia madisonensis* Foerste, from Indiana. Journal of Paleontology 40:1,127–1,130, pl. 146.

———. 1967a. A new polyactinal sponge from the Antelope Valley Formation (Ordovician) in the Toquima Range, Nevada. Journal of Paleontology 41:511–515, 4 fig.

———. 1967b. Two new Early Paleozoic sponges and the sponge-like organism *Gaspespongia basalis* Parks, from the Gaspé Peninsula, Quebec. Journal of Paleontology 41:766–775, 5 fig., pl. 101–102.

———. 1969. A new Middle Cambrian hexactinellid sponge from western Utah. Journal of Paleontology 43:125–128, 2 fig., l pl.

———. 1970a. Two new upper Devonian hexactinellid sponges from Alberta. Journal of Paleontology 44:7–16, 2 fig., pl. 3–4.

———. 1970b. *Ellesmerespongia feildeni,* a new Permian sponge from the Canadian Arctic. Journal of Paleontology 44:1143–1145, 2 fig.

———. 1971. Sponges of the Ordovician Cat Head Member, Lake Winnipeg, Manitoba. *In* D.C. McGregor, F. H. Cramer, Rousseau H. Flower, & J. K. Rigby, Contributions to Canadian Paleontology, Fossils of the Ordovician Red River Formation (Cat Head Member), Manitoba. Geological Survey of Canada Bulletin 202:35–78, 6 pl.

———. 1974. *Vaurealispongia* and *Twenhofelella,* two new brachiospongid hexactinellid sponges from the Ordovician and Silurian of Anticosti Island, Quebec. Canadian Journal of Earth Sciences 11:1,343–1,349, 3 fig.

———. 1975. Some unusual hexactinellid sponge spicules from the Cambrian Wilberns Formation of Texas. Journal of Paleontology 49:412–415, 2 fig.

———. 1976a. A new Devonian heteractinid sponge from southwestern Ellesmere Island, Arctic Canada. Canadian Journal of Earth Sciences 13:120–125, 8 fig.

———. 1976b. Some observations on occurrence of Cambrian Porifera in western North America and their evolution. Brigham Young University Geology Studies 23(2):51–90, 6 fig.

———. 1977a. Two new Middle Ordovician sponges from Foxe Plain, southeastern District of Franklin. Geological Survey of Canada Bulletin 269:121–129, 2 pl.

———. 1977b. *Constellatospongia,* a new heteractinid astraeosponge from the Upper Ordovician Churchill River group, Manitoba. *In* Geology of Ordovician rocks, Melville Peninsula and region, southeastern District of Franklin. Geological Survey of Canada Bulletin 269:131–137, 2 fig., l pl.

———. 1977c. A new Middle Ordovician sponge from western Newfoundland. Canadian Journal of Earth Sciences 14:2,662–2,668, 8 fig.

———. 1977d. A new chiastoclonellid sponge fauna from the Devonian of Michigan. Journal of Paleontology 51:1,215–1,219, 3 fig.

———. 1978. Two wewokellid calcareous sponges in North America. Journal of Paleontology 52:705–716, 3 fig., 1 pl.

———. 1979. The genus *Ensiferites,* a Devonian astraeosponge of North America. Journal of Paleontology 53:475–493, 2 fig., 3 pl.

———. 1980. *Ichnospongia perplexa,* a new sponge from the Mississippian Chainman Shale near Eureka, Nevada. Journal of Paleontology 54:1,278–1,281, 3 fig.

———. 1981. The sponge fauna of the Eocene Castle Hayne Limestone from east-central North Carolina. Tulane Studies in Geology and Paleontology 16:123–144, 4 fig., 3 pl.

———. 1983a. Sponges of the Middle Cambrian Marjum Limestone from the House Range and Drum Mountains of western Millard County, Utah. Journal of Paleontology 57:240–270, 11 fig.

———. 1983b. Heteractinida. *In* T. W. Broadhead, ed., Sponges and Spongiomorphs: Notes for a Short Course. University of Tennessee Department of Geological Sciences, Studies in Geology 7:70–89, 12 fig.

———. 1984. Permian sponges from western Venezuela. Journal of Paleontology 58:1,436–1,462, 4 fig.

———. 1986a. Sponges of the Burgess Shale (Middle Cambrian) British Columbia. Palaeontographica Canadiana 2:105 p., 27 fig., 20 pl.

———. 1986b. Late Devonian sponges of Western Australia. Geological Survey of Western Australia, Report 18:vii + 59 p., 14 fig., 7 pl.

———. 1986c. The sponge fauna from the Mississippian Heath Formation of central Montana. *In* J. T. Dutro, Jr. & H. W. Pfefferkorn, eds., Neuvième Congrès International de Stratigraphie et de Géologie du Carbonifère. Compte Rendu, vol. 5, Paleontology, Paleoecology, Paleogeography. Southern Illinois University Press. Carbondale & Edwardsville. p. 443–456, 5 fig., 2 pl.

———. 1987a. Early Cambrian sponges from Vermont and Pennsylvania, the only ones described from North America. Journal of Paleontology 61:451–461, 4 fig.

———. 1987b. Phylum Porifera. *In* R. S. Boardman, A. H. Cheetham, & A. J. Rowell, eds., Fossil Invertebrates. Blackwell Scientific Publications. Palo Alto, California. p. 116–139, 21 fig.

———. 1991a. Evolution of Paleozoic heteractinid calcareous sponges and demosponges—patterns and records. *In* Joachim Reitner & Helmut Keupp, eds., Fossil and Recent Sponges. Springer-Verlag. Berlin & Heidelberg. p. 83–101, 15 fig.

———. 1991b. The new Devonian (Givetian) heteractinid sponge *Gondekia* from Ontario, Canada, and evolution of the astraeospongiids and eiffeliids. Journal of Paleontology 65:38–44, 4 fig.

———. 1994. Well-preserved specimens of the sponges *Gondekia* (Heteractinida) and *Pseudohydnoceras* (Hexactinellida), Middle Devonian of New York State. Journal of Paleontology 68:727–734, 4 fig.

Rigby, J. K., & W. I. Ausich. 1981. Lower Mississippian sponges from the Edwardsville Formation, southern Indiana. Journal of Paleontology 55:370–382, 2 fig., 1 pl.

Rigby, J. K., & T. N. Bayer. 1971. Sponges of the Ordovician Maquoketa Formation in Minnesota and Iowa. Journal of Paleontology 45:608–627, 9 fig., pl. 69–70.

Rigby, J. K., & R. B. Blodgett. 1983. Early Middle Devonian sponges from the McGrath Quadrangle of west-central Alaska. Journal of Paleontology 57:773–786, 4 fig.

Rigby, J. K., C. K. Chamberlain, & B. A. Black. 1970. Mississippian and Pennsylvanian sponges from the Ouachita Mountains of Oklahoma. Journal of Paleontology 44:816–832, 12 fig., 3 pl.

Rigby, J. K., & B. D. E. Chatterton. 1989. Middle Silurian Ludlovian and Wenlockian sponges from Baillie-Hamilton and Cornwallis Islands, Arctic Canada. Geological Survey of Canada Bulletin 391:69 p., 8 fig., 10 pl.

———. 1999. Silurian (Wenlock) demosponges from the Avalanche Lake area of the Mackenzie Mountains, southwestern District of Mackenzie, Northwest Territories, Canada. Palaeontographica Canadiana 16:43 p., 2 fig., 10 pl.

Rigby, J. K., & C. R. Clement. 1995. Demosponges and hexactinellid sponges from the Lower Devonian Ross Formation of west-central Tennessee. Journal of Paleontology 69:211–232, 12 fig.

Rigby, J. K., & D. H. Collins. 2003. Sponges of the Middle Cambrian Burgess and Stephen Shale Formations, British Columbia. Royal Ontario Museum Bulletin In press.

Rigby, J. K., & A. Desrochers. 1995. Lower and Middle Ordovician demosponges of the Mingan Islands, Gulf of St. Lawrence, Quebec. Paleontological Society Memoir 41(Journal of Paleontology supplement) 60(4):35 p., 14 pl., 5 fig.

Rigby, J. K., & O. A. Dixon. 1979. Sponge fauna of the Upper Silurian Read Bay Formation, Somerset Island, District of Franklin, Arctic Canada. Journal of Paleontology 53:587–627, 14 fig., 3 pl.

Rigby, J. K., Patrick Embree, & Michael Murphy. 1996. An unusual Upper Cretaceous (Santonian) hexactinellid sponge from the Great Valley Sequence, western Sacramento Valley, northern California. Journal of Paleontology 70:713–717, 3 fig.

Rigby, J. K., Fan Jiasong, & Zhang Wei. 1989a. Sphinctozoan sponges from the Permian reefs in South China. Journal of Paleontology 63:404–439, 20 fig.

———. 1989b. Inozoan calcareous Porifera from the Permian reefs in South China. Journal of Paleontology 63:778–800, 13 fig.

Rigby, J. K., Fan Jiasong, Zhang Wei, Wang Shenghai, & Zhang Xiaolin. 1994. Sphinctozoan and inozoan sponges from the Permian reefs of South China. Brigham Young University Geology Studies 40:43–109, 15 pl.

Rigby, J. K., & J. K. Gilland. 1977. A new fossil sponge from the Ordovician Garden City Limestone of southeastern Idaho. Great Basin Naturalist 37:475–480.

Rigby, J. K., & Q. H. Goodbody. 1986. *Malluviospongia,* a new Devonian heteractinid sponge from the Bird Fiord Formation of southwestern Ellesmere Island, Northwest Territories, Canada. Canadian Journal of Earth Sciences 23:344–349, 2 fig.

Rigby, J. K., & T. C. Gosney. 1983. First reported Triassic lyssakid sponges from North America. Journal of Paleontology 57:787–796, 5 fig.

Rigby, J. K., & R. C. Gutschick. 1976. Two new Lower Paleozoic hexactinellid sponges from Utah and Oklahoma. Journal of Paleontology 50:78–85, 1 pl.

Rigby, J. K., & D. R. Harris. 1979. A new Silurian sponge fauna from northern British Columbia, Canada. Journal of Paleontology 53:968–980, 3 fig., 2 pl.

Rigby, J. K., R. D. Horrocks, & J. M. Cys. 1982. A new hexactinellid brachiosponge from the Upper Permian of west Texas. Journal of Paleontology 56:315–323, 2 fig., 1 pl.

Rigby, J. K., & Hou Xian-Guang. 1995. Lower Cambrian demosponges and hexactinellid sponges from Yunnan, China. Journal of Paleontology 69:1,009–1,019, 5 fig.

Rigby, J. K., R. Keyes Jr., & A. Horowitz. 1979. Two new Mississippian sponges from northeastern Alabama. Journal of Paleontology 53:709–719, 4 fig., 1 pl.

Rigby, J. K., J. E. King, & L. F. Gunther. 1981. The new Lower Ordovician protosponge, *Asthenospongia,* from the Phi Kappa Formation in central Idaho. Journal of Paleontology 55:842–847, 1 fig., 1 pl.

Rigby, J. K., & E. I. Leith. 1989. *Tiddalickia manitobensis,* a new dictyosponge, and an unusual specimen of the lithistid sponge, *Aulocopella winnipegensis* Rauff, from the Ordovician of Manitoba. Journal of Paleontology 63:550–553, fig. 1–2.

Rigby, J. K., & A. C. Lenz. 1978. A new Silurian astylospongid sponge from Baillie-Hamilton Island, Canadian Arctic Archipelago. Canadian Journal of Earth Sciences 15:157–162, 5 fig.

Rigby, J. K., G. Lindner, & C. H. Stevens. 2004 (in press). A new occurrence of the "hydrozoan" *Radiotrabeculopora reticulata* Fan, Rigby, & Zhang, 1991, in the Permian of California. Journal of Paleontology 77:000–000, 2 fig.

Rigby, J. K., & B. J. Maher. 1995. Age of hexactinellide beds of the Roberts Mountains Formation, Snake Mountains, Nevada, and additions to the Silurian sponge fauna. Journal of Paleontology 69:1,020–1,029, 4 fig.

Rigby, J. K., & W. L. Manger. 1994. Morrowan lithistid demosponges and hexactinellids from the Ozark Mountains of northwestern Arkansas. Journal of Paleontology 68(4):734–746, 7 fig.

Rigby, J. K., & R. H. Mapes. 2000. Some Pennsylvanian and Permian sponges from southwestern Oklahoma and north-central Texas. Brigham Young University Geology Studies 45:25–67, 6 fig., 6 pl.

Rigby, J. K., & Dorte Mehl. 1994. Middle Devonian sponges from the northern Simpson Park Range, Nevada. Brigham Young University Geology Studies 40:111–153, 18 fig.

Rigby, J. K., & R. W. Moyle. 1959. Some Mississippian and Pennsylvanian sponges from Utah. Journal of Paleontology 33:399–403, 1 fig., 1 pl.

Rigby, J. K., & M. A. Murphy. 1983. *Gabelia,* a new late Devonian lyssakid protosponge from the Roberts Mountains, Nevada. Journal of Paleontology 57:797–803, 5 fig.

Rigby, J. K., & M. H. Nitecki. 1975. An unusually well preserved heteractinid sponge from the Pennsylvanian of Illinois and a possible classification and evolutionary scheme for the Heteractinida. Journal of Paleontology 49:329–339, 3 fig., 1 pl.

Rigby, J. K., M. H. Nitecki, C. M. Soja, & R. B. Blodgett. 1994. Silurian aphrosalpingid sphinctozoans from Alaska and Russia. Acta Palaeontologica Polonica 39:341–391, 14 fig.

Rigby, J. K., A. Pisera, T. T. Wrzolek, & G. Racki. 2001. Upper Devonian (Frasnian) sponges from the Holy Cross Mountains, Central Poland. Palaeontology 44(3):447–499, 4 fig., 9 pl.

Rigby, J. K., & T. L. Pollard Bryant. 1979. Fossil sponges from the Mississippian Fort Payne Chert in northeastern Alabama. Journal of Paleontology 53:1,005–1,012, 1 fig., 1 pl.

Rigby, J. K., & A. W. Potter. 1986. Ordovician sphinctozoan sponges from the eastern Klamath Mountains, northern California. Journal of Paleontology (Memoir 20) 60:1–47, 11 fig.

Rigby, J. K., A. W. Potter, & R. B. Blodgett. 1988. Ordovician sphinctozoan sponges of Alaska and Yukon Territory. Journal of Paleontology 62:731–746, 5 fig.

Rigby, J. K., G. Racki, & T. Wrzolek. 1982. Occurrence of dictyid hexactinellid sponges in the Upper Devonian of the Holy Cross Mountains (Poland). Acta Geologica Polonica 31:163–168, 4 pl.

Rigby, J. K., D. Schumacher, & S. J. Meader. 1979. The genus *Ensiferites,* a Devonian astraeosponge of North America. Journal of Paleontology 53:475–493, 2 fig., 3 pl.

Rigby, J. K., & Baba Senowbari-Daryan. 1995. Permian sponge biogeography and biostratigraphy. *In* P. A. Scholle, T. M. Peryt, & D. S. Ulmer-Scholle, The Permian of Northern Pangea, vol. 1, Paleogeography, Paleoclimates and Stratigraphy. Springer-Verlag. Berlin. p. 153–166, 6 fig.

———. 1996a. Upper Permian inozoid, demospongid, and hexactinellid sponges from Djebel Tebaga, Tunisia. The University of Kansas Paleontological Contributions (new series) 7:130 p., 81 pl.

———. 1996b. *Gigantospongia,* new genus, the largest known Permian sponge, Capitan Limestone, Guadalupe Mountains, New Mexico. Journal of Paleontology 70:347–355, 5 fig.

Rigby, J. K., Baba Senowbari-Daryan, & Liu Huaibao. 1998. Sponges of the Permian Upper Capitan Limestone, Guadalupe Mountains, New Mexico and Texas. Brigham Young University Geology Studies 43:19–117, 19 fig., 13 pl.

Rigby, J. K., & R. J. Stuart. 1988. Fossil sponges from the Silurian-Devonian Roberts Mountains Formation in northeastern Nevada. *In* D. L. Wolberg, ed., Contributions to Paleozoic Paleontology and stratigraphy, in honor of Rousseau H. Flower. New Mexico Bureau of Mines and Mineral Resources Memoir 44:129–137, 3 fig.

Rigby, J. K., & F. M. Terrell. 1973. Permian sponges from western Ellesmere Island, Arctic Canada. Canadian Journal of Earth Sciences 10:1,431–1,443, 17 fig.

Rigby, J. K., & A. T. Washburn. 1972. A new hexactinellid sponge from the Mississippian-Pennsylvanian Diamond Peak Formation in eastern Nevada. Journal of Paleontology 46:266–270, 3 fig., 1 pl.

Rigby, J. K., & B. D. Webby. 1988. Late Ordovician sponges from the Malongulli Formation of central New South Wales, Australia. Palaeontographica Americana 56:1–147, 24 fig., pl. 1–44.

Rigby, J. K., T. T. Wrzolek, & G. Racki. 2001. Upper Devonian (Frasnian) sponges from the Holy Cross Mountains, Central Poland. Palaeontology 43:447–488, 4 fig., pl. 1–9.

Rigby, J. K., Wu Xichung, & Fan Jaisong. 1998. Triassic hexactinellid sponges from patch reefs in north-central Sichuan, People's Republic of China. Brigham Young University Geology Studies 43:119–165, 11 pl.

Ringueberg, E. N. S. 1884. New fossils from the four groups of the Niagara period of western New York. Proceedings of the Academy of Natural Sciences of Philadelphia 1884:144–150, 3 pl.

Robinson, P. D., & S. K. Haslett. 1995. A radiolarian dated sponge microsclere assemblage from the Miocene Dos Bocas Formation of Ecuador. Journal of South American Earth Sciences 8(2):195–200, 2 fig., 1 pl.

Roemer, C. F. 1848. Über eine neue Art der Gattung *Blumenbachium* (Koenig) und meher unzweifelhafte Spongien in obersilurischen Kalkschichten der Graftschaft Decatur im State Tennessee in Nord-America. Neues Jahrbuch für Mineralogie, Geologie, und Paläontologie 1848:680–686, pl. 9.

———. 1852. Lethaea geognostica, oder Beschreibung and Abbildung der für die Gebirgs-Formationen bezeichnendsten Versteinerungen, 3rd ed., theil II. E. Schweizerbart'sche Verlagshandlung. Stuttgart.

———. 1854. Palaeo-Lethaea: Kohlen Periode (Silur-Devon-, Kohlen- und Zechstein Formation), vol. 1, no. 2. *In* H. G. Brönn & F. Roemer, eds., Lethaea Geognostica. E. Schweizerbart. Stuttgart. 788 p.

———. 1860. Silurische Fauna des westlichen Tennessee. Eine paläontologische Monographic. Edvard Trewendt Verlag. Breslau. 97 p.

———. 1861. Die fossile fauna der Silurischen Diluvial-Geschiebe von Sadewitz bei Oels in Nieder-Schlesien. Breslau. p. 1–15, pl. 1–3; p. 55–56, pl. 7.

———. 1876–1880. Lethaea geognostica oder Beschreibung und Abbildung der für die Gebirgs-Formationen bezeichnendsten Versteinerungen. Part 1, Lethaea palaeozoica. E. Schweizerbart'sche Verlagshandlung. Stuttgart. 668 p., 2 pl.; atlas, 62 pl.

———. 1883. Notiz über die Gattung *Dictyophyton.* Zeitschrift der Deutsche Geologischen Gesellschaft 35:704–708.

———. 1885. Lethaea erratica oder Aufzählung und Beschreibung der in der norddeutschen Ebene vorkommenden Diluvial-Geschiebe nordischer Sedimentär-Gesteine. Palaeontologische Abhandlungen herausgeber von Dames und Kayser, vol. 2, part 5. Georg Reimer. Berlin. 173 p.

———. 1887. *Trochospongia,* eine neue Gattung silurischer Spongien. Neues Jahrbuch für Mineralogie, Geologie und Palaeontologie 2:174–177, pl. 6.

Roemer, F. A. 1839. Nachtrag zu Versteinerungen der Norddeutschen Oolithen Gebirges. Hannover.

———. 1840–1841. Die Versteinerungen des norddeutschen Kreidegebirges. Hahn'schen Hofbuchhandlung. Hannover. Lieferung 1, p. 1–48, pl. 1–8, 1840; Lieferung 2, p. i–iv, 49–145, pl. 8–16, 1841.

———. 1864. Die Spongitarien des norddeutschen Kreidegebirges. Palaeontographica 13(1–2):1–64, 19 pl.

Rogers, W. S., M. Jackson, & K. McKinney. 1964. A new genus of sponge from the Middle Ordovician. Journal of Paleontology 38:135–137, 5 fig.

Roll, Artur. 1934. Form, bau, und entstehung der Schwammstotzen im südeutschen Malm. Palaeontologische Zeitschrift 16:197–246, 18 fig.

Romanenko, Y. W. 1968. Kembriyskiye gubki utryada Heteractinellida Altya [Cambrian sponges of the order Heteractinellida in the Altay]. Paleontologicheski Zhurnal 2:134–137, 3 fig. Translated in Paleontological Journal 2:271.

Rothpletz, A. 1900. Über einen neuen jurassischen Hornschwamm und die darin eingeschlossenen Diatomeen. Zeitschrift Deutsche Geologische Gesellschaft 52:154–160, 3 fig.

Row, R. W. H. 1909. Reports on the marine biology of the Sudanese Red Sea. XIII. Report on the sponges collected by Mr. Cyril Crossland in 1904–5. Part 1, Calcarea. Journal of the Linnean Society, Zoology 31:182–214, pl. 19–20.

Ruedemann, Rudolf. 1921. Report on fossils from the so-called Trenton and Utica beds of Grand Isle, Vermont. Vermont State Geologist Twelfth Report 1919–1920:90–100, 1 pl.

———. 1925. The Utica and Lorraine Formations of New York: Part 2, Systematic paleontology; No. I, Plants, sponges, corals, graptolites, crinoids, worms, bryozoans, brachiopods. New York State Museum Bulletin 262:1–171, 75 fig., pl. 1–13.

———. 1934. Palaeozoic plankton of North America. Geological Society of America, Memoir 2:141 p., 6 fig., 26 pl.

Russo, Franco. 1981. Nuove spugne calcaree triassiche di Campo (Cortina d'Ampezzo, Belluno). Bollettino della Società Paleontologica Italiana 20(1):3–17, 4 pl.

Rützler, K. 1965a. Substratstabilität als ökologischer Faktor im marinen Benthos, dargestellt am Beispiel adriatischer Poriferen. Internationale Revue der gesamten Hydrobiologie 50:281–292.

———. 1965b. Systematik und Ökologie der Poriferen aus Litoral-Schattengebieten der Nordadria. Zeitschrift für Morphologie und Ökologie der Tiere 55:1–82, 41 fig.

———. 1971. Bredin-Archold-Smithsonian Biological survey of Dominica; burrowing sponges, genus *Siphonodictyon* Bergquist, from the Caribbean. Smithsonian Contribution, Zoology 77:1–37, 11 fig., 9 pl.

———. 1973. Clionid sponges from the coast of Tunisia. Bulletin de l'Institut National Scientifique et Technique d'Oceanographie et de Peche de Salammbô 2(4):623–636, 7 fig.

———. 1974. The burrowing sponges of Bermuda. Smithsonian Contributions to Zoology 165:1–32, 26 fig.

Rützler, K., & G. Rieger. 1973. Sponge burrowing: Fine structure of *Cliona lampa* penetrating calcareous substrata. Marine Biology 21:144–162.

Salomon, Dorte. 1990. Nomenklatur und taxonomischer status der fossilen Gattung *Cribrospongia* d'Orbigny, 1849 (=*Tremadictyon* Zittel, 1877) (Hexactinosa, Hexactinellida). Berliner geowissenschaftliche Abhandlungen (Reihe A) 124:35–41, 1 fig., 1 pl.

Salter, J. W. 1861. Descriptions and lists of fossils, Appendix. *In* Geology of the neighbourhood of Edinburgh. Memoirs of the Geological Survey of Great Britain, London, Sheet 32:132–151, pl. 2.

———. 1864. On some new fossils from the Lingula-Flags of Wales. Quarterly Journal of the Geological Society of London 20:233–241, 3 fig., pl. 13.

Saper, J., & W. E. White. 1958. Amino-acid composition of scleroprotein of the sponge *Hippospongia equina.* Nature 181(4,614):285–286.

Sarà, Michele. 1970. Competition and cooperation in sponge populations. *In* W. G. Fry, ed., Biology of the Porifera. Zoological Society of London Symposium 25:273–284, 4 fig.

Sarà, Michele, & E. Manara. 1991. Cortical structure and adaptation in the genus *Tethya* (Porifera, Demospongiae). *In* J. Reitner & H. Keupp, eds., Fossil and Recent Sponges. Springer-Verlag. Berlin & Heidelberg. p. 306–312, 1 fig.

Sarà, Michele, & J. Vacelet. 1973. Écologie des demosponges. *In* P. Grassé, ed., Traité de Zoologie, vol. 3. Masson et Cie. Paris. p. 462–576.

Sars, G. O. 1872. Spongiae. *In* Kongelige Norske Universited, ed., On some remarkable forms of animal life from the great depths off the Norwegian coast. I, Partly from posthumous manuscripts of the late Professor Michael Sars. Brøgger & Christie. Christiana, Norway. p. 62–82.

Schäfer, P., & B. Senowbari-Daryan. 1981. Facies development and paleontologic zonation of four Upper Triassic patch-reefs, Northern Calcareous Alps near Salzburg, Austria. *In* D. F. Toomey, ed., SEPM Special Publication 30:241–259, 10 fig.

Schiller, Wolfgang. 2000. Feinstratigraphische Untersuchungen der Kernbohrung Enspel 1991 (2) unter besonderer Berucksichtigung der kiesel Microfossilien. Mainzer Naturwissenschaftliches Archiv 38:39–91, 84 fig.

von Schlotheim, E. F. 1820. Die Petrefactenkunde auf ihrem jetzigen Standpunkte durch die beschreibung seiner Sammlung versteinerter und fossiler Überreste des Thier- und Pflanzenreichs der vorwelt erläutert. Becher'schen Buchhandlung. Gotha. 437 p.

Schlüter, C. A. 1868. Über die jungsten Schichten der untern Senon-Bildung en und deren Verbreitung. Sitzungberichte der Niederrcinischen Gesellschaft für Natur- und Heilkunde zu Bonn 1868:92–93.

———. 1870. Über die Spongitarien-Banke der unteren Mucronaten- und oberen Quadraten-Schichten, und über *Lepidospongia rugosa* inbesondere. Sitzungberichte der Niederrheinischen Gesellschaft für Natur- und Heilkunde zu Bonn 8:139–141.

———. 1872. Über die Spongitarien-Banke der oberen Quadraten und unteren Mukronaten-schichten des Münsterlandes. Festschrift für 20 Hauptvers de Deutschen Geologischen Gesellschaft zu Bonn. 38 p., 1 pl.

———. 1885. Über eine Spongie des reinischen Devon, *Octacium rhenanum* n. g., et. sp. Sitzungberichte der Niederrheinische Gessellschaft, Naturalhistorische und Heilkunde, Verhandlungen, Bonn 42:151–152.

Schmidt, Oscar. 1862. Die Spongien des adriatischen Meeres. Wilhelm Engelmann. Leipzig. 88 p., 7 pl.

———. 1864. Supplement der Spongien des adriatischen Meeres. Enthaltend die Histologie und systematische Ergänzungen. Wilhelm Engelmann. Leipzig. iv + 48 p., pl. 1–4.

———. 1868. Die Spongien der Küste von Algier. Mit Nachträgen zu den Spongien des Adriatischen Meeres (Drittes Supplement). Wilhelm Engelmann. Leipzig. vi + 44 p., 5 pl.

———. 1870. Grundzuge einer Spongien-Fauna des atlantischen Gebietes. Jena. Leipzig. iv + 88 p., 6 pl.

———. 1879. Die Spongien des Meerbusen von Mexico und des Caraibischen Meeres, part 1. Verlag von Gustav Fischer. Jena. 32 p., 4 pl.

———. 1880. Die Spongien des Meerbusen von Mexico (und des Caraibischen Meeres), part 2. Verlag von Gustav Fischer. Jena. p. 33–90, pl. 5–10.

———. 1890. Die pliocänen und glacialen Bildungen am Nordabhang des Monte San Salvatore, vol. 2, no. 1. Eclogae geologicae Helvetiae. Mittheilung der Schweizerische Geologisches Gesellschaft. Lausanne. p. 56.

Schrammen, Anton. 1899. Beitrag zur Kenntnis der obersenonen Tetractinelliden. Mitteilungen aus dem Roemer Museum, Hildesheim 10:9 p., 3 pl.

———. 1901. Neue Kieselschwämme aus der oberen Kreide der Umgebung von Hannover und von Hildesheim. Mitteilungen aus dem Roemer Museum, Hildesheim 14:26 p., 5 pl.

———. 1902. Neue Hexactinelliden aus der oberen Kreide. Mitteilungen aus dem Roemer Museum, Hildesheim 15:26 p., 4 pl.

———. 1903. Zur Systematik der Kieselspongien. Mitteilungen aus dem Roemer Museum, Hildesheim 19:21 p.

———. 1910. Die Kieselspongien der oberen Kreide von Nordwestdeutschland, I Teil, Tetraxonia, Monaxonia und Silicea incert. sedis. Palaeontographica, Supplement 5(1):1–175, 8 fig., 24 pl.

———. 1912. Die Kieselspongien der oberen Kreide von Nordwestdeutschland, II Teil, Triaxonia (Hexactinellida). Palaeontographica, Supplement 5(2):177–385, fig. 9–15, pl. 25–45.

———. 1924a. Die Kieselspongien der oberen Kreide von Nordwestdeutschland, III und letzter Teil. *In* W. Soergel, ed., Monographien zur Geologie und Palaeontologie, serie I, heft 2. Verlag von Gebrüder Borntraeger. Berlin. 159 p., 17 pl.

———. 1924b. Zur Revision der Jura-Spongien von Süddeutschland. Jahresbericht und Mittheilungen Oberrheinischen Geologie (new series) 13:125–154.

———. 1936. Die Kieselspongien des oberen Jura von Süddeutschland. A. Vorwort und Allgemeiner Teil. Palaeontographica (Abt. A) 84:149–194, pl. 14–23 (1–10).

———. 1937 [1936]. Die Kieselspongien des oberen Jura von Süddeutschland. B. Besonderer Teil. Palaeontographica (Abt. A) 85:1–114, pl. 1–17 (11–27).

Schröder, R. 1963 [1962]. Vertikalverteilung des Zooplanktons und thermokline. Archiv fuer Hydrobiologie, supplement 25(4):401–410, 9 fig.

Schulze, F. E. 1880. Untersuchungen über den Bau und die Entwicklung der Spongien. IX Mittheilungen. Die Plakiniden. Zeitschrift für Wissenschaftliche Zoologie 34:407–451, pl. xx–xxii.

———. 1885. The Hexactinellida. *In* T. H. Tizard, H. M. Moseley, J. Y. Buchanan, & J. Murray, eds., Report on the Scientific Results of the Voyage of H.M.S. 'Challenger,' 1873–1876. Narrative 1(1):437–451.

———. 1887a. Report on the Hexactinellida collected by H. M. S. Challenger during the years 1873–1876. Reports of the Scientific Results of the Voyage of H.M.S. Challenger, Zoology 21:513 p., 104 pl., 1 map.

———. 1887b. Über den Bau und das System der Hexactinelliden. Abhandlungen der Königlichen preussischen Akademie der Wissenschaften zu Berlin (Physikalisch-Mathematische Classe) 1886:1–97.

———. 1897. Revision des Systems de Asconematidae und Rossellidae. Konigliche Preussiche Akademie der Wissenschaften, Berlin, Sitzungberichte 26:520–558.

———. 1899. Amerikanische Hexactinelliden nach dem Materials der Albatross-Expedition. Verlag Gustav Fischer. Jena. 126 p., 19 pl.

———. 1904. Hexactinellida. Wissenschaftliche Ergebnisse der Deutschen Tiefsee-Expedition auf dem Dampfer "Valdivia" 1898–1899, vol. 4. Gustav Fisher. Jena. 266 p., 52 pl.

Schulze, F. E., & R. Kirkpatrick. 1910a. Preliminary notice on Hexactinellida of the Gausse-Expedition. Zoologischer Anzeiger 35:293–302.

———. 1910b. Die Hexactinelliden der deutschen Südpolar-Expedition 1901–1903 (Aus. d. Engl. übers.). Deutsche Südpolar-Expedition 1901–1903, G. Reimer, Berlin 12(1):1–62, 10 pl.

Schweigger, A. F. 1819. Beobachtungen auf naturhistorischen Reisen von August Friedrich Schweigger. George Reimer. Berlin. 127 p., 8 pl., 12 tables.

———. 1820. Handbuch der Naturgeschichte skeletloser ungegliedeter Thiere. Leipzig. 766 p.

Sdzuy, Klaus. 1969. Unter- und mittelkambrische Porifera (Chancelloriida und Hexactinellida). Paläontologische Zeitschrift 43(3/4):115–147, 9 fig., pl. 14–16.

Seeley, H. M. 1885. A new genus of Chazy sponges, *Strephochetus*. American Journal of Science (series 3) 30:355–357, 3 fig.

———. 1902. Some sponges of the Chazy Formation. *In* G. H. Perkins, The Geology of Grand Isle. Report of the State Geologist on the mineral industries and geology of certain areas of Vermont, Report 3:151–161, 1 pl.

Seilacher, A. 1962 [1961]. Die Sphinctozoa, eine Gruppe fossiler Kalkschwämme. Akademie der Wissenschaften und der Literatur in Mainz, Abhandlungen der Mathematisch-Naturwissenschaftlichen Klasse jahrgang 1961, 10:721–790, 7 fig., 9 pl.

Selenka, E. 1879. Über einen Kieselschwamm von achtstrahligen Bau, und über Entwickslung der Schwamm-knospen. Zeitschrift für Wissenschaftliche Zoologie 33:467, pl. 27–28.

Semper, C. 1868a. (On *Hyalonema schultzei* and on *Eurete*). Einige neue Kieselschwämme der Philippien. Annals and Magazine of Natural History (series 4) 2:372–373.

———. 1868b. Über einige neue Kieselschwämme der Philippinen. Verhandlungen der physikalisch-medizinische Gesellschaft Würzburg 1:29–30.

———. 1874. Die Stammverwandschaft der Wirbelthieve und Virbelosen. Arbeiten aus dem Zoologisch Zootomischen Institut in Würzburg (new series) 2:25–76, pl. 3–5.

Senowbari-Daryan, Baba. 1978. Neue Sphinctozoen (segmentierte Kalkschwämme) aus den "oberrhätischen" Riffkalken der nördlichen Kalkalpen (Hintersee/Salzburg). Senckenbergiana lethaea 59:205–227.

———. 1980. Neue Kalkschwämme (Sphinctozoen) aus ober-triadischen Riffkalker von Sizilien (Beitrage zur Paläontologie und Microfazies der obertriadischen Riffe des alpin-mediterranen, Gebietes 15). Mitteilungen Gesellschaft der Geologie, Bergbaustudenten 26:179–203.

———. 1981. Zur paläontolgie einer kleinen riffes innerhalb der Amphyclinen-Schichten (Localität: Huda Juzna, Slowenien). Rozpravi IV. Razpreda Slovenska Akademija Znabistu in Umetnosti 23(3):99–118, 1 fig., 10 pl.

———. 1989. Spicula in segmentierten Schwämmen. Berliner Geowissenschaftliche, Abhandlungen (Reihe A) 106:473–515, 4 fig., 14 pl.

———. 1990. Die systematische Stellung der thalamiden Schwämme und ihre Bedeutung in der Erdgeschichte. Münchner Geowissenschaftliche Abhandlungen (Reihe A, Geologie und Paläontologie) 21:1–325, 70 fig., pl. 1–63.

———. 1991. "Sphinctozoa:" an overview. *In* J. Reitner & H. Keupp, eds., Fossil and Recent Sponges. Springer-Verlag. New York, Berlin. p. 224–241, 8 fig.

———. 1994a. Segmentierte Schwämme ("Sphinctozoen") aus der Obertrias (Nor) des Taurus-Gebirges (S-Türkei). Jahrbuch der Geologischen Bundesanstalt 50:415–446.

———. 1994b. Mesozoic sponges of the Pucará Group, Peru. Palaeontographica (Abt. A) 233:57–74, 3 fig., 12 pl.

Senowbari-Daryan, B., & Pietro Di Stefano. 1988. Microfacies and sphinctozoan assemblage of some Lower Permian breccias from the Lercara Formation (Sicily). Rivista Italiana di Paleontologia e Stratigrafia 94:3–34, 8 pl.

Senowbari-Daryan, B., & Theo Engeser. 1996. Ein beitrag zür Nomenklatur sphinctozoider Schwämme (Porifera). Paläontologische Zeitschrift 70(1/2):269–271.

Senowbari-Daryan, Baba, & D. C. Garcia-Bellido. 2002. Fossil 'Sphinctozoa:' chambered sponges (polyphyletic). *In* J. N. A. Hooper & R. W. M. Van Soest, eds., Systema Porifera: A guide to the classification of sponges. Kluwer Academic/Plenum Publishers. New York. p. 1,511–1,533, 13 fig.

Senowbari-Daryan, Baba, & Rucha Ingavat-Helmcke. 1994. Sponge assemblage of some Upper Permian reef limestones from Phrae province (Northern Thailand). Geologija 36:3–59, 13 pl.

Senowbari-Daryan, Baba, & R. P. Reid. 1987. Upper Triassic sponges (Sphinctozoa) from southern Yukon, Stikinia terrane. Canadian Journal of Earth Sciences 24:882–902, 5 fig., 7 pl.

Senowbari-Daryan, Baba, & J. K. Rigby. 1988. Upper Permian segmented sponges from Djebel Tebaga, Tunisia. Facies 19:171–250, 15 fig., pl. 22–40.

Senowbari-Daryan, Baba, & P. Schäfer. 1979. Neue Kalkschwämme und ein Problematikum (*Radiomura cautica* n. g., n. sp.) aus Oberrhät-Riffen südlich von Salzburg (Nördliche Kalkalpen). Mitteilungen der Österreichischen Geologische Gesellschaft 70:17–42, 2 fig., 7 pl.

———. 1986. Sphinctozoen (Kalkschwämme) aus den norischen Riffen von Sizilien. Facies 14:235–284, pl. 44–53.

Senowbari-Daryan, Baba, P. Schäfer, & B. Abate. 1982. Obertriadische Riffe und Rifforganismen in Sizilien. Facies 6:165–184, 4 fig., 3 pl.

Senowbari-Daryan, Baba, Kazem Seyed-Emami, & Ali Aghanabati. 1997. Some inozoid sponges from Upper Triassic (Norian–Rhaetian) Nayband Formation of central Iran. Revista Italiana di Paleontologia e Stratigrafia 103(3):293–321, 9 pl.

Senowbari-Daryan, B., & G. D. Stanley. 1988. Triassic sponges ("Sphinctozoa") from Hells Canyon, Oregon. Journal of Paleontology 63(3):419–423, 3 fig.

———. 1992. New thalamid sponges from the Triassic Luning Formation of Nevada. Journal of Paleontology 66:183–193.

Senowbari-Daryan, B., & Detlef Wurm. 1994. *Radiocella prima* n. g., n. sp., der erste segmentierte Schwamm mit tetracladinen Skelett aus den Dachstein-Riffkalken (Nor) des Gosaukammes (Nördliche Kalkalpen, Österreich). Festschrift zum 60, Geburstag von Erik Flugel. Abhandlungen der Geologischen Bundesanstalt (Austria) 50:447–452, 4 fig., 1 pl.

Senowbari-Daryan, Baba, Rainer Zühlke, Thilo Bechstädt, & Erik Flügel. 1993. Anisian (Middle Triassic) buildups of the Northern Dolomites (Italy): The recovery of reef communities after the Permian/Triassic crisis. Facies 28:181–256, 17 fig., pl. 40–65.

Sherborn, C. D. 1922–1932. Index Animalium, sive index nominum qua ab A. D. MDCCLVIII, generibus et speciebus animalium imposita sunt. Cambridge University Press. Cambridge. Sectio Primo, MDCCLVIII–MDCCC, p. i–lix, 1–1,195; Sectio Secundo, MDCCI–MDCCL, parts 1–28, p. i–cxxxii, 1–7,056; part 29, Epilogue, additions to bibliography, additions and corrections, index to trivialia under genera, p. cxxxiii–cxlvii, 1–1,098.

Sieber, R. A. 1937. Neue Untersuchungen über die Stratigraphie und Ökologie der alpinen Triasfaunen. I. Die Fauna der nordalpinen Rhätriffkalke. Neues Jahrbuch für Mineralogie, Geologie und Paläontologie 78:123–188.

Siemiradzki, J. R. 1913. Die spongien der Polnischen Juraformation. Beiträge zur Paläontologie und Geologie Österreich-Ungarns und des Orients: Mitteilungen des Geologischen und Paläontologischen Institutes der Universität Wien 26:163–211, 8 pl.

Simon, L. 1953. Über die Spezifität der Nadeln und die Variabilität der Arten bei den Spongilliden. Zoologische Jahrbücher für Allgemeine Zoologie 64:207–234, 19 fig.

Simonowitsch, S. 1871. Beiträge zur Kenntniss der Bryozoen des Essener Grünsande. Verhandlungen der Naturhistorischen vereins der preussische Rheinland und Westfalens (series 3) 8:27–34, 1 pl.

Simpson, T. L. 1963. The biology of the marine sponge *Microciona prolifera* (Ellis and Solander). I. A study of cellular function and differentiation. Journal of Experimental Zoology 154:135–151, 3 fig., 2 pl.

Sintzova, I. 1878. O melovykh gubkakh Saratovskoi gubernii (Dopolnenie k state "O yurskikh i melovykh okamenelostyakh Saratovskoi gubernii") [Cretaceous sponges of the Saratov region. Supplement to the article "Jurassic and Cretaceous fossils of the Saratov region"]. Zapiski Novorossiiskogo Obshchestva Estestvoispytatelei 6(1):1–40.

———. 1879. On calcareous sponges from the government of Saratow (Russia). Transactions of the New Russian Society of Natural History 6:1–40, pl. 1–6.

Listed in Rauff, 1893, as Kalkspongien des gouvernements Saratow (Russich). Mémoire, Société Nouvelle Russie, Odessa 6:1–40.

Siribelli, Lycia. 1961. Differenze nell' aspetto sterno e nello scheletro fra *Axinella verrucosa* O.S. e *Axinella damicornia* (exper.) O. S. (Demospongiae). Analli Institute e Museo Zoologica, Universitario Napoli 13(5):1–23, 3 pl.

Sirková, A. 1938a. Houby z kopiri ivnického tithonu na Morave. Rozpravy Ceské Akademie Ved A Umeni, Trída 2, 48(36):1–31, pl. 7.

———. 1938b. Die Schwämme aus dem Koprivicer Tithon in Mahren. Bulletin International de l'Académie Tchèque des Sciences, Prague 39:181–191, 1 pl.

Smith, J. 1911. Carboniferous limestone rocks of the Isle of Man. Transactions of the Geological Society, Glasgow 14(2):119–164.

Smith, J. Toulmin. 1847. On the Ventriculidae of the Chalk; including the description of peculiar characteristics of structure observed in their tissues. Annals and Magazine of Natural History (series 1) 20:74–97, 176–191, pl. 7–8.

———. 1848. On the Ventriculidae of the Chalk; their classification. Annals and Magazine of Natural History (series 2) 1:36–48, 203–220, 279–295, 352–372, pl. 13–16.

van Soest, R. W. M. 1980. Marine sponges from Curaçao and other Caribbean localities. Part II. Haplosclerida. Studies on the Fauna of Curaçao and other Caribbean Islands 62(191):1–173.

———. 1991. Demosponge higher taxa classification re-examined. *In* J. Reitner and H. Keupp, eds., Fossil and Recent Sponges. Springer-Verlag. Berlin. p. 54–71.

Sokolova, M. 1964. Nekotory e zakonomerst raspredeleniya pishchevykh gruppiravok glubokovodnog e bentosa. Okeanologiya 4(6):1,079–1,088.

Sokolova [also Sokoleva], V. Z. 1962. [Feeding resources of benthos in Lake Syamozep]. *In* S. V. Gerd, Transactions of the Syamozer Complex Expedition, vol. 2. ANSSR. Leningrad. p. 36–55, 2 fig.

Sollas, W. J. 1873. On the coprolites of the upper Greensand Formation and on flints. Quarterly Journal of the Geological Society of London (new series) 29:76–81.

———. 1875. Sponges. Encyclopedia Britannica, 9th ed. Adam and Charles Black. Edinburgh. p. 427–446, 26 fig.

———. 1876a. On *Eubrochus clausus,* a vitreo-hexactinellid sponge from the Cambridge "Coprolite" bed. Geological Magazine (new series, Decade II) 3:398–403, pl. 14.

———. 1876b. On the foraminifera and sponges of the Cambridge Upper Greensand. Proceedings of the Cambridge Philosophical Society 1876:299–300.

———. 1877a. On *Pharetrospongia strahani* Sollas, a fossil holoraphidote sponge. Journal of the Geological Society (London) 33:242–255.

———. 1877b. On *Stauronema,* a new genus of fossil hexactinellid sponges, with a description of its two species, *S. carteri* and *S. lobata.* Annals and Magazine of Natural History (series 4) 19:1–25, pl. 1–5.

———. 1878. On the structure and affinities of the genus *Catagma*. Annals and Magazine of Natural History (series 5) 2:353–364.

———. 1880a. On the structure and affinities of the genus *Protospongia*. Quarterly Journal of the Geological Society of London 36:362–367, 2 fig.

———. 1880b. The sponge-fauna of Norway; a report on the Rev. A. M. Norman's collection of sponges from the Norwegian coast. Annals and Magazine of Natural History; Zoology, Botany and Geology (series 5) 5:130–144, pl. 6–7, 11–12, 17.

———. 1880c. On sponge spicules from the Chalk of Trimmingham, Norfolk. Reports of the British Association for the Advancement of Science 1880:586–587.

———. 1880d. On the flint nodules of the Trimmingham Chalk. Annals and Magazine of Natural History; Zoology, Botany and Geology (series 5) 6:384–395, pl. 19–20.

———. 1881. On *Astroconia granti,* a new lyssakine hexactinellid from the Silurian formation of Canada. Quarterly Journal of the Geological Society, London 37:254–260, fig. 1–11.

———. 1883. Descriptions of fossil sponges from the Inferior Oolite, with a notice of some from the Great Oolite. Quarterly Journal of the Geological Society of London (new series) 39:541–554, pl. 20–21.

———. 1884. On the development of *Halisarca lobularis.* Quarterly Journal of Microscopical Science 24:603–621, fig. 1–39, pl. 37.

———. 1885a. A classification of the sponges. Annals of Natural History (series 5) 16:395.

———. 1885b. On *Vetulina stalactites* (O.S.) and the skeleton of the Anomocladina. Proceedings of the Royal Irish Academy of Science (series 2) 4(4):486–492, pl. 3–4.

———. 1886. Preliminary account of the tetractinellid sponges dredged by the 'Challenger' 1872–1876. Scientific Proceedings of the Royal Dublin Society (new series) 5:177–199.

———. 1887. Sponges. Encyclopaedia Britannica, 9th ed., vol. 22. Adam and Charles Black. Edinburgh. p. 412–429, 26 fig.

———. 1888. Report on the Tetractinellida collected by H. M. S. Challenger during the years 1873–1876. Report on the Scientific Results of the voyage of H. M. S. Challenger during the years 1873–76, Zoology, vol. 25. London, Edinburgh, & Dublin. clxvi + 458 p., 44 pl., 1 map.

Solle, Gerhard. 1938. Die ersten Bohr-Spongien im europäischen Devon und einige andere Spuren. Senckenbergiana 20:154–178, 22 fig.

Sorokin, J. I. 1964. Quantitative study of the microflora in the Central Pacific Ocean. Journal Conseil 29:25–60.

Spreisterbach, J. 1935. Beitrage zur Kenntnis der Fauna des rheinischen Devon. Jahrbuch, Preussische Geologische Landesanstalt 1935:447–483, 2 fig., 2 pl.

Stearn, C. W., & J. W. Pickett. 1994. The stromatoporoid animal revisited: building the skeleton. Lethaia 27:1–10.

Steiner, M. D., D. Mehl, J. Reitner, & B.-D Erdtmann. 1993. Oldest entirely preserved sponges and other fossils from the lowermost Cambrian and a new facies reconstruction of the Yangtze platform (China). Berliner Geowissenschaftliche Abhandlungen (series E) 9:293–329, 13 fig., 7 pl.

Steinmann, Gustav. 1878. Über fossile Hydrozoen aus der Familie der Coryniden. Paläontographica 25(3):101–124.

———. 1881. Über *Prototetraclis linki.* n. f., eine Lithistide des Malms. Neues Jahrbuch 2:154–163, pl. 9.

———. 1882. Pharetronen-Studien. Neues Jahrbuch für Mineralogie, Geologie und Palaeontologie 2:139–191, pl. 6–9.

———. 1913. Pharetronen. *In* F. Toula, Die Kalke von Jägerhause unweit Baden (Rauchstallbrunnengraben) mit nordalpiner St. Cassianer Fauna. Jahrbuch der Kaiserlich-Königliche Geologischen. Reichsanstalt, Wien 63:86–89, fig. 1–4.

Stempien, M. F. Jr. 1960. The nucleic acids of sponges. Annals of the New York Academy of Science 50:910–912.

Stewart, C. 1870. On a new sponge, *Tethyopsis columnifer.* Quarterly Journal of Microscopical Science, London (new series) 10:281–282.

Stoehr (Stöhr), P. A. 1880. Die Radiolarienfauna der Tripoli von Grotte Provinz Gergenti in Sicilien. Palaeontographica 26 (1879–1880):69–124.

Stoppani, Antonio. 1858. Les Petrifications d'Ésino ou description des fossiles appartenant au dépot Triasique superieur des environs d'Ésino en Lombardie. Milan. p. 126–131, pl. 29–31.

———. 1860. Risultai paleontologici dedotti dallo studio dei petrefatti d'Esino: 1, Del parallelismo del deposito di Esino con altri fuori de Lombardia. 2, Del posto occupato dal deposito de Esino nella serie stratigraphica de Lombardia. Atti della Societa Italiana di Scienze naturali, Milano 2(1859–1860):65–92.

Strand, E. 1928. Miscellanea nomenclatorica zoologica et palaeontologica. Archiv für Naturgeschichte, Berlin 72A:31–36.

Stuckenberg, A. 1895. Korallen und Bryozoen der Steinkohlenablagerungen des Ural und des Timan. Mémoires du Comité Géologique, St. Petersbourg 10(3):244 p., 24 pl.

Tanaka-Ichihara, Keiko, & Yoko Watanabe. 1990. Gametogenic cycle in *Halichondria okadai. In* Klaus Rutzler, ed., New perspectives in sponge biology, Smithsonian Press. Washington, D.C. & London. p. 170–174, 3 fig.

Tate, Ralph. 1865. On the correlation of the Cretaceous formations of the north-east of Ireland. Quarterly Journal of the Geological Society of London 21:15–44, pl. 3–5.

Taylor, S., & J. G. Wells. 1975. Cowries. T. F. H. Publications Inc., Ltd. Neptune City, New Jersey. 288 p., pl.

Teichert, Curt. 1958. Cold- and deep-water coral banks. American Association of Petroleum Geologists Bulletin 42(5):1,064–1,082.

Termier, Henri, & Geneviève Termier. 1955. Contribution à l'étude des Spongiaires permien du Djebel Tebaga (Extrème Sud Tunisien). Bulletin de la Société Géologique de France (series 6) 5:613–630.

———. 1973. Stromatopores, sclérosponges et Pharétrones: les Ischyrospongia. Annales des Mines et la Geologie, Tunis 26:285–297.

———. 1974. Sponges permiens du Djebel Tebaga (sud Tunisien). Comptes Rendus de l'Académie des Sciences, Paris (series D) 279:247–249.

———. 1977a. Paléontologie des Invertébrés. *In* Henri Termier, Geneviève Termier, & D. Vachard, Monographie paléontologique des affleurements Permiens du Djebel Tébaga (sud Tunisien). Palaeontographica (Abt. A) 156:25–99.

———. 1977b. Structure et évolution des spongiaires hypercalcifiés du Paléozoïque supérior. Mémoires de l'Institut Géologique de l'Université de Louvain 29:57–109.

———. 1980. Stromatopores, trépostomes et tabuliatomorphes du Paléozoïque d'Afrique du nord. Annales de Paléontologie (Invertébrés) 66(1):1–16, 4 pl.

———. 1981. Descriptions of species. *In* H. Termier, G. Termier, & H. H. Tsien, Spongiaires des calcaires récifaux du Frasnien de l'Ardenne. Bulletin de la Société belge de Géologie 90(4):290–292, fig. 3, pl. 3.

Termier, Henri, Geneviève Termier, & H. H. Tsien. 1981. Spongiaires des calcaires récifaux du Frasnien de l'Ardenne. Bulletin de la Société belge de Géologie 90(4):287–298, fig. 3, pl. 3.

Termier, Henri, Geneviève Termier, & J. Thibieroz. 1990. Hexactinella Lyssakida liassiques de la bordure sud-est des Cevennes. Bulletin Trimestriel de la Société Géologique de Normandie et des Amis du Muséum du Havre 77(3–4):5–17, 1 fig., 4 pl.

Termier, Henri, Geneviève Termier, & Daniel Vachard. 1977a. Monographie paléontologique des affleurements permiens du Djebel Tebaga (sud Tunisien). Palaeontograpica (Abt. A) 156:1–109, 52 fig., 18 pl.

———. 1977b. Étude comparative de quelques ischryosponges. Géologie Méditerranéenne 4(2):139–180.

Thomas, A. O. 1922. Some new Paleozoic glass-sponges from Iowa. Proceedings of the Iowa Academy of Science 29:85–91, pl. 1.

Thomson, C. W. 1868. On the vitreous sponges. Annals and Magazine of Natural History (series 4) 1(2):114–132, pl. 4.

———. 1873a. The depths of the sea. Macmillan and Co. London. 527 p., 84 fig.

———. 1873b. Notes from the 'Challenger,' I–VII. Nature 8:28–30, 51–53, 109–110, 246–249, 266–267, 347–349, 400–403.

———. 1877. The voyage of the 'Challenger,' The Atlantic, a preliminary account of the general results of the exploring voyage of H.M.S Challenger during the year 1873 and the early part of the year 1876. Macmillan and Co. London. vol. 1., p. 1–424; vol. 2, p. 1–396.

Tiwari, Meera. 1997. *Nabaviella acanthomorpha* n. sp., a sponge spicule from the Precambrian-Cambrian boundary interval in the Tethys sequence of northwestern Kashmir. Journal Geological Society of India 50:655–658, 2 fig.

———. 1999. Organic-walled microfossils from the Chert-phosphorite Member, Tal Formation, Precambrian-Cambrian boundary, India. Precambrian Research 97:99–113, 4 fig.

Tiwari, Meera, C. C. Pant, & V. C. Tewari. 2000. Neoproterozoic sponge spicules and organic-walled microfossils from the Gangolihat Dolomite, Lesser Himalaya, India. Current Science 79(5):651–654.

Toomey, D. F. 1964. Ellenburger (Lower Ordovician) sponge beds of central Texas. Tulsa Geological Society Digest 32:98–111, 2 fig., 3 pl.

———. 1970. An unhurried look at a Lower Ordovician mound horizon, southern Franklin Mountains, west Texas. Journal of Sedimentary Petrology 40:1,318–1,334, 15 fig.

Toomey, D. F., & R. M. Finks. 1969. The paleoecology of Chazyan (lower Middle Ordovician) "reefs" or "mounds" and Middle Ordovician (Chazyan) mounds, southern Quebec, Canada, a summary report. New York State Geological Association Guidebook, 41st Annual meeting, Plattsburgh, New York. State University of New York, Department of Geology. Brockport. p. 93–134.

Toomey, D. F., & M. H. Nitecki. 1979. Organic buildups in the Lower Ordovician (Canadian) of Texas and Oklahoma. Field Museum of Natural History, Fieldiana, Geology (new series) 2:1–181, 85 fig.

Topsent, Émile. 1892. Contribution à l'étude des Spongiaires de l'Atlantique Nord, fasc. II. Résultats des Campagnes scientifiques accomplies sur son yacht par Albert I, Prince souverain de Monaco. Publiés sous sa direction, avec the concours du Baron Jul. de Guerne. Monaco. 165 p., 11 pl.

———. 1898. Introduction à l'étude monographique des Monaxonides de France, Classification des Hadromerina. Archives de Zoologie expérimentale et générale 4(3):91–113.

———. 1901a. Les Spongiaires de l'expedition antarctique belge et la biopolarité des faunes. Comptes Rendus, Académie des Sciences 132(3):168–169.

———. 1901b. Considérations sur la Faune des Spongiaires des Côtes d'Algérie: Éponges de la Calle. Archives de Zoologie expérimentale et générale (series 3) 9:327–370.

———. 1902. Spongiaires. Expédition Antarctique Belge, Résultats du voyage du S. Y. Belgica en 1897–1898–1899, Rapports Scientifiques, Zoologie. J. E. Buschmann. Anvers. 54 p., 6 pl.

———. 1904. Spongiaires des Açores, fasc. XXV. Résultats des Campagnes Scientifiques accomplies sur son yacht par Albert I., Prince souverain de Monaco. Publiés sous sa direction, avec the concours du M. Jules Richard. Monaco. Imprimerie de Monaco. p. 1–280, 18 pl.

———. 1913. Spongiaires provenant des Campagnes scientifiques de la *Princess-Alice* dans les Mers du Nord (1898–1899–1906–1907). Résultats des Campagnes Scientifiques du Prince de Monaco 45:1–67, 5 pl.

———. 1920a. Spongiaires du Musée zoologique de Strasburg. Monaxonides. Bulletin de l'Institut océanographique, Monaco 381:1–36.

———. 1920b. *Tethya aurantium* et les *Tethya* de Lamarck. Bulletin du Muséum d'Histoire Naturelle, Paris 1920:640–646.

———. 1922. Les mégasclères polytylotes des Monaxonides et la parenté de Latrunculines. Bulletin du l'Institut océanographique, Monaco 415:1–8.

———. 1928a. Une Mycaline productrice de desmes, *Desmatiderma arbuscula,* n. g., n. sp. Bulletin de l'Institut Océanographique, Monaco 519:1–8.

———. 1928b. Spongiaires de l'Atlantique et de la Mediterranée. Résultats des Campagnes Scientifiques du Prince Albert 1st de Monaco 74:1–376, 11 pl.

———. 1928c. Sur deux Eurétides du Japan (Note préliminaire). Bulletin de l'Institut océanographique, Monaco 74:1–4.

Trammer, Jerzy. 1979. Some aspects of the biology of fossil solid-branching demosponges, exemplified by *Reiswigia ramosa* gen. n., sp. n., from the Lower Oxfordian of Poland. Acta Geologica Polonica 29(1):39–49, fig. 1–5, pl. 1–3.

———. 1989. Middle to Upper Oxfordian sponges of the Polish Jura. Acta Geologica Polonica 39(104):49–91.

Trautschold, H. 1870. Palaeontologischer Nachtrag zu der Abhandlung des Fürsten P. Krapotkin über die geognostischen Verhaltnisse des Kreises Meschtschowsk im Gouvernement Kaluga. Bulletin de la Société Impérial des Naturalistes de Moscou 42 (année 1869, no. 4):230–233, 1 pl.

Trébougoff, G. 1942. Contribution à la connaissance des larves planctoniques d'éponges. Archives de Zoologie Expérimentale et Générale 82:357–399, 7 fig., pl. ix–x.

Trejo, M. 1967. La esponja fosil *Rhaxella sorbyana* (Blake) y su significacion estratigrafica. Boletin de la Asociacion Mexicana de geologos petroleros 19:33–38, 7 pl.

Trest'yan, G. N. 1972. New Upper Cretaceous glass sponges (Hyalospongiae) from the Dniester region. Paleontologicheskii Zhurnal 1972(2):32–41, 4 fig. In Russian, translated into English in 1972, Paleontological Journal 6(2):171–179.

van Trigt, H. 1919. A contribution to the physiology of the fresh-water sponges (Spongillidae). Leiden Tijdschrift Nederlands Dierk. Verhandelingen (series 2) 17:1–220, 6 pl.

Tschernyschev, T. 1898. Über die Artinsk- und Carbon-Schwämme vom Ural und vom Timan. Bulletin of the Academy of Science, St. Petersbourg (series 5) 9:1–36, pl. 1–5.

Tschernyschev, T., & R. Stepanov. 1916. Obercarbon Fauna von Konig Oskars und vom Heibergs Land. *In* Report of the Second Norwegian Arctic Expedition in the "Fram" 1898–1902, vol. 4. Videnskabs-Selskabet i Kristiana (Society of Arts and Sciences of Kristiana). A. W. Brogger. Oslo. p. 1–67, 12 pl.

Tuzet, Odette. 1973a. Introduction et place des spongiaires dans la classification. *In* P.-P. Grassé, ed., Traité de Zoologie. Anatomie, Systématique, Biologie, III. Spongiaires. Masson et Cie. Paris. p. 1–26, fig. 1–10.

———. 1973b. Éponges calcaires. *In* P.-P. Grassé, ed., Traité de Zoologie. Anatomie, Systématique, Biologie, III. Spongiaires. Masson et Cie. Paris. p. 27–132, fig. 12–89.

Tuzet, O., & R. Connes. 1962. Spicules anormaux d'une variété écologique d'*Ephidatia fluviatilis* Linné. Vie et Milieu 13:467–470, 1 fig.

Tuzet, O., & J. Paris. 1964. La spermatogenèse, l'ovogenèse, la fécondation et les premiers stades du developpement chez *Octavella galangaui* Tuzet et Paris. Vie et Milieu 15:309–327, 9 fig.

Tuzet, O., & M. Pavans de Ceccatty. 1959. La spermatogenèse, l'ovogenèse, la fecondation et les premiers stades du développement d'*Hippospongia communis* Lmk. (=*H. equina* O.S.). Bulletin Biologique 92:1,331–1,348, 5 fig.

Twenhofel, W. H. 1928. The geology of Anticosti Island. Memoir of the Geological Survey of Canada 154:481 p., 60 pl.

Ulbrich, Hans. 1974. Die Spongien der Ilsenburg-Entwicklung (Oberes Unter-Campan) der Subherzynen Kreidemulde. Freiberger Forschungshefte (Paläontologie) C291:121 p., 19 pl.

Ulrich, E. O. 1878. Descriptions of new genera and species of fossils from the Lower Silurian about Cincinnati. Journal of the Cincinnati Society of Natural History 2:8–30.

———. 1879 [1878]. Descriptions of some new species of fossils from the Cincinnati group. Journal of the Cincinnati Society of Natural History 1:92–100.

———. 1889. Preliminary description of new lower Silurian sponges. The American Geologist 3:233–248, fig. 1–7.

———. 1890a. American Paleozoic sponges. Illinois Geological Survey (Paleontology of Illinois, part 2, section 3) Bulletin 8:209–241.

———. 1890b. Sponges of the Devonian and Carboniferous systems. Illinois Geological Survey (Paleontology of Illinois, part 2, section 4) Bulletin 8:242–253.

Ulrich, E. O., & Oliver Everett. 1890. Lower Silurian sponges. Illinois Geological Survey (Paleontology of Illinois, part 2, section 5) Bulletin 8:255–282.

Ushakov, P. V. 1955. Polychaetes of the far-eastern seas of the Soviet Union. Ed. ZIN, Akademiya Nauk SSSR (Academy of Science, USSR), Tabulation and Analysis of the fauna of the USSR 56:445 p., 164 fig.
In Russian.

Vacelet, J. 1961. Quelques éponges remarquables de Méditerranée. Revue des Travaux, Institut des Pêches Maritimes 25:351–354.

———. 1962. Existence de formations de réserve chez une éponge calcaire pharétronide. Comptes Rendus, Académie des Sciences, Paris 254:2,425–2,426, 3 fig.

———. 1965 [1964]. Étude monographique de l'éponge calcaire pharétronide de Méditerranée, *Petrobonia massiliana* Vacelet et Lévi. Les pharétronides actuelles et fossiles. Recueil des Travaux aux de la Station Marine d'Endoume 50(34):1–125.

———. 1967a. Descriptions d'éponges pharétronides actuelles des tunnels obscurs sous-récifaux de Tuléar (Madagascar). Recueils et Travaux de la Station Marine d'Endoume 6:37–62, 6 fig., 13 pl.

———. 1967b. Quelques éponges pharétronides et "silico-calcaires" des grottes sous-marines obscures. Recueils et Travaux de la Station Marine d'Endoume 58(42):121–132.

———. 1969. Éponges de la roche du large et de l'étage bathyal de Méditerranée. Muséum National d'Histoire Naturelle (Paris), Mémoires (series A) 59:145–219, 54 fig., 12 pl.

———. 1977a. Éponges pharétronides actuelles et sclérosponges de Polynésie française, de Madagascar et de la Réunion. Bulletin du Muséum National d'Histoire Naturelle (Paris), Zoologie 307(444):345–367, 7 fig., 2 pl.

———. 1977b. Une nouvelle relique du secondaire: un répresentant actuel des éponges fossiles sphinctozaires. Comptes Rendus Hebdomaires des Séances de l'Académie des Sciences, Paris 285(5):509–511, 1 pl.

———. 1979. Description et affinités d'une éponge sphinctozoaire actuelle. *In* C. Lévi & N. Boury-Esnault, eds., Biologie des Spongiaires, Colloques Internationaux du Centre National de la Recherche Scientifique 291:483–493.

———. 1980. Squelette calcaire facultatif et corps de régénerations dans le genre *Merlia,* éponges apparentées aux chaetétides fossiles. Comptes Rendus Hebdomadaires des Séances, Académie des Sciences, Paris (series D) 290(3):227–230.

———. 1981. Éponges hypercalcifiées ('pharétronides', 'sclerosponges') des cavitiés des récifs coralliens de Nouvelle-Caledonie. Bulletin, Muséum National d'Histoire Naturelle (Paris) Zoologie, Biologie et Ecologie Animales 3(2):313–351.

———. 1983. Les éponges hypercalcifiées, reliques des organismes constructeurs de récifs du Paléozoïque et du Mesozoïque. Bulletin de la Société Zoologique de France 108:547–557.

———. 1985. Coralline sponges and the evolution of Porifera. *In* S. Conway-Morris, ed., Organisms and Relationships of Lower Invertebrates. Systematics Association Special Volume 28:2–13.

———. 1991. Recent Calcarea with a reinforced skeleton ("Pharetronids"). *In* J. Reitner and H. Keupp, eds., Fossil and Recent Sponges, Springer-Verlag. Berlin. p. 252–265.

———. 1994. Porifera. *In* C. Juberthie & V. Decu, eds., Encyclopaedia Biospeologica, vol. 1. Société de Biospéologie. Saint-Girons, France. p. 36–38.

Vacelet, J., N. Boury-Esnault, & J.-G. Harmelin. 1994. Hexactinellid Cave, a unique deep-sea habitat in the scuba zone. Deep Sea Research 1, 41(7):965–973.

Vacelet, Jean, & Claude Lévi. 1958. Un cas de survivance, en Méditerranée, du groupe d'éponges fossiles des Pharétronides. Comptes Rendus hebdomadaires des Séances de l'Académie des Sciences, Paris 246(2):318–320, 3 fig.

Vacelet, J., & M. J. Uriz. 1991. Deficient spiculation in a new species of *Merlia* (Merliida, Demospongiae) from the Balearic Islands. *In* J. Reitner & H. Keupp, eds., Fossil and Recent Sponges. Springer-Verlag. Berlin. p. 170–178.

Vacelet, J., & P. Vasseur. 1971a. Éponges des récifs coralliens de Tuléar (Madagascar). Tethys, Supplement 1:51–126, 77 fig., 4 pl.

———. 1971b. Spongiaires des grottes et surplombs des récifs de Tuléar (Madagascar). Recueils et Travaux de la Station Marine d'Endoume (Fascicule Hors Serie) Supplement 4:71–123.

Vachard, D., & C. Montenat. 1981. Biostratigraphie, micropaléontologie et paléogéographie du Permien de la région de Tezak (Montagnes Centrales d'Afghanistan). Palaeontographica (Abt. B) 178:1–88, 15 pl.

Vandercammen, Antoine. 1950. Contribution à l'étude des spongiaires hétéractinellides. Bulletin de l'Institut Royal des Sciences de la Société Naturelle de Belgique 26 (19):1–23, 1 fig., 6 pl.

Van de Graaf, W. J. E. 1969. Carboniferous Sphinctozoa from the Cantabrian Mountains, Spain. Leidse Geologische Mededelingen 42:239–257, 2 pl.

Vanuxem, Lardner. 1842. Geology of New York. Part 3. Comprising a survey of the third geological district. New York Geological Survey. Albany. 306 p.

Van Weel, P. B. 1949. On the physiology of the tropical fresh water sponge *Spongilla proliferans.* I. Ingestion, digestion and excretion. Physiologia Comparata Oecologia 1:110–126.

Vasserot, Jean. 1961. Caractère hautement spécialisé du régime alimentaire chez les astérides *Echinaster sepotus* et *Henricia sanguinolenta,* prédateurs de spongiaires. Bulletin de la Société Zoologique de France 86(6):796–809, 5 fig.

Vasseur, G. 1880. Reproduction asexuelle de la *Leucosolenia botryoides.* Archives de Zoologie expérimentale et générale, Paris 8:59–65.

Vaughan, T. W., & J. W. Wells. 1943. Revision of the suborders, families, and genera of the Scleractinia. Geological Society of America Special Paper 44:363 p., 39 fig., 51 pl.

Veizer, V., & J. Wendt. 1976a. Mineralogy and chemical composition of Recent and fossil skeletons of calcareous sponges. Neues Jahrbuch für Geologie und Paläontologie, Monatshefte 1976(9):558–573, 8 fig.

———. 1976b. The Southern Bükk (N. Hungary) Triassic revisited: The Beruavölgy limestone. Annales Section Geologique 27:16–65, 15 pl.

Vernadsky, W. J. 1934 [1933]. Ozeanographie und Geochemie. Mineralogische und Petrographische Mitteilungen 44(2–3):168–192.

Verrill, A. E. 1907. Porifera of the Bermuda Islands. New Haven, Connecticut, Transactions of the Academy of Arts and Sciences 12:330–344.

Vinassa de Regny, P. 1901. Trias-spongien aus dem Bakony. Resultatae der Wissenschaften Erforschung der Balatonsees, I, Palaeontologie der Umgebung des Balatonsees, vol. 1. Wien. 22 p., 7 fig., 3 pl.

———. 1911a. Trias-Spongien aus dem Bakony. *In* Resultate der wissenschaftlichen Untersuchungen des Balaton (Plattensee), Anhang I, Band 1911, no. 2. Wien. p. 1–22, 3 pl.

———. 1911b. Neue Schwamme, Tabulaten und Hydrozoen aus dem Bakony. *In* Resultate der wissenschaftlichen Untersuchungen des Balaton (Plattensee), Anhang l, Band 1911, no. 3. Wien. p. 1–18, 4 pl.

———. 1915. VIII. Triadische Algen, Spongien, Anthozoen und Bryozoen aus Timor. Paläontologie von Timor, vol. 4, no. 8. M. Wanner. Stuttgart. p. 75–118, pl. 63–72.

Vinogradov, A. P. 1953. The elementary chemical composition of marine organisms. Sears Foundation for Marine Research, Memoir 2:176–193.

Vinogradov, M. E. 1959a. Über die quantitative Verbreitung des Tiefseeplanktons im nordwestlichen Teil des Stillen Ozeans. Internationale Revue der Hydrobiologie 44(2):217–225, 3 fig.

———. 1959b. On the vertical distribution of deep-sea plankton in the west part of the Pacific Ocean. Proceedings of the International Congress of Zoology 15:223–225.

Vinogradov, N. G. 1958. Vertikalnoe raspredelenie glubokovodnogo donnai faune okeana. Trudy Institut Okeanologii Akademii Nauk SSSR, 1958:86–122.

Volkmer, C. 1963. *Spongilla jewelli* n. sp. from freshwater sponge at Brazil. Anais Academia Brasileira de Ciencia 35(2):271–273.

Volkmer-Ribeiro, C. 1970. *Oncosclera*—a new genus of freshwater sponges (Porifera-Spongillidae) with redescription of two species. Amazoniana 2(4):435–442.

Von Brand, Theodor. 1939 [1938]. Further experiments on the decomposition and regeneration of nitrogenous organic matter in sea water. Biological Bulletin 77:285–296.

Von der Marck, 1876 [1873]. Neue Beiträge zur Kenntniss der fossilen Fische und anderer Theirreste. Palaeontographica 22:55–74, pl. 2,*10.*

Vosmaer, G. C. J. 1880. The sponges of the Leyden Museum, I, The Family of the Desmacidinae. Notes from the Leyden Museum [Rijksmuseum voor de Geschiedenis der Natuurwetenschappen, Leyden] 2:99–136.

———. 1882. Spongien (Porifera). *In* H. G. Brönn, ed., Die Klassen und Ordnungen des Thierreichs, part 1. H. G. Brönn. Leipzig & Heidelberg. p. 1–32, pl. 1,2,4.
Translated by A. Dendy, Annals and Magazine of Natural History (series 5) 19:249–260.

———. 1883. Spongien (Porifera). *In* H. G. Brönn, ed., Die Klassen und Ordnungen des Thierreichs, part 2. H. G. Brönn. Leipzig & Heidelberg. p. 33–64, pl. 5–6.
Translated by A. Dendy, Annals and Magazine of Natural History (series 5) 19:249–260.

———. 1884. Spongien (Porifera). *In* H. G. Brönn, ed., Die Klassen und Ordnungen des Thierreichs, parts 3–6. H. G. Brönn. Leipzig & Heidelberg. p. 65–176, pl. 3,7,8.
Translated by A. Dendy, Annals and Magazine of Natural History (series 5) 19:249–260.

———. 1885. Spongien (Porifera). *In* H. G. Brönn, ed., Die Klassen und Ordnungen des Thierreichs, parts 7–10. H. G. Brönn. Leipzig & Heidelberg. p. 177–320, pl. 19–25.
Translated by A. Dendy, Annals and Magazine of Natural History (series 5) 19:249–260.

———. 1887. Spongien (Porifera). *In* H. G. Brönn, ed., Die Klassen und Ordnungen des Thierreichs, parts 11–16. H. G. Brönn. Leipzig & Heidelberg. p. 321–496.
Translated by A. Dendy, Annals and Magazine of Natural History (series 5) 19:249–260.

Vosmaer, G. C. J., & H. P. Wijsman. 1904. Ober den Bouw van sommige Kiezelspicula bij Sponzen. I. De styli van *Tethya lyncurium.* Koninklijke Akademie van Wetenschappen te Amsterdam. Verslag van de gewone Vergaderingen der Wisen Natuurkundige Afdeeling. Published by the society. Amsterdam. p. 733–748, 1 fig.

Waagen, W., & J. Wentzel. 1888. Salt-Range Fossils. Productus Limestones: Coelenterata, Amorphozoa, Protozoa. Memoirs of the Geological Survey of India 1(4\5):854–998.

Wagner, Wolfgang. 1963. Die schwammfauna der Oberkreide von Neuberg (Donau). Palaeontographica (Abt. A) 122:166–250, 5 pl.

———. 1964. Kalkschwämme aus dem Korallenkalk des oberen Malm von Laisacker bei Neuburg a. d. Donau. Mitteilungen der Bayer. Staatssammlung für Paläontologie und historische Geologie 4:23–36, 7 fig., pl. 5–7.

Wähner, F. 1903. Das Sonnwendgebirge im Unterinntal, ein Typus eines alpinen Gebirgsbaues. F. Deuticke. Leipzig-Wien. 356 p., 19 pl., 96 fig.

Walch, J. E. I. 1768–1776. Recueil des monuments des catastrophes que le globe de la terre a éssuiées contenant des pétrefactions et d'autres pierres curieuses, (commencé) par G. W. Knorr (et continué par J. E. I. Walch), vol. II–IV. Nuremburg.

Walcott, C. D. 1879. Fossils of the Utica Shale. Transactions of the Albany Institute 10:18–19, fig. 16–18, pl. 2.

———. 1886. Second contribution to the studies of the Cambrian faunas of North America. United States Geological Survey Bulletin 30:369 p.

———. 1892. Preliminary notes on the discovery of a vertebrate fauna in Silurian (Ordovician) strata. Bulletin of the Geological Society of America 3:153–172, pl. 3–5.

———. 1912. Notes on fossils from limestones of Steeprock series, Ontario, Canada. Canada Department of Mines, Geological Survey Branch, Memoir 28:16–23, pl. 1–2.

———. 1919. Cambrian geology and paleontology, IV. Middle Cambrian algae. Smithsonian Miscellaneous Collections 67(5):217–260.

———. 1920. Cambrian Geology and Paleontology. IV. no. 6. Middle Cambrian Spongiae. Smithsonian Miscellaneous Collections 67(6):261–364, fig. 4–10, pl. 60–90.

Walker, W. R., & R. K. Bambach. 1974. Analysis of communities. Sedimenta 4. Principles of benthic community analysis. *In* A. M. Ziegler, K. R. Walker, E. J. Anderson, E. G. Kauffman, R. N. Ginsberg, and N. P. James. Sedimenta 4, Principles of benthic community analysis, Notes for a short course. Division of Marine Geology, Rosenstiel School of Marine and Atmospheric Sciences, University of Miami. Miami, Florida. p. 2.1–2.20, 10 fig.

Wallace, S. J. 1878. On the "geodes" of the Keokuk Formation and the genus *Biopalla,* with some new species. American Journal of Science (series 3) 15:366–370.

Walther, J. 1893–1894. Einleitung in die Geologie als historische Wissenschaft, I Bionomie des Meeres, II. Die Lebenweise der Meeresthiere. G. Fischer. Jena. xxx + 531 p.

———. 1904. Die Fauna der Solnhofener plattenkalke. Denkschriften der Medecinisch-naturwissenschaftlichen Gesellschaft, vol. 11. Festschrift E. Haeckel. Jena. p. 161–163.

Warburton, F. E. 1960. Influences of currents on form of sponges. Science 133:89, 1 fig.

Watanabe, Y. 1957. Development of *Tethya serica* Lebwohl, a tetraxonian sponge. Observations on external changes. Natural Science Report, Ochanomizu University 8:97–104.

Webby, B. D. 1969. Ordovician stromatoporoids from New South Wales. Palaeontology 12:637–662, pl. 117–129.

Webby, B. D., & Lin Baoyu. 1988. Upper Ordovician cliefdenellids (Porifera: Sphinctozoa) from China. Geological Magazine 125:149–159, 6 fig.

Webby, B. D., & J. K. Rigby. 1985. Ordovician sphinctozoan sponges from central New South Wales. Alcheringa 9:209–220, 10 fig.

Webby, B. D., & J. Trotter. 1993. Ordovician sponge spicules from New South Wales, Australia. Journal of Paleontology 67:28–41, 7 fig.

Weidlich, O., & B. Senowbari-Daryan. 1996. Late Permian "sphinctozoans" from reefal blocks of the Ba'id area, Oman Mountains. Journal of Paleontology 70:27–46.

Weller, J. M. 1930. Siliceous sponge spicules of Pennsylvanian age from Illinois and Indiana. Journal of Paleontology 4:233–251, pl. 15–20.

Wells, J. W. 1957. Corals. *In* J. W. Hedgpeth, ed., Treatise on Marine Ecology and Paleoecology. Volume l, Ecology. Geological Society of America Memoir 67:1,087–1,104, 1 fig.

Welter, O. A. 1911 [1910]. Die Pharetronen aus dem Essener Grünsand. Verhandlungen des Naturhistorischen Vereins der preussischen Rheinlande und Westfalens 67:1–82, 10 fig., 3 pl.

Wendt, J. 1974. Der Skelettbau aragonitischer Kalkschwämme aus der alpinen Obertrias. Neues Jahrbuch für Geologie und Palaeontologie, Monatshefte 1974:498–511, fig. 1–9.

———. 1979. Development of skeletal formation, microstructure, and mineralogy of rigid calcareous sponges from the Late Palaeozoic. Colloque Internationaux, Centre National de la Recherche Scientifique 291:449–457.

Whitelegge, T. 1901. Report on sponges from the coastal beaches of New South Wales. Records of the Australian Museum 4(2):55–118, pl. 10–15.

Whitfield, R. P. 1881. Remarks on *Dictyophyton* and descriptions of new species of allied forms from the Keokuk beds, at Crawfordsville, Indiana. American Museum of Natural History Bulletin 1:10–20.

———. 1886. Notice of a new fossil body, probably a sponge related to *Dictyophyton.* Bulletin of the American Museum of Natural History 1(8):346–348, pl. 35.

———. 1905. Descriptions of new fossil sponges from the Hamilton group of Indiana. American Museum of Natural History Bulletin 21:301–303.

Wiedenmayer, F. 1974. Recent marine shallow-water sponges of the West Indies and problems of speciation. *In* Contributions to the geology and paleoecology of the Carribean and adjacent areas. Verhandlungen der Naturforschenden Gesellschaft in Basel 84(1):361–376.

———. 1977a. Shallow-water sponges of the western Bahamas. Experientia, Supplementum 28:1–287, 43 pl.

———. 1977b. The Nepheliospongiidae Clarke, 1900 (Demospongea, Upper Devonian to Recent), an ultraconservative, chiefly shallow-water sponge family. Eclogae Geologicae Helvetiae 70(3):885–918.

———. 1994. Contributions to the knowledge of post-Paleozoic neritic and archibenthal sponges (Porifera). Kommission der Schweizerischen Palaeontologischen Abhandlungen 116:5–140.

Wilckens, O. 1937. Beiträge zur Paläontologie des Ostindischen Archipels, XIV. Korallen und Kalkschwämme aus dem obertriadischen Pharetronenkalk von Seran (Molukken). Neues Jahrbuch für Mineralogie, Geologie und Paläontologie (Abt. B) 77:171–211, pl. 6–13.

Wilson, A. E. 1948. Miscellaneous classes of fossils, Ottawa Formation, Ottawa-St. Lawrence Valley. Canada Department of Mines and Resources, Geological Survey Bulletin 11:116 p., 4 fig., 28 pl.

Wilson, B. R., & D. S. Gillette. 1971. Australian shells; illustrating and describing 600 species of marine gastropods found in Australian waters. Reed. Sydney. 168 p.

Wilson, H. V. 1904. Sponges. Reports on an exploration off the west coasts of Mexico, Central and South America, and off the Galapagos Islands, in charge of Alexander Agassiz, by the U. S. Fisheries Commission Steamer "Albatross" during 1891. Memoir of the Museum of Comparative Zoology, Harvard College XXX(1):161 p., 26 pl.

———. 1925. Siliceous and horny sponges collected by the U. S. Fisheries steamer "Albatross" during the Phillipine Expedition, 1907–1910. U.S. National Museum Bulletin 100, vol. 2(4):273–506, 16 pl.

Wilson, J. L. 1950. An Upper Cambrian pleospongid (?). Journal of Paleontology 24:591–593, 1 fig., pl. 80.

Winchell, N. H., & Charles Schuchert. 1893. Sponges, graptolites and corals from the Lower Silurian of Minnesota. Minnesota Geological Survey Final Report 3(1):55–95, pl. F–G.

Wintermann, G. 1951. Entwicklungs-physiologie Untersuchungen an Susswasserschwämmen. Zoologische Jahrbücher, Abteilung für Anatomie und Ontogenie der Tiere 71(4):428–486, 38 fig.

Wisniowski, A. 1885. Staniworka gabek w systematyce zwierat wedlug Marshall [Systematic position of sponges according to Marshall]. Wszechswiat 4:563.

Wisniowski, T. 1888. Wiadomosc o krzemieniach jurajskich okolicy Krakao. Kosmos Rocznicki 13:175–185.

———. 1889a [1888]. Przyczynek do poznania microfauny krzemieni jurasjaskich okolic Krakowa. Kosmos Rocznicki, vol. 13 [Beitrag zur Kentniss der Mikrofauna aus den oberjurassischen Feuersteinknollen der Umgegend von Krakau]. Jahrbuch der Kaiserlich-Königlichen Geologischen Reichsanstalt 38:657–702.

———. 1889b. Nowy przycznek do zhajomosci górnojurajskich Monactinellidów i Tetractinellidów. Kosmos Roczniki 14:185–189, 230–237.

Wolfenden, E. B. 1959. New sponges from Lower Carboniferous reefs of Derbyshire and Yorkshire. Journal of Paleontology 33:566–568.

Wood, R. 1990. Reef-building sponges. American Scientist 78:224–235.

———. 1991. Non-spicular biomineralization in calcified demosponges. *In* J. Reitner & H. Keupp, eds., Fossil and Recent Sponges. Springer-Verlag. Berlin. p. 322–340, 9 fig.

Wood, R., & J. Reitner. 1988. The Upper Cretaceous "chaetetid" demosponge *Stromatoaxinella irregularis* n.g. (Michelin) and its systematic implications. Neues Jahrbuch für Geologie und Paläontologie, Abhandlungen 177:213–224.

Wood, R., J. Reitner, & R. R. West. 1989. Systematics and phylogenetic implications of the haplosclerid stromatoporoid *Newellia mira* nov. gen. Lethaia 22:85–93.

Worthen, A. H. 1875. *Cnemidium? trentonensis. In* F. B. Meek & A. H. Worthen, Descriptions of invertebrates. Illinois Geological Survey 6:491.

Wu S. K. 1965. Comparative functional studies of the digestive systems of the muricid gastropods *Drupa ricina* and *Morula granulata.* Malacologia 2:211–233, 5 pl.

Wu Xichun. 1989. Late Triassic Carnian strata in western Sichuan Basin and a new sponge family. Acta Palaeontologica Sinica, Beijing 28(6):766–771, 1 pl.
In Chinese, with English summary.

———. 1990 [1989]. Late Triassic Lychniscosa fauna in northwestern Sichuan. Acta Palaeontologica Sinica 29(3):349–363, 3 pl.
In Chinese, with English summary.

Wu Xichun, & Xiao Rongwu. 1989. Discovery of Late Triassic sponge fauna in northwestern Sichuan. Journal of Kunming Institute of Technology 14(1):12–21, 1 pl.
In Chinese.

Wu Xichun, & Zhang L. 1982. Late Triassic (Carnian) sponge patch reefs in northwestern Sichuan basin. Scientia Geologica Sinica 10:379–385.
In Chinese with English abstract.

Wu Ya Sheng. 1991. Organisms and communities of Permian reef of Xiangbo, China. International Academic Publishers. Beijing. 192 p., 26 pl.

Wu Ya Sheng, & Fan Jiasong. 2002. Permian-Triassic history of reefal thalamid sponges: evolution and extinction. Acta Palaeontologica Sinica 41(2):163–177, 1 fig.
In Chinese and English.

Wulff, J. L. 1994. Sponge feeding by Caribbean angelfishes, trunkfishes, and filefishes. *In* R. W. M. Van Soest, T. M. G. Van Kempen and J.-C. Braekman, eds., Sponges in Time and Space, Proceedings of the 4th International Porifera Congress, Amsterdam. A. A. Balkema. Rotterdam. p. 265–271, 1 fig.

Wyatt, D. J. 1979. Carbonate mud mounds from the Lower Ordovician Wah Wah Limestone of the Ibex area, western Millard County, Utah. Brigham Young University Geology Studies 26(2):101–114.

Xiao Shuhai, Jie Hu, Xunlai Yuan, R. L. Parsley, & Ruiji Cao. 2002 (MS). Articulated sponges from the Early Cambrian Hetang Formation in southern Anhui, South China: Their age and implications for significant gaps in the early sponge record. Palaeogeography, Palaeoclimatology and Palaeoecology, v. 00, 49 p., 10 fig.

Xiao Shuhai, Xunlai Yuan, M. Steiner, & A. H. Knoll. 2002. Macroscopic carbonaceous compressions in a terminal Proterozoic shale: a systematic reassessment of the Miaohe Biota, South China. Journal of Paleontology 76(2):347–376, 11 fig.

Yabe, Hisakatsu, & Toshio Sugiyama. 1934. *Amblysiphonella* and *Rhabdactinia gen.* and *sp. nov.* from the Upper Palaeozoic Limestone of Mimikiri, near Sakawa-mati, Tosa Province, Sikoku, Japan. Japanese Journal of Geology and Geography 11:175–180, pl. 20–22.

———. 1939. *Marindiqueia mirabilis,* gen. et sp. nov., a sponge-like fossil from the Eocene limestone of Marinduqu Island, Phillippine Islands. Transactions and Proceedings of the Palaeontological Society of Japan 15:60.

Yochelson, E. L. 1968. Biostratigraphy of the Phosphoria, Park City, and Shedham Formations. U.S. Geological Survey Professional Paper 313-D:571–600.

Young, J., & J. Young. 1876. *Acanthospongia smithii. In* John Armstrong, John Young, & David

Robertson, Catalogue of western Scottish fossils. British Association for the Advancement of Science. Blackie and Son. Glasgow. p. 38.

———. 1877. On a Carboniferous *Hyalonema* and other sponges from Ayrshire. Annals and Magazine of Natural History (series 4) 20:425–432, pl. 14–15.

Yuan, Xunlai, Shuhai Xiao, R. L. Parsley, Chuanming Zhou, Zhe Chen, & Jie Hu. 2002. Towering sponges in an Early Cambrian lagerstätte: disparity between nonbilaterian and bilaterian epifaunal tierers at the Neoproterozoic-Cambrian transition. Geology 30(4):363–366, 4 fig.

Yurewicz, D. A. 1977a. Evolution of the Capitan massive limestone (Permian) of Guadalupe Mountains, New Mexico and West Texas. American Association of Petroleum Geologists, Bulletin 61(5):843–844.

———. 1977b. The origin of the massive facies of the lower and middle Capitan Limestone (Permian), Guadalupe Mountains. *In* M. E. Hileman & S. J. Mazzullo, eds., Upper Guadalupian facies, Permian reef complex, Guadalupe Mountains, New Mexico and West Texas. Society of Economic Paleontologists and Mineralogists, Permian Basin Section, 1977, Field Conference Guidebook, Publication 76-16:45–92.

Zahálka, Cenek. 1900 [1899]. Pásmo IX–Brezezenské-kridového útvaru v Poohri. (Die IX (Preisner) Etage der Kreideformation im Egergebeite). Sitzungsberichte der Kaiserlichen Böhmischen Gesellschaft der Wissenschaften, 1899, article 4:103 p., 6 pl.

———. 1901 [1900]. Pásmo X-Tepliché-kridového útvaru v Poohri. (Die X (Teplitzer) Etage der Kreideformation im Egergebeite). Sitzungsberichte der Kaiserlichen Böhmischen Gesellschaft der Wissenschaften, 1900, article 9:51 p., 3 pl.

Zangerl, Rainer, & E. S. Richardson, Jr. 1963. The paleoecological history of two Pennsylvanian black shales. Fieldiana, Geological Memoir 4:352 p., 55 pl.

Zeise, O. 1897. Die Spongien der Stramberger Schichten. Palaeontologische Studien über die Grenzschichten der Jura- und Kreideformation im Gebiete der Karpathen, Alpen, und Apeninen, VIII. Palaeontographica, Supplement 2:i–iv, 289–342, pl. 19–21.

Zenker, J. C. 1836. Historisch-topographisches Taschenbuch von Jena und seiner Umgebung, besonders in naturwissenschaftlicher und medicinischer Bezeihung. F. Fromman. Jena. 388 p.

Zenkovitch, B. 1963. Biology of the Seas of the U.S.S.R. Interscience Publishers of John Wiley & Sons & George Allen & Unwin, Ltd. New York & London. 955 p., 427 fig.

Zhang Wei. 1983. Study on the sphinctozoans of Upper Permian Changxing Formation from Lichuan area, West Hubei, China. *In* A collection of Theses for Master's Degree (1981). Institute of Geology, Academia Sinica. Beijing. p. 1–11.

———. 1987. A new genus *Neoguadalupia* with notes on connections of interrelated genera in Sebargasiidae, Sphinctozoa. Scientia Geologica Sinica 7:231–238, 5 fig.

Zhang Xiaolin, & Zhang Wei. 1990. Paleoecology of reef-building sponges in Kefeng Reef, Longlin, Quangxi. Shiyou Yu Tianrangi Dizhi 11(4):427–535.
In Chinese with English summary.

Zhuravleva, I. T. 1962. *Tscheryschevo-Stuckenbergia*. *In* P. D. Rezvoi, I. T. Zhuravleva, & V. M. Koltun, Class Porifera (Spongia). Osnovy Paleontologii [Fundamentals of Paleontology], Porifera, Archaeocyatha, Coelenterata, Vermes Akademii Nauk SSSR. Moscow. p. 63.
English translation, p. 72, A. Mercado and H. Mills, Israel Program for Scientific Translations Ltd., 1971, available from U. S. Department of Commerce, National Technical Information Service.

———. 1967. *Girphanovella* Zhuravleva, *gen. nov. In* I. T. Zhuravleva, N. M. Zadorozhnaya, D. V. Osadchaja, N. V. Pokrovskaya, N. M. Rodionova, & V. D. Fonin, Fauna Nzyhnego Kembryja Tuvy (Opornyy Razrez r. Shivelig-Khem) [Lower Cambrian fauna of Tuva (the Reference Section of the Shivelig-Khem River)]. Akademii Nauk SSSR. Moscow. p. 107, pl. 59.

Zhuravleva, I. T., & I. A. Pyanovskaya. 1995. Fossil description. *In* I. A. Pyanovskaya & I. T. Zhuravleva, Biostratigraphy and new forms of Lower Paleozoic fossils of the Bogambir Ridge (North Nuratau Range, South Tien Shan). Geologiya i Geofizika (Russian Geology and Geophysics) (Novosibirsk) 36(3):36–44, fig. 1–4.

Ziegler, Bernhard. 1962. Beobachtungen an hexactinelliden spongien. Eclogae Geologicae Helvetiae 55:573–586, 2 fig., 3 pl.

Ziegler, Bernhard, & Siegfried Rietschel. 1970. Phylogenetic relationships of fossil calcisponges. *In* W. G. Fry, ed., The Biology of the Porifera, Zoological Society of London Symposium 25:23–40, 4 fig.

von Zittel, K. A. 1876. Über *Coeloptychium*. *In* Beitrag zur Kenntniss der Organization fossiler Spongien. Abhandlungen der mathematisch-physikalischen Classe der Königliche bayerischen Akademie der Wissenschaften (München) 12(3):1–80, 7 pl.

———. 1877a. Beitrage zur Systematik der fossilen Spongien. Neues Jahrbuch für Mineralogie, Geologie und Paläontologie 1877(1):337–378.

———. 1877b. Studien über fossile Spongien, 1 Abt., Hexactinellidae. Königlisch Bayerischen Akademie der Wissenschaften, Mathematisch-Physikalischen Klasse, Abhandlungen 13(1):1–63, pl. 1–4.
Translated in Annals and Magazine of Natural History (series 5) 20:257–273, 405–424, 501–517.

———. 1877c. *Protachilleum kayseri*. *In* E. Kayser, Über primordial und Untersilurische Fossilien aus der Argentine Republik. Beiträge Geologie und Paläontologie der Argentine Republik 2(1):22–23.

———. 1878a. Studien über fossile Spongien, II, Lithistidae. A. Allgemeiner Theil. Abhandlungen der kaiserische Bayerischen Akademie der Wissenschaften 13(1):67–154, pl. 1–10.
Translated in Annals and Magazine of Natural History (series 5) 2:113–135, 235–248, 324–341, 385–394, 467–482.

———. 1878b. Studien über fossile Spongien, Dritte Abtheilung: Monactinellidae, Tetractinellidae und Calcispongiae. Abhandlungen der kaiserische Bayerischen Akademie der Wissenschaften 13(2):1(93)–48(138), pl. 11–12.
Translated in Annals and Magazine of Natural History (series 5) 3:304–312, 364–379, and vol. 5:61–73, 120–135.

———. 1878c. Handbuch der Palaeontologie: Protozoa, Coelenterata, Echinodermata und Molluscoidea, vol. 1, no. 1, Class Spongia, Seeschwämme. *In* Paläozoologie, 1876–1880. R. Oldenbourg. Munchen & Leipzig. p. 128–202.

———. 1878d. Beiträge zur Systematik der fossilen Spongien, II Theil. Neues Jahrbuch für Mineralogie, Geologie und Paläontologie

———. 1879. Beiträge zur Systematik der fossilen Spongien, Dreitter Theil. Neues Jahrbuch für Mineralogie, Geologie und Paläontologie, 1879:1–40, 3 pl.

———. 1884. Über Astylospongidae und Anomocladina. Neues Jahrbuch für Mineralogie, Geologie und Paläontologie 2:75–80, pl. 1–2.

———. 1895. Grundzüge der Paläontologie (Paläozoologie), Abteilung 1, Invertebrata. R. Oldenbourg. München & Leipzig. p. viii + 1–971.

———. 1903. Grundzüge der Paläontologie (Paläozoologie), ed. 2., Abteilung 1, Invertebrata. R. Oldenbourg. München & Berlin. 558 p., 1,405 fig.

INDEX

[redacted]

[redacted]